Phlebology '95

Phlebology '95

Proceedings of the XII World Congress
Union Internationale de Phlébologie

London 3 – 8 September 1995

Editors

David Negus
Georges Jantet
Philip D. Coleridge-Smith

Volume 1

ISBN 978-3-540-19999-1   ISBN 978-1-4471-3095-6
DOI 10.1007/978-1-4471-3095-6

First Published in two volumes by Springer-Verlag London Ltd as
Supplement 1 (1995) to the journal *Phlebology*.

Originally published by the Venous Forum of the Royal Society of Medicine and
Societas Phlebologica Scandinavica in 1995

Union Internationale de Phlébologie

Executive Committee
Comité Executif

President:	André Davy
Vice Presidents:	Georges Jantet, Jean-Pierre Kuiper, Hugo Partsch, Pauline Raymond-Martimbeau, J Leonel Villavicencio
General Secretary:	Pierre Wallois
Deputy General Secretary:	Ivan Staelens
Treasurer:	Jaques Follereau
Honorary Members:	Henrik R van der Molen, Wilhelm Schneider, Jean van der Stricht, Maryse Horáková, Saul Umansky

Congress President/
Président du Congres: *Georges Jantet*

Organising Committee
Comité Organisateur

Chairman:	John Scurr
Scientific Chairman:	Vaughan Ruckley
Treasurer:	Gilbert Carswell
Editor:	David Negus
Assistant Editors:	Georges Jantet, Philip Coleridge-Smith

Norman Browse, Kevin Burnand, Frank Cockett, Richard Corbett, John Dormandy, Brian Gilchriot, Jean-Pierre Kuiper (UIP representative/Représentant de l'UIP), Charles Michel, Peter Mortimer, representatives of major sponsors/Représentants des principales compagnies parrainant le Congrès.

International Advisory Committee

Message de Bienvenue du Président

Depuis quarante ans, le traitement des maladies du système veineux a beaucoup évolué.

Pendant trop longtemps divers aspects de ces maladies étaient traités par des disciplines différentes, travaillant indépendamment qu'il s'agisse de dermatologues, médecins généralistes, chirurgiens et même dans certains pays, vénérologues.

C'est grâce surtout à l'école française que la Phlébologie a gagné ses titres de noblesse en tant que spécialité à part entière, mais chaque pays a trop longtemps travaillé de façon isolée.

Le développement des réunions internationales a favorisé des échanges enrichissants entre pays aboutissant à la création de l'Union Internationale de Phlébologie en 1959.

Les nouvelles méthodes d'exploration, l'application d'une méthode scientifique et épidémiologique étendue ainsi que d'autres études dans divers domaines, ont développé le champ d'intérêt vers de nouvelles spécialités; c'est ainsi que Radiologues, Gynécologues, Angéiologues, Physiologistes, Lymphologistes, Epidémiologistes et autres ont été amenés à jouer un rôle dans le traitement des maladies veineuses, au profit de la Phlébologie qui est devenue une spécialité réellement multidisciplinaire nécéssitant un travail d'équipe.

Le contrôle des dépenses est à l'ordre du jour et la récente prise de conscience du taux élevé des maladies veineuses, avec la morbidité et même la mortalité qui les accompagnent, a conduit à mettre l'accent sur la prévention compte tenu du coût élevé du traitement des complications tant sur le plan humain que financier.

L'UIP est le forum des Phlébologues de tous les pays et le "Venous Forum" de la Royal Society of Medicine (foyer des Phlébologues du Royaume Uni) est très conscient de l'honneur qui lui est fait et de l'importante responsabilité qui lui incombe d'organiser le XIIe Congrès Mondial à Londres en 1995, exactement dix ans après avoir organisé la première réunion au Royaume Uni du Chapitre Européen de L'UIP.

Le Comité d'Etude/Organisation travaille de façon assidue depuis quatre ans afin que ce Congrès présente un intérêt pour toutes les disciplines représentées, soit d'un très haut niveau scientifique et offre un programme socio-culturel exceptionnel.

En tant que Président du Congrès de 1995, je suis haureux de vous accueillir à ce Congrès – membres de toutes les disciplines liées à la phlébologie, qu'elles soient médicales ou non: vous y trouverez tous un intérêt.

Faire un exposé oral, ou sous forme de posters ou vidéo, est bien sûr très important, mais nous attachons aussi une très grande importance à la participation de l'auditoire. Ceci nous paraît être un des meilleurs moyens de favoriser les échanges. Ainsi donc, que vous fassiez ou non un exposé, permettez-moi, au nom du "Venous Forum", de vous assurer de notre accueil chaleureux.

Georges Jantet
Président du Congrès

Welcome from the President

The management of diseases of the venous system has evolved very rapidly during the past 40 years.

For too many years different aspects of these diseases were managed by different disciplines working independently e.g. Dermatologists, General Physicians, Surgeons and, in some countries, Venereologists.

The French School was responsible to a large extent for the development of Phlebology as a specialty in itself but for too long each country worked rather in isolation.

With the development of international meetings, cross fertilisation between different countries became a reality, culminating in the creation of the Union Internationale de Phlébologie in 1959.

New methods of investigation, application of the scientific method and large-scale epidemiological and other studies have broadened the field of interest to further and new specialties; thus it is that Radiologists, Gynaecologists, Angiologists, Physiologists, Lymphologists, Epidemiologists and others have become involved in the management of venous diseases to the enrichment of Phlebology which has become a truly multidisciplinary specialty requiring proper teamwork.

Cost containment is now the order of the day and the recent realisation of the high incidence of venous disease with its accompanying morbidity and even mortality, has led to an emphasis on prevention, given the very high cost, in human and financial terms, of the management of the complications.

The UIP is the forum for Phlebologists the world over and the Venous Forum of the Royal Society of Medicine (the home of Phlebologists in the United Kingdom) is very conscious of the great honour and responsibility bestowed upon it to organise the XII World Congress in London in 1995, just 10 years after it organised the first meeting in the United Kingdom of the European Chapter of the UIP.

The Planning/Organising Committee has been working very hard over the past 4 years to ensure this Congress is of interest to all the disciplines involved, is of the highest scientific standard and is well balanced by an outstanding social/cultural programme.

As President of the 1995 Congress, it is my privilege and pleasure to welcome you from all of the involved disciplines, whether medical or non medical. You will all find something of interest.

Making a presentation, either orally or by poster or by video, is very important, of course, but we are attaching great importance also to audience participation as one of the best ways of fostering cross exchanges. Thus, whether you are making a presentation or not, allow me to say, on behalf of the Venous Forum, how warmly welcome you all are.

Georges Jantet
Congress President

Welcome Address

On behalf of the Organising Committee it gives me great pleasure to welcome you to the XII World Congress of the Union Internationale de Phlébologie, London. I hope you enjoy the Meeting and that you take full advantage of the very extensive scientific programme and the excellent social programme.

To those of you that have contributed to the scientific sessions I am most grateful. Our decision to hold plenary sessions during the morning to stimulate discussion on important issues in Phlebology, resulted from your own comments and feedback following previous Meetings. This has resulted in fewer papers being presented orally, but at the same time has improved the standard of poster presentations. I hope this formula proves successful and will be repeated at future Meetings.

We are particularly fortunate to have a number of unique events in our social programme. By including these events in the overall registration of the Meeting we hope to encourage all of you to stay for the whole week to make new friends and, above all, enjoy yourselves.

We are, of course, extremely grateful to our sponsors without whom none of these events could have taken place. I have, throughout the organisation of this Meeting, been supported by an extremely dedicated Organising Committee. Their contribution has been enormous. Without their support and the help of our Conference Organisers, this Meeting could not have occurred.

If during your stay in London we can do anything to make the Meeting more enjoyable for you, we will be delighted to do so. Once again, it is a great pleasure to welcome you to London to yet another very successful UIP Meeting.

John H. Scurr FRCS
Chairman of the Organising Committee

Acknowledgements

We are most grateful to the following
for their generous support

Hoechst

Sigvaris ® Worldwide/Ganzoni & Cie AG
St Gallen/Switzerland and St Louis/France

Beiersdorf & Jobst

HNE Healthcare
– a division of Huntleigh Technology plc

Aethoxysklerol Sclerosing Agent
from Kreussler Pharma

Kendall International

Laboratoire Negma

Lohmann GmbH & Co KG

Medi Bayreuth

Rhone-Poulenc Rorer

Venosan Medical Stockings
from Salzmann AG

Groupe de Recherche Servier, France

STD Pharmaceutical

Thuasne SA

Zyma SA

Contents - Volume 1

Epidemiology and Socio-economics

Anatomy/Pathology

Basic Science
Anatomy

Physiology

The Investigation of Venous Disorders

Non Invasive Investigations
Duplex Scanning

Angiology

Angioscopy, Intravascular Ultrasound

The Treatment of Varicose Veins and Telangiectasias

Surgery

Injection Sclerotherapy

Pharmacology

Compression

Epidemiology and Socio-economics

Worldwide

E x c e l l e n c e   i n   P h l e b o l o g y

AUSTRALIA
Beiersdorf Australia Ltd.
112-118 Talavera Road
AUS – North Ryde/
NSW 2113
Tel: 2 / 888 09 77

BELGIUM
SA Beiersdorf NV
Boulevard Industriel 30
B – 1070 Bruxelles
Tel: 2 / 562 52 11

CHILE
BDF Chile SA
Camino lo Espejo 501
Maipú
RCH – Santiago de Chile
Tel: 2 / 557 36 66

GERMANY
Beiersdorf AG
Unnastr. 48
D – 20245 Hamburg
Tel: 040 / 569 - 0

JOBST GmbH
Beiersdorfstr. 1
D – 46446 Emmerich
Tel: 02822 / 607 - 0

GREAT BRITAIN
BDF UK Ltd.
Yeomans Drive,
Blakelands
GB – Milton Keynes,
Bucks, MK 14 SLS
Tel: 908 / 21 13 33

ITALY
Beiersdorf SpA
Casella postale 17094
I – 20170 Milano
Tel: 2 / 257 72 - 1

JAPAN
Beiersdorf Japan K.K.
Hiroo SK Bldg.
2-36-13 Ebisu – Shibuya-ku,
J – Tokyo 150
Tel: 3 / 54 21 46 70

NETHERLANDS
Beiersdorf NV
Bolderweg 2
Bedrijvenpark "De Vaart"
NL – Almere 1332 AT
Tel: 3653 / 891 00

SPAIN
Beiersdorf S.A.
Carretera de Mataró
a Granollers, Km. 5,4
E – 08310 Argentona
(Barcelona)
Tel: 3 / 758 3300

SWEDEN
Beiersdorf AB
Box 10056
S – 43421
Kungsbacka
Tel: 300 / 550 00

USA
Jobst Institute, Inc.
P.O. Box 471048
USA – Charlotte,
NC 28247-1048
Tel: 800 / 332 7573

Phlebology '95, D. Negus et al. (eds.). Phlebology (1995) Suppl. 1: 3-5

PI/1.5

Social Factors and the Healing of Venous Ulcers

P.J. Franks[1]*, N. Bosanquet[2], M. Connolly[2], M.I. Oldroyd[1], C.J. Moffatt[1]*, R.M. Greenhalgh[1] and C.N. McCollum[3]

[1] Department of Surgery, Charing Cross & Westminster Medical School, London
[2] Health Policy Unit, St Marys Medical School, London
[3] Department of Surgery, University Hospital of South Manchester
* current address Centre for Research & Implementation of Clinical Practice, London

INTRODUCTION

Leg ulceration is a common source of morbidity in the elderly with estimated population prevalence in the UK of 0.15-0.18% [1,2]. Delivery of care is chiefly through community services, particularly community nurses who treat more than 50% of patients in their home. Recent innovations in venous ulcer care include the four layer bandage (4LB) which provides sustained compression for at least one week [3,4]. This standardised approach to treatment allows for an in-depth investigation of the factors which are likely to prolong ulcer healing. Although information on clinical risk factors have been investigated in a number of studies, it is possible that other social factors may also influence the healing process. This study is an investigation of the relationship between socio-economic factors and the healing of venous ulceration.

METHODS

Patients for this study presented to five of six community leg ulcer clinics within Riverside Community NHS Trust over the first six months from the start of each clinic. In patients with significant arterial disease (ankle brachial pressure index <0.8), high compression is contra-indicated and the 4LB was not used. Only patients considered to have venous ulceration were treated using this method, and these made up the study population. At the first attendance at a community clinic patients were interviewed about their lifestyle, including questions on income, housing and social interaction. After 12 weeks of treatment patients were re-interviewed and their legs examined for areas of ulceration. Statistical analysis was performed using logistic regression analysis, the dependent variable being failure of the ulcer to complete healing. Patients with bilateral ulceration were only considered to have healed if both limbs were completely free of areas of ulceration after 12 weeks. The results of the regression analysis are given as odds ratios (95% confidence intervals) with a low odds ratio (<1.0) indicating a greater chance of healing over the 12 week period, whilst OR>1.00 indicating a prolonged healing time.

RESULTS

In all, 168 patients presented to the clinics within the first six months and were considered suitable for the 4LB, with a mean age of 76 (sd=11) years of whom 106 (64%) were women. Most (140/168, 82%) had areas of ulceration <10cm^2 with a median ulcer duration of seven months (range=one week to 63 years). Most patients were retired, with the most frequent source of income a combination of social security and private pensions (42%). Only 22% were owner occupiers, but of the total, half had central heating in their accomodation. One half of all patients lived alone, but more than three quarters saw family and friends every day.

Of the 168 patients in this study 87 (52%) had complete ulcer healing after 12 weeks of treatment. Table I gives some of the key factors which were associated with venous ulcer healing.

Table I. Univariate comparison of selected social factors in prolonged venous ulcer healing.
(*=significant at p<0.05)

Factor	Healed	Unhealed	Odds ratio (95% CI)
Marital Status			
married	29 (33%)	16 (20%)	1.00
single	17 (20%)	26 (32%)	2.77 (1.15,6.69) *
widowed	34 (39%)	34 (42%)	1.81 (0.82,3.99)
div/sep	7 (8%)	5 (6%)	1.29 (0.34,4.88)
Social Class			
I & II	18 (21%)	6 (7%)	1.00
IIIN+M	42 (48%)	43 (54%)	3.07 (1.09,8.67) *
IV + V	27 (31%)	31 (39%)	3.44 (1.17,10.14) *
Central Heating			
yes	52 (60%)	33 (41%)	1.00
no	34 (40%)	48 (59%)	2.22 (1.18,4.18) *

It can be seen that being single, having a low social class and not having central heating were significantly associated with prolonging ulcer healing. When adjusted for the known clinical risk factors for venous ulcer healing (ulcer size, ulcer duration and mobility) only lack of central heating remained significantly different (OR=2.27, 95%CI 1.11, 4.55). Factors which produced high OR>2.0 indicating poor healing were unemployment (OR=4.27, 95%CI 0.37, 49.33) and social contact <weekly

(OR=5.33, 95%CI 0.91, 31.09), whilst factors with low OR<0.5 indicating improved healing were income from work (OR=0.48, 95%CI 0.12, 1.99), and council house tenancy (OR=0.42, 95%CI 0.16-1.09). Stepwise analysis with adjustment for clinical risk factors gave improvements in healing in patients living in council housing (OR=0.63, 95%CI 0.43, 0.92) whilst social contact <weekly was associated with prolonged healing (OR=2.34, 95%CI 1.00,5.73).

DISCUSSION

It is often felt that leg ulceration is a problem of the socially deprived, but with little evidence to support this. In this study we have identified a number of social factors which on univariate analysis appear to be associated with venous ulcer healing in a group of 168 patients. However, following adjustment for clinical risk factors these factors are no longer significant, though the odds ratios are of similar magnitude. There is clearly scope for further work to confirm these results and to determine whether these associations are chance, or a causative effect on the healing of chronic venous ulceration.

REFERENCES

1. Callam MJ Ruckley CV Harper DR Dale JJ: Chronic ulceration of the leg: extent of the problem and provision of care. Br Med J 1985;i: 1855-6

2. Cornwall JV Dore CJ Lewis JD: leg ulcers: epidemiology and aetiology. Br J Surg 1986; 73: 693-6

3. Blair SD Wright DDI Backhouse CM Riddle E McCollum CN. Sustained compression and healing of chronic venous ulcers. Br Med J 1988; 297: 1159-61

4. Moffatt CJ Franks PJ Oldroyd M Bosanquet N Brown P Greenhalgh RM McCollum CN
Community leg ulcer clinics and impact on healing. Br Med J 1992; 305: 1389-92

Phlebology '95, D. Negus et al. (eds.). Phlebology (1995) Suppl. 1: 6-7

P103

Males with a Standing Profession: Epidemiology of Chronic Venous Insufficiency

R.M.A. Krijnen[1], E. M. de Boer[1], H.J. Adèr[2] and D.P. Bruynzeel[1]

Depts of [1]Dermatology and [2]Epidemiology, Free University Hospital, Amsterdam, The Netherlands

INTRODUCTION

It has been long recognized that not only hereditary, but also environmental factors are associated with the occurrence of venous disorders in a population. Prolonged standing for many years is indicated as one of these risk factors [1,2].
In the Netherlands the prevalence of venous disorders in an occupational population was never studied; as in many countries venous disorders are registered only in case of hospital admittance.
The aim of the present study was to investigate the presence of venous disorders in workers with a standing profession. Furthermore, the correlation between the presence of CVI (chronic venous insufficiency) and the presence of subjective complaints of the legs and several presumed risk-factors was investigated.

MATERIALS AND METHODS

A field trial was performed in 14 factories, mainly in the meat and shoe industry. Included were all males with a standing profession, present at the time, resulting in 387 subjects, mean age 37 years. All subjects were examined for the presence of CVI by physical examination, Doppler ultrasound investigation and light reflection rheography. This was accomplished near the working place. Furthermore, subjects were asked for the presence of complaints of the legs, such as a tired and heavy feeling or pain in the legs.
Major CVI was defined as the presence of skin changes (Widmer II-III) or signs of deep venous insufficiency or stem varicosis.
Minor CVI was defined as beginning corona paraplantaris phlebectatica, ankle flare or side branch varicosis without dermal changes.
Individuals with only intracutaneous varices were considered not to have CVI.

RESULTS

'Major CVI' was found in 42 subjects (11%) and 'Minor CVI' was found in 70 subjects (18%). In 275 (71%) subjects no signs of CVI, or only a few hemodynamicly unimportant varicose veins, were present.

In 47 subjects (12%) an insufficiency of one or both saphenal veins was found.

A tired feeling or pain in the legs was reported by 79% of the individuals with CVI but also by 61% of the individuals without CVI. A tired feeling was the most reported complaint. It was reported by 70% of the subjects having CVI and by 53% of the healthy subjects. Pain in the legs was reported by 46% of the subjects having CVI and by 27% of the healthy subjects.

A statistically significant positive correlation of CVI with age, with weight and with duration of standing work was demonstrated.

CONCLUSIONS

CVI occurred in over a quarter of male workers with a standing position at work. The prevalence increased with age and with years of standing profession (after correction for age). Complaints of the legs, such as a tired feeling were common findings, also without objective signs of CVI.

ACKNOWLEDGEMENTS

This study was made possible by a grant of the Praeventiefonds.

REFERENCES

1. Mekky S, Schilling RSF, Walford J. Varicose veins in Women Cotton Workers. An Epidemiological Study in England and Egypt. BMJ 1969(2):591-595.

2. Guberan E, Widmer LK, Glaus L, Muller R, Rougemont A, Da Silva A, et al. Causative factors of Varicose veins: Myths and Facts. An Epidemiological Study of 610 women. VASA 1973;2:115-120.

Phlebology '95, D. Negus et al. (eds.). Phlebology (1995) Suppl. 1: 8-9

P109

Chronic Venous Insufficiency (CVI) and Saphenofemoral and/or Saphenopopliteal Incompetence

J. Gawrychowski, A. Romanski and B. Lazar-Czyzewska

Varicose Vein Clinic, 41-800 Zabrze ul. Gen.de Gaulle'a 60, Poland

INTRODUCTION

In middle European countries about 20% of adult people are affected with chronic venous insufficiency (CVI) (4,6). An underlying cause of CVI is venous insufficiency in lower extremities, both within superficial and deep systems, resulting in hemodynamic disorders (1,2,7). Insufficiency of saphenofemoral and/or saphenopopliteal junction seems to be particularly important (3,5,6).

The aim of this study was retrospective analysis of a series of 983 patients (1966 legs) with varicose veins.

METHODS

Using the Doppler-probe we have examined vena saphena (v.s.) magna and v.s. parva. We have studied CVI using digital photo-plethysmography (D-PPG).

RESULTS

281 out of 983 patients (28.6%) had saphenofemoral junction [232 (23.6%) one -, and 49 (5%) two-sided] and 9 (2.7%) had saphenopopliteal incompetence. Out of 330 legs with saphenofemoral or saphenopopliteal incompetence, D-PPG has revealed no CVI in 109 (33%); I° CVI in 55 (16.7%); II° in 154 (46.7%) and III° in 12 (3.6%). No deep veins thrombosis have been found. Out of 1404 legs (702 patients) with saphenofemoral or saphenopopliteal competence we observed no CVI in 1123 (80%); I° CVI in 187 (13.3%); II° CVI in 47 (3.3%) and III° CVI also in 47 (3.3%).

CONCLUSIONS

CVI has been observed significantly ($p < 0,001$) more frequently in patients with saphenofemoral and/or saphenopopliteal incompetence than in those with competence.

REFERENCES:

1. Bergqvist D., Bergentz S.E.: Diagnosis of deep vein thrombosis. World J. Surg. 1990, 14, 679-687

2. Blazek V., Schultz-Ehrenburg U., Kerner J., Schmitt H.J.: Two new non-invasive computer-aided measuring systems for vascular diagnoses: digital photo-plethysmography (D-PPG) and computer-aided gravimetric plethysmography (CGP).

3. Blazek V., Schultz-Ehrenburg U.: Fortschritte in der computerunterstutzten nichtinvasiven Gefassdiagnostik. Phlebol. 1991, 20, 169-175

4. Hachen H.J., Lorenz P.: Klinische und photopletysmographische Doppelblind-Studie der Wirkung Calciumdobesilat bei Patientinnen mit peripheren Storungen der Mikrozirkulation. Angiology 1982, 33, 480-486

5. Holmgren K., Jacobsson H., Johnsson H., Lofsjogard-Nilsson E.: Thermography and photoplethysmography, a non-invasive alternative to venography in the diagnosis of deep vein thrombosis. J. Int. Med. 1990, 228, 29-33

6. O'Donnel T.F.jr, McEnroe C.S., Heggerick P.: Chronic venous insufficiency. Surg. Clin. North Am. 1990, 70, 159-180

7. Weindrof N., Schultz-Ehrenburg U.: Der Wert der Venenverschlussplethysmographie in der phlebologie. Akt. Dermatol. 1984, 10, 83-87

Phlebology '95, D. Negus et al. (eds.). Phlebology (1995) Suppl. 1: 10-13

P247

Recurrent Varicose Veins; A Doppler and Computerized Air Pletysmography Study

U. Alonzo and A. Garavello

Servizio di Flebologia Chirurgica, Azienda Ospedaliera S.Filippo Neri, Roma, Italy

Recurrent varicose veins are a common clinical problem; the incidence is variable, but data from Centers worldwide report a recurrence rate (after 5 years) in the order of 20%, and this probably increases with time.

At today the real efficacy of the stripping operation is under debate, but results of C.H.I.V.A. operation or simple flush ligation with avulsion of distal varicosities seems to give no better results.

Studies performed to identify the reasons for this high recurrence rate found an inadequate primary operation in most cases; the aim of our study was to analize the hemodynamic pattern of recurrent varicose veins to find the mechanisms underlying this pathology.

Diagnostic study was performed with Doppler C.W. and Air Pletysmography (AP); AP evaluated the hemodynamic role of recurrent varicose veins mapped with Doppler C.W.

Patients and Methods

The present study was undertaken to evaluate the problem of recurrence of varicose veins after surgery.

Between the January 1993 and January 1995 we evaluated 70 patients (59 female 11 male, mean age 53) affected by recurrent varicose veins; technique of previous operation, complication occurred, timing of recurrence and related symptoms were recorded.

In all the patients Doppler C.W. and Air Pletysmography (AP) were performed (single operator). In AP (our technique) a rubber cuff is posed around the calf and then inflated to 24-30 MmHg.. The patient (in the erect position) is asked to perform a standard exercise (20 tip-toes) to activate the calf muscle pump; the exercise is repeated after tourniquet application at the upper thigh and then over and below the knee (Perthes test)..

In superficial venous system incompetence a significant decrease of calf volume at Perthes test may be observed; impaired deep outflow (postphlebitic syndrome e.g.) shows a lack of lowering of the PA trace and an hypertensive rebound with tourniquet apposition.

Results

General results are summarized in the following table I and II;

Table I

Etiology of varices	Diagnostic studies before first operation	First operation	Result of first operation
Idiopathic 32 pts.	Not performed 36 pts.	Stripping 56 pts.	Improved 51 pts.
Pregnancy 25 pts.	Performed 34 pts.	Muller op. 11 pts.	Status quo 13 pts.
Postphlebitic 13 pts.		Flush ligation 3 pts.	Worsened 6 pts.

Table II

Follow-up of first operation	Timing of recurrence after operation	Symptoms of recurrence
DVT 8 pts.	Early (1 year); 24 pts.	Severe 25 pts.
Pregnancy 8 pts.	1-2 years; 13 pts.	Moderate 34 pts.
SVT 1 pt.	2-5 years; 17 pts.	Mild 11 pts.
	>5 years; 16 pts.	

Causes of recurrence were the following;

- in 23 pts; reflux from the groin.
- in 12 pts; deep veins obstruction (postphlebitc or "secondary" varices).
- in 10 pts; Great Saphenous vein incompetence (undiagnosed or developed after first operation).
- in 9 pts.; thigh perforator/s incompetence.
- in 6 pts.; impaired deep outflow caused by muscular compression on the popliteal vein (soleus arch syndrome or related).
- in 5 pts; Saphena Parva incompetence (undiagnosed or developed after first operation)
- in 5 pts.; leg perforator/s incompetence

Discussion

Glauco Bassi identified two tipes of recurrences;

-TRUE recurrences when the first therapy was correct,

-FAKE recurrences when the first therapy was inappropriate.

We believe this is the correct point of view to face the problem.

The recurrence is not just a cosmetic problem; in our experience 25 pts. experienced severe symptoms like leg edema or hypodermitis and ulcers were not uncommon.

A striking fact is that in half of the patients no preoperative diagnostic tests were performed; the operation was planned only on clinical examination, insufficient for a correct assessment of saphenous veins incompetence. Another fact is that in 19 pts. the operation was unsuccessful, sometime worsening the symptoms.

In our experience, like for others Authors, reflux from the groin was the commonest cause of recurrence, 23 pts. Generally the reoperation showed unresected tributaries joining the saphenous stump; frequently the previous groin incision had been either too low or too short for adequate exposure.

It has been recently reported that neovascularization of the saphenous stump may develop through newly formed vessels; we also observed cases of "pseudoangioma" of the groin nourishing recurrent varicose veins of the thigh. Raju emphasized the importance of the femoral vein insufficiency (FVI) in the etiopahogenesis of varicose veins; Folse noted this is also true for recurrence. In our experience an early recurrence (2 years) was observed in 9 of 12 pts. with VFI; its our opinion that FVI may play a major role in groin recurrences.

In 10 pts. the recurrence followed simple avulsion of varicosities (Muller operation); Saphena Magna incompetence was invariably present. Inappropriate preoperative diagnostic examination was mainly responsible for these failures.

In 9 pts. a thigh perforator incompetence caused the recurrence; errors in preoperative mapping of the varicosities (Hunterian perforators join the saphenous trunk directly, others must be separately ligated) can explain some of this cases but not at all. It's possible that stripper trauma can predispose thigh perforators to incompetence; furthermore one has to consider that incompetence might develop in the ensuing years.

In 5 pts. recurrent reflux occurred from leg perforating veins (in 2 cases "veine fosse poplitèe"), probably missed at the first operation - residual varices.

In 12 pts. the recurrent varicosities were "postphlebitic"; in 3 pts. the Deep Venous Thrombosis (DVT) was preoperative (and the recurrence early!), in others the DVT had occurred following the operation. Postphlebitic secondary varicosities must be differentiated from primary varicosities. We believe that the first step in varicose veins therapy is the assessment of deep veins patency; silent DVT is probably more frequent than commonly believed and any suspect must be screened with AP.

In 5 pts. Saphena Parva (SP) incompetence was present; a careful Doppler (sometimes Ecodoppler when doubts exist) examination of SP before surgical treatment is always mandatory.

In 6 pts. deep outflow obstruction was caused by soleus arch compression on the popliteal vein; in these patients the superficial venous system acts like a veno-venous by-pass and then became insufficient. Stenosis may be fixed or intermittent; PA shows a calf volume hypertension at Perthes test (and sometimes in basic conditions), subsiding with knee flexion. Suggestive symptoms are calf pain, leg edema, "ascending varices" with early recurrence after operation; varicosities became evident at a young age (in our study a patient was stripped at 12!). Doppler examination is not specific; ascending flux in the erect position on the saphenous trunk is sometime found.

Conclusions

A significant proportion of the surgery carried out for varicose veins is inadequate; faulty surgical technique is the main source of recurrences, particularly when saphenous vein tributaries are not correctly ligated.

Accurate preoperative diagnostic assessment of sites of reflux is critical to achieve a good result and in preventing early recurrences; combined SP and SM incompetence must be kept in mind. Clinical examination alone is insufficient and Doppler CW evaluation is mandatory in every patient. If a deep venous obstruction is suspected (either postphlebitic or muscular compression) PA is highly desirable; we routinely perform preoperative PA to assess deep veins patency.

Reflux from recurrent perforators remains an unsolved problem; excluding "missed" perforator or errors in preoperative mapping we believe that further studies will be necessary to fully understand perforators pathophysiology.

References

Negus D.
Recurrent varicose veins: a national problem
Br.J.Surg. 1993; 80: 823-824

Olivier C.
Le traitment chirurgical des recidives après operation pour varices essentielles.
J.Chir. 1975; 109: 565-574

Christopoulos D., Nicolaides A.N., Szendro G.
Venous reflux: quantification and correlation with the clinical severity of chronic venous disease.
Br.J.Surg. 1988; 75:352-356

Sarin S., Scurr J.H., Coleridge Smith P.D.
Assessment of stripping the long saphenous vein in the treatment of primary varicose veins.
Br.J.Surg. 1992; 79: 889-893

Redwood N.F.W., Lambert D.
Patterns of reflux in recurrent varicose veins assessed by duplex scanning.
Br.J.Surg. 1994; 81:1450-1451

Glass G.M.
Neovascularization in recurrence of the varicose great saphenous vein following transection.
Phlebology 1987; 2: 81-91.

Nabatoff R.A.
Reasons for major recurrence following operations for varicose veins.
Surg.Gyn.Obst. 1969; 128: 275-278

Phlebology '95, D. Negus et al. (eds.). Phlebology (1995) Suppl. 1: 14-16

PI/1.6

Legal Action Following Treatment for Varicose Veins

W.G. Tennant and C.V. Ruckley

Vascular Surgery Unit, Royal Infirmary, Edinburgh, UK

Introduction.

Procedures carried out for the treatment of varicose veins are among the commonest in surgical practice. Major complications are rare but may be life threatening. Minor complications are common, and may give rise to grievances particularly in the light that patients often seek treatment for cosmetic reasons. These considerations combined with the increase in medicolegal action in the UK prompted this investigation of litigation following treatment of varicose veins in Britain.

Patients.

Each of the three medical defence organisations in the U.K. were asked to provide details of legal action initiated in the last ten years in connection with varicose vein treatment.

Results.

349 cases were notified; (mean age 42 years, 75% female).

Arterial or Venous Damage (48 Patients)

The superficial femoral artery (SFA) or common femoral artery (CFA) was damaged by ligation or division in 6 cases and the popliteal artery in 1 case. In one case of damage to the CFA, repair is documented, but there were no reports of repair in the remaining cases.

The "femoral vein" is named as injured by division or ligation in 25 cases. In the remaining 15 cases the damage was non-occlusive. Immediate repair is documented in just over half the cases of occlusive damage, most going on to a successful outcome.

Nerve Damage (55 Patients)

Two cases of alleged sciatic nerve damage occurred and one injury to the sural nerve. In 10 cases the nerve damage was unspecified.

Damage to the femoral (2 cases) and saphenous nerves (14 cases) occurred in the femoral triangle during groin dissection and stripping of the saphenous vein. The commonest cause of peroneal nerve damage (26 cases) was injudicious use of the avulsion technique in the region overlying the head and neck of the fibula. Dissection of the short saphenous vein in the popliteal fossa accounts for the others.

General Complications

General surgical complications included those events which might occur after any surgical operation, such as pulmonary embolus (5 cases) and deep venous thrombosis (3 cases). The category of "specific complications" contains all cases where the complication was specific to the performance of an operation for varicose veins. This class is most interesting, containing 2 cases of retained vein strippers. The other causes of litigation in this category are diverse.

Recurrence and Cosmesis

These are detailed as causes for legal action in a total of 36 cases. It is impossible in any of these cases to determine the degree of the problem or if further surgery was undertaken for recurrence.

Anaesthetic complications

Anaesthetic complications ranged from minor dental damage (2 cases) to temporary peripheral nerve palsies and three allegations of intra-operative anoxia, one believed to have caused the death of the patient.

Communication or Miscellaneous problems

Poor communication was cited as a cause of dispute in 9 cases. Each case was different, but examples included failure to provide adequate counselling pre-operatively and failure to provide adequate follow-up. Operations were classified as "unsatisfactory" where action was initiated for complaints not otherwise specified nor included in any other of the categories.

Complications of Sclerotherapy (48 Patients)

As might be expected, the major proportion of these are ulceration (25 cases) and discolouration at injection sites (14 cases). Cases of intra-arterial injection and saphenous nerve damage are reported - emphasising the importance of correct technique.

Grade of Medical Staff

The grade of medical staff against whom litigation was taken in each case is detailed in **table 1.** The incidence of litigation against consultant grade staff is the highest of all the groups in this study, probably made falsely high by the institution of NHS indemnity in 1991.

Discussion.

The injuries leading to litigation in this study have all been reported by other authors as complications of varicose vein surgery or sclerotherapy [1 2] and figure regularly in the annual reports of the Medical Defence Organisations [3 4 5 6].

It is noteworthy that legal actions are being initiated for recurrence of varicose veins and for 'dissatisfaction'.

The pattern of major vascular injuries reported in this study emphasises the need for a thorough display of the anatomy when undertaking varicose vein surgery.

These include:

1. Improvements in surgical training in venous anatomy and surgical technique.

2. Pre-operative assessment should include routine use of continuous wave Doppler and judicious use of Duplex sonography and phlebography.

3. Surgery performed by experienced surgeons with a specific training in the field.

4. Patients should be carefully counselled as to the likely outcome of treatment.

Insitution of these measures would not only reduce the number of legal actions but it would transform the care of this long neglected condition.

Table 1

Nature of Injury	CON	SR	REG	SHO	OTHER	
Arterial Damage	4	0	2	1	1	
Venous Damage	13	0	14	11	2	
Neural Damage	17	4	12	7	15	
Wrong leg	2	1	0	0	0	
Wrong operation	5	0	1	0	3	
Recurrence in same leg	6	3	2	1	1	
Poor cosmesis	15	0	5	1	2	
"Unsatisfactory"	24	1	4	3	10	
Local wound complications	24	2	5	2	13	
General surgical complications - DVT, PE	7	0	1	0	5	
Specific complications	14	1	3	4	8	
Anaesthetic	6	0	2	1	1	
Poor communication	7	0	0	0	2	
Sclerotherapy complications	22	1	3	6	16	
TOTAL	166	13	54	37	79	349

[1] Cockett FB. Arterial complications during surgery and sclerotherapy of varicose veins. Phlebologie 1986;1:3-6

[2] Tera H. Emergency repair of femoral vein accidentally divided at operation for varicose veins. Acta. Chir. Scand. 1967;133:283-287

[3] Editorial. Hazards of Varicose Vein Treatment. S.A. Medical Journal 1977; 956

[4] Keddie NC. Medico-legal problems from the treatment of varicose veins. Journal of the Medical Defence Union. Winter 1987:8-9

[5] Medical Defence Union Annual Report. Lost in the Femoral Triangle. 1986:43

[6] Medical Defence Union Annual Report. Popliteal Nerve avulsed. 1988:57

Phlebology '95, D. Negus et al. (eds.). Phlebology (1995) Suppl. 1:17-19

OP/12.1

Perceived Health in a Randomised Trial of Single and Multilayer Bandaging for Chronic Venous Ulceration

P.J. Franks[1], N. Bosanquet[2], D. Brown[3], J. Straub[3], D.R. Harper[3], and C.V. Ruckley[3]

[1] Centre for Research & Implementation of Clinical Practice, London
[2] Health Policy Unit, St Mary's Hospital Medical School, London
[3] Lothian & Forth Valley Leg Ulcer Study, Royal Infirmary of Edinburgh, Edinburgh and Falkirk & District Royal Infirmary, Falkirk

INTRODUCTION

It is widely acknowledged that chronic leg ulceration has a major impact on patients quality of life, yet few studies have investigated this effect [1,2]. Patients may experience benefits of treatment above and beyond the simple healing of their ulcer, which may in turn lead to improved psychological state and changes to social and physical functioning. As part of a randomised trial of two bandage regimens, we have examined patients perceived health, both at the start, and following 24 weeks of treatment to determine changes in perceived health status associated with healing of the leg ulceration whilst examining the potential benefits of the two bandage systems which were offered.

METHODS

The trial was conducted in patients with current leg ulceration, present for a minimum of two months and a minimum diameter of 10 mm. Patients underwent non invasive investigation of the cause of their ulceration and were classified according to the cause of their ulcer into i. simple venous ulceration, ii.complex venous ulceration or iii.arterial ulceration. Only patients with ankle brachial pressure index (ABPI) >0.8 were included in the pure venous arm of the trial. The design was factorial with patients being randomised to different bandage, dressing and drug regimens simultaneously. This paper refers to the comparison of bandage treatments in the trial, the clinical results of which are reported elsewhere. The two treatments were the four layer bandage (4LB) which consists of Velband (Johnson & Johnson), crepe (Smith & Nephew), Elset (Seton) and Coban (3M) [3] compared with the Granuflex Adhesive compression bandage (ACB, Convatec Ltd). Patients had both dressings and bandages changed weekly, or more frequently if required.

At the first visit (or within one week) of starting treatment patients were interviewed about their lifestyle. Included in this was the Nottingham Health Profile, a well validated assessment of patients perceived health recommended for use in patients with chronic disease [4]. The NHP produces scores for energy, pain, emotion, sleep, social functioning and mobility over a range from zero (no interfernce in activity) to 100 (total interference in activity). At the end of the trial period of 24 weeks, patients were reinterviewed irrespective of whether their ulcer had healed. For this analysis of perceived health, only patients who presented to the 24 week visit with legs free of all areas of ulceration were considered to have healed.

Comparison of the change in NHP over 24 weeks was performed initially using Mann-Whitney U-test, whilst comparison of the change in scores between the two groups was performed using ANOVA with adjustment for presenting NHP scores.

RESULTS

Of the 200 patients randomised the mean age was 69.6 years (range=35-94) of which 132 (66%) were women. The randomisation was 98 to the 4LB compared with 103 to the Granuflex bandage, both groups receiving similar proportions of randomised dressings and randomised drug treatment. Over the 24 weeks analysis revealed that there were significant improvements in perceived health as determined by reductions in scores for energy (mean difference [d]=6.8 ,p=0.005) emotion (d=5.3, p<0.001) sleep (d=13.7, p<0.001) and mobility (d=4.9, p<0.001) and a corresponding significant reduction in pain (d=20.1, p<0.001) for the total group after 24 weeks of treatment.

Table I. Changes in NHP scores between the two randomised bandage groups.

Subscore	4LB	ACB	F	p	F(adj.)	p
n	98	103				
Energy	-12.0	-1.7	4.61	0.033	4.26	0.040
Pain	-22.5	-17.8	1.65	0.201	0.88	0.348
Emotion	-6.0	-4.6	0.35	0.555	0.38	0.541
Sleep	-14.1	-13.2	0.05	0.824	0.03	0.874
Social	-1.8	-1.4	0.06	0.806	0.04	0.850
Mobility	-7.1	-2.7	3.55	0.061	4.04	0.046

From the table it is clear that following adjustment for the presenting scores the 4LB was able to demonstrate significantly greater reductions in energy and mobility scores, implying that patients on the 4LB experienced improvements in both of these. The difference in energy could be explained by differences between healed and unhealed ulcers (-11.5 versus -0.3, p=0.023), but could not explain the difference in mobility which was similar between healed and unhealed ulcers (-6.1 versus -3.2, p=0.228).

DISCUSSION

Following a period of 24 weeks in a clinical trial patients suffering from leg ulceration improved in all scores of the Nottingham Health Profile. These changes were largely associated with healing the areas of leg ulceration. The 4LB appeared to demonstrate greater improvements in both energy and mobility compared with the ACB. Although the energy score may have been related to increased healing rate in patients randomised to 4LB, it could not explain the improvements in mobility which were only weakly associated with complete ulcer healing.

This trial has demonstrated that the Nottingham Health profile is sensitive to the detection of changes in patients perceived health status over 24 weeks of treatment in a clinical trial. It has also proven capable of detecting differences in scores between patients whose ulcer heals and those in whom it does not. The use of the 4LB has led to demonstrable improvements in patients perceived health state above those experienced using the ACB, which has since been withdrawn.

REFERENCES

1. Franks PJ Moffatt CJ Connolly M Bosanquet N Oldroyd M Greenhalgh RM McCollum CN: Community leg ulcer clinics: effect on quality of life. Phlebology 1994; 9: 83-86

2. Hyland ME Ley A Thompson B : Quality of life of leg ulcer patients: questionnaire and preliminary findings. J Wound Care 1994; 3: 294-298

3. Blair SD Wright DDI Backhouse CM Riddle E McCollum CN: Sustained compression and healing of chronic venous ulcers. Br Med J 1988; 297: 1159-61

4. Hunt SM McEwan J McKenna J. Measuring Health Status. London: Croom Helm, 1986.

Phlebology '95, D. Negus et al. (eds.). Phlebology (1995) Suppl. 1: 20-22

P088

A Comparison of the Cost Effectiveness in Venous Ulcer Management Between Current Methods in the Community and Treatment with a Multilayer Compression Bandage System

A.D. Fox, D. Trkulja, M.S. Whiteley, J.S. Budd and M. Horrocks

University Department of Surgery, Royal United Hospital, Combe Park, Bath, UK

INTRODUCTION

Chronic venous ulceration is a debilitating and socially isolating disease affecting approximately 150,000 people at any one time in the United Kingdom [1]. The prevalence increases with age and women are affected twice as commonly as men. Greater than 100,000 patients will be receiving treatment at any one time and a proportion of these remain unhealed [2]. Continuing treatment for this disorder costs the NHS £300 - 400 M per year [3]. Treatment remains largely community-based [1,4] but this can be ineffective due to staff variation, financial constraints and dressing inconsistencies. The aim of this study was to compare the cost effectiveness of current community treatment with a multilayer compression bandage system recently introduced into our own dedicated hospital-based ulcer clinic.

PATIENTS AND METHODS

Inclusion criteria included the presence of venous ulceration for greater than 2 months and the absence of arterial disease (ankle brachial pressure index greater than 0.8). A multilayer compression bandage system (Figure 1) was applied twice weekly and a computerised planimetry system was used to assess ulcer healing. Following full resolution of the ulcer patients were referred for venous duplex and definitive management. Analysis of the community treatment to date was performed retrospectively (costs accurate February 1994) and compared with the multilayer treatment system.

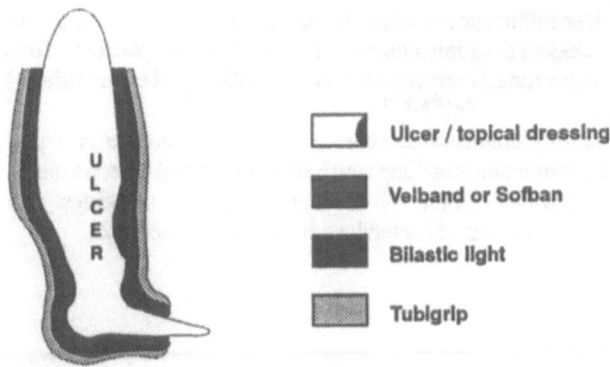

Figure 1   **Multilayer Compression Bandage System**

RESULTS

Eleven patients (9 females, 2 males) with chronic venous ulcers were recruited directly from the outpatient clinic. Their mean age was 67.7 years. Two patients had a past history of deep vein thromboses and 4 patients had undergone superficial venous surgery. Six patients had a history of arthritis (4 rheumatoid). All patients had generalised lipodermatosclerosis but only one patient had an established infection (staphylococcus aureus). All patients still had active ulceration at the time of recruitment to the dedicated hospital clinic despite a median duration of community treatment of 75 weeks (range 8 - 160 weeks). The comparative data for community and hospital treatment is shown in table 1.

PARAMETERS PER PATIENT	COMMUNITY	DEDICATED CLINIC
DURATION OF ULCER (weeks)	75 (8 - 160)	6 (2 - 18)
NO. OF VISITS	129	8
SALARY (£)	774	48
COST OF DRESSINGS(£)	642	48
TOTAL COST/ PATIENT (£)	1416	96

Table 1   Individual patient data comparing community treatment and the dedicated clinic using the multilayer compression bandage system

The dedicated ulcer clinic successfully healed all the ulcers in a mean time of 6 weeks (range 2 - 18 weeks) at a total cost of £1,056 (£96 per patient). Total community treatment cost at the time of recruitment was £15,576 (£1,416 per patient) but no ulcers had fully healed.

Duplex sonography demonstrated deep venous incompetence in 3 patients and the remainder had a combination of long and short superficial venous incompetence. All patients were prescribed Grade II graduated support stockings following ulcer resolution and remain healed at a median follow-up of 6 months.

DISCUSSION

Community venous ulcer treatment appears to be costly and ineffective and in this small series a potential saving slightly in excess of £14,000 could have been achieved by introduction of the dedicated ulcer clinic at an earlier stage. Providing the clinic offers elective compression a dramatic improvement in patient outcome may be achieved. Current community problems include lack of continuity, variation in types of dressings, inadequate compression and poor follow-up which may result in early recurrence particularly in the absence of graduated stockings following ulcer resolution.

Healing rates of 74% within 12 weeks have been reported [5,6] and similar results have been achieved in community pilot studies using the same techniques [7]. However, early referral for further investigation, definitive treatment and prescription of appropriate graduated support stockings is essential [8,9]. We recommend standardisation of services by a dedicated team, which may be hospital- or community-based, using a multilayer compression bandage system.

REFERENCES

1   Freak L, McCollum CN. The effective management of venous ulceration. Vascular Medicine Review. 1992; 3: 53 - 62.
2   Callam MJ, Ruckley CV, Harper DR, Dale JJ. Chronic ulceration of the leg: extent of the problem and provision of care.
     BMJ 1985; 290: 1855 - 1856
3   Editorial. King's Fund Grant to help treat leg ulcers. BMJ 1988; 297: 1412
4   McIntosh JB. Decision making on the district.
     Nursing Times 1979; 75 (29): 77 - 80
5   Moffatt CJ, Dickson D. The Charing Cross high compression four-layer bandage system. J Wound Care 1993; 2(2): 91 - 94
6   Moffatt CJ, Stubbings N. The Charing Cross approach to venous ulcers.
     Nursing Standard 1990; 5(12): 6 - 9
7   Callum MJ, Ruckley CV, Dale JJ, Harper DR. Hazards of compression treatment of the leg: an estimate from Scottish Surgeons.
     BMJ 1987; 295:1382
8   Browse N, Burnand KG. The causes of venous ulceration.
     Lancet 1982; 2: 243 - 245
9   Editorial. Venous ulceration. Lancet 1982; 2: 247 - 248

Phlebology '95, D. Negus et al. (eds.). Phlebology (1995) Suppl. 1: 23-24

P116

Chronic Venous Insufficiency of the Lower Limbs and its Socio-Economic Significance

M. de Castro Silva

Mineira Academy of Medicine, Belo Horizonte, Minas Gerais, Brazil

INTRODUCTION

The importance of Chronic Venous Insufficiency (CVI) is not restricted to its medical aspect. Its socio-economic repercussions have roused increasing interest in view of its cost to the various services of social security, thus reflecting on the national economy. To confirm this we have only to consider the fact that nine tenths of all vascular diseases that affect the lower limbs consist of phlebopathies. In the specific cases of CVI, studies made in other countries have shown the prevalence of various forms of this syndrome. In Denmark, stasis ulcers have been observed in 3.9% of individuals, of which 1.9% were men and 5.9% women. In Switzerland, 1% of individuals in full professional activity were afflicted and in the United States 0.2% of the women and 0.1% of the men. The number of lost workdays per year caused by the morbidity is estimated at 500,000 in England and Wales, and 2,000,000 in the United States. Only now has the social and economic importance of CVI been taken under serious consideration in Brazil. An epidemiological study made by the Health School Center of Botucatu, São Paulo, Maffei found predominance of varicose veins in a group of individuals above 15 years of age-pregnant women not included-who had come to the Center for a routine examination or because they needed medical advice. In this group, 47.6% had varicose veins, 37.9% being men and 50.9% women. The predominance of serious or moderate varicose veins was 21.2%.

MATERIAL AND METHODS

A report published by the Brazilian Ministry of Health Insurance in September 1984 lists the incidence according to their frequency of the 50 principal diseases that cause temporary absence from work, as well as the social benefits paid due to this situation. The reference years was 1983 (Table 1).

Despite the high incidence of CVI, 14th place, these numbers do not reflect entirely Brazilian reality. In a total of 25,062,000 contributors, 65% were men and 35% were women. The social security benefity is exclusively accorded to the worker, his family not being included. Nonetheless, although women represent only a third of the working force in Brazil, they form a larger contingent of beneficiaries: 3,825 women, 3,717 men (Table 2). These numbers indicate clearly how serious the problem is. The Brazilian population between 25 and 69 years of age is expected to soar up to 64,463,000 by 1994 (source: IBGE-Brazilian Institute of Geography and Statistics), of which 31,037,000 (48.1%) are women, among whom the incidence of varicose veins and its consequent complications is much higher as compared with men.

Table 1. Incidence of the principal diseases that caused absence from work and the granting of benefit (illness assistance) by order of frequency: year 1983. (Source: Brazilian Ministry of Health Insurance. Published in September 1984).

Diseases	Incidence in 1,000 benefits granted	No. of cases
1. Postoperative	278.8	208,167
2. Neurotic disorders	69.9	52,258
3. Essential hypertension	50.3	37,556
4. CID 724	39.3	29,403
5. Osteoarthrosis and related disorders	27.5	20,570
6. Pulmonary tuberculosis	19.5	14,580
7. Psychosis other than organic	19.4	14,544
8. Hypertensive cardiac diseases	16.1	12,024
9. Fracture of the cubitus and the radius	14.7	11,008
10. Epilepsy	13.3	9,956
11. Schizophrenic psychosis	12.6	9,468
12. Viral hepatitis	10.8	8,104
13. Alcohol dependency syndrome	10.7	7,994
14. Chronic venous insufficiency	10.0	7,542
15. Duodenal ulcer	9.0	6,731
16. Cardiac insufficiency	8.0	5,974
17. Rheumatoid arthritis	6.9	5,219
18/22.		
23. Bronchial asthma	6.3	4,712
24/33.		
34. Diabetes mellitus	4.8	3,594
35/42.		
43. Renal infections	3.5	2,669

Table 2. Number of contributors and illness assistance: year 1983. (Source: Brazilian Ministry of Health Insurance, 1984).

	No.	(%)
Number of contributors	25,062,000	
Men 16,290,300		(65)
Women 8,771,700		(35)
Illness assistance (IA) granted	746,556	
Chronic venous insufficiency	7,542	
Men 3,717		
Women 3,825		

REFERENCES

1. Maffei FHA. Contribuição para o conhecimento da epidemiologia das varizes e da insuficiência venosa crônica de membros inferiores. Estudo em adultos atendidos no Centro de Saúde Escola de Botucatu. Tese de Livre-Docência, Faculdade de Medicina de Botucatu, UNESP, Botucatu, 1982: 131.
2. Castro-Siva M. Insuficiência venosa crônica: diagnóstico e tratamento clínico. In: Maffei FHA, ed. Doenças vasculares periféricas. Rio de Janeiro: MEDSI, 1987.

Phlebology '95, D. Negus et al. (eds.). Phlebology (1995) Suppl. 1: 25-28

PI/1.1

The Epidemiology of Venous Disease

R. Beaglehole

Department of Community Health, Faculty of Medicine and Health Science, University of Auckland, New Zealand

INTRODUCTION

From a public health perspective, venous diseases (varicose veins and venous ulcers) remain neglected and under-researched, despite contributing considerably to the burden of disease experienced by individuals and health services. Venous disease are among the commonest chronic diseases in rich countries. Fortunately much of the burden of varicose veins is mild; only "severe" varicose veins are associated with major health service costs [1].

The epidemiological situation is frustrating since there is convincing evidence that venous diseases are preventable [2].

This paper reviews the epidemiological literature on varicose veins and venous ulcers identified from computerised data-bases and concentrates on the higher quality papers from the period 1985-1994.

METHODOLOGICAL CONSIDERATIONS

The epidemiological literature on venous diseases leaves much to be desired. As a consequence, precise and unbiased estimates of the prevalence and incidence of venous diseases and their association with possible causal factors are still rare. Common methodological problems include: highly selected study populations, inadequate case definition, inadequate measurement of venous disease and the risk factors, and lack of standardisation within and between studies.

From an epidemiological point of view it is not difficult to design and conduct high quality studies of venous diseases. The sample size of studies is not usually an issue since venous disease, at least in adults in rich countries, are very common and it is relatively easy to study large numbers of people. A well designed study would be based on a clearly defined geographic population, although it might be justifiable to study an occupationally defined population for the purpose of testing a specific hypothesis; no further studies of hospital patients, either in- or out-patients, are justified.

Future studies should use clearly defined criteria in a standardised manner with detailed attention to the reduction of observer bias. Unfortunately there is no objective criteria for venous diseases. According to Callam [3], the best criteria for varicose veins are those used in the Basle study based on the clinical appearance of the lower leg [4]. Cross national or cross-cultural studies have the potential to increase our understanding of venous disease; if these studies use different observers, particular attention must be given to minimising observer variation. An important responsibility rests with both funding agencies and journal editors to encourage good studies.

In this paper, the minimal criteria for a high quality epidemiological study of varicose veins are that it be community-based, used clearly specified criteria, described the methods used to reduce observer bias, and achieved a high response rate. The presentation of age specific or age standardised data are important since the prevalence of venous disease is strongly age related.

THE EPIDEMIOLOGY OF VARICOSE VEINS

Prevalence and incidence

Of the seven studies identified from Medline searches in the last decade, only one met the minimal criteria for high quality in that randomly chosen community samples were studied by direct examination and explicit criteria for varicose veins were used. Unfortunately, the Framingham studied presented data only on incidence which cannot be compared with the more common measures of prevalence, and no mention is made of efforts to reduce observer variation in this study [5]. Only one study mentioned training of observers, and this was a clinic based study [6]. All of the studies which used objective criteria used the criteria proposed by Arnoldi [7]; two studies utilised self assessment alone [8, 9]. The community based Finnish study, which validated self-reports of varicose veins in a small sub sample, did not specify the criteria used [8]. It is of interest, however, that the validation exercise found a high degree of both sensitivity and specificity for self reports of varicose veins. A study based on a sclerotherapy practice has not been included [10].

The prevalence of varicose veins from the recent literature varies from 15% to almost 50%. The two studies with the highest prevalence used a very broad definition of varicose veins which included all venous abnormalities and/or included people over the age of 70 years [6], [11]. From the data, both recent and old, it is likely that in middle aged men and women in rich countries the prevalence of various veins is about 14% in men and about 20% in women.

The only incidence study found a two-year incidence rate of 40 and 52 per 1,000 in men and women respectively. Interestingly, there was no relationship between age and the incidence of varicose veins; the higher prevalence in older people is simply a reflection of the fact that this is a non fatal condition.

Risk factors for various veins

Age and female sex are the best documented non-modifiable risk factors for the prevalence of varicose veins. There is also good evidence that people living in environments not overly exposed to "western" influences have a much lower prevalence. Of the modifiable risk factors, obesity is the best documented, but usually only in women which suggests that obesity may not be a causal factor; it may however, increase the severity of the condition. Multiparity is usually associated with a higher prevalence of varicose veins but it is not clear that this is a modifiable risk factor. The evidence on other modifiable factors, such as occupation, prolonged standing or sitting, is not sufficient for firm conclusions to be drawn. More information is required on the associations of diet and physical inactivity with varicose veins.

It is reassuring that varicose veins are not themselves associated with an increased risk of cardiovascular disease, after controlling for other causal factors [5]. Regrettably, there are no controlled studies on the impact of weight control and physical activity on the development of varicose veins.

The Epidemiology of Venous Ulceration

Venous ulcers are the most severe form of venous disease and in comparison with varicose veins are more easily measured, although their prevalence is much lower.

Well-designed community based studies have been published recently from Sweden [12] and Australia [13]. Both studies identified cases by referral from health care professionals and the Australian study also received self-referrals. Venous ulcers comprise about half of all chronic ulcers in the community and the prevalence is approximately 1%, the age specific prevalence is similar in men and women [13] and increases with age. Obesity is associated with the prevalence of venous ulcers [12] although formal aetiological studies have yet to be reported. Most venous ulcers are a result of deep vein thrombosis [12, 13].

Conclusion

The epidemiology of venous disease has not progressed over the last decade. There is a need for collaborative multinational studies, especially of varicose veins, to explore specific hypotheses relating to the aetiology of this common condition. Our understanding of the epidemiology and prevention of other chronic diseases, including coronary heart disease, is increasing rapidly from such studies. The methods used in these studies could easily be applied to venous diseases.

REFERENCES

1.   Biland L, Widmer LK. Varicose Veins (VV) and Chronic Venous Insufficiency (CVI). Acta Chir Scand 1988;Suppl 544:9-11.
2.   Beaglehole R. Epidemiology of varicose veins. World Journal of Surgery 1986;10:898-902.
3.   Callam MJ. Epidemiology of varicose veins. Br J Surg 1994;81:167-173.
4.   Widmer LK. Peripheral Venous Disorders Basell III. Bern, Hans Huber, 1978.
5.   Brand FN, Dannenberg AL, Abbott RD, Kannel WB. The Epidemiology of Varicose Veins: The Framingham Study. Prev Med 1988;4:96-101.
6.   Maffei FHA, Magaldi C, Pinho SZ, Lastoria S, Pinho W, et al. Varicose veins and chronic venous insufficiency in Brazil: Prevalence among 1755 inhabitants of a country town. Int J Epidemiol 1986;15:210-217.
7.   Arnoldi CC. The aetiology of primary varicose veins. Dan Med Bull 1957;4:102-107.
8.   Laurikka J, Sisto T, Auvinen O, Tarkka M, Laara E, et al. Varicose veins in a Finnish population aged 40-60. J Epidemiology and Community Health 1993;47:355-357.

9.   Franks PJ, Wright DDI, Moffatt CJ, Stirling J, Fletcher AE, et al. Prevalence of Venous Disease: A Community Study in West London. Eur J Surg 1992;158:143-147.

10.   Sadick NS. Predisposing Factors of Varicose and Telangiectatic Leg Veins. Phlebology 1992;18:883-886.

11.   Hirai M, Naiki K, Nakayama R. Prevalence and risk factors of varicose veins in Japanese women. Angiology 1990;3:228-232.

12.   Nelzen O, Bergqvist D, Lindhagen A. Venous and non-venous leg ulcers: clinical history and appearance in a population study. Br J Surg 1994;81:182-187.

13.   Baker SR, Stacey MC, Jopp-McKay AG, Hoskin SE, Thompson PJ. Epidemiology of chronic venous ulcers. Br J Surg 1991;78:864-867.

Phlebology '95, D. Negus et al. (eds.). Phlebology (1995) Suppl. 1: 29-30

PI/4.2

Classification of Recurrent Varicose Veins

P.A. Stonebridge ChM FRCS Edin

Department of Surgery, Ninewells Hospital and Medical School, Dundee, Scotland

Recurrence of varicose veins following operation is in the order of 20-30%[1-3]. Classification of this relatively common problem should preferably serve two functions; firstly to identify and if possible highlight the cause of the recurrence and secondly to indicate the nature of the intervention required to rectify the problem. A widely accepted classification would also allow the comparison of series, audit and be of use in surgical training. There are two recently published classifications. The first by Darke[2] identified three types of recurrence based on a series of 76 limbs with recurrent varicose veins.

Type 1   recurrence without demonstrable incompetence in either long or short saphenous systems (28% of the series).
Type 2   recurrence derived from the ipsilateral saphenous system not previously operated on (9% of the series).
Type 3 -   recurrence in either long or short saphenous systems following previous operation on that system (63% of the series).

More recently the author published the results of a varicographic analysis of the pattern of recurrent varicose veins in 128 limbs. This study concemtrated on the important clinical question 'do I need to re-explore the saphenofemoral junction?', which put another way is 'was the first operation done properly?' The classification which developed from this pragmatic approach can be seen in figure 1. The presence of the long saphenous vein in the thigh was also noted.

The classification relates solely to recurrence following operation on the long saphenous vein. It is therefore less encompassing than Darke's classification but is more specific. Type 1 recurrence equates with a technically inadequate primary operation requiring re-exploration of the saphenofemoral junction. Type 2 recurrence does not require groin re-exploration but may involve a retained long saphenous vein in the thigh.

In the latter study 66% were due to unsatisfactory primary surgery (type 1), with 60% of recurrence being associated with a retained thigh segment of long saphenous vein. Sixty three per cent of recurrence in Darke's study were Type 3 and therefore also related to inadequate primary operation.

30

Both classifications identify and highlight the cause of the recurrence and interesting had virtually identical findings, which, if either, is the more useful clinical tool will only be answered by time.

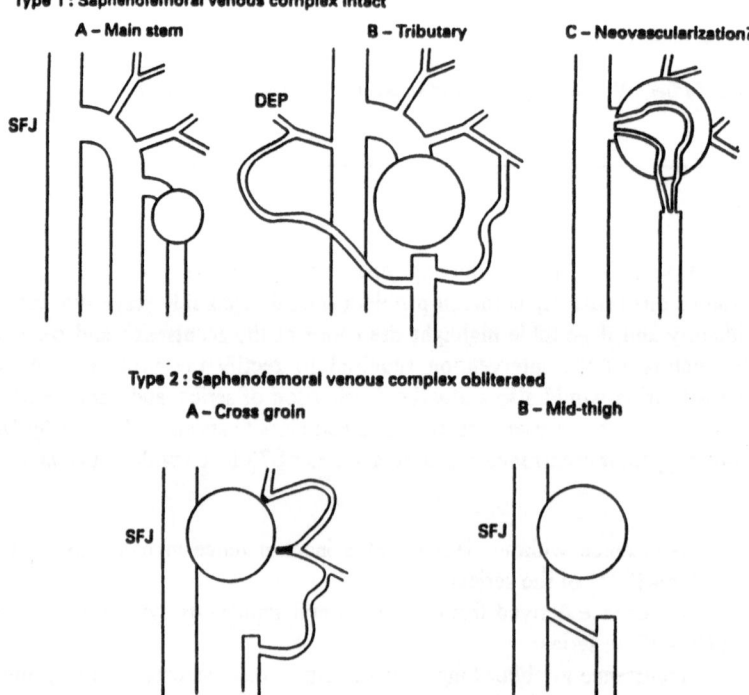

Type 1 : Saphenofemoral venous complex intact

A – Main stem   B – Tributary   C – Neovascularization?

SFJ   DEP

Type 2 : Saphenofemoral venous complex obliterated

A – Cross groin   B – Mid-thigh

SFJ   SFJ

SFJ = saphenofemoral junction; DEP = deep external pudendal; ◯ = scar tissue

References

1. Royle JP. Recurrent varicose veins. World J Surg 1986; 10: 944-53.

2. Darke SG. Recurrent varicose veins and short saphenous insufficiecy. In: Venous Disorders, Bergan JJ, Yao JST 9eds). Philadelphia: WB Saunders, 1991: 217-32.

3. Stonebridge PA, Chalmers N, Beggs I, et al. Recurrent varicose veins: a varicographic analysis leading to a new practical classification. Brit J Surg 1995; 82: 60-2.

PI/4.1

RECURRENT VARICOSE VEINS OF THE LOWER LIMB

Naga R, El-Sheikh S

Departments of Surgery and Radiology, Faculty of Medicine, Alexandria University, Egypt.

Objective: To review the etiology and pathogenesis of recurrent primary varicosity of the lower limbs.
Design: Retrospective review of consecutive case.
Patients: Out of 432 stripping for primary varicose veins in the lower limbs during one year, 25 limbs presented with recurrent varicosities after an average period of 2.5 years from the initial operation.
Intervention: All limbs were operated to excise the recurrent varicosities and explore the anatomical and pathological causes of recurrence.
Measurements: Doppler venous studies were carried before and after surgery.
Results: Stripping and/or excision of recurrent varicosity results in a better functional limbs than the conservative measures.
Conclusions: Major recurrent varicosities should be properly excised. Minor recurrence without haemodynamic abnormality can be managed conservatively.

P104

QUALITY OF LIFE OF PATIENTS ATTENDING A SPECIALIZED LEG ULCER CLINIC

Vandongen YK,* Stacey MC,* French, D✝

Department of Surgery* and Department of Psychology✝, University of Western Australia, Perth, Australia.

Objective: To assess the quality of life (QoL) of a group of patients receiving treatment for chronic leg ulceration and to determine whether the use of a less restrictive dressing type affects quality of life.

Design: Non randomized pre-test post-test design.

Patients: 26 patients with chronic ulceration of the leg. Comparison group subjects were matched for age and sex.

Measurements: Schedule for the Evaluation of Individual Quality of Life (SEIQoL) was administered by one interviewer on presentation to the clinic and repeated after 12 weeks.

Results: Mean QoL index scores improved significantly in the patient group in the 12 weeks treatment period (57.37-68.14) (paired t-test, p=0.02). These however did not reach the levels of the comparison group which remained unchanged during the same period of time (78.92-82.17) (paired t-test, p>0.05). The improvement in the patient group was not dependent on whether complete healing had occurred (repeated meaures ANOVA, p>0.05). Dressings which allowed better mobility did not significantly alter QoL compared with the standard treatment of compression bandaging (repeated measures ANOVA, p>0.05).

Conclusions: Improvement in QoL began with the commencement of treatment for ulceration however it did not reach the levels of the comparison group. Improved QoL was not dependent on complete healing or on the type of dressing used.

Pl/1.4

THE TRUE PREVALENCE OF LEG ULCERS HAS BEEN UNDERESTIMATED

Nelzén Q, Bergqvist D*, Lindhagen A.
Depts of Surgery, Kärnsjukhuset Skövde and *University Hospital Uppsala, Sweden

Objective: To assess the true prevalence of leg ulcers in the community.

Design: A retrospective review of the combined results from three consecutive surveys assessing leg ulcer prevalence in a defined population.

Patients: In the Skaraborg study(i) 827 patients with current ulcers known to health care professionals were identified out of a population of 270 800. The Skaraborg & Malmö study(ii) produced prevalece estimates based on a questionnaire to 12 000 randomly selected persons from the general population, aged 50-89 years. In the Skövde study(iii) the prevalence calculations were based on a selected sample of 2 785 industry employees, aged 30-65 years.

Results: By comparing the age-specific point prevalences registered in the studies we found that assessment through health care severely undervalued the true prevalence below the age of retirement(65 years). This indicated a high proportion of self treatment for leg ulcers in younger people. In people of old age most appeared to be known to health care professionals. Overall about half of all people with open ulcers appear to be self-caring, most of them being of working age. Only 1/10 aged 30-49(i,iii), 1/7 aged 50-59(i,ii,iii) and 1/3 aged 60-69 years(i,ii,iii) seems to recieve professional treatment. One out of three with history of leg ulceration appears to have an open ulcer at any one time. Every second ulcer has a venous cause.

Conclusions: The point prevalence of open leg ulceration in Skaraborg county was 0.6% of total population, only half of which is known to health care. Leg ulcer prevalence based on patients known to health care professionals underestimates the true prevalence.

P187

COMPUTERISED ORGANISATION OF A CENTRE OF PHLEBOLOGY

Botta G, Mancini S
III Department of General Surgery and Pabisch Centre of Phlebology - Director Prof. Sergio Mancini
Faculty of Medicine and Surgery, University of Siena, Italy

Objective: To organise and manage a computerised modern Centre of Phlebology.

Design: To analyse and refer the experience acquired in the organisation and management of the computerised "W Pabisch" Centre of Phlebology of the University of Siena.

Patients: All the phlebologic patients seen and treated by surgeons of the Pabisch Centre, articulated for sex, age, city, week, month and years of clinical observation.

Characteristics: A local network of Personal Computer has been created with the Windows NT 3.5 Server operative system, with the possibility of connecting both the token-ring type of network which links the Polyclinic of Siena, and Computer Vax of the Calculus Centre of the University of Siena, which is integrated in the Internet.

Results: Simplification of the bureaucratic procedures of admittance to the Centre; computerised management of the bookings for first visits, controls, non-invasive instrumental diagnostic examinations, medications for varicose ulcers, keeping of list of patients for surgical treatment or for sessions of sclero and pressotherapy with automated recording of the effected operation and follow-up. Connection through modem/fax to other Institutes, Academies and Scientific Societies for exchange of clinical data, protocols and scientific works on phlebologic topics.

Conclusions: To evaluate in real time, in great number, the results obtained in reference to the illness of the patients.

P188

AMBULATORY PHLEBECTOMY AND COMBINED TREATMENTS

Sánchez, C.; Altmann-Canestri, E.; Tropper, U.; Tkach, E.; Clar, D.

Clínica de Flebología "Dr. César F. Sánchez"
Avellaneda 198 - (1642) San Isidro, Buenos Aires, Argentina.

Objective: To prove the advantages of combining different techniques with ambulatory phlebectomy in order to avoid and solve its inconveniences with excellent esthetic results.

Design: 3,580 interventions carried out since 1986.

Intervention: Iconographic sequences are shown of the indications, techniques and results of each of the associated methods.

Measurements: Sequences of photographs before and after the patients were subject to ambulatory phlebectomy and combined treatment.

Results: Iconographic sequences are shown of patients treated with these techniques which show excellent therapeutic and esthetic results obtained in some ambulatory cases of the 3,580 interventions made.

Conclusions: The impossibility to obtain a definitive cure of the varicose disease, genetically determined, induces us to offer our patients simple, practical, economic and esthetic treatments, and these are the advantages offered by the ambulatory phlebectomy and combined treatments.

P110

INCREASING MORTALITY DUE TO VENOUS DISEASES IN HUNGARY

Sándor T, Molnár L, Acsády Gy, Monos E

Dept.of Surgery Teaching Hospital, Institute of Behavioural Sciences, Cardiovascular Surgical Clinic, Experimental Research Department - 2nd Inst. of Physiology, Semmelweis University of Med.Budapest,Hungary

In 1970 in Hungary venous mortality caused mainly PE and DVT was 9/100 000 for males,12,5/100 000 for females and the incidence of fatal venous diseases was 1116.WHO data indicated medium incidence of mortality caused by vein diseases compared to the economically more developed European countries.

Unfortunately, from the early seventies a continuous deterioration of health of the Hungarian population has been observed.Especially a striking increase in mortality due to myocardial infarction under the age of 40, suicide, alcoholism and drug has been recorded.Parallelly with the increasing number of the arteficial abortions the number of the whole Hungarian population has decreased.In 1980 it was 10,7 M and in 1990 only 10,4 M. By 1990 mortality due to venous diseases had doubled actually: 16,8/100 000 for males and 22,8/100 000 for females. That means that the incidence of fatal venous diseases was 2064.

During this 20 years period due to effective preventive programs the mortality rate caused by vein pathology decreased in several European countries, e.g. by 80% in Austria and by 50% in Finland.The mortality rate between Finland and Hungary is 1:7 for males and 1:5 for females. Even the relatively high mortality in Switzerland comprises only 50% of that observed in Hungary.In the contrary many Hungarian surgeons ignored the high risk of venous thromboembolism and harboured fears and doubts about using modern thromboprophylaxis with LDH or LMWH.

Recently a research project has been initiated at the Semmelweis Univ. of Medicine in Budapest which focuses on factors contributing to the high incidence of vein pathology and morbidity in Hungary in order to reduce their occurence.The study uses methodology and experiences of clinical physiology, phlebology, vascular surgery and population health related sociology.At the same time large scale activity has been started for using up-to-date thromboprophylactic methods.

OP/12.3

A PROSPECTIVE AUDIT OF SURGERY FOR VARICOSE VEINS

T. LEES, S. SINGH, P. SPENCER, C. RIGBY, J. COOPER, J BEARD

Rotherham DGH & Royal Hallamshire Hospital, Sheffield, South Yorkshire

Recurrence rates of up to 77% have been reported following varicose vein surgery and this has been attributed to inadequate surgery. This study assesses the adequacy of surgery for varicose veins by comparing pre and post operative sites of venous incompetence.

30 limbs in 21 patients with varicose veins taken from the waiting list of 6 surgeons were examined by duplex scanning preoperatively and 6 weeks postoperatively. Sites of venous incompetence were recorded, and the patients were assessed postoperatively for the clinical result of the surgery.

Prior to surgery 23 limbs had venous reflux in the long saphenous vein at the saphenofemoral junction and 7 limbs had reflux in the short saphenous vein at the saphenopopliteal junction. Postoperatively residual incompetence at these sites was present in 6 limbs at each site. The reasons for persistent incompetence were failure in preoperative diagnosis (1 limb), inadequate surgery (10 limbs), and inappropriate surgery (1 limb). 77% of varicose vein surgery was performed by Higher Surgical Trainees or Staff Grade Surgeons who were responsible for 83% of the persistent venous incompetence. 25 limbs had a good or moderate clinical result when assessed by the patient.

This audit has confirmed that incomplete surgery for varicose veins is common and implies a need to improve this surgery. Further follow up of these patients will assess if incomplete surgery is associated with the development of recurrent varicose veins.

P112

THE TREATMENT OF VARICOSE VEINS WITHIN A HEALTH REGION

T.A. LEES, J.D. HOLDSWORTH

Wansbeck Hospital, Woodhorn Lane, Ashington, Northumberland, NE63 9JT

A high recurrence rate following varicose vein surgery has been attributed to inappropriate or inadequate initial surgery[1]. We have surveyed how varicose veins are assessed and treated by Surgeons in a single Health Region.

A postal questionnaire was distributed to 86 surgeons. 83% responded with 60 Surgeons (85%) treating varicose veins. It was estimated that 4307 limbs were treated in the Region annually of which 58% were treated by Surgeons with a vascular specialist interest. Initial assessment of varicose veins by clinical examination was accompanied by tourniquet testing and by hand held doppler examination in 75% and 37% respectively. In the surgical treatment of long saphenous vein incompetence all surgeons performed high saphenous ligation, 67% stripped the long saphenous vein to the knee and 23% stripped it to the ankle, with 88% performing multiple avulsions. For short saphenous vein incompetence 92% of surgeons performed saphenopopliteal ligation, with 28% localising the saphenopopliteal junction by investigation, 13% stripping the short saphenous vein and 80% undertaking multiple avulsions;

Considerable variation in varicose vein management continues. Standardising the assessment and surgical treatment may help to reduce recurrence rates.

1. Negus D. Recurrent varicose veins: a national problem. Br J Surg 1993; **80**: 823-4.

P192

COMPUTERISED MANAGING PROGRAM IN PHLEBOLOGICAL DISEASE BY A NEW CLINICAL CHART

Lorenzo Tessari
Ambulatorio Flebologico "Dr Bassi" Via Cormeo 16 Trieste-Casa di Cura "Dr Pedarzoli, Peschiera d/gVr - ITALY

Objective: A doctor dealing with patients suffering from varicose disease needs fast access to:

- fast medical anamnesis of the patient
- a list of surgical procedures and elastic compression bandages
- the number and kind of sclerotherapy done during the phlebological treatment.

Quick access to this information is possible only with a computerised database clinical chart, easy to see also for someone not adept with computers.

This approach with computerised clinical charts might a useful approach in the phlebological field. Therefore it could also represent a useful attempt for the unification of the management of the phlebological medical rooms.

Conclusion: We propose to work out a suitable clinical chart in order to supply a computerised management program of the veins disease stages.

OP/12.6

PSYCHOLOGICAL IMPLICATIONS OF SYNDROME ULCEROVARICOSUM: A POSSIBLE KEY FOR AN EARLY REHABILITATION

Grivceva-Panovska V, Pavlova Lj, Nikolovska S, Starova A, Dervendzi D
Dept of Dermatology, Medical Faculty, Skopje, Macedonia

Objective: To identify, classify and evaluate psychological implications in patients with syndrome ulcerovaricosum.

Design: Randomized parallel group open clinical trial.

Patients: 40 patients (30 out- and 10 in-patients, 28 female, 12 male, age range 28-70 yrs) vs control groups of matching-pairs of healthy volunteers and sy ulcerovaricosum patients, conventionaly treated.

Intervention: standard angiological treatment was enriched by means of a psychotherapeutical approach and liason therapy.

Measurements: MMPI, SAT9 applied by a trained dermatologist and psychiatrist, and clinical outcome by scoring system designed and applied by a dermatologist-angiologist, documented at the zero-time, and every 3 months during the 24-months follow up period.

Results: All patients were depressive, but in terms of subclinical extent. They were deeply worried for the further evolution of the disease. Despite continuous education they were reluctant to use elastic badages vs surgical intervention, accepted as a "final solution". All patients were more concerned for the complications and visible aspects of the syndroma vs physical symptoms (pain, edema, etc.). There is not a positive correlation between the angiological score and the degree of alexithymity as measured by SAT9, but in terms of the clinical outcome time needed for epithelization is up to 30 % shorter in patients with lower SAT9 score, and up to 43 % shorter in psychotherapeutically treated group. All patients stated a pre-treatment decision to change a working place for a new one, far enough from the judgemental eyes of "healthy" others.

Conclusions: Our results indicate that patients with Sy ulcerovaricosum are "poor communicators" and more than "skin deep" affected. The suggested complex diagnostic and therapeutically approach might offer a key for better understanding and a more efficient early rehabilitation of our patients.

OP/12.2

Cost effectiveness of oxpentifylline in venous ulcer healing

N Bosanquet, PJ Franks D Brown J Straub DR Harper CV Ruckley

Health Policy Unit, St Mary's Medical School, London, Lothian & Forth Valley Leg Ulcer Study, Royal Infirmary of Edinburgh, Edinburgh and Falkirk & District Royal Infirmary , Falkirk UK

Objective: To determine the cost effectiveness of healing venous ulceration in a randomised double blind trial of oxpentifylline.

Design: Randomised double blind placebo (p) controlled parallel groups trial of oxpentifylline (ox) , the main clinical results being presented in another communication. Cost of treatment assessed by patient questionnaire and clinical notes to assess frequency of visits to or by health professionals. Related to complete ulcer healing on the reference limb after 24 weeks of treatment.

Patients: Patients attending leg ulcer clinics with venous ulcer present for longer than two months, with a diameter >1cm .

Measurements: Questionnaires to patients with leg ulceration at baseline and after 24 weeks together with clinic notes to assess frequency of visits to or by health care professionals.

Results: 200 patients were randomised (101 ox, 99 p) and followed up for 24 weeks. Patients were well matched for presenting variables. Preliminary analysis showed that the total cost of ulcer care was greater per healed ulcer on placebo (£1,787) compared with oxpentifylline (£1,524). Estimates of costs for delivering care in the normal situation will be presented together with past treatment costs.

Conclusions: Oxpentifylline healed more ulcers and was more cost effective than placebo.

OP/3.3

Patterns of venous reflux in 317 patients with obvious varicose veins.

J. Jérôme GUEX MD, Michel PERRIN MD, Bernard HILTBRAND MD, J. Marc BAYON MD, Franck HENRI MD , François A. ALLAERT MD,PhD.

32 Bd Dubouchage, NICE. Clin. du Grand Large, DECINES. Labos Beaufour, 24 Rue Erlanger, PARIS. Département de Biostatistiques et Informatique Médicale, CHU du Bocage, DIJON.

Objective: The aim of this study was to determine symptomatology, clinical class, and topographic pattern of varicose veins in a consecutive series of patients with venous complaints.

Methods: From Dec 1, 1992 to May 31, 1993, we examined 498 lower limbs in 317 patients with obvious varicose veins and no previous treatment. We looked for a reflux in 25 different venous segments in each limb, and collected clinical and historical data to determine anatomo-clinical correlations.

Results:

- *Population:* M: 70 (22.1 %), Fem: 247 (77.9 %) out of 317 patients. Age: Mean = 48.9 ± 14.6 y. Height: Mean = 1.654 ± 0,083 m. Weight: Mean = 66 ± 13.2 Kg.

Paternal heredity in 66 (20.8%), maternal in 167 (52.7%) of subjects. Bilateral Varicose veins: 192 (60.6%). Patient-known prior Deep Venous Thrombosis: 15 (4.7 %).

Lipodermatosclerosis: 36 out of 498 legs (7.2%), Skin changes: 14 ulcers out of 498 legs (2.8%) and 28 varicose eczema (5.6%).

Classes of chronic venous insufficiency (CVI) out of 498 legs:

grade 0: 117 (23.5%); gr 1: 310 (62.2%); gr 2: 47 (9.4%); gr 3: 24 (4.8%).

- *Topography:* Duplex derived reflux was found in the greater saphenous vein territory (junction or trunk or related perforator or main tributary) in 423 limbs (85.3%) but sapheno-femoral junction was incompetent only in 342 legs (68.7%). A reflux was found in lesser saphenous vein territory in 100 limbs (20.1%) and in sapheno-popliteal junction in 92 (18.5%). Strictly non saphenous origin of varicosities was found in 31 limbs (6.2%). A deep venous incompetence was found in 48 legs (9.6%).

Conclusions:

These findings yield data on distribution and occurence of the venous lesions of the lower limbs in varicose disease. They show that a complete duplex-scan allows appropriate management of the various lesions and preservation of main saphenous trunks.

PI/12

Early and preclinical signs of genuine varices - results of a longitudinal epidemiological study with children and juveniles (Bochum Study I-III)

U. Schultz-Ehrenburg*, N. Weindorf and H. Hirche

Department of Dermatology, Medical Center of Berlin-Buch, Berlin (Germany)

The Bochum Study I-III had been performed as a longitudinal study (1982-1991) on pupils of 11 secondary schools (Gymnasium) of the city of Bochum, when being in the 5th and later in the 9th and 13th classes. A questionnaire, physical examination as well as Doppler sonographic and photoplethysmographic investigations had been done with all the pupils explored in this study.

A significant correlation between familial predisposition and the occurrence of varivose veins could not be found even in the Bochum Study III. The girls predominated with respect to reticular varices and telangiectasias, the boys with respect to trunk varices, branch varices and incompetent perforators. Photoplethysmography showed shorter refilling times for children and adolescents than for adults, but was not suitable to distinguish between pathological and normal findings at these age stages.

	Bo I n = 740 10-12 ys.	Bo II n = 518 14-16 ys.	Bo III n = 459 18-20 ys.
Saphenous refluxes	2 %	12 %	20 %
Large varices	0 %	2 %	6 %
Small varices	10 %	30 %	40 %

Large varices appeared later than small varices and were preceded by refluxes of the long and short saphenous veins (see Table). Thus, saphenous refluxes can be regarded as the earliest preclinical sign. It clearly identifies individuals at risk.

P194

MALPRACTICE IN PHLEBOLOGY

J TKACH-E, TKACH-C SANCHEZ

Centro de Flebologia y Estética Integral. Brown 591- Moron (1708). Buenos Aires, REPUBLICA ARGENTINA

Objective: Demonstration of a case of malpractice in phlebology with a great aesthetic damage.
Patient: A 53 year old woman without any clinical records attends a Phlebology consultancy room following her medical insurance booklet. She was diagnosed with ambulatory segmentary phlebectomy with extrasaphenous stretches with predominance of the lower right limb.
Results: The patient reports great pain and burning during the infiltration with the "alleged" local anaesthetic being calmed by her doctor who carries out stripping in the corresponding stretches through crochet.
Conclusions: The pain went on for 12 hours. 24 hours later ecchymotic halo close to the microincision was observed. A week later, there were 14 ulcers on the ecchymosis which measured 120 cm^2 (see poster), with pseudo purulent exudate, from intense skin sclerosis to aponeurosis, fibrin, necrotic remains, tissue of granulation with vessels of neoformation, accumulation of haemosiderin and inflammatory infiltration (histopathology).
The cause: Sclerosing detergent was mistakenly used as a local anaesthetic. We consider this an important case, a big contribution to phlebology since it actually happened. Something to be taken into consideration when choosing the procedure, preparation and presentation of substances, laboratories and safety systems to be used.

OP/12.5
THE COST EFFECTIVENESS OF THE DIAGNOSIS OF DEEP VEIN THROMBOSIS IN SYMPTOMATIC PATIENTS

Hull RD, Feldstein W, Pineo GF, Raskob GE;

University of Calgary, Calgary, Alberta

Objective: To determine the most cost effective approach to the diagnosis of deep vein thrombosis (DVT in symptomatic patients.

Design: An economic evaluation to identify the most cost effective approach to diagnosis of DVT based on a prospective study of over 500 patients referred to a Regional Thromboembolism Program with a first episode of clinically-suspected DVT. Real costs in 1992 Canadian dollars (CAN) and charges in 1992 U.S. dollars (US) were used. The diagnostic approaches (clinical diagnosis), Doppler ultrasound with B-mode imaging (U/S) and impedance plethysmography (IPG) were ranked from least to most cost effective, with effectiveness (health benefit) being defined as the number of patients with DVT correctly identified by objective testing, or the number in whom treatment was correctly withheld.

Results: Clinical diagnosis was not cost effective ($2,590, 784.00 CAN, $2, 624,200.00 US). Outpatient diagnosis using non-invasive testing was most cost effective. Serial U/S was more costly ($618,265.00 CAN, $1,326,180 US) than IPG ($527,165.00 CAN, $1,052,880.00 US). Initial U/S with serial IPG for patients with negative U/S was less costly than serial U/S ($551,065.00 CAN, $1,124,580.00 US).

Conclusion: Non-invasive leg testing for the diagnosis of DVT is more cost effective than clinical diagnosis, with serial IPG being less costly than serial U/S. The combined approach of initial U/S followed by serial IPG combines the advantage of the more sensitive U/S with the less costly serial IPG.

P195
A COMPUTERISED PHLEBOLOGY RECORD

George M Robb
New Westminster Vein Clinic, 501-625 5th Avenue, New Westminster BC, Canada V3M 1X4

Objective: To describe a system for development of a computerised phlebological record.

Results: A system that has been in use in this clinic for the past four years is described. Important features are the method of recording the location of varices, and relating this to the results of treatment.

P177

VARICOSE DISEASE EPIDEMIOLOGICAL STUDY

Figueirôa CLS, Figueirôa ES, Soares MV

Hospital STA Izabel - Escola da Medicina - Salvador - Bahia

Summary: The AA, in isolated collections 410 under aged 5-10 years, and 1952 adults studied the prevalence of manifestations of the varicose disease and influence of the risk factors.
The analysis of the material permitted to conclude the existence of the initial manifestations of the varicose disease in the infantile population presenting the predominance of the masculine sex, and the initial forms, a factor of the maxim importance is the maternal heredity, we can observe a tendency to the progression of the clinical manifestations with the advance of the ages, remaining for the definitive conclusions evaluations that must be realised in 4 and 8 years in the future.
In the collected material among individuals more than 11 year olds, existed a significant parcel of the feminine sex maintaining the maternal positive heredity as an expressive risk factor, and an injured predominance for the left inferior limb.
The gestations were confirmed as accelerating the eclosions factors observing the gestational trimonthly with injured specifications, and the third gestation pontificating as the bigger accelerator.
It is discussible, in spite of statistically demonstrated in our material, if the prevalence of the most severe lesions are dependent on the age or a characteristic of the duration factor of the disease.

P184

EVALUATING THE SF 36 AS A MEASURE OF QUALITY OF LIFE IN PATIENTS WITH CHRONIC LEG ULCERS.

Price P E, Harding K G

Wound Healing Research Unit, University of Wales College of Medicine, Cardiff,U.K.

Objective: To evaluate a generic measure of Quality of Life (SF-36, Ware 1989) in measuring QoL in patients with chronic wounds.

Design: A longitudinal, single cohort study.

Patients: 74 patients with open leg ulcers attending a specialist unit, minimum duration of wound = 3 months.

Measurements: A self-reported generic measure of QoL (SF-36) was administered within 2 weeks of referral to the unit (time point 1) and again 4 months later (time point 2).

Results: The demographic features of this group suggest that they are representative of the population of patients with leg ulcers (mean age = 70.3 years, M:F ratio = 1:2). At time point 1 these patients rated themselves significantly lower than age-equivalent norms on 7 of the 8 sub-scales of the SF-36 (Two sample unrelated t-test, $p < 0.001$). Improvement in ratings at time point 2 was only significant for the Social Functioning sub-scale (paired t-test, $p < 0.05$).

Conclusion: Using the SF-36 with this patient group confirms clinical impressions that they are functioning at a much poorer level that age-equivalent norms, however, this measure may not be sufficiently sensitive to highlight changes in patient condition over time.

Anatomy/Pathology

HNE

The following HNE products will be shown on stand No. 18

* A new range of garments for the Flowtron^R Intermittent Compression System

* The Flowtron DVT^R system for use in the prevention of Deep Vein Thrombosis

* The complete range of Vascular Dopplers including Mini, Super and Multi Dopplex^R II, Printa and Reporter Software package

* The Lymphotron system for use in the treatment of Lymphatic disorders

* New Diagnostic and Assessment products

A sound approach to quality care

HNE, 310-312 Dallow Road, Luton, Bedfordshire, LU1 1TD
Tel : 01582 413104 Fax : 01582 459100

Anatomy/Pathology

Basic Science

Anatomy/Pathology

Basic Science
Anatomy

Phlebology '95, D. Negus et al. (eds.). Phlebology (1995) Suppl. 1: 45-47

OP/16.2

The Anatomy of the Saphenofemoral Junction - A Vascular Network with Therapeutic Implications

H.G. Kluess, E. Rabe, G. Gallenkemper, P. Mulkens and H.W. Kreysel

Dermatologic Clinic and Policlinic of the University of Bonn, Germany

INTRODUCTION

True recurrence of varicose veins following treatment usually occurs only as a refilling of temporarily occluded veins after sclerotherapy [1]. At the level of the saphenofemoral junction (SFJ) this is almost inevitable, whereas sidebranches, reticular varices and telangiectases may be treated with increasing durable success rates. Following surgical treatment with ligation of the SFJ and stripping of the greater saphenous vein (GSV) removed segments definitely may not recur. Cases of persistent or new varicosities at similar sites should be termed "pseudo-recurrences". They may be due to several reasons, related to the great anatomic variability of the venous system [2] and inadequately performed surgi-cal procedures [3,4].

OBJECTIVES

According to the anatomic site and required treatment recurrences after operation of the SFJ with stripping of the GSV could be divided into four groups [1,5]:

1. Persisting of the GSV after ligation (and stripping) of a sidebranch to the SFJ or a doubled GSV instead of the insufficent GSV itself. This occasionally happens in adipose patients and as a result of an inexperienced surgeon, insufficient preparation and/ or lack of anatomic knowledge. In these cases usually there is no benefit for the patient immediately after the procedure.

2. Remaining of a long trunk of the SFJ instead of having placed ligature plain at the wall of the common femoral vein (CFV). In these cases tributaries originated in this trunk connect to the lower limbs superficial vein system. Time and extent of recurrence de-pend on incompetence of the valves in these vessels.

3. Sidebranches directly flowing into the CFV at about the level of the SFJ, which have not been noticed at the time of the operation. Some of these may lead to reflux into the lower limbs superficial vein system, as there are collaterals. Time and extent of reflux ist not to be predicted.

4. The remaining recurrences occure as diffuse vessel networks in and around the groin. possibly leading to reflux pathways. In cases of connections to insufficient sidebranches of the lower limb, clinical relevant varicosities may be seen.

Treatment should aim on a correctly placed ligature of the proximal point of insufficiency (e.g. the SFJ) [6]. This easily might be done by operation in group 1 ("untouched" region), whereas surgery could be difficult in group 2 and 3, depending on the extension of scar-tissues in the groin. In group 4 (repeated) sclerotherapy is neccessary. Best to avoid these instances will be to perform surgery comprehensively at first [3,6].

METHODS

Intended to reveal the anatomic variability near the SFJ, in 40 consecutively operated limbs of 32 patients the sidebranches were extensively preparated up to 10 cm distally to the SFJ. All divisions, subdivisions and connections were noted, ligated and cut. The SFJ was ligated in the plain of the wall of the CFV in all cases. Ventral, medial and lateral sidebranches direct to the CFV up to 2 cm distant to the SFJ were exposed and ligated seperately. The GSV was partially stripped to above or below the knee in all cases, predo-minantly using the atraumatic technique of invagination. Percutaneous miniphlebectomy was added if neccessary. The procedures were performed either in general anesthesia or after local infiltration of anesthetics. Preoperative planning included photoplethysmogra-phy, continuous-wave-doppler and color-coded-duplex-sonography. Venography in most of the cases was dispensable.

RESULTS

In 9 cases sidebranches directly to the to the femoral vein were found. Doubling and tripling of the GSV was seen in 3 resp. 2 limbs. 2 to 8 primary branches of the SFJ (avg.: 4,7) ended in 4 to 21 ultimate branches (avg.: 11,1). 2 to 6 divisions (avg.: 3,2) were noted. In 9 cases shunt-like circuits between different tributaries were detected. The true anatomic situation is illustrated in Fig.1.

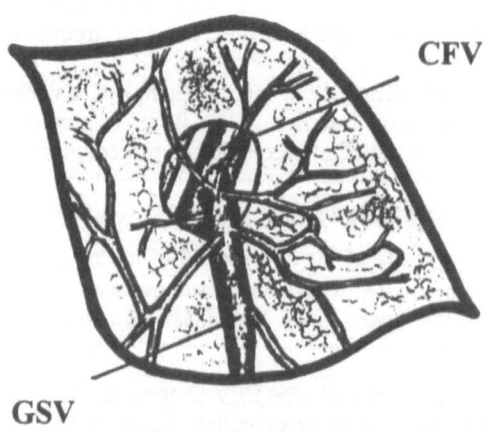

Fig.1 Anatomic site of the right saphenofemoral junction
       (CFV = common femoral vein, GSV = greater saphenous vein)

CONCLUSIONS

In order to reduce recurrence rates after surgical treatment of varicosities of the GSV liga-tion of the SFJ has to be done at the level of the wall of the CFV (while being cautious not do narrow the deep veins lumen) [6]. Direct branches of the CFV must be looked for and ligated separately. As we could demonstrates in addition there are numerous divisions and subdivisions of the tributaries of the SFJ, which easily can be reached through a small incision in the groin. We recommend to cut them at least beyond the second to third bi-furcation, as those could lead to reflux-pathways into epifascial veins in a diffuse manner. This concept is illustrated in Fig.2.

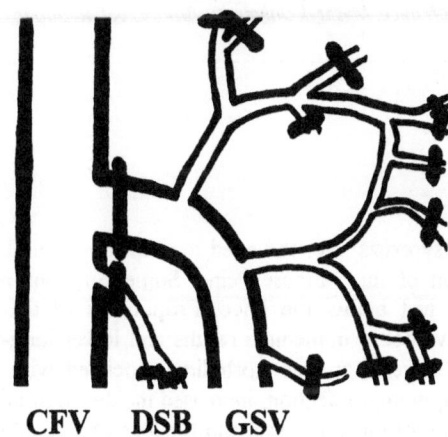

CFV   DSB   GSV

Fig.2   Operation of the saphenofemoral junction
(CFV = common femoral vein, GSV = greater saphenous vein, DSB = directly into femoral vein flowing sidebranch, ⬤▬ = correctly placed ligatures)

Doubling of the GSV should not be overlooked. Ligation of the posteromedial vein is to be obtained as this may avoid postoperative bleeding and furthermore prevents distal re-currences. Using this strategy recurrence rates following the operation of varicosities of the GSF should be minimized, leaving the question of neo-vascularisation in the area of the ligated SFJ open to further investigation [7].

REFERENCES

1.   Vin F. La Sclérothérapie Écho-guidée dans les Récidives Variqueuses Post-opéra-toires. Phlébologie 1995; 48: 25-29
2.   Somjen GM. Anatomy of the Superficial Venous System. Dermatol Surg 1995; 21: 35-45
3.   Goldman MP, Weiss RA, Bergan JJ. Diagnosis and treatment of varicose veins: a review. J Am Acad Dermatol 1994; 31: 393-413
4.   Thibault PK, Lewis WA. Recurrent Varicose Veins - Evaluation Utilizing Duplex Venous Imaging. J Dermatol Surg Oncol 1992; 18: 618-624
5.   Hach W. Phlebographie der Bein- und Beckenvenen. Konstanz: Schnetztor 1985
6.   Bergan JJ, Kistner RL (eds.). Atlas of venous surgery. Philadelphia: Saunders 1992
7.   Couffinal JC, Bouchereau R. Angiogenèse et insuffisance veineuse superficielle. La malaide induite. Phlébologie 1990; 43: 609-613

Phlebology '95, D. Negus et al. (eds.). Phlebology (1995) Suppl. 1: 48-50

P027

Relations of the Long Saphenous Vein and Saphenous Nerve From the Point of View of Varicose Vein Surgery

L. Veverková[1], M. Ruzicka[2] and J. Kalac[1]

[1] 1st Surgical Clinic Masaryk University, Brno, Czech Republic
[2] Department of Surgery Bohunice, Masaryk University, Brno, Czech Republic

INTRODUCTION

Considerable controversy exists over the need to strip the whole Long Saphenous Vein (LSV) during operation of the varicose veins. Some surgeons recommend that LSV should be stripped to just below the knee. Proponents of the stripping argue that stripping procedure have better immediate results and lower long-term recurrence rate. Opponents argue that there is greater morbidity associated with stripping of the vein owing to bleeding, pain, wound infection, increased incidence of injury to the Saphenous Nerve (SN) and stripping of the LSV results in the loss of a possible conduit suitable for arterial or venous reconstruction.

As we are proponents of the stripping of the whole LSV (from the ankle to the groin) by the operation of the stem varicose veins we wanted to clarify anatomic relations between LSV and SN in the region of the shin.

Description of the anatomical course of the SN isn't identical in different anatomical atlases and differences are especially in :

1. the level of the branching of the SN
2. presence of one or more branches of the SN
3. position of the nerve in relation to LSV (anterior or posterior, lateral or medial)

MATERIAL

86 of the lower extremities of the cadarvers were dissected in the region from ankle to the knee. LSV and SN were visualised and meticulously examined. The course of the both structures and their relations were studied and documented by photographs.

RESULTS

1. Saphenous nerve wasn't branching in 12 legs (13.9 %). SN was devided into two branches in proximal third of the shin in 24 legs (27.9 %), in the middle third in 36 legs (41.8 %) and distal third in 14 legs (16.3 %).
2. Anatomic position of the branch of the SN accompaniing LSV :

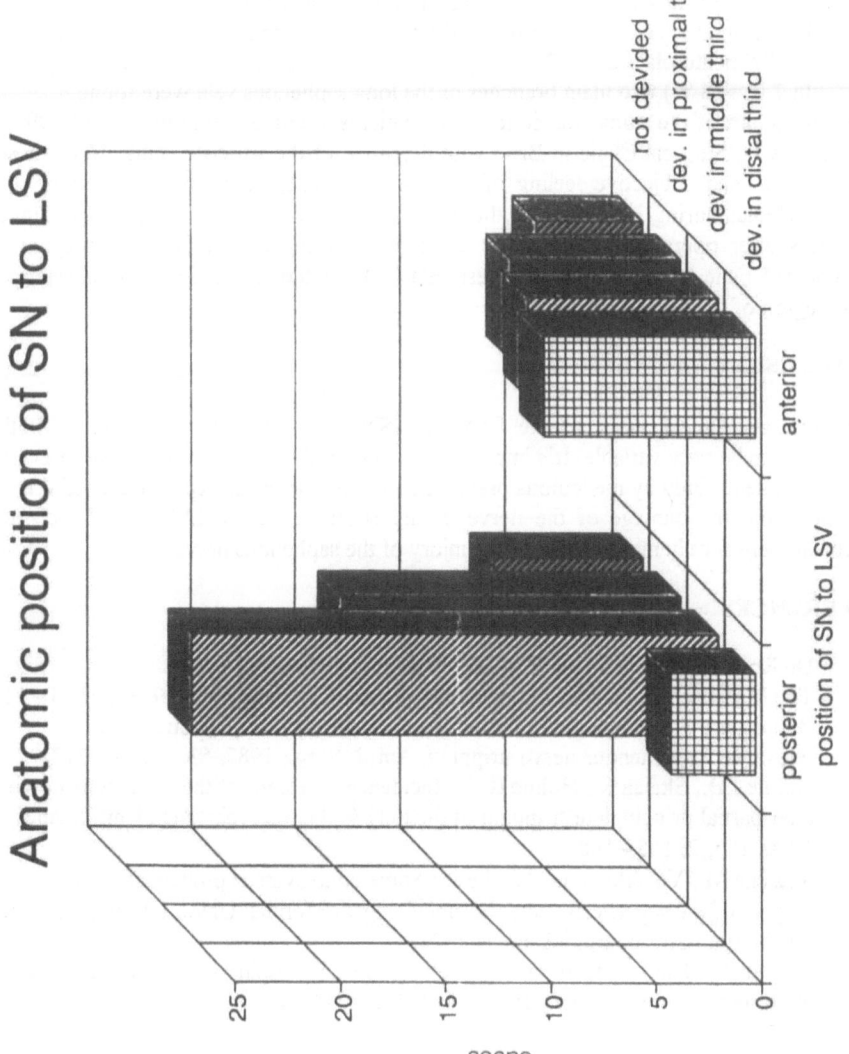

a - not branching nerve was on the posterior aspect of the LSV in 7 cases, on the anterior aspect of the LSV in 5 cases.

b - branch of the SN divided in proximal third of the shin was found on the posterior aspect of the LSV in 16 cases, on the anterior aspect in 16 cases.

c - branch of the SN divided in the middle third of the shin was on the posterior aspect of the LSV in 25 cases and on the anterior aspect in 9 cases.

d - SN divided in the distal third of the shin was on the posterior aspect of the LSV in 4 cases and on the anterior aspect in 10 cases.

3. In proximal third of the shin where the SN was not divided was the position of the SN posterolateral in 62 cases (72 %) and anterolateral in 24 cases (28 %).

4. In 7 legs (8 %) two main branches of the long saphenous vein were found.

The clinical part of the study was done in 215 patients, operated on in the period 1992 - 1994 at the 1st Surgical Clinic in Brno with diagnosis of the varicose veins. 22 of these patients have had subjective feeling of anesthesia or paresthesia in the region of the medial maleolus during 3 weeks after the operation. All of these patients were examined 6 months after operation. 15 of these 22 patients were without previous subjective symptoms. 5 patients had only mild paresthesia (2.4 %) and two patients had anesthesia in the region of the medial maleolus (0.9 %).

CONCLUSION

From this anatomical study of the LSV and SN is apparent that relation of both structures can be very variable. It's impossible to predict the course of the nerve in every operated patient. Only by meticulous preparation the operator can separate the LSV and SN and avoid the damage of the nerve during stripping of the LSV. With such an approach there is only minimal risk of the injury of the saphenous nerve.

REFERENCES

1. De Roos K.- P., Neumann H.A.M. : Treatment of varicose veins
   The Maastricht perspective in phlebology. Praktická flebologie, 1993, 2, 1, 5 - 11
2. Ramasastry S.S., Dick G.O., Futrell J.W. : Anatomy of saphenous nerve :
   relevance to saphenous nerve stripping, Am. J. Surg., 1987, 53, 5, 274 - 277
3. Holme J.B., Skajaa K., Holme K. : Incidence of lesions of the saphenous nerve
   after partial or complete stripping of the long saphenous vein, Acta Chir. Scand.,
   1990, 156, 2, 145 - 148
4. Růžička M., Veverková L., Wotke J. : Some controversial problems of the
   surgery of the superficial veins, 10 eme Congres Mondial Union Internationale de
   Phlebologie Strasbourg, Abstracts p. 233
5. Creton D. : The results of internal saphenous vein stripping under local anesthesia
   in outpatient care, Phlebol., 1991, 44, 2, 303 - 311

Phlebology '95, D. Negus et al. (eds.). Phlebology (1995) Suppl. 1: 51-53

P024

The Developmental Anomalies of the Inferior Vena Cava Could be Considered Truncular Vascular Malformation?

C. Setacci and G. Sozio

Chair of Vascular Surgery, University of Siena, Italy

Introduction

Anomalies of the inferior vena cava (I.V.C.) are often discovered incidentally during imaging studies performed for other reasons, since patients with these abnormalities are usually asymptomatic. The congenital absence of (I.V.C.) is uncommon (1,2). In these persons, venous drainage from structures below the diaphragm is diverted to collateral channels, including the axygous system, vertebral vessels (3), and superficial vessels in the abdominal wall (4). Although anomalous IVC occurs in 0.6% of patients with heart diseases (5) and 1% of patients undergoing cardiac catheterization (6), the incidence of absent IVC is unknown.

Case report

We observed a case of Atresia of the inferior vena cava in May 1993. A 50-years old man was admitted as an emergency after compressive thoraco-abdominal trauma. Physical examination on current admission showed dyspnea and remarkable emithoracic pain on the right side.

Chest radiography showed a pneumothorax associated with right pleural effusion and multiple ribs fractures. A percutaneous thoracic drainage was performed under antibiotic therapy.

The patient remained 25 days at the hospital and then he was descharged with the pneumotorax and the pleural effusion completely healed. After a week the patient returned at our observation complaing fewer and right inguinal pain. At the physical examination a moderate edema was observed on the right leg. An antibiotic and heparin therapy was administered and the patient was submitted to phlebography that showed thrombosis of the internal iliac vena on the left side, thrombosis of the common, internal and external iliac vena on the right side. Than the patient underwent computed tomography that confirm the bilateral thrombosis of the iliac axis, and also showed the atresia of the infrarenal part of the inferior vena cava. Blood from the legs was carried towards the heart through the lumbar venous system and the abnormaly enlarged azygos and hemiazygos venis.

The anticoagulant therapy replaced the heparin one stabilizing the Quick time on 45 per cent. After 13 days, when there were no more symptoms of disease, the patient was discharged suggesting to continue anticoagulant therapy at home and to wear elastocompressive stockings on the right leg in upright position.

Discussion

According to the Embryonic School's theories about the etiology of congenital vascular malformations (7,8), in the first ten weeks of embryonic life, three stages can be distinguished in the developmen of the vascular system: stage I, the undifferentiated or capillary network stage; stage II, the retiform stage with the beginning of organization into larger channels; and stage III, the truncular stage in which recognizable main vascular truncus appear.

Any cessation or disturbance of this development can be the source of vascular malformations, and the stage in which the aberrations occur will determine the very nature of the malformation.

If the dysontogenetic processes take place at the truncular stage, they are superimposed upon a predetermined vessel whose structure and anatomic course is altered, thereby leading to the so-called truncular vascular malformation.

The inferior vena cava develops in a complex series of events, beginning with the appearance of plained posterior cardinal veins that drain the dorsal and inferior portions of the embryo these vessels anastomose with the subcardinal veins that is situated medially when the posterior cardinal veins regress the subcardinal veins became the route if preferred drainage.

The right hepatic vein is formed by hepatic sinusoids abd becames the hepatic right hepatic vein is formed by hepatic sinusoids and becames the hepatic segment of the inferior vena cava.

The right subcardinal vein anastomoses with the right hepatic vein and persists to became the prerenal segment of the inferior vena cava. Supracardinal veins also develop medial to the posterior cardinal veins and dorso lateral to the subcardinal veins. The supracardinal system is the last system to develop, it forms the azygos and hemiazygos vein. The right supracardinal vein also forms the post renalm segment of the inferior vena cava. The renal segment of the inferior vena cava develops from anastomoses between the supracardinal and subcardinal veins.

In the case that we present we have observed an atresia of the infrarenal part of the inferior vena cava. This is the resuls of an anomaly of the development of the right supracardinal vein.

Dysontogenetic effects can be the result of genetic errors or environmental factors:

The last ones are more frequently involved in producing anomalies of development at truncular stage: in particular fetal disease like allantoic vessel thrombosis (9) and embryonic infection by cytomegalovirus or toxoplasmosis (10) have been implicated in cases of truncular venous malformations.

Of course, prolonged hemodynamic effects may contribute considerably to the clinical appearance and progression of venous dysplasias (11,12) even after many years from the birth.

In the case that we observed we were able to make diagnosis of the atresia of the infrarenal part or the inferior vena cava for the presence of a concomitant disease, not directly related to the vascular abnormality, that forced the patient to recumbent position for a long period, creating the conditions to make such anomaly evident.

References

1. Chuang V.P., Mera C.E., Hoskins P.A.: Congenital anomalies of the inferior vena cava: review of embryogenesis and presentation of a simplified classification. Br.J.Radiol. 1974; 47, 206-213.

2. Stewart G., Farmer G.: Sturge-Weber and Klipper-Trenauny syndromes with absence of the inferior vena cava. Arch. Dis. Child. 1990; 65, 446-447.

3. Mayo J., Gray R., StLouis E., Grosman H., McLoughlin M., Wise D.: Anomalies of the inferior vena cava. Am.J.Roentgenol. 1983; 140, 339-345.

4. Datta D.V., Saha S., Samanta A.K.S. et al.: Chronic Budd-Chiari syndrome due to obstruction of the infrahepatic portion of the inferior vena cava. Gut 1972; 13, 372-378.

5. Anderson R.C., Adams P., Buerke B.: Anomalies of the inferior vena cava with azygous continuation (intrahepatic interruption of the inferior vena cava). J.Pediatr. 1961; 59, 370-383.

6. Muelheims G.H., Mudd J.G.: Anomalous inferior vena cava. Am.J.Cardiol. 1962; 9, 945-952.

7. Rienhoff W.F.: Congenital arteriovenous fistula. Bull. Johns Hopkins Hosp. 1924; 271-284.

8. Schobinger R.A.: Periphere Angiodysplasien. Bern, Stuttgart, Wien, Hans Huber, 1977.

9. Benirschke K., Driscoli E.: The pathology of the human placenta. In: Uehlinger E. (ed.), Handbuch der speziellen pathologischen anatomie und histologie. Vol.7, Berlin, Springer, 1967; p.98.

10. Kloss K., Vogel M.: Pathologie der perinatalperiode. Stuttgart, Georg Thieme, 1974.

11. Hach W.: Atiologie und pathogenese der primären varicose. Dtsch. Med. Wochenschr. 1967; 92, 1400.

12. Martin A., Odling-Sinee W.: Pressure changes in varicose veins. Lancet 1976; 1, 768.

Phlebology '95, D. Negus et al. (eds.). Phlebology (1995) Suppl. 1: 54-56

P022

Double Inferior Vena Cava: A Case Report

J.E. Amorim, L.C.U. Nakano, M. Nunes, A. Lourenço, R.P. Nunes Filho, M. Reicher, M.C.J. Perez and E. Burihan

Division of Vascular Surgery, Department of Surgery - Escola Paulista de Medicina, Universidade Federal de São Paulo, São Paulo, Brazil

INTRODUCTION

The congenital anomalies of the inferior vena cava have a great importance to the vascular surgeon when he needs an access to the abdominal aortic infra-renal. Among these anomalies, double vena cava and left vena cava have been pointed out, although the second one is a rarity. (1,2,3)

There was only one case of double inferior vena cava observed in our service in the recent five years. We have done 1063 ascending phlebographies in which have been visualized the iliac and inferior vena cava, even if a punction of femoral vein is needed to be done the cavography.

The objective of this work is to describe a rare ocurrence of these anatomic anomalies, such as to emphasize the necessity to have ascending phlebography and inferior cavography when the diagnostic is not so clear.

RELATE OF THE CASE

White female, 32 years old, taking oral contraceptive for several months, clinically presenting varicose veins in left inferior extremity for 3 years, worsening with pain and edema in the last year. It was determined a phlebographic study to help in a probably diagnostic of postphlebitic syndrome or Cockett syndrome.

Rabinov-Paulin modified technique was used appearing no suggestive images of chronic or acute deep venous thrombosis. It was observed suggestive image of obstruction in the left common iliac vein, having proximal flow through a large paravertebral colateral. As the diluition of contrast medium damaged the quality of the image, an inferior cavography was firstly done by left femoral vein puncture and subsequently to the right femoral. (Figures 1 and 2) The diagnostic was only possible after injection of contrast medium through bilateral femoral puncture. In the left side the inferior vena cava joins up left renal vein. The left renal vein joins up right inferior vena cava in the right side of vertebral column. The vein pressure in iliacs was measured having the same result in both sides.

 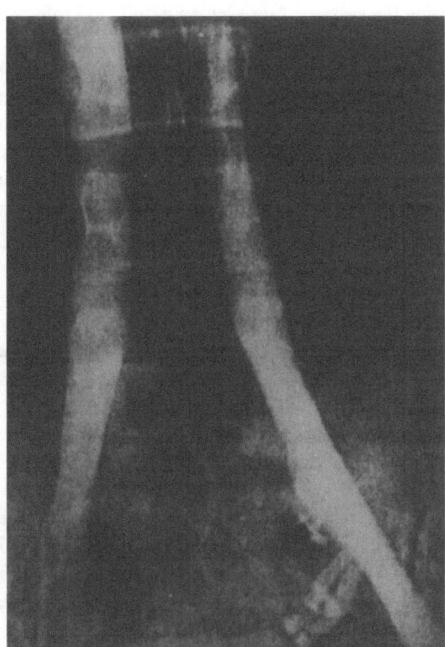

Fig. 1   Fig. 2

Figures 1 and 2: (1) Cavography by left femoral puncture, showing left inferior vena cava that joins up left renal vein. (2) Cavography by bilateral femoral injection, showing double inferior vena cava.

DISCUSSION

The inferior vena cava embriological formation is complex. Each segment originates from the fusion of several venous segments when the size of the embryo exceds 15 mm. The anatomics anomalies occur in the fusion and regression of the subcardinals, supracardinals and posterior cardinals veins. Simplifying, the hepatic and pre-renal segment are formed by subcardinal vein. The renal segment is the result of anastomosis of supracardinal and subcardinal veins and, at last, the postrenal segment is formed from supracardinal and posterior cardinal veins. It is frequent anatomic alterations in the position of the right ureter, staying in the back side of the inferior or surrounding the inferior vena cava, carrying the obstruction to the ureteropelvic junction or distal ureter. By the way, the symptoms of double or left inferior vena cava can be from urologic origin, which were observed between third or fourth decade.(5)

The ocurrence of anatomic anomalies of inferior vena cava is between 2,0% and 3,0% of necropsy studies (6), being more frequent double inferior vena cava (2,0%) than left inferior vena cava (0,2% and 0,3%).(1,2) These incidences are not found in clinical descriptions. It is also rare the descriptions of inferior vena cava anomalies which are diagnostified through noninvasive evaluation like tomography, ultra-sound and duplex scan.(7)

In our research, however, in amount of 1063 phlebography which were performed in recent five years, we observed only this case of double inferior vena cava (0,09%)

In the literature is found reports of one case of inferior vena cava anomalies with reference to duplication of vena cava or its position on the left, suggering that its frequency in clinical pratice has been less frequent than in autopsies. The reports relate urologic symptons or observations

56

during the surgery, however we didn't find references which show any relation between anomalies of inferior vena cava and chronic venous stasis.

We agree with the Wagner & Boguch idea in investigating an accurate way, the presence of double vena cava by noninvasive evaluation or cavography, in patients with indication to vein interruption to prophylaxis for pulmonary embolism.

The difference between the incidences of inferior vena cava anomalies observed in necropsies and clinical pratice suggests that the patients have less symptons or this diagnostic is less estimated by the examinator, despite having accurate noninvasive evaluation to the diagnostic like ultra-sound, tomography, duplex scan and magnetic ressonance.

REFERENCES

1.   Chuang, V.P., Mena, C.E., Hoskins, P.A.. Congenital Anomlies of the infferior vena cava.Review of embryogenesis and presentation of a simplified classification. Br. J. Radiol 1974; 47:206-13.

2.   Knodtzon J., Svane S. Left-sided inferiior vena cava. A peroperatively demonstrated case. Acta Chir Scand 1986; 152:547-9.

3.   Giuffrida, G.F., Trimarchi, S., Miani,S., Lovaria, A., Giorgette, P.L., Giordanengo F. Chirurgia ricostruttiva dell'aorta sottorenale com vena cava inferiore sinistroposta. Minerva Cardio Angiol 1993; 41(1-2):43-7.

4.   Redman J.F., Aguilar-Gusman O.F. Ureteropelvic junctoin obstruction caused by accessory renal vessels in association with peruretral vena cava and vena caval duplication. Urology 1992; 40(4):362-7.

5.   Kenawi, M.M., Williams, D.I. Circumcaval ureter: a report of 4 cases in children with a review of the literature and a new classification. Br J Urol 1976; 48:183-6.

6.   Baldridge, E.D., Canos, A.J. Venous anomalies encountered in aortoiliac surgery. Arch Surg 1987; 122: 1184-7.

7.   Sener, R.N. Nonobstructive right circumcaval ureter associated with double inferior vena cava. Urology 1993; 41(4):356-60.

8.   Wagner, J., Bogusch G. An abnormal pattern of blood vessels in the retroperitonial space with a duplicated inferior vena cava in an adult: a case report. Surg Radiol Anat 1993; 15:201-5.

Phlebology '95, D. Negus et al. (eds.). Phlebology (1995) Suppl. 1: 57-60

P017

Anatomical Study of the Small Saphenous Vein (Saphena Parva): Types of Termination

E. Burihan[1] and J.C.C. Baptista-Silva[1]

Division of Vascular Surgery, Department of Surgery[1], Escola Paulista de Medicina, Universidade Federal de São Paulo, São Paulo, Brazil

INTRODUCTION

The termination of the small saphenous vein (SSV) is of fundamental and great practical importance in the diagnosis and surgical treatment of varicose veins. There is wide divergence of opinion between different authors on this subject.

METHODS

We have dissected 200 legs of 100 adult cadavers. Sixty were male (60%) and forty female (40%); 44 were white and 56 non-white. With cadaver in prone position both the small saphenous veins were dissected beginning at the lateral malleolus; a catheter was inserted into the vein and contrast media was injected and serial filming was performed. After the phlebographic examination, we dissected the small saphenous throughout its length from the lateral malleolus to its termination.

RESULTS

We found two groups of termination of the small saphenous vein.

First group
To the deep veins (popliteal and femoral); types A - principal deep vein of the leg 55/200 (27.5%); A_1 - two different veins 10/200 (5%); A_2 - after given one collateral to principal deep

vein, it continues to the great saphenous vein (saphena magna) 4/200 (2%); A_3 - through two branches, one to the principal deep vein and the other to the great saphenous vein 30/200 (15%); A_4 - through two branches, one to the principal deep vein and other continues as the femoropopliteal vein 11/200 (5.5%); A_5 - through two branches, one to the principal deep vein and the other to the muscular vein of the 1/3 inferior of the thigh 9/200 (4,5%); A_6 - through these two branches, one to the principal deep vein and the other which lies subcutaneously without definite termination 11/200 (5,5%); A_7 - through two branches, one to the principal deep vein and other to the great saphenous vein 4/200 (2%); A_8 - through two branches, one to the principal deep vein and the other to the perforating vein of the 1/3 inferior of the thigh 9/200 (4,5%); A_9 - through two branches, one to the principal deep vein and the other to the vein which closely follows the squiatic nerve 1/200 (0,5%). A_9 - to the great saphenous vein (saphena magna) 4/200 (2%).

Second group
To other collector veins (not to principal deep veins). B - to the great saphenous vein (saphena magna) at the high thigh 13/200 (6,5%); C - to the cutaneous vein of the posterior part of the thigh 6/200 (3%); D - to the muscular veins of the 1/3 inferior of the thigh 4/200 (2%): E - through two branches, one to the great saphenous vein and the other to the perforating vein of the 1/3 inferior of the thigh 5/200 (2,5%); F - through two branches, one to the great saphenous vein and other to the muscular vein of the 1/3 inferior of the thigh 4/200 (2%); G - to the perforating vein of the 1/3 inferior of the thigh 4/200 (2%); H - to the vein which closely follows the sciatic nerve of the 1/3 inferior of the thigh 2/00 (1%); H_1 - to the deep femoral vein after crossing the adductor magnus muscle 2/200 (1%); I - to the inferior gluteal vein of the 1/3 inferior of the thigh 3/200 (1,5%); J - to the gastrocnemius veins and to the great saphenous vein below the popliteal area 9/200 (4,5%).

DISCUSSION

Since the classic abstract of the Giacomini [2], we find that the term "termination" is not always used adequately. In our article we have found that termination of the small saphenous vein is different from the classical descriptions of Giacomini [2], Kosinski [3] , Blomquist [4], Mullarky [5] and Lazaro da Silva [6]. In the first group of our results, type A was more frequently as those encountered by Haeger [7]. In the second group we have encountered many variations in the termination of the small saphenous vein compared to those encountered by Carrol [8], Bueno Neto & Zupo [9], Allegra & Machini [10].

CONCLUSIONS

1. The termination of the small saphenous vein is very variable; a single termination in the deep principal vein of the limb was encountered most frequently (27,5%).

2. From the types of termination of the small saphenous vein that show more than one termination, the predominant type has one communication with the principal deep vein of limb (popliteal or femoral veins) and another termination with the great saphenous vein (15%).

3. From the types of termination of the small saphenous vein that do not have any communication with the principal deep vein of limb, the most frequent type was termination in the great saphenous vein in the thigh (6,5%).

4. Of all types of termination of the small saphenous vein, at least one had communication with the principal deep vein of the limb which we encountered in 144 to 200 limbs dissected (72%).

REFERENCES

1. Burihan E. Estudo anatômico de veia safena parva: tipos de terminações. Rev Bras Cardiovasc 1974; 10(2): 89-99.

2. Giacomini C. Osservazioni anatomiche per service allo studio della circulazioni venosa delle extremitá inferiori. Torino: Tip V Vercellino, 1873.

3. Kosinski G. Observations on the superficial venous systems of the lower extremity. J Anat 1926; 60:131-42.

4. Blomquist H E Variability in the terminations of the short saphenous vein in the Finns. ann Chir Gynecol Finn 1968; 57:55-8

5. Mullarky R E Termination of the small saphenous vein. Northwest Med 1963; 62: 878-80

6. Lazaro da Silva A. Contribuição ao estudo da "veia saphena parva". De suas afluentes e anastomoses ao nivel da "fossa poplitea". Sua aplicação cirúgica. Tese dedoutoramento. Faculdade de Medicina da Universidade de Minas Gerais, Belo Horizonte, 1965.

7. Haeger K. The Surgical anatomy of the saphenous-femoral and the saphenous-popliteal junctions. J Cardiovasc Surg 1962; 3:420-7.

8. Carrol W W Varicosities of the lesser saphenous vein. Arch Surg 1949; 59:578-87.

9. Bueno Neto J, Zupo M. Varizes dos membros inferiores. Rev. Med (S. Paulo) 1956; 40:18-39.

10. Allegra G. Machini M. Le varicidella piccola safena. Sintomi, diagnosi e terapia radicale. Il Policlinico (Ser Chir) 1958; 65: 123-40.

Phlebology '95, D. Negus et al. (eds.). Phlebology (1995) Suppl. 1: 61-64

P131

Complications du Traitement de la Saphène Externe Liées à la Présence de L'Artère Petite Saphène

F. Chleir, R. Rettori and F. Vin

Paris, France

Complications of the Treatment of the Short Saphenous Vein Associated with the Presence of the Short Saphenous Artery

SUMMARY

The authors stress the complexity of the anatomy of the popliteal fossa and the difficulties of sclerotherapy and surgery in this region. A plea is made for the use of Doppler screening before any intervention in this area.

COMPLICATIONS DU TRAITEMENT DE LA SAPHENE EXTERNE LIEES A LA PRESENCE DE L'ARTERE PETITE SAPHENE

F. Chleir, R. Rettori, F. Vin (Paris, France)

INTRODUCTION

L'anatomie du creux poplité, spécialement celle veineuse, est complexe et variable. On connait les variations du système veineux profond avec dédoublement plus ou moins complet de la veine poplitée du fait de la terminaison à différents niveaux des veines jambières. De même la terminaison de la crosse de la saphène externe dans la veine poplitée, qui est plus ou moins haute, se situe le plus souvent entre 2 et 5 cm au dessus de l'interligne du genou ; elle peut être double ; elle reçoit souvent une collatérale supérieure, se terminant sur sa convexité et semblant prolonger la veine saphène externe à la cuisse.

La cure de l'insuffisance veineuse superficielle conduit à des gestes thérapeutiques, notament sur la saphène externe. Sa grande variabilité anatomique rend ce traitement difficile et le résultat aléatoire. A partir d'un cas clinique en rapport avec une atteinte de la saphène externe, nous avons cherché à préciser les rapports entre la veine saphène externe et l'artère petite saphène ; ensuite nous avons recherché les différentes descriptions faites des incidents liés au traitement par sclérothérapie, ainsi que les cas d'incidents chirurgicaux.

RAPPEL ANATOMIQUE

L'existence d'une petite artère détachée d'une artère jumelle qui "s'accole au nerf saphène externe l'accompagnant jusqu'à mi-jambe" (L. Testut) ou "descendant satellite de la veine saphène externe" (Ch. Dujarrier) est signalée dans les classiques traités d'anatomie de ces auteurs.

G. Paturet (1) est plus précis décrivant cette artère petite saphène comme une collatérale constante, généralement grèle, naissant de la face postérieure de l'artère poplitée ou même souvent d'un tronc commun avec une artère jumelle généralement la jumelle interne. Elle descend en dedans de la veine poplitée et du nerf sciatique poplité interne, puis elle s'accole à la veine saphène externe qu'elle accompagne pour aller se distribuer aux téguments du mollet et au nerf saphène externe. Les constatations opératoires (2) comme celles échographiques (3) montrent que l'artère petite saphène entre ainsi en contact direct avec la paroi de la saphène externe à un niveau variable (crosse ou petite terminale du tronc), l'abordant en règle par en dedans, mais pouvant longer son bord postérieur ou superficiel.

COMPLICATIONS DE LA SCLEROTHERAPIE

Plusieurs centaines de milliers d'injections sont effectuées chaque année en France et les accidents sont rares mais parfois dramatiques. P. Wallois (4) en 1972 dans : "La sclérose des varices" de R. Tournay a consacré un chapitre aux complications de la sclérothérapie et a décrit les risques au niveau du creux poplité lors d'injections au niveau de l'artère petite saphène ou de branches de l'artère jumelle. J. Natali (5) en a recensé 39 en intra-artérielle, 46 en sous-cutanée et 8 accidents généraux entre 1975 et 1991. A. et M. Avramovic (6) ont rapporté une étude statistique des complications de la sclérothérapie à propos de 7 200 cas, ils ont retrouvè 2,9 / 1 000 cas d'ulcérations. Au niveau du creux poplité, la présence de structures vasculaires artérielles et veineuses dans un espace confiné rend la sclérothérapie délicate même entre les mains d'un médecin expérimenté, le risque d'injection intra-artérielle étant lié à la présence toute proche de l'artère petite saphène et des artères jumelles ou de leurs collatérales. La complication majeure est la nécrose, soit simplement cutanée, s'il s'agit d'une petite branche superficielle, soit musculaire avec risque de rétraction et d'importantes séquelles fonctionnelles.

La symptomatologie est constante avec une douleur extrèmement vive dès la fin de l'injection, en aval, rapidement suivie d'une incapacité fonctionnelle quasi complète. Ces symptomes s'aggravent après quelques heures et nécessitent parfois le recours à des antalgiques majeurs. Les traitements actuels sont essentiellement faits de vaso-dilatateurs, d'anti-inflammatoires associés à de l'héparine. Il faut reconnaitre que ces traitements sont peu ou pas efficaces et l'on a souvent recours à la chirurgie d'exérèse. On peut espérer une amélioration de ces résultats grâce à l'utilisation du puissant vaso-dilatateur et anti-agrégant qu'est l'Iloprost.

COMPLICATIONS DE LA CHIRURGIE

Nous ne reviendrons pas sur les différents types de complications de la chirurgie veineuse superficielle, celles-ci ont été bien décrites et sont heureusement rares. Notre objectif est de rapporter un cas original qui nous a semblé particulièrement intéressant . Il s'agit dd'une femme de 45 ans, ayant 5 enfants, opérée il y a dix ans de sa saphène externe à droite, qui vient consulter pour un ulcère malléollaire externe persistant depuis plus d'un an. L'examen retrouve un ulcère à contours nets, indolore, peu bourgeonnant, assez propre, d'environ 16 cm^2. Au plan local, il existe une importante varicose non systématisée, un notable oedème de toute la jambe, avec un aspect boudiné des orteils, le caractère un peu dystrophique du lymphoedème évoquant un syndrome post-thrombotique. L'écho-Doppler nous a montré une absence d'élément en faveur de séquelles de thrombose veineuse profonde, mais une masse échogène de la terminaison de la saphène externe avec au Doppler pulsé un flux pulsatile de type artériel. Devant ce tableau, nous avons fait pratiquer une artériographie du membre inférieur droit avec prise de clichés tardifs pour étudier le retour veineux. Cet examen a confirmé l'existence de 3 fistules artério-veineuses dont 2 provenaient de l'artère fémorale superficielle et une de la poplitée. Devant ce tableau, il a été décidé d'une embolisation sélective de ces branches. Seules les 2 branches supérieures ont pu être embolisées, la dernière communiquant avec la poplitée étant trop sinueuse ; la fistule artério-veineuse lui correspondant peut être expliquée par la pose d'une ligature unique intéressant à la fois la crosse de la saphène externe et l'artère petite saphène. L'apparition secondaire de telles fistules après ce type de ligature "en masse" est bien connue en particulier au niveau du pédicule rénal.

Environ 6 semaines après l'embolisation, l'ulcère s'est complètement fermé ; l'aspect dystrophique a nettement diminué, la patiente continuant à porter une compression.

COMMENTAIRES

La situation de l'artère petite saphène au contact de la saphène externe la rend difficile à individualiser lors d'une crossectomie de la saphène externe ; elle rend également la

64

sclérose délicate. Toutefois, si les chirurgiens individualisent les nerfs afin de ne pas les léser, la plupart ne s'intéressent pas à cette artère et disent la lier en même temps que la saphène externe. Ceci amène deux réflexions : d'abord on peut s'interroger sur les risques, lors d'une sclérose, de la crosse saphénienne externe, du passage du produit dans l'artère petite saphène. La destination de cette artère est cutanée, mais, en cas d'introduction du liquide caustique dans sa lumière, ce n'est pas uniquement la peau qui est intéressée, il y a aussi une atteinte et une amyotrophie du jumeau interne. Ce qui se passe est facile à imaginer : on exerce au moment de l'injection une contre-pression qui renvoie le produit en amont. Ensuite du fait du courant artériel, ce produit passe dans la jumelle interne et le muscle jumeau en même temps qu'il diffuse en territoire cutané. Cela explique l'infarctus musculaire. Si par malchance la petite saphène naît directement de la poplitée, le produit se déverse dans cette artère, c'est alors qu'on observe la mise en cause de la vitalité de la jambe, heureusement exceptionnelle.

L'autre question est celle de connaître la fréquence et l'importance des fistules artério-veineuses après chirurgie de la veine saphène externe et la nécessité d'isoler l'artère petite saphène pour la lier isolément, indépendamment de la section - ligature de la crosse.

CONCLUSION

La présence de l'artère petite saphène ainsi que la diversité du niveau d'abouchement de la veine saphène externe rendent la sclérothérapie délicate et la chirurgie difficile. Les complications sont rares mais graves en sclérothérapie, les inconvénients de la chirurgie sont moins importants qualitativement mais probablement plus fréquents. Seul un bon repérage par écho-Doppler pulsé avec une cartographie précise, autorise un traitement efficace et sûr de l'insuffisance de la saphène externe, la sclérothérapie guidé par échographie ayant permis de faire un grand pas dans ce sens. L'étude de ce cas a permis de redire à quel point le creux poplité est complexe ; il a également permis de rappeler que certains ulcères peuvent être secondaires à une fistule artério-veineuse.

BIBLIOGRAPHIE

1 - PATURET G. Traité d'anatomie humaine tome II. Masson et Cie Editeur, Paris 1951 : 966- 67

2 - OUVRY P. , DAVY A., GUENNEGUEZ H. L'artère saphène externe. Phlébologie 1980 ; 33 : 307 - 12

3 - SOMER -LEROY R. de, WANG A., OUVRY P. Echographie du creux poplitéé ; recherche d'une artériole petite saphène avant échographie. Phlébologie 1991 ; 44 : 69 - 78

4 - WALLOIS P. Accidents de la sclérose. In La sclérose des varices.Expansion édit Paris 1992 : 273 - 283.

5 - NATALI J. Les complications de la sclérothérapie dans le traitement des varices. XI congrès mondial de Phlébologie. Montréal, John Libbey éditeur 1992, 859 - 860.

6 - AVRAMOVIC A. et M. Complications of sclerotherapy : a statistical study. XI congrès mondial de Phlebologie. Montreal. John Libbey 1992, 859 - 860.

Phlebology '95, D. Negus et al. (eds.). Phlebology (1995) Suppl. 1: 65-68

OP/18.4

Morphology of Arteriovenous Anastomoses (AVAs) and Their Influence on the Vein Wall and its Endothelial Lining

L. Schalin and G. Moberger

Department of Surgery, Sabbatsberg Hospital, Stockholm, Sweden. & Department of Pathology, Huddinge Hospital, Huddinge, Sweden

Keywords: AV-anastomoses, higher varicose skin temperature, pulsatile flow, shear stress, endothelial damage, Nitric oxide.

Objective: A light-microscopic study of varicose specimens and the inside surface of the varicose vein wall. To determine what appeared obvious during microsurgical dissection of varicose veins 1) arteriovenous anastomoses(AVAs) connecting to varicose convexities, 2) and from flow study experience the influence of shear stress on the vein wall and its endothelium caused by the speed and force of arterial flow transmitted through the AVAs.
Design: Morphologic study of varicose vein segments embracing the site where the postulated anastomoses run into the varicose veins and the interior vein wall opposite this entrance. Patients aimed for varicose surgery underwent microsurgial dissection. Excised specimens examined by routine pathology. The material(14 samples) too small for statistics.
Material: Thirteen patients with 2-4 fairly well defined varicosities of moderate to large but not extensive size at the calf or thigh(mean age 57.2 ± 13.5). A primary collection of specimens(n=100) in 46 patients were stored but lost due to the break down of the freezer.
Interventions: Localization of AVAs by thermography depicting the increased skin heat overlying varicosities. Microsurgical dissection performed under spinal or general anesthesia.
Main outcome measures: Varicose segments(n=14) with connecting, identified and differently marked venous tributaries and AV-anastomoses obtained by meticulous microsurgical dissection were fixed in 4 % formalin, stained with Hematoxylin-Eosin or van Gieson and serially sectioned for light microscopy.

Introduction

Our light microscopy study is based on the repeated observation of pulsating tributaries connecting to varicose veins and the assumption that AV-shunting is important for their development, a statement supported by indisputable but neglected observations well substantiated[1,2,3,4,5] and called attention to since 1953[2] when they were properly described. Calculations[6] where varicose meanders were transferred to a coordinate system showed them to be sinous curves indicating that pathological flow(through perforators) appeared before the varicose shape. As no correlation between varicose skin heat and the site of perforating veins could be shown in a study[7] aimed to reproduce the promising findings of a thermographic method[8] used to localize perforating veins and secondly retrograde flow according to fluid dynamic principles did not explain varicose tortuosity an *alternative hypothesis* was launched. Varicose meandering (tortuosity) was claimed[9,10] to be due to 1) increased arterial flow in varicose calves and seg-ments[11], 2) transmitted through small sized arteriovenous anastomoses connecting to varicose convexities[10,12], 3) the influence of shear stress forces on the endothelium and the the vein wall. Another prerequisite appeared to be 4) a probably hereditary constitutional defect with degenerative changes of the vein wall with marked increase of connective tissue elements[13], loss of contractility due to disturbance of muscle cell collagen/elastic fibre balance[14] resulting in increased vein wall distensibility[15,16,17]. The varicose development includes a metabolic disorder due to biochemical processes in the vein wall revealed by measurable amounts of connective tissue degradation products from the wall[18], excreted and recorded in the urin. 5) Endothelium is sensible to flow induced shear stress and pulsatile flow[19], stated to release NO=Nitric oxide [20] [NO=EDRF, endothelial derived relaxing factor[21]. Pulsatile flow transmitted through AVAs could be that factor that initiates varicose dilatation. Listed five factors(1-5) constitute the theoretical basis for flow dependant development of VARICOSE MEANDERS similar to the pertubation theory valid for river meanders. A very small disturbance of flow causes the wall first to bulge in a sinuos manner successively elongate and finally to coil between locking tributaries of venous or arteriovenous character.

Flow studies(Schalin & Liepsch 1987-93) in translucent one-to-one, true-to-scale varicose vein models showed that the arterial 'jet' transmitted through simulated AVAs hit the opposite vein wall at such low flow rates as 1-2 ml/min. continuos flow. When the flow was pulsatile the wall was hit alreadey at a lower flow rate of 0.5-1.0 ml/min. Disturbed flow and whirls appeared quite close to the inside vein wall where flow separation was recorded[22]. Possible damage of the endothelium opposite the entrance of the AVCs is fully in accordance with such observations in specimens from varicose surgery[13]. Repeated influence(over years) of hydrodynamic shear forces and release of NO might be the initial step of phlebecasies and subsequent dilatation, elongation and varicose meandering(=definition of varicose veins).

Aim of our light microscopy study - to consider the existence of AV-anastomoses connecting to varicose veins and evaluate their possible role concerning vein wall changes and endothelial patterns in the vein wall opposite and adjacent to the entrance of the suggested AVA.

Results

Evidence of AVAs In 7/14 specimens. Negative outcome(n=7) due to discoloration, artefacts, lack of important sections. 1) The arterial character of the efferent tributary and its transition through the vein wall proven in 6/7 specimens. 2) Three effects due to shear stress and pulsatile flow either a) thickening of the vein wall(n=4), b) dilatation of the vein with thinning of the wall opposite the entrance opening of the AVA(n=2) accompanied by c) endothelial cell degeneration appearing as enlarged nuchlei forming chromatin clumps with distraction of the cytoplasm.

Remarks on Light Microscopy findings.

A. Four major observations

1) the *arterial character* of the efferent vessel outside the varicose vein wall *confirmed;*
2) the *break through site of the AVAs* through the vein wall *identified* and *loss of muscle cells and elastic membranes during its wall transit,* 3) the often *very small size of the wall transition,*
4) *endothelial changes* of the vein wall in the vicinity or *opposite to the entrance of the AVA.*
B. Three effects which could be associated with the biofluid mechanic influence of shear stress and pulsatile flow acting on the vein wall[10] and **due to the penetration of the AVAs.**
1. *Increased wall thickness* of the vein distally to the AV-entrance through the vein wall.
The thickened wall presented smooth muscle cell hyperplasia and hypertrophy, increase
of collagen and appearance of elastic fibrils normally not part of the vein wall.
In other parts of the vein were observed:
2) dilatation of the vein with more or less pronounced thinning of the wall opposite the entrance opening of the AVA, muscle cells being reduced in number and replaced by collagen. The thinning paralleled the degree of dilatation with separation of the muscle fibres and increasing infiltration of collagenous fibrous tissue. 3) *Degeneration of endothelial cells* opposite the AV-shunt entrance. The endothelial cells showed degenerative changes with enlarged nuchlei forming chromatin clumps(multinuchleated cells) and with distraction of the cytoplasm. At the end remained only the thinned out vein wall.

Discussion

Since1953[2]the role of AVAs for varicose development has been suggested but disregarded. The morphological demonstration of arteriovenous anastomoses connecting to varicose veins and endothelial changes at the interior vein wall represent novelties. The findings of apparant structural and endothelial changes anatomically correlating to a nearby AVA calls for a mechanical interpretation. One such is the the turbulence caused by the AVA-transmitted arterial 'jets' visualized in experimental stream path studies[10,22]. There is no better explanation to endothelial changes, varicose meandering and irregular dilatation, phlebectasies, nor the increased skin heat overlying varicosities. Afferent AVAs and venous tributaries anchors the vein so that dilatation and elongation caused by the flow dependant shear stress action(and NO?) takes place

68

between these locking tributaries. At first there is a bulge, then comes elongation forcing the vessel to coil. This explains why the AVAs and venous tributaries in the great majority were observed to connect to the convexities of the meandering varicosities.

Conclusion

1) *AVAs to varicose veins proven in morphological sections* present an anatomical route to explain 2) arterial maintenance of varicose veins with hot blood *explaining the increased skin heat over-lying varicosities*, 3) supports the launched theory of varicose coiling. 4) There is no alternative explanation to endothelial changes opposite the AV-shunt orifice.

REFERENCES

[1] Pratt GH.: Arterial varices, a syndrome. Am J Surg 1949;77: 456-460.

[2] Piulachs P & Vidal-Barraquer F.:Pathogenic study of varicose veins. Angiology 1953;4: 59-100.

[3] Hœger K.: Arteriovenous connections in the calf as a cause of pain and walking difficulties. The J of Cardiovascular Surg 1963;4: 673-680.

[4] Gius JA.: Arteriovenous anastomoses and varicose veins. Arch.Surg. 87; 299-310, 1960.

[5] Schalin L.: Arteriovenous communications localized by thermography and identified by operative microscopy. Acta Chir Scand 1981;147: 409-420.

[6] Nylander G.: a) Meander phenomenon in veins. Current Medical Digest 1970; 1185-1194.: b) The meandering phenomenon in the pathogenesis of varices. Angiology, 1969; 20:592-600.

[7] Schalin L. Revaluation of incompetent perforating veins . In Superficial and deep venous disease of the lower limbs(Ed. M. Tesi). Edizioni Minerva Medica 1984; 98-104.

[8] Patil KD, Williams JR & Lloyd Williams K. Thermographic localization of incompetent perforating veins in the leg. Br Med J 1970;1: 195-197.

[9] Schalin L.: Varicose meandering and arteriovenous communications.. Abstr Int Symp on Venous Diseases of the Lower Limbs. March 10-12, 1982 Florence. O.I.C. Medical Press. Via G. Modena 19, 50121 Florence.

[10] Schalin L & Liepsch D.: Varicose meandering an indirect sign of locally increased arterial flow transmitted by arteriovenous communications. In Liepsch D.(Ed): Proc. 2nd Int Symp. on Biofluid Mechanics, München.Springer Verlag 1990; 257-274,

[11] Schalin L. Arterial resting flow with respect to the varicose and non-varicose condition of the patients calf In: de'Oliveira F(Ed): Advances in vascular surgery, Proc. IInd Int Meeting on Vascular Pathology, Coimbra Spain 1983: 245-264.

[12] Schalin L. Small arteriovenous communications to varicose and so called "Incompetent perforating veins". Proc 10th Annual Congr.1986; 88-100. The Phlebology Society of America, 5530 Wisconsin Av., NW Washington DC 20815;

[13] Thulesius O, Ugaily-Thulesius L, Gjöres J & Neglen P. The varicose saphenous vein, functional and ultrastructural studies, with special reference to smooth muscle. Phlebology1988;3: 89-95,

[14] Rose S & Ahmed A. Some thoughts on the œtiologi of varicose veins. J Cardiovasc Surg 1986;27: 534-543

[15] Zsotér T & Cronin RFP. Venous distensibility in patients with varicose veins. The Can Med Ass. Le Journal 1966;94: 1293-1297.

[16] Prerovsky I, Kruszevska A, Linhart J & Hlavová A. Distensibility of the forearm veins in patients with primary varicose veins. Angiologica 1969;6: 354-361.

[17] Thulesius O, Ekman L & Gjöres JE. Functional study of isolated venous valves, with comments on the etiology of varicose veins. In: William E (Ed) Venous Diseases: Medical and Surgical Management 1974; 74-76.

[18] Niebes P, Engels E & Jegerlehner ML. Studies on normal and varicose human saphenous veins. Differences in the composition of collagen and glucose-aminoglucans. Bibl Anatomic. 1977; 16 (part 2): 301-303.

[19] Davies P. How do vascular endothelial cells respond to flow? NIPS 1989; 4: 22-25.

20 Collier J & Vallance P. Second messenger role for NO widens to nervous and immune systems. TIPS 1989;10: 427-431.

21 Palmer RMJ, Ferrige AG & Moncada S. Nitric oxide release accounts for the biological activity of endothelium-derived relaxing factor. Nature, 1987;327: 524-526.

22 Schalin L & Moberger G. Endothelial and vein wall response to turbulent shear stress from arteriovenous communications, In Liepsch D(Ed): Biofluid Mechanics, Proc 3rd Int Symp Fortschr.-Ber. VDI Reihe 17 Nr 107 Düsseldorf: VDI Verlag 1994: 53-65.

Phlebology '95, D. Negus et al. (eds.). Phlebology (1995) Suppl. 1: 69-72

P241

Teleangectasia Studied by Means of Optic Probe Video-Capillaroscopy

S.B. Curri

Center of Molecular Biology, Milan, Italy

INTRODUCTION

Teleangectasia was never studied by means of classic capillaroscopic methods on account of the technical difficulties related to their anatomo-topographical seat, so that the finest morphological peculiarities of the altered smallest skin blood vessels remain an unsolved problem. Only after the introduction of the Optic Probe Video-Capillaroscopy (OPVC: 1,2,3), a combination of the optic fibers technics with informatic, it was possible to examine the structural patterns of the skin and mucosal microcirculation of the whole body surface (4). Teleangectasia is generally considered as a local benign inesthetism of the lower limbs skin, without clinical implications. On the contrary, teleangectasia must be interpreted as the first objective sign of a microcirculatory maldistribution (5), during the so called preclinical stages of venous stasis and/or insufficency (6). Aim of our study was : 1. to define the pathological changes of the smallest skin blood vessels in macroscopically "unaffected" skin areas, and 2. to explain the apparently bizarre architectonic distribution of teleangectasia on the skin of the lower limbs.

MATERIAL AND METHODS

Comprehensively 48 female patients (mean age 27 +/- 6.4 y.) suffering from light and desultory symptoms of venous stasis, as heaviness of the legs, ankle-oedema, sometimes paresthesia and nocturnal cramps, but without any instrumental sign of venous insufficiency (so called "Hypotonic Phlebopathy" acc. to Andreozzi, 7) were selected. All the subjects show teleangectasic areas, generally of small extension and localized on the tight, knee or leg skin. Teleangectasia and with nude eye unaffected, 10-20 cm distant skin areas were examined by means of OPVC (Scopeman Moritex MS 504, Meisei, Japan), using contact objectives 50 x and 200-400 x. The capillaroscopic images were then registered (YV Videoreg. CV300E, Sony, Japan) and printed (colour printer VY 150E, Hitachi, Japan).

OBSERVATIONS

At lower magnification (50 x) the dilated small venules of teleangectasia are easily detectables and may be followed in their whole course up to the finest ramifications. The single venular segments show an irregular, serpiginous course; the wall diameter is discontinuous, with segmentary dilations and restrainings (Fig.1 A). The teleangectasic venules are sorroundend by a fine network of capillaries. The red blood cells mouvements are not appreciables at this magnification. At 200 - 400 x the sludge phenomens and the remarkable slackening of the blood flow become more evident. Sometimes the dilated venules appears in form of irregular, polygonal meshes (Fig.1 B) oriented parallely to the skin surface. In unaffected skin areas, 10-20 cm from the teleangectasia, there are severe pathological changes of the skin microcirculation. The normal microangiotectonic is narrowed. Capillary dilation, microaneurisms, disjoining of the meshes are constant findings (Fig. 1 C-D).

CONCLUSIONS

As reported by Weindorf et al. (8) and repeatedly by us emphasized (3,4,5), microvascular changes are present in macroscopically unchanged skin in patients suffering from chronic venous insufficiency. These changes initiate long time before the chronicization of the clinical symptoms of venous stasis and/or insufficiency. A precocious alteration of the vein wall microcirculation in the first stages of the process appears highly probable. Teleangectasia and the related microcirculatory changes are perhaps the top of an iceberg, whereas the main damage play their dangerous roll in the vein wall. Concerning the apparently capricious and casual anatomo-topographical distribution of teleangectasia, our observations on the microangiotectonics of the whole body surface in healthy young males and females show that there are in the lower limbs great differences of the spatial distribution of capillaries and small venules. Teleangectasia may appear only in body seats, where the capillary and venular meshes are parallely oriented to the skin surface (5). On the contrary, where the capillary loops are ascending from the deeper skin layers and appears as points or small commas, whereas the small venules are localized more profoundly, the "classic" teleangectasia in their polymorphous aspects cannot to manifest one's self. The appearance of teleangectasia is therefore related to the specific seat-bound skin microangiotectonics (2).

REFERENCES

1. Thulesius O, Capillaroscopy with fibre optic microscope. Vasa 1992; 21:6.8.
2. Curri SB, Anatomia del microcircolo cutaneo: nuove acquisizioni morfo-funzionali sulla microangiotettonica distrettuale delle diverse regioni della superficie cutanea indagata con Video-Capillaroscopia a Sonda Ottica. Flebologia 1992;3: 247-259.
3. Curri SB, Nouvelles tendences en microcirculation: angiobiotopie des maladies veineuses. Artéres et Veines 1994;13:16-22.
4. Curri SB, Relationship between Laser-Doppler Signal and Skin Microangiotectonic in Chronic Venous Stasis, with Special Reference to Teleangectasia, Proc Eur Congr int Union Phlebology, Budapest Sept 6-10 1993. Multiscience Publ Co Ltd Essex 1994; 185-194.
5. Curri SB, Aspects microcirculatoires des téleangectasies étudiés par vidéo-capillaroscopie a sonde optique (VCSO). Phlébologie 1994;47/4: 325-328.

6. Curri SB, Stase veineuse chronique et microcirculation. Artéres et Veines 1991; 10: 337-344.
7. Andreozzi GM, La Flebopatia Ipotonica, in: S.Bilancini and M.Lucchi (Eds) Le varici nella pratica quotidiana. Minerva Medica Publ Co Turin 1991: 403-408.
8. Weindorf NG, Schultz-Ehrenburg U, Capillary Microscopic Findings in Patients with C.V.I. under Different Hydrostatic Conditions. Eur Congr int Union Phlebology, Budapest Sept 6-10 1993: Abstr Book 134.

72

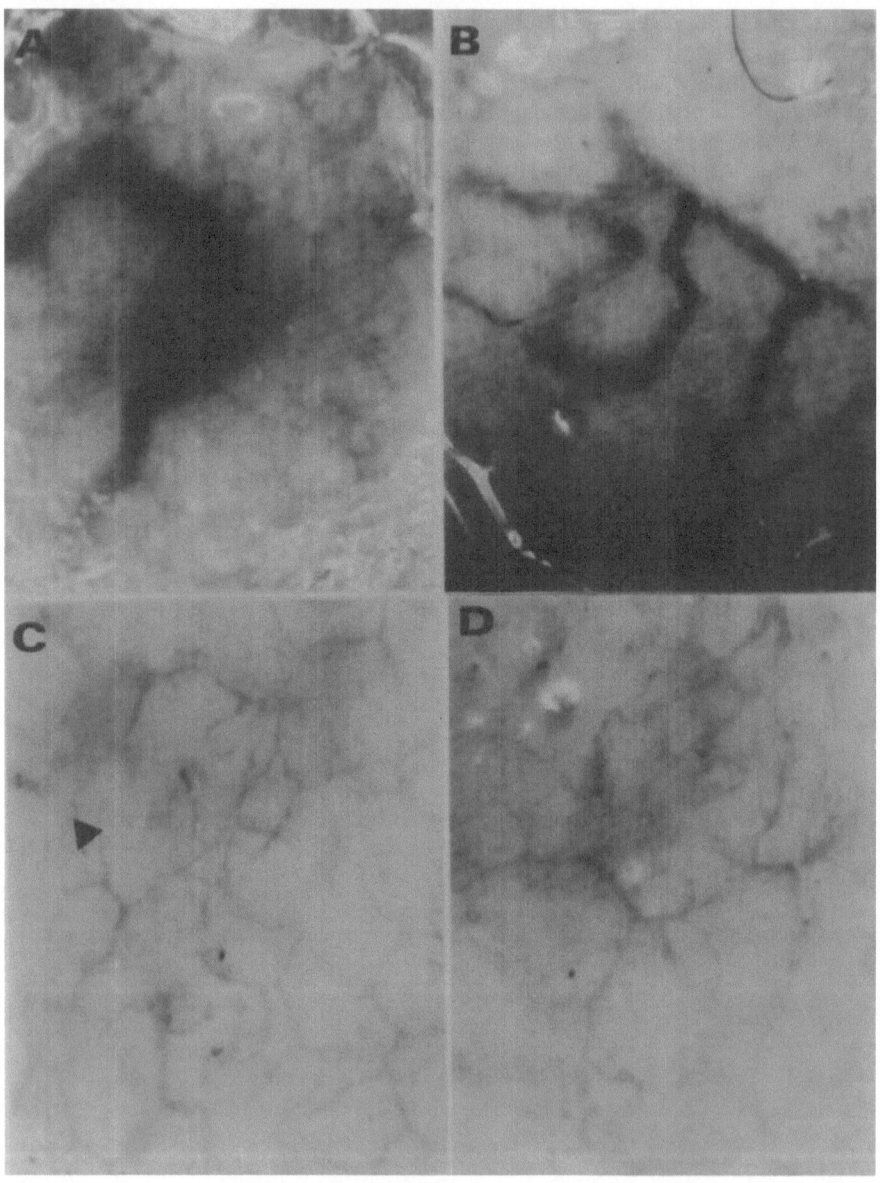

Fig.1

In A: irregular, serpiginose course of a teleangectasia with segmentary dilations and restrainings. In B: polymorphous meshes formed by dilated small venules. In C-D: narrowing and pathological changes of the microvessels in unaffected skin areas, 10-20 cm from the teleangectasia. Contact objective 50 x.

Phlebology '95, D. Negus et al. (eds.). Phlebology (1995) Suppl. 1: 73-75

P298

Anatomical and Physio-Pathological Fundamentals of the "Venous Buffer Circuit" - Modifications to Cockett's and Linton's Operations

E.A. Enrici[1] and H.S. Caldevilla[2]

[1] Phlebology & Lymphology Department, Post-Graduate School of Health Sciences, Argentine Catholic University, Buenos Aires, Argentina
[2] Argentine College of Venous & Lymphatic Surgery, Buenos Aires, Argentina

In 1956, Cockett pinpointed the difference in the genesis of trophic disorders triggered by direct and indirect perforants in the leg.

Indirect perforants conform to variable topography in the upper and mediumm portions of the leg, showing 2 anatomical characteristics, fundamental in their physiopathological implications: (a) the existence of the muscular vein inserted in its subaponeurotic route, acting as buffer to tensions produced, and (b) meeting of large calibre vessels at the superficial venous system level which, when flow reverses, easily neutralizes the retrograde hypertensive waves and reinstates them into the deep venous system.

By means of phlebographic and anatomical dissections, authors verified a fundamental anatomical detail: Leonard's vein does not end isolated in the retromalleolous, as previously believed, but continues in an internal submalleolous manner from the lower perforant up to anastomote with inner saphenous. The authors named his arc, which links Leonard's, lower perforant and inner saphenous veins, the "Lower Anastomotic Arc", as published by the Argentine Society of Surgeons in 1974. Besides, they found that at the arc's convex face's middle a vein is born which, in its distal route, splits into 2 branches: one ends at the retromalleolous capillaries and Léjar's sole, the other at the inner plantar veins. Due to its physiopathological implications, this vein has been called the "Fourth Perforant".

As per authors' understanding, the indurative hypodermitis is caused by a yielding "Venous Buffer Circuit" in the lower inner half of the leg (Fig.1). This circuit consists of: inner saphenous and Leonard's veins and the lower anastomotic arc at the upper superficial level, admitting the 3 inner direct and the 4th perforants from the posterior tibial group.

74

1st Cycle

Perforant
valvular
forcing

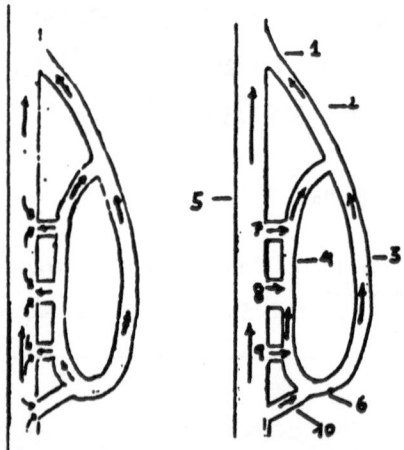

2nd Cycle

Direct per-
forant in-
sufficiency

4th Cycle

Global in-
sufficiency
of the buf-
fer venous
circuit

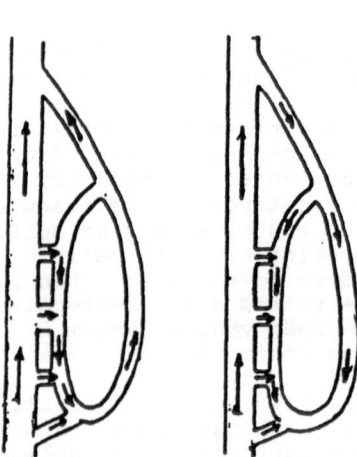

3rd Cycle

Direct per-
forant and
Leonard's
vein insuf-
ficiencies

Fig. 1. VENOUS BUFFER CIRCUIT

1. External saphenous vein arch
2. Inner saphenous vein, thigh
3. Inner saphenous vein, leg
4. Leonard's vein
5. Deep venous system

6. Lower anastomotic arc
7. Upper direct perforant
8. Medium direct perforant
9. Lower direct perforant
10. 4th perforant

In the posthrombotic syndrome, in the authors' concept, the physiopathological sequence leading to the aforementioned circumstance, occurs as follows:

In a first stage, the soleo-gemelli muscular pump (Barrow's peripheal core) compensates, by its hypertrophy, the deep venous insufficiency and the patient shows a swollen calf and a moderate, or absent, ædema. But also in this same stage the powerful short waves, due to strong muscular contractions and the "blow-out" mechanism, are capable of turning the perforrants' valves insufficient. In turn, the direct perforants expand more than the rest as they meet resistance at flow's running into lesser calibre veins, like Leonard's vein.

In the second stage, the flow's reversal is channelled upwards through Leonard's vein thanks to its valve sufficiency, following a centripetal route towads the inner saphenous vein. But very soon Leonard's vein will yield and the flow will revert towards the malleolous region to reach the lower anastomotic arc, onset of the inner saphenous and 4th perforant veins.

Based on the above, the authors implemented the following modifications to the Cockett and Linton operations:

COCKETT'S OPERATION:

1. An incision bordering the indurative hypodermitis-cellulitis region, adaping it to tissue vitality and reaching Achilles' tendon, without hindering any other surgical stages. In its lower part, the incision is retro- and sub-malleolous, allowing approach to the lower and valvulated von Limborg's perforants (1974). 2. The incision continues surrounding inner malleolus underneath inner saphenous vein (1976) to resect the lower anastomotic arc and ligate 4th perforant and marginal veins that approach said arc from the venous sole. 3. Dissection of the posterior tibial group from upper to lower perforant, allowing ligation of upper, medium, lower and valved von Limborg's perforants in the original region that generates the retrograde hypertensive wave, but without leaving recesses which might develop thrombotic areas. 4. The pinpointing of the arterial perforant which usually accompanies upper and medium venous perforants, assuring higher skin vitality after surgery. 5. Ligation of outer perforants in the posterior flap at the poeroneal wall. 6. Radical surgery to the superficial venous system and inner and/or outer saphenectomy and ectomy of varicose routes feasible due to graded small incisions, according to he type of pathology and its ætiology.

LINTON'S OPERATION

Direct perforants sector: identically to Cockett's operation. At the indirect perforants level, dissection of the deep venous collectors is not extended upwards so as not to damage aponeurotic insertion of the muscular pump.

Resorting to these modifications the authors, and members of their team, have thus treated more than 2500 patients since 1974.

OP/16.1

THE VEINS OF THE M. SOLEUS
AN ATTEMPT TO AN ANATOMICAL "MODEL"

I Staelens MD

Lab. d'Anatomie ULB B1070 Brussels, Belgium

Objective: To develop an anatomical model for the veins of the M Soleus.

Design: Dissection of the veins of the M Soleus after injection of the venous and arterial system of cadaver legs with a mixture of gelatine and pigments. The vascular system of fresh cadaver legs was previously rinsed with water.

Material: 50 fresh cadaver legs not considering their previous medical history, except for the absence of surgical treatment for venous insufficiency.

Results: 1) Four groups of small, paired, horizontal veins were found at the external side of the muscle, three groups on the medial side.

2) Two groups of large paired vertical veins, one above each other, in the centre of the muscle draining directly into the tibio-peroneal trunks with a short large communicating vein.

Conclusion: The soleus muscle is drained by the association of a peripheral, small sized horizontal venous system and a central large sized vertical system.

Anatomy/Pathology

Basic Science
Pathology

Phlebology '95, D. Negus et al. (eds.). Phlebology (1995) Suppl. 1: 78-80

P011

The Influence of Blood Flow on Neointima Formation in PTFE Grafts in Venous System

K. Ziaja[1], M. Blaszczynski[1], M. Zabski[1], P. Karczewski[2], B. Bialas[2], T. Drazkiewicz[1], and M. Simka[1]

[1] 1-St Department of General and Vascular Surgery   [2] Department of Patomorphology
Silesian Medical School, Katowice, Poland

INTRODUCTION

In our previous studies we observed that there were some differences in neointima formation between prostheses in venous and arterial systems [1]. This could be due to: different chemical and/or physical parameters of blood flow in veins and arteries, different histological structure or differences in blood flow pattern [1,2,3,4,5].

The aim of this study is to determine the causes of differences in neointima formation between grafts in arteries and in veins and to evaluate the influence of blood flow on neointima formation.

MATERIAL AND METHODS

Experiments were performed on albino New Zealand rabbits divided into 4 groups (5 animals in each group):

Group I - we replaced an 8 mm segment of the aorta below the renal arteries with PTFE;

Group II - we replaced an 8 mm segment of the vena cava below the renal veins with exactly the same prosthesis as in group I;

Group III - as group II plus arterio-venous fistula distally between iliac vessels;

Group IV - prosthesis as above sutured as an arterio-venous fistula between abdominal aorta and vena cava inferior below the level of renal vessels.

The observation period was 3 weeks.

Specimens were taken for microscopic examination. The specimens were divided longitudinally at the mid-line of the grafts, fixed in 10% buffered formaldehyde solution at room temperature, embedded in paraffin, sectioned longitudinally and stained with haematoxilin-eosin. We used computer aided image analyser VIDS-IV for morphometrical analyses.

We estimated the thickness of the neointima at the caudal and cranial anastomoses and in the middle area of the graft. Also neointima ingrowth at the proximal and distal end of the prosthesis was assessed.

RESULTS

In all the prostheses in group 2, 3 prostheses in group 3 and 2 prostheses in group 4 neointima formation was completed - all these prostheses were completely covered with neointima. In other cases there was no neointima in the middle part of the prosthesis. Detailed morphometric measurements (mean for each group) are shown in tables 1 and 2.

Table 1. Detail thickness of the neointima (in μm) at the: inflow, outflow and middle areas of the graft.

Group	Thickness of neointima (in μm) at the:		
	inflow area	middle area	outflow area
1	72,31	0	152,21
2	67,66	175,41	76,68
3	68,97	79,63	100,12
4	90,87	16,23	172,11

Table 2. Detail neointima ingrowth (in mm) from the caudal and cranial anastomoses. Only cases where neointima formation was not completed.

Group	Neointimal ingrowth from the:	
	outflow end	inflow end
1	0,89	1,47
2	All the prostheses were covered with neointima	
3	0,9	1,85
4	1,18	1,5

DISCUSSION

Most authors agree that there are marked differences between neointima formation in prostheses used in arterial and venous systems [1,2,3,4,5]. The

80

causes of this fact remain unknown. In this experiment we investigated the same type of prosthesis in different models. This allowed us to specify the blood flow as the cause of marked differences between neointima formation in prostheses used in arterial and venous systems. Two facts confirming this hypothesis should be emphasized:

a) In all groups the neointima is thicker at the outflow end of the prosthesis than at the inflow end of the prosthesis.

b) In animals where neointima formation is not completed, the ingrowth of neointima is more extensive at the inflow end, than at the outflow end.

CONCLUSIONS

1. Neointima formation is more rapid and more complete with prostheses used in veins than in arteries.

2. There are marked differences between neointima formation in prostheses used in arterial and venous systems. This seems to result mainly from differences in blood flow pattern.

REFERENCES

1. Błaszczyński M., Ziaja K., Żabski M., Białas B., Karczewski P.: Experimental Comparisment in the Healing Process of the Politetrafluoroethylene Vascular Prosthesis Used in Arterial and Venous Systems. Proceedings of the European Congress Of The International Union Of Phlebology. London 1993;.

2. Fillinger M.F., Reinitz E.R., Schwartz R.A., Resetarits D.E., Paskanik A.M., Bruch D., Brendenberg C.E.: Graft geometry and venous intimal - medial hyperplasia in arteriovenous loop grafts. J. Vasc. Surg., 1990,11,556-566.

3. Kohler T.R., Kirkman T.R., Kraiss L.W., Zierler B.K., Clowes A.W.: Increased blood flow inhibits neointimal hyperplasia in endothelialized vascular grafts. Circ. Res., 1991,69,1557-1565.

4. Kraiss L.W., Kirkman T.R., Kohler T.R., Zierler B.K., Clowes A.W.: Shear stress regulates smooth muscle proliferation and neointimal thickening in porous polytetrafluoroethylene grafts. Arterioscler. Thromb., 1991,11,1844-1852.

5. Watase M., Kambayashi J., Itoh T.,Tsuji Y., Kawasaki T., Shiba E., Sakon M., Mori T., Yashika K., Hashimoto P.H.: Ultrastructural analysis of pseudo-intimal hyperplasia of polytetrafluoroethylene prostheses implanted into the venous and arterial systems. Eur. J. Vasc. Surg. 1992,6,371-380.

Phlebology '95, D. Negus et al. (eds.). Phlebology (1995) Suppl. 1: 81-84

PI/3.4

Expression of the Adhesion Molecules ICAM-1, VCAM-1, LFA-1 and VLA-4 in the Skin with Chronic Venous Insufficiency

M. Peschen, A. Weyl, J.M. Weiss, T. Lahaye, E. Schöpf and W. Vanscheidt

Department of Dermatology, University of Freiburg, Germany

INTRODUCTION

Pathophysiologically decisive for the occurrence of chronic venous insufficiency (CVI) is "ambulatory venous hypertension". The characteristic changes of CVI result in corona phlebectatica, hyperpigmentation, stasis dermatitis, dermatoliposclerosis and chronic ulcus cruris. The effects of chronic hypertension in the area of terminal capillary system are largely unknown. One current hypothesis postulates altered leukocyte rheology [1].

Adherent leukocytes present in the capillary area may result in occlusion of the vessel and thus to reduced perfusion. Extravasation of T-cells and granulocytes may result in a release of mediators which cause damage to the endothelium and increased permeability of the capillaries [1,2]. The migration of leukocytes through the endothelial layer of the terminal capillary system is decisive for inflammatory processes such as occur in patients with CVI [3,4,5].

A very close interaction between leukocytes (T-lymphocytes, granulocytes, monocytes, etc.) is necessary for these migratory processes. The contact is mediated by cellular adhesion molecules (CAM) which are expressed on the surface of endothelial cells and leukocytes [3,4,5]. The adhesion of the leukocytes to the endothelium is currently seen as divided into various stages. First, there is a weak adhesion of the leukocytes to the endothelial cells. This weak binding is mediated by molecules of the selectin group, such as E-selectin on the endothelial cells and their ligands sialyl Lewis X. Tight binding of the leukocytes to the endothelial cell is mediated by ICAM-1 and VCAM-1 to the endothelial cells and their ligands LFA-1 and VLA-4 to the T-cells. The tight binding is a direct prerequisite for the transmigration of the leukocytes to the neighboring tissue.

Aim of our investigation was therefore to determine the behavior of the expression of ICAM-1, VCAM-1, VLA-4 and LFA-1 in the earlier stages of CVI.

METHODS

Patients

Six mm punch biopsies were obtained from 10 healthy patients and 40 patients suffering from CVI of the lower limb. The clinical diagnosis was confirmed by doppler sonography and plethysmography. A concomitant arteriosclerosis of the extremities was excluded by an ankle-arm index > 0.9.

Immunhistochemistry

Frozen skin specimen were embedded in Tissue Freezing Medium (Fa. Jung, Nussloch, FRG) and 5μm serial cryostat frozen sections were stained using a 4-step immunohistochemical staining protocoll (ABC technique, Dako Hamburg, FRG):
1. primary mAb (mouse IgG1 kappa); 2. Biotin-conjugated-goat-anti-mouse IgG
3. peroxidase conjugated streptavidin; 4. diaminobenzidine as chromogenic substrate.
Sections were counterstained with hemalaun. Stainings were evaluated by three independent observers in a blinded fashion using a Zeiss Axioskop, equipped with a MC 100 camera system.

Evaluation of slides

In each specimen, capillary loops and pericapillary infiltrate were evaluated separately for the expression of ICAM-1, VCAM-1, VLA-4 and LFA-1 respectively (fig.1). The expression of adhesion molecules on endothelial cells was determined by counting all capillary loops in each specimen. Thereafter, the percentage of capillary loops which expressed ICAM-1 and VCAM-1 was calculated. Furthermore, pericapilllary infiltrating leukocytes and lymphocytes were assessed for VLA-4 and LFA-1 positive staining and the percentage of capillary loops with LFA-1 and VLA-4 positive infiltrate was determined.

RESULTS

The expression of ICAM-1 and VCAM-1 on endothelial cells was slightly increased in patients with corona phlebectatica. Specimen of hyperpigmentation, stasis dermatitis and dermatoliposclerosis showed a significant increased expression of ICAM-1 and VCAM-1 when compared with healthy skin (Fig.1).
Furthermore, a marked perivascular infiltration of LFA-1 and VLA-4 positive leukocytes and lymphocytes was observed in patients with hyperpigmentation and stasis dermatitis, in specimen of corona phlebectatica and dermatoliposclerosis only a slight perivascular infiltration was observed in comparison to healthy skin (Fig.1).

Adhesion molecule	Healthy Skin	Corona phlebec-tatica	Hyperpig-mentation	Stasis dermatitis	Dermato-liposcle-rosis
ICAM-1	37% ± SD 3.7 n=241	40% ± SD 4.5 n=253	49% ± SD 4.2 n=521	53% ± SD 3.6 n=317	50% ± SD 4.0 n=631
VCAM-1	14% ± SD 3.1 n=195	20% ± SD 3.5 n=174	28% ± SD 3.9 n=376	40% ± SD 4.6 n=284	47% ± SD 4.2 n=602
VLA-4	27% ± SD 4.4 n=195	30% ± SD 4.2 n=125	56% ± SD 3.4 n=269	73% ± SD 3.3 n=303	42% ± SD 4.7 n=502
LFA-1	29% ± SD 3.4 n=179	34% ± SD 4.7 n=239	50% ± SD 4.0 n=357	62% ± SD 4.6 n=250	35% ± SD 4.3 n=341

Fig.1: Percentage and ± SD of ICAM-1- and VCAM-1- positive capillary loops and percentage and ± SD of LFA-1- and VLA-4- positive pericapillary infiltrate compared to unstained loops (n= total number of capillary loops).

DISCUSSION

Recently, our team was able to demonstrate a markedly altered expression pattern of ICAM-1/VCAM-1 and LFA-1/VLA-4 in chronic venous ulcers [6]. In the perimeter of the Ulcera cruris, compared to healthy lower-calf tissue, there is a pronounced immunohistochemical expression of ICAM-1 and VCAM-1 on the endothelial cells. The ICAM-1 and VCAM-1 ligands LFA-1 and VLA-4 are also very prevalent on cells of the perivascular inflammatory infiltrate. The chronic ulcers examined to date represent a late stage of CVI. In our study an increased expression of ICAM-1/VCAM-1 and LFA-1/VLA-4 in earlier stages of CVI was observed again. These findings indicate that inflammatory reactions are controlled in the vascular terminal capillary system and maintained by increased expression of these adhesion molecules.
We speculate that in patients with CVI the increased expression of the above mentioned adhesion molecules could contribute to an augmented leucocyte number both interstitially [7]. and intracapillarily [2]. These leukocytes cause endothelial cell damage, which in turn promotes diapedesis of erythrocytes with subsequent iron overloading [8] and pericapillary deposition of fibrinogen [9].
Our investigations suggest that the downregulation of the increased expression of adhesion molecules in stasis dermatitis could be a goal of further pharmacological studies to prevent the development of the clinical signs of CVI [10].

REFERENCES
1. Thomas PRS., Nash GB., Dormandy JA.: White cell accumulation in dependent legs of patients with venous hypertension: A possible mechanism for trophic changes in the skin. Br Med J 1988;296:1693-1695.
2. Vanscheidt W., Laaf H., Weiss JM., Schöpf E.: Immunohistochemical investigation of dermal capillaries in chronic venous insufficiency. Acta Dermatol Venerol 1991;71:17-19.
3. Cavender DE. Lymphocyte adhesion to endothelial cells in vitro: Models for the study of normal lymphocyte recirculation and lymphocyte emigration into chronic inflammatory lesions. J Invest Dermatol 1989;93:88S-95S.

4. Pardi R, Inverardi L, Bender JR. Regulatory mechanism in leucocyte adhesion: Flexible receptors for sophisticated travelers. Immunol Today 1992;13:224-230.
5. Shimizu Y., Newman W., Tanaka Y., Shaw S.: Lymphocyte interactions with endothelial cells. Immunol Today 1992;13(3):106-112.
6. Weyl A., Vanscheidt W., Weiss JM., Peschen M., Schöpf E., Simon JC.: Expression of the Adhesion Molecules ICAM-1,VCAM-1, ELAM-1 and their Ligands VLA-4 and LFA-1 in chronic venous leg ulcers. (JAAD in Print)
7. Scott HJ, Coleridge Smith PD, Scurr JH. Histological study of white blood cells and their association with lipodermatosclerosis and venous ulceration. Br J Surg 1991;78:210-211.
8. Ackermann Z, Seidenbaum M, Loewenthal E. Overload of iron in the skin of patients with varicose ulcers. Arch Dermatol 1988;124:1376-1378.
9. Falanga V, Moosa HH, Nemeth AJ. Dermal pericapillary fibrin in venous disease and venous ulceration. Arch Dermatol 1987;123: 620-623.
10. Weiss JM., Vestweber D., Weyl A., Peschen M., Schöpf E., Vanscheidt W., Simon JC.: Inhibition of TNFα-induced T-cell adhesion to endothelial cells by Pentoxifylline. J Invest Dermatol 102: 599, 1994.

PI/3.6

CLINICAL HISTOCHEMICAL AND IMMUNOHISTOCHEMICAL INVESTIGATION OF THE CAPILLARY BASAL MEMBRANE IN CHRONIC VENOUS INSUFFICIENCY

Vanscheidt W, Peschen M, Weiss JM, Weyl A, Schöpf E, Department of Dermatology, University of Freiburg, Germany

Objective: Regarding the pathogenesis of chronic venous insufficiency (CVI) the adhesion of leukocytes to microvascular endothelial cells and the pericapillary halo are of central interest. Aim of our investigation was therefore to examine the correlation between thickness of pericapillary type IV collagen layer, basal membrane alterations and transcutaneous oxygen tension (TcPO2).

Design: Histochemical and immunohistochemical investigation of the capillary basal membrane was performed and compared to the TcPO2.

Patients: 15 biopsies from normal controls, as well as 30 patients with CVI stage I and CVI stage III (Widmer classification).

Measurements: Thickness of pericapillary type IV collagen layer was measured in each specimen microscopically with 10x magnification. TcPO2 was measured just prior to biopsy procedures in exactly the same area the specimen was subsequently excised.

Results: The microscopically measured thickness of the collagen IV layer and thickness of the basal membrane was increased significantly in patients with CVI. Specimens from normal controls showed a collagen IV layer thinner than 0.1mm. Patients with CVI stage III revealed strong collagen IV depositions between 0.2 and 0.3mm. Comparison between TcPO2 and histological findings in the measured areas showed oxygen pressure varying from 72mmHg (SEM 15mmHg) in normal controls to 14mmHg (SEM 12mmHg) in patients with thick collagen IV layers.

Conclusions: In patients with CVI a thick collagen IV layer is associated with decreased TcPO2 possibly contributing to trophic alterations.

P097

THE CAUSE OF VARICOSE VEINS

Fegan G

Trinity College Dublin

True understanding of the cause of varicose veins can only follow a new approach to the concept of venous return from the lower limb. It is necessary to study in depth the effect of one single walking step, to compare the transport of blood in the many paired chambers of the venous system in the leg with the flow of blood in the heart in a single heart beat. We must consider why the venous capacity of the leg is so great compared with the venous systems returning blood from other organs and how the cycle in the leg veins involves a collecting phase and a boosting phase.

In each step the collecting time is similar to the collecting time of each heartbeat and the boosting time through the open atrio-ventricular valve is very similar to the boosting time through the valve in the perforating vein.

When a part of this system fails we must look at the result, both in the short term leading to physiological compensatory hypertrophy of the superficial veins and to the subsequent pathological changes resulting from turbulent blood flowing in the wrong direction.

Taking this view of the function of venous return from the lower limb supports the restoration approach to the treatment of varicose veins and explains how you can obtain resolution of apparently complex varicose veins after destroying one incompetent perforation vein.

Corrective treatment should be aimed towards restoration of correctly operating pumps rather than eradication.

P028

SUPERFICIAL FEMORAL VEIN LEIOMYOSARCOMA
Diagnostic and therapeutic management

L CASTELLANI; R MARTINEZ; D GARCES; S ROUCHET; A de MURET; G CALAIS; P ROSSET

Service di Chirurgie Cardio-Vasculaire CHR TROUSSEAU

Vascular leiomyosarcomas are rare tumours of the soft tissues. The great majority of cases occurred in the inferior vena cava.
Only a few cases of peripheral leiomyosarcomas have been reported.
At this time neither limited margins' resection, nor the benefits of post-operative radiotherapy and chemotherapy are well established.
To try to analyse progress in diagnostic and therapeutic management; we report the case of a woman of 65 years of age who underwent surgical treatment for a leiomyosarcoma of the superficial femoral vein at our institution by an entire resection of a ten centimetre tumour with vascular reconstruction. The histological examination confirmed the vascular nature of the fusiform leiomyosarcomatous cell proliferation (stage II).
Phlebography, computed tomography and magnetic resonance imaging can support the preliminary approach but only the pathology examination of the tumour could definitely establish the diagnosis.
The disease prognosis is dominated by the local tumour recurrence and the high metastatic risks that could be reduced by post-operative radiotherapy and chemotherapy.
It is very difficult to draw conclusions about the management of this infrequent and rare tumour, but the experience of this case and opinions of various authors suggest that adjuvant therapy protocol still needs some working on and that the optimal therapeutic modality of leiomyosarcoma of the superficial femoral vein remains surgical.

P115

EVOLUTION HISTOLOGIQUE DE LA MALADIE VARIQUEUSE

COCET J.M. - MILLIEN J.P. - CREUSY C.

Cabinet d'Angéiologie 61 Rue de Turenne 59000 LILLE FRANCE

Objectif : Etude de l'évolution histologique de la varicose chez des sujets opérés.
Projet : Etude rétrospective comparant les résultats anatomopathologiques de varices uni et contro-latérales, opérées chez le même individu, à plusieurs années d'intervalle.
Patients : 40 patients (33 femmes - 7 hommes) avec une insuffisance saphène interne primitivement uni-latérale uniquement, puis secondairement controlatérale. Délai moyen entre deux interventions : 7 ans.
Méthodes : Etude histologique systématique de toutes les varices opérées. Observation des lésions pariétales élémentaires après coloration au Safran, bleu d'Alcian et Orcine, selon le même protocole histopathologique. Classification histologique d'après les critères du Pr. SAOUT.
Résultats : on retrouve exactement dans 70 % des cas une classification histologique strictement identique entre la première et la deuxième intervention. Il existe par ailleurs des anomalies pariétales similaires entre les deux groupes dans 92,5 % des cas. Cette étude permet en outre de comparer l'intensité des lésions observées chez le même individu à distance.
Conclusion : l'analyse de ces résultats plaide en faveur d'une maladie génétique.

P115

HISTOLOGICAL EVOLUTION OF VARICOSE DISEASE

COGET JM - MILLIEN J P - CREUSY C

61 Rue du Turenne 59000 Lille - FRANCE

Objective: To study the histological evolution of a varicose vein, treated surgically, within the same population.
Project: Retrospective study comparing pathological results of uni and controlateral varicose veins, operated in the same patient, with an interval of several years.
Patients: 40 patients (33 women - 7 men) with an insufficient of long saphenous veins, originally only unilateral, then controlateral. The operation was indicated in all cases. Average delay between two surgical interventions: 7 years.
Measurements: Systematic histological study of all varicose veins which were operated. Observation of the elementary parietal lesion after colouring technique with Safran, blue of Alcian Orceine, using the same protocol. Histological classification according to the criteria of Prof. Saout. Comparison of the intensity of the observed lesion as well.
Results: In exactly 70% of the cases, we can notice a strictly identical histological classification between the first and the second surgical intervention. More over there are similar parietal anomalies between the two groups in 92,5% of the cases.
Conclusion: The analysis of those results seem to confirm that varicose veins are really a genetic disease.

Anatomy/Pathology

Clinical Research

Anatomy/Pathology

Clinical Research
Microcirculation

Phlebology '95, D. Negus et al. (eds.). Phlebology (1995) Suppl. 1:90-92

K/1

Activated Leukocytes and Endothelium in Chronic Venous Insufficiency

Geert W. Schmid-Schönbein PhD

Department of Bioengineering and Institute for Biomedical Engineering, University of California, San Diego, La Jolla, CA 92093-0412, USA

INTRODUCTION

Evidence from molecular, biochemical, biophysical, microcirculatory, pharma-cological, and epidemiological studies suggests that activation of leukocytes and interaction with the endothelium may lead to cardiovascular disorders with progressive tissue injury. Ischemia and reperfusion or physiological shock leads to acute accumulation of leukocytes attachment to the endothelium in vital organs, including heart, liver, skeletal muscle, lung, intestine, and brain (1). In many of these acute conditions, neutropenia, suppression of leukocyte adhesion, or cellular downregulation has served to explore the role of circulating leukocytes and their interaction with the endothelium as a contributor to progressive organ failure. There is also evidence to suggest that activation of leukocytes and endothelial cells may play a role in chronic diabetic angiopathy, hypertension as well as venous disease.

LEUKOCYTE PROPERTIES

Leukocytes (neutrophils and monocytes) impose a higher resistance in the microcirculation than red cells or platelets. Leukocyte diameters exceed the average capillary diameter, and their cytoplasm is made up of a dense network of actin with stiff viscoelastic properties. Pseudopods which are rich in actin fibers exhibit even more rigid viscoelastic properties. Thus after pseudopod formation, leukocytes impose an even higher resistance in the capillary network than without pseudopods. Neutrophils granules are carriers of lysosomal enzymes and adhesion glycoproteins. Both endothelial cells and leukocytes serve as source for oxygen free radicals.

THE LEUKOCYTE-ENDOTHELIAL INTERACTION

Adhesion requires membrane contact area formation. This can be achieved in part by passive cell spreading due to attractive membrane forces between leukocyte and substrate, and in part by active spreading with actin polymerization and rearrangement of cell cytoplasm. Neutrophil and monocyte rolling on endothelium is mediated by L-selectin and P-selectin, a family of glycoproteins that are either constitutively expressed on neutrophils/monocytes and endothelial cells, respectively. L-selectin is involved in attraction of monocytes to local inflammatory sites via adhesion to the endothelium. In neutrophils and monocytes, a family of heterodimer glycoproteins mediate adhesion, known as integrins. Leukocyte integrins CD11a,b,c/CD18 share a common β_2 chain (CD18) that is monocovalently associated with each α-chain. Monocyte adhesion is mediated by $a_4\beta_1$ integrin while spreading requires β_2 integrins (2). Leukocyte stimulation leads to transfer of integrins from an intracellular pool to the plasma membrane as well as shedding of L-selectin. The intercellular adhesion molecule 1 (ICAM-1, CD54) serves as counterreceptor for CD11b/CD18. An inducible cell ligand sLe^x, possibly in form of clustered saccharide patches, is counterreceptor for P-selection, and heparin-like chains for L-selectin. ICAM-1 can also be found on parenchymal cells that are subject to lethal attack by neutrophils. There is evidence to suggest that alternative adhesion pathways exist, especially in situations with chronic peroxidation of cell membranes.

Neutrophil transendothelial migration may require integrins (e.g. CD11b/CD18). Neutrophils which establish adhesion with the endothelium and stop rolling are stimulated into an active state. They actively penetrate the endothelium and the basement membrane.

LEUKOCYTE MICROVASCULAR ENTRAPMENT

Leukocyte entrapment in microvessels may lead to complete obstruction of the capillary lumen, but may more frequently be associated with only partial blockade of the vessel lumen. Immuno-protection of leukocyte membrane integrins (CD18) or adhesion molecules on the endothelium with monoclonal antibodies serves to reduce the membrane adhesion stress and to abolish the capillary no-reflow. Individual neutrophils are encountered, which are deformed inside the capillaries into cylindrical shapes with close membrane abutment to the endothelium. Neutrophil stiffening and expression of membrane adhesion proteins lead to enhanced entrapment. Fully obstructed capillaries contain high numbers of red cells and platelets, but in acute no-reflow both cell types are passively trapped due to the leukocyte capillary obstruction. Attachment of neutrophils to post-capillary venules is prominent in most forms of ischemia and reperfusion and raises the hemodynamic resistance. During reperfusion, when leukocytes are pushed out of an obstructed capillary, all red cells and platelets are washed out of the capillary and flow is restored

An alternative mechanism to entrap leukocytes in the capillary network is to elevate the venous pressure. Elevation of the venous pressure serves to raise the capillary pressure, thus distending the capillaries, but it compromises at the same time the pressure drop along the length of the capillaries. Reduction of the pressure drop in capillaries to push plasma and blood cells along the lumen leads to a reduction of the fluid layer thickness between leukocytes and the endothelium, as well as a reduction of the shear rate in post capillary venules, so that entrapment of leukocytes is favored. In the leg of humans, such microvascular mechanism can be directly observed. Leukocyte are trapped in the peripheral circulation during elevated venous pressure (in a leg down position) but are washed out of the microcirculation in a leg up position with reduced venous pressure (3).

Entrapment of leukocytes in the microcirculation has a number of implications. It promotes local organ injury and causes capillary no-reflow. At the same time it contributes to immune suppression, since microvascular entrapment is accompanied by a reduction of circulating leukocyte numbers, and entrapped leukocytes have reduced capacity to carry out phagocytosis since there is less likelihood for contact formation with microorganisms than in freely circulating leukocytes. Attachment to a single site on the endothelium may initiate transformation of the leukocytes (mostly neutrophils and monocytes) into an active state. The cells then project pseudopods and migrate across the endothelium into the interstitium.

Attachment of leukocytes to the postcapillary endothelium is frequently associated with oxygen free radical formation. Free radical formation is prominent in the gap between leukocytes and endothelial membranes mediated by integrins (CD11b/CD18) (4). In the absence of antithrombotic agents, leukocytes also serve to trigger platelet aggregate formation with adhesion to the endothelium and fibrin deposition. The presence of leukocytes in platelet thrombi can be mediated by P-selectin (5).

VENOUS ULCERS IN THE LEG

Patients with chronic venous insufficiency and lesion formation exhibit microvascular regions in the skin with no apparent blood flow while at the same time circulating leukocytes become trapped in the microcirculation of the skin and subcutaneous tissue of the leg (3, 6). The entrapment of leukocytes is enhanced when the venous pressure is raised by placing the patient in a sitting position and can be reduced when the leg is raised in a supine position (7). Immunohistochemistry of skin biopsies from the site of tissue lesions shows a rich perivascular accumulation of T lymphocytes and macrophages as dominant infiltrating cells, a situation that is not observed in controls. In extreme cases, the cell infiltration is accompanied by enhanced expression of the endothelial adhesion receptor ICAM-1 together with other molecular markers of inflammation. The entrapment of neutrophils is accompanied by release of proteolytic enzymes due to degranulation, including enzymes like elastase that are known to break down connective tissue components (8).

The entrapment of leukocytes is caused by a loss of the normal pressure gradient across the microcirculation due to venous pressure elevation and the adhesion of leukocytes to microvascular endothelium. The abnormal entrapment in microvessels may further be aggravated by central activation of leukocytes and endothelial cells. Application of compression stockings is one of the relative few treatment modalities to reduce the lesions. Compression stockings enhance the removal leukocytes from regions with skin lesions, probably by improving the microvascular perfusion and elevation of local shear stresses in the microcirculation (9).

REFERENCES

1. Granger ND, Schmid-Schönbein GW. Physiology and Pathophysiology of Leukocyte Adhesion. ed. New York: Oxford University Press 1995.
2. Luscinskas FW, Kansas GS, Ding H, Pizcueta P, Schleiffenbaum BE, Tedder TF, et al. Monocyte rolling, arrest, and spreading on IL-4-activated vascular endothelium under flow is mediated via sequential action of L-selectin, β_1-integrins, and β_2-integrins. J Cell Biol 1994; 125: 1417-1427.
3. Moyses C, Cederholm-Williams SA, Michel CC. Haemoconcentration and accumulation of white cells in the feet during venous stasis. Int J Microcirc Clin Exp 1987; 5: 311-320.
4. Suematsu M, Schmid-Schönbein GW, Chavez-Chavez RH, Yee TT, Tamatami T, Miyasaka M, et al. In vivo visualization of oxidative changes in microvessels during neutrophil activation. Am J Physiol 1993; 264: H881-H891.
5. Palabrica T, Lobb R, Furie BC, Aronovitz M, Benjamin C, Hsu YM, et al. Leukocyte accumulation promoting fibrin deposition is mediated in vivo by P-selectin on adherent platelets. Nature 1992; 359(6398): 848-851.
6. Thomas PRS, Nash GB, Dormandy JA. White cell accumulation in dependent legs of patients with venous hypertension: a possible mechanism for trophic changes in the skin. Brit Med J 1988; 296: 1693-1695.
7. Edwards J, McMullin GM, Scott HJ, Wilkinson L. White blood cell distribution in chronic venous insufficiency. In: Microcirculation in Venous Disease. Eds. Coleridge Smith PD. Austin: R.G. Landes Company, 1994; pp 113-128.
8. Shields D. White cell activation. In: Microcirculation in Venous Disease. Eds. Coleridge Smith PD. Austin, TX: R.G. Landes Comp., 1994; pp 129-143.
9. Abu-Own A, Sarin S. Compression treatment in venous disease. In: Microcirculation in Venous Disease. Eds. Coleridge Smith PD. Austin, TX: R.G. Landes Comp., 1994; pp 145-173.

Phlebology '95, D. Negus et al. (eds.). Phlebology (1995) Suppl. 1: 93-95

K/4

Lymphatic Microcirculation of the Skin in Venous Disease

U.K. Franzeck

Department of Medicine, Angiology Division, University Hospital, Zürich, Switzerland

INTRODUCTION

Extensive investigations have been performed in the past to evaluate the microangiopathy of blood capillaries in chronic venous insufficiency (CVI), but recent studies are looking also at the involvement of the lymphatic system in CVI. Especially the microlymphatics of the skin are involved in chronic venous disease.

METHODS

Fluorescence Microlymphography

Unlike the blood capillaries of the skin the cutaneous microlymphatics cannot be visualized *in vivo* without fluorescent dyes. Therefore a subepidermal injection of 0.01 ml FITC (fluorescein-iso-thio cyanate)-dextran with a molecular weight of 150'000 in a 25 % solution is performed at the medial ankle region by means of a steel microcannula with an outer diameter of 0.2 mm. From an initial deposit of the fluorescent dye the lymphatic capillaries are stained. The lymphatic capillaries form a network, which is recorded by fluorescence video microscopy on video tape (1, 2).

The apparatus consits of a incident light fluorescence microscope (Leica AG, Herbrugg, Switzerland) with a mercury vapor lamp (HBO 100W, Osram, F.R.G.), a 3 CCD video camera (DXC-930, Sony, Tokyo, Japan), a video timer (For-A-Company, Chiba, Japan), a video scale marker (For-A-Company), a video monitor (Philipps, Eindhoven, The Netherlands) and a video tape recorder (S-VHS, Panasonic, Japan). The microscope and the camera a mounted on a special support (Leica AG, Glattbrugg, Switzerland), which permits optimal adjustment of the microscope to the skin (3).

RESULTS

Mild CVI

In cases with CVI without trophic changes no changes of lymphatic capillary morphology were observed. However, the propagation of the fluorescent dye was significantly enhanced (see Table 1). The enlarged extension of the microlymphatic network is likely due to the increased lymphatic drainage caused by the enhanced leakage of the blood capillaries.

Table 1.

Extension of FITC-dextran 150'000 in the superficial lymphatic capillary network in healthy subjects (n = 17 studies) and patients with CVI (n = 32 studies). Mean values and standard deviations are given in mm.

Extension	Controls	Patients with CVI	Level of significance
proximal	3.5 ± 2.5	16.9 ± 24.4	$p < 0.05$
distal	2.9 ± 2.1	9.3 ± 9.4	$p < 0.01$
ventral	3.9 ± 2.9	14.8 ± 9.4	$p < 0.001$
dorsal	7.1 ± 3.2	17.4 ± 9.6	$p < 0.001$

Moderate and severe CVI

In patients with trophic changes in the medial ankle region the superficial network of the lymphatic capillaries was damaged. The meshes which form a regular network in healthy controls and in patients without trophic skin changes, are interupted, only partially filled or completely obliterated. Fragments of lymphatic capillaries may be filled relatively far away from the deposit. In cases with no filling of capillaries at all the dye moved diffuse into the interstitial space.

DISCUSSION

The morphology of lymphatic capillaries is still normal in the initial disease. However, the extension of the FITC-stained network is already significantly increased compared with control subjects (10). This is caused mainly by the increased capillary leakage which consequently leads to an increase in lymph fluid.

In the more severe stage of the disease lymphatic microangiopathy progresses and the lymphatic capillary network becomes destructed and obliterated (10). The diffusion of FITC-dextran 150'000 is enhanced. In severe cases only fragments of lymphatic capillary can be found. These changes demonstrate an additional lymphatic component in edema formation in CVI.

Venous ulceration is obviously caused finally by localized microvascular ischemia, which results from several factors. Further research is necessary to gain more direct evidence of the importance of white cells in the cascade of events leading to ulcer formation.

Recent observations suggest a lymphatic involvement in nonhealing venous ulcers. Fluorescence microlymphography performed in such a patient revealed a lymphatic fistula in the ulcer zone. This implies an additional lymphatic component, which may be in part responsible for the delayed healing process.

References

1. Bollinger A, Jäger K, Sgier F, Seglias J: Fluorescence microlymphography. Circulation 1981;64:1195-1200.
2. Franzeck UK, Isenring G, Frey J, Jäger K, Mahler F, Bollinger A. Eine Apparatur zur dynamischen intravitalen Videomikroskopie. VASA 1983;12:233-238.
3. Bollinger A. Microlymphatics of human skin. Int J Microcirc: Clin Exp 1992;12:1-15.
4. Franzeck UK, Isenring G, Bollinger A: Lymphatic microangiopathy in chronic venous insufficiency (CVI). In The Initial Lymphatics. New Methods and Findings. Edited by A Bollinger, H Partsch and JHN Wolfe. New York, Thieme-Stratton Inc., 1985, pp 171-177.

Phlebology '95, D. Negus et al. (eds.). Phlebology (1995) Suppl. 1: 96-98

P039

Study of Morphoquantitative Leukocytic District Behaviour in Chronic Venous Disease: Relationship Between Haematic Values and Leukocytic Integrin

E. Capodicasa[1], F. Corazzi[1], C. Muscat[1], F. De Bellis[1], R. Tognellini[1], G. Venturini[2], G. Lolli[3] and R. Bisacci[3]

[1] Institute of Internal Medicine and Oncological Science, [2] Laboratory of Chemical Clinical Analysis and [3] Department of Surgery and Vascular Surgery, University of Perugia, Italy

INTRODUCTION

In the last few years numerous and detailed publications have indicated the leukocyte as potentially beging in a position to play a pathogenous rather than a defensive role in many circumstances. Nowadays particular interest is being paid to the pathogenous role thatwhite blood cells can have in the cardiovascularfield, and particulary at a microcirculatory level. A good number of epidemiological studies have already underlined the existence of a significant correlation between the leukocyte count and the incidence of ischemic diseases, such as ictus and coronary diseases. Many anatomic-pathological and clinical studies of an experimental nature indicate the leukocyte,and particularly the neutrophil granulocytes the principal mediator of the damage so-called post-ischemia riperfusion.

The pathogenous action of white blod cells seems to come about in such circumstances both because of the theological properties of leukocytes and the funcional attributes of these cells, whic can be activated, adhere to the endothelium migrate across the vascular wall and liberate numerous lithic and toxic products. Whe have therefore undre taken a study withthe preliminarypurpose of evaluating the behaviour of the ematic cells from a morpho-quantitative and functional point of view both in the systemic venous blood and in the reluxblood from the foot,after the subjects had been kept for a predetermined time il lyng position with the limbs in horizontal position or sitting with the limbs lowered.

PATIENTS AND METHODS:

15 subjects have been studied (4 men and 11 women; average age 48 +- 16DS) whit chronic venous insufficiency in the lower limbs,hospitalised for a varycectomy operation. They have been chosen for this study on a consecutive basis, following their informed consent; a venous blood specimen to be analysed was whitdrown from each of the subjects. Blood was drown from both the upper and the lower limbs,respectively from the

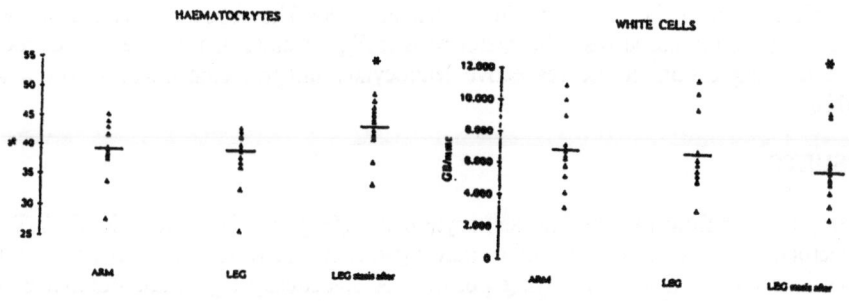

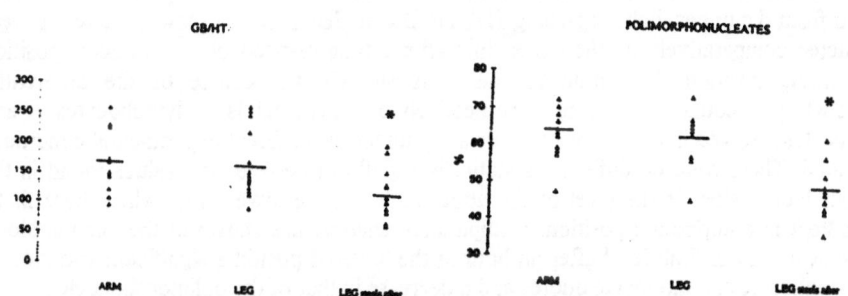

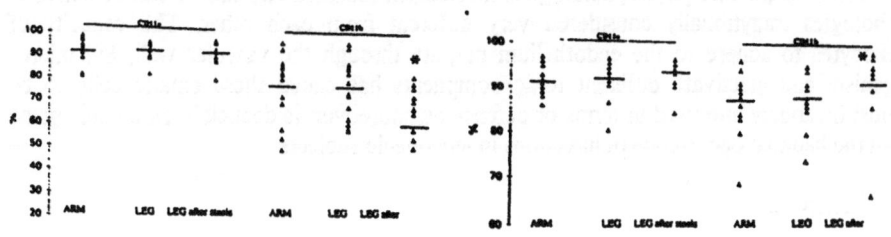

anterior cubital veins and from the big safene vein, next to the medial malleolus. The patients were at first invited to lie down on their backs with their streched out on the bed for two hours. At the end of this period the first peripheral venous blood specimens to be anallysed were drown (time 0) both from the upper and the lower limbs using the technique of double withdrawal without stasis , at the level of both the upper and the lower limbs. After this the patients were kept sitting ith legs down fo an hour. At the end of this per5iod the second set of venous blood specimens were taken from the safene vein on the same side, again in the region of the medial malleolus.The following parametres were evalueted from the blood tests: haematocrytes (HT), absolute and differential leukocyte count and expression of the respective leukocytary integrin component CD11 a and CD11b.

RESULTS

As may be seen from the absolute leukocyte count (GB) and from the ratio GB/HT the haematocrit values (HT) do not differestiate significantly between the upper limb and the lower limb after a time in the lying position. A statistically significant difference was noticed, however, in the comparative analysis between the values observed at this level and those found in the blood taken from the lower limb after a long period in the lowered position. No statically significant difference was observed in the composition of venous blood from the upper limbs regarding HT, NGB and GB/HT,when these parameters were evalueted comparatively on the upper limb after a time respectively in horizontalposition and sitting position. Inaddition to this it is shoown the course of the differential leukocytary count referring respectivelyto neutrophils, lymphocytes and monocytesobserved in the various venous areas under the different experimental conditions indicated. There were no differences statically significant between the values found in the venous blood taken at the level of the upper limb and the lower limb, when these limbs were kept in a horizontal position. Comparative analysis has shown in the venous blood taken at the lower limb level after an hour in the lowered positin a significant variation in the lymphocyte and monocyte quotes and a decrease in that of the polimorphonucleates.

DISCUSSION

One of the most campelling aspects in the medical-biological field is undoubtedly represeted by the fact physiophatological mechanism fudamentally similar can be active in pathologies tradytionally considered very different from each other. The capacity of leukocytes to adhere to the endothelium migrate through the vascular wall, kill micro-organism and inactivate different toxic compnents has caued these ematic cells to be almost invariably interpred in terms of defence as, moreover, is deducible on a clinic plane from the habitual occurrence of infections in leucopenic subjects.

REFERENCES

1. Edzard E., Dale E. et aa. :"Leukocyte and the risk of ischemic disease". JAMA 1987; 257(17):160-171
2. Welbourn C.R.B., Goldamn G. et aa.:"Pathophysiology of ischaemia reperfusion iniury: central role of the neutrophyl",Br.J.Surg. 1991; 13:58-68.
3. Etman M.L.,Lloyd M. et aa."Inflammation in the course of early myocardial ischemia". FASEB J. 1991; 5: 2529-2537.
4. Moyses C.,Cederholm-Williams S.A. and Michel C.C.:"Haemoconcentration and accumulation of white cells in the feet during venous stasis"I.J.Microcirc.Clin.E.1987

Phlebology '95, D. Negus et al. (eds.). Phlebology (1995) Suppl. 1: 99-101

P041

Plasma Levels of Elastase-Alfa 1-Proteinase Inhibitor in Dependent Legs of Patients with Venous Disease: A Preliminary Report

F. de Bellis[1], R. Bisacci[2], A. Villa[3] F. Gregorio[4], R. Biondi[4], and E. Capodicasa[1]

[1] Institute of Internal Medicine and Oncological Science, [2] Department of Surgery, [3] Laboratory of Chemical Clinical Analysis and [4] Department of Clinical Medicine, Pathology and Pharmacology, University of Perugia, Italy

INTRODUCTION

Polymorphonuclear leukocyte elastase (PMN-elastase) is a potent serine protease stored in azurophilic granules of neutrophils. During activation or lysis of polymorphonuclear cells, PMN-elastase, primarily responsable for intracellular protein breakdown, is liberated in intercellular compartments. The concentration of PMN-elastase in plasma determined by enzyme immuno-assay in a complex with α-1-proteinase inhibitor (Ela/α 1-PI complex), has proved one of the most sensitive and specific markers for early diagnosis of inflammation and neutrophil activation [1]. Diverse evidence suggests that leukocytes, and neutrophils in particular, are important mediators of vascular injury and could by involved in the pathogenesis of tissue damage in venous disease [2;3].

Recently, a significantly raised plasma elastase level was found not only in patients with active venous ulceration or lipodermatosclerosis (LDS) but also in patients with uncomplicated varicose veins [4]. There was no difference in the neutrophil count or erythrocyte sedimentation rate (ESR) in patients and normal controls in this study. Although increased plasmatic concentration of PMN-elastase in peripheral blood might be due to an inflammatory response and neutrophil activation in patients with venous ulceration and LDS, this cannot account for the increased levels observed in patients with uncomplicated varicose veins.

It is known that the Ela/α1-PI complex, once formed in the plasma is subject to clearance by the endothelial reticulum system (ERS). Analytical variability in the methods employed for the Ela/α1-PI complex is relatively low (c.v.< 5 %). Biological variability is much more consistent and normal reference range is relatively wide (4- 41 ng/ml in our laboratory). Finding a concentration gradient amoung the different venous sectors is better way of giving indications about liberation of PMN elastase at loco-regional level in individual patients.

This preliminar report, gives the results of our study designed to evaluate the Ela/α1-PI complex levels in human plasma from peripheral venous blood and dependent legs of patients with chronic venous insufficiency and venous hypertension complicated or not with varicophlebitis.

PATIENTS AND METHODS

Two groups of patients, hospitalized and awaiting varicetomy, entered the study in consecutive order and after informed consens. Patients were excluded from the study if suffering from any condition that could result in neutrophil activation or if they had taken any medication known to alter white cell activity.

The first group (2 males and 14 females, mean age 36 years ± 7,2 DS) were patients awaiting surgery for uncomplicated varicose veins. The second group (2 males and 6 females, mean age 43 years ± 6,4 DS) were patients suffering from varicophlebitis. All patients underwent a clinical examination, colour duplex ultrasonography and photopletismography to confirm the presence of venous disease, estabilish the extent of venous disease and identifity the site and severity of venous reflux.

Patients were asked to remove any support bandaging or stockings on the day before the study and all subjects were asked to fast and remain supine with legs flat on the bed for two hours at room temperature. After the two hours, samples of venous blood were taken for analysis using the double blood sampling without stasis technique, from anterior cubital veins and varicose veins of the great saphena. An automized method with a Coulter Counter Analizer was used to determine the haematocrit (HCT), absolute and differential leukocytaric counts in blood anticolagulated with EDTA. Ela/α1-PI dosage in plasma was done using a commercially available immuno-assay (IMAC, Merck, Darmstad, Germany), according to kit instruction [5].

RESULTS AND DISCUSSION

Figure 1 shows that significantly different concentrations of Ela/α1-PI levels in plasma from venous blood of dependent legs and from the arm are only found in patients with varicophlebitis.

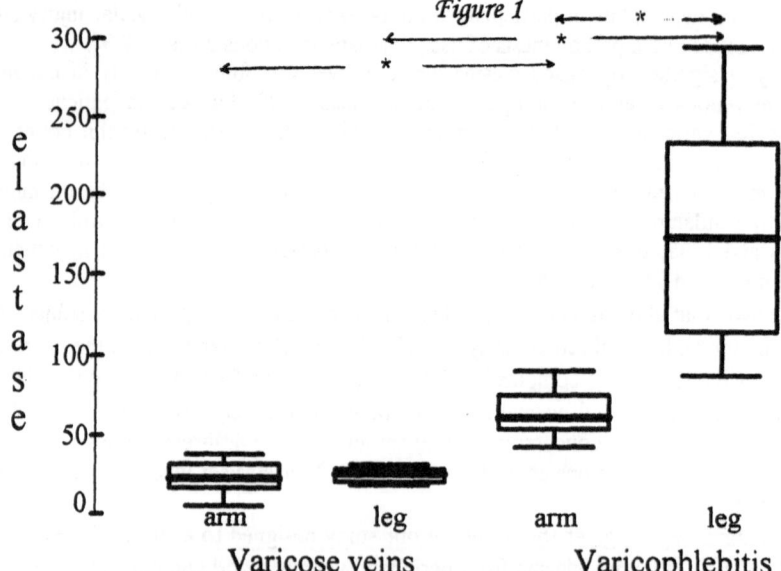

Plasma neutrophil elastase concentration (ng/ml) from venous blood of the dependent legs and from the arm in the two patient groups (mean, 95% confidence interval; *p <0.05)

Once more, a significantly elevated plasma Ela/α1-PI concentrations was found, both in the venous blood from dependent legs and from the arm, in the group of patients with varicophlebitis compared with the group of patients with uncomplicated varicose veins. Haematocrit values, absolute and differential leukocytaric counts in all subjects studied were in the normal reference range and no significant intra-patient differences in these blood parameters were found in the two venous regions considered in this study (data not given).

DISCUSSION

In recent years more numerous and detailed studyes agree that leukocytes are able, under different conditions, to produce a pathogenic rather than defence reaction. This pathogenic action can presumibly occur, also with certain limits, in case of chronic venous disease and its complications.

Though activation of PMN can intervene in chronic venous disease our experimental data suggest that this activation is not accompained by loco-regional degranulation of these leukocytes, great enough to increse the regional plasma levels of the the Ela/α1-PI complex, unless there is a manifest inflammatory complication, such as phlebitis, in the involved venous region.

These results provide evidence of increased neutrophil loco-regional degranulation in patients with venous disease and suggest the possible utilization of PMN-elastase as a sensitive and rapid responsive laboratory indicator of acute inflammatoy activity in dependent legs of patients with chronic venous disease .

This study is ongoing and we are incresasing the casuistry and evaluating regional behavior of PMN elastase and other haematologic-leukocytaric parameters, under different experimental conditions, in both venous disease patients and artificially induced venous hypertension in normal subjects.

REFERENCES

1. Froeschle MC, Goetz WA. Elastase - a new marker for inflammatory diseases. GIT VELARG . Darmstdt (Germany) 1985.
2. Thomas PRS, Nash GB, Dormandy JA.White cell accumulation in dependent legs of patients with venous hypertension: a possible mechanism for trophic changes in the skin.BMJ 1988;296:1693-5.
3. Cheatle TR, Sarin S, Coleridge Smith PD, Scurr JH. The pathogenesis of skin damage in venous disease: a review. Eur J Vasc Surg 1991; 5: 115-23.
4. Shields DA, Andaz SK, Sarin S, Scurr JH, Coleridge Smith PD. Plasma elastase in venous disease. Br J Surg 1994; 81:1496-99.
5. Diagnostica Merck: PMN Elastase. Now a routine determination: Faster with IMAC technology. E. Merck , Darmstadt, Germany, 1994.

Phlebology '95, D. Negus et al. (eds.). Phlebology (1995) Suppl. 1: 102-104

P242

Study of Microcirculation of Patients with Varicose Veins of Lower Limbs

A. Grasso, C. Costanzo, S. Luca, C. Germiglio and S. Romeo

Chair and School of Specialization of Vascular Surgery - Dir. Prof. S. Romeo, University of Catania, Italy

INTRODUCTION

Varicose veins of lower limbs have always caused, because of their rate, highest social spendings. Those patients who have trophic disturbances often need surgical treatment even after surgical intervention. This condition becomes worse in Southern Europe countries because of environmental and climatic factors.
It is therefore necessary to improve diagnostic methods in order to prevent trophic disturbances.
In order to reach this target, our Institute studies all patients affected by varicose disease not only by doppler c.w. but also by microcirculatory methods, such as Videocapillaroscopy and Reflected Light Rheography on medial malleolus.

MATERIALS AND METHODS

Thirty patients (24 females and 4 males) suffering from varicose veins of lower limbs at 2nd or 3rd stage of Widmer classification, were studied by Reflected Light Rheography and Videocapillaroscopy on medial malleolus of both legs. In order to avoid any interference, capillaroscopy was ever made before RLR; in the same way, no drugs during the period of study were administered.
In the patients with only one affected leg, the other leg gave useful comparison parameters.

RESULTS

In the group of 30 patients the following parameters were evaluated: Venous networks refilling time (T0), venous drain capability (dr), capillaries density and morphology. About rheographic parameters we found T0 equal to 12+-3 seconds and dr equal to 100+-50 mV (mean values). In the group of patients at 2nd Widmer stage, T0 was equal to

16+-3 seconds and   dr was equal to 120+-30 mV, while in
the patients at 3rd Widmer stage T0   and dr were found,
respectively, equal to 8+-2 seconds   and to 50+-15 mV.
About capillaroscopic parameters,   we   found normal
capillaries density in all subjects; "reticular" and "halo
formation" pictures, respectively, in 26 (76,6%) and 4
(13,3%) patients.

CONCLUSIONS

The study of microcirculation by Capillaroscopy and
Rheography , in the patients suffering from varicose veins
of lower limbs, was with no doubt useful to demonstrate
that any stage of venous disorders causes different
microcirculatory alterations.
In fact, we found "halo formation" pictures in patients
with   more   pathological rheographic parameters (T0 less
than 10 seconds) and nearly all these patients suffered
from   varicose   disease   at   3rd   stage   of   Widmer
classification. Furthermore, microcirculatory alterations
were more evident in patients with trophic disturbances and
elder venous disease.

REFERENCES

1) S.B. CURRI: " LE MICROANGIOPATIE" ed. Inverni della
Beffa 1986- pagg. 221-230
2) V. WIERNET: " DIAGNOSTIC PHLEBOLOGIE FONCTIONELLE PAR
RHEOGRAPHIE A REFLEXION DE LUMIERE" Phlebologie 1984, vol.
3- pag. 303- 308
3) W. GOOR: " UN METODO NUEVO PARA EL FLEBOLOGO: LA
REOGRAFIA POR REFLEXION LUNINOUSA" Angiologia 1985, vol.2 ,
pag.89-90
4) G. SPINELLA, A. GRASSO, S. ROMEO: " LA
FOTOPLETISMOGRAFIA A LUCE RIFLESSA NELLA DIAGNOSTICA DELLE
FLEBOPATIE DEGLI ARTI INFERIORI : UTILITA' E LIMITI NELLA
NOSTRA ESPERIENZA " Atti del 3 congresso italiano di
flebologia- Catania 1986- Monduzzi Editore - pag. 261-264
5) S.B. CURRI : "CHANGES OF THE SKIN, ADIPOSE TISSUE AND
MUSCLE   MICROCIRCULATION   IN   VENOUS   DISORDERS"
Proc.Eur.Cong.Union Phlebology, Budapest 1993,, pag.91-100

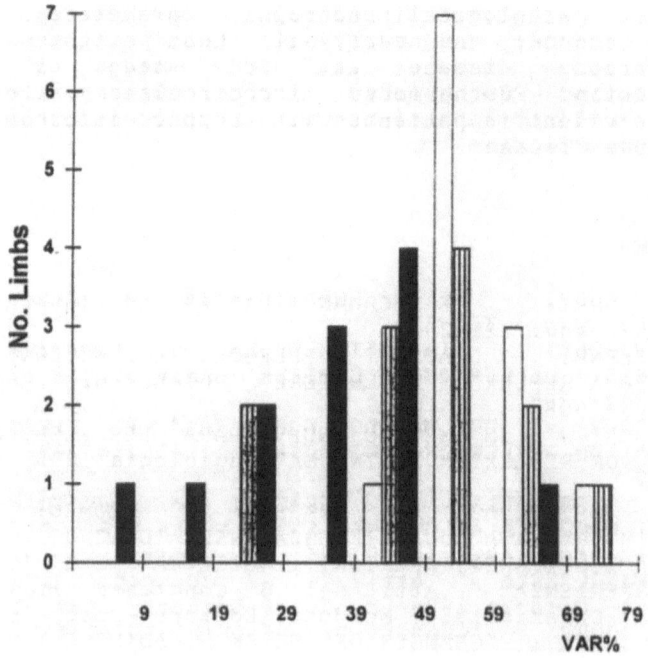

Phlebology '95, D. Negus et al. (eds.). Phlebology (1995) Suppl. 1: 105-107

PI/7.6

Capillary Response to Stretching Skin in Venous Disease and Normal Subjects

A.M. Shutt, S.R. Dodds, A.R. Cowan and A.D.B. Chant

Department of Vascular Surgery, Royal South Hants Hospital, Graham Road, Southampton SO9 4PE, UK

INTRODUCTION

Skin capillary blood flow measured with laser Doppler in the legs of subjects with chronic venous insufficiency (CVI) is known to be different from normal subjects [1]. Many factors affect skin capillary blood flow including temperature, exercise, recent food or drug ingestion, as well as postural changes. With changes in posture there is both a change in limb volume and distribution of flow. The distribution of flow is affected by centrally and locally mediated events. At the capillary level, this is promoted by an increase in the venous pressure above 25mmHg. This vasoconstriction has been variously ascribed to a reflex [2] or some kind of local autoregulation [3,4]. Whether or not this "reflex" in venous disease persists is still a moot point [5].

Stretching the skin causes it to blanch as external forces occlude the blood supply to the area. If the applied force is sufficiently large and prolonged, tissues may become ischaemic. The lower limb is known to swell when adopting the upright position as confirmed by plethysmographic techniques. This swelling will cause the skin to stretch and may transiently reduce skin blood flow. To test just how great the effect of stretching is we have investigated the effect of a 30% uniaxial extension of the skin in the gaiter region of normal subjects and subjects with CVI.

METHOD

A laser Doppler probe was placed between the arms of a uniaxial skin extensometer (Dermatronics Ltd, UK.), connected to an MBF3 Laser Doppler Fluximeter (LDF)(Moor Instruments, UK.), on a subject lying supine in a temperature and humidity controlled environment. Guidelines set out by the European Society of Contact Dermatitis for measurement of cutaneous blood flow using laser Doppler were followed [6].

The feet of the extensometer were applied using ethylcyanoacrylate cement (Powabond 100. Staident Ltd. UK) 1 cm apart in the medial gaiter region of the lower limb, on either side of the LDF probe. The probe was held lightly in place with adhesive tape. The tape was applied in such a fashion as not to exert undue pressure on the probe and occlude the signal. No decrease in flux was seen after surrounding the probe with the cemented feet.

A 30 % uniaxial extension was then undertaken and the stress relaxation of skin followed for two minutes. Stress relaxation is the phenomenon seen when skin is stretched and held at constant length. With time a reduction in force required to maintain skin at that length is observed. The output from the extensometer went to a flat bed chart recorder calibrated to 1 Kgf per 1 volt. This was later converted to Newtons by multiplying by a factor of 9.8 . Great care was taken not to apply any inward compressive pressure on the extensometer arms to indent the skin as this would have affected LDF signals. At the end of this time the extensometer arms were closed to their original position, whilst continuing to monitor the LDF signal for a further short period. Mean flux or its percentage change from a reference level were measured. Zero level was set at the instrument zero. Biological zero was not measured. Way points throughout the extensometer cycle were identified and then compared. These included:

A. Resting mean baseline flux prior to stretching skin;
B. Minimum Flux during Skin Extension: the minimum flux recorded at peak extension;
C. Maximum Flux during Skin Extension: the maximum flux recorded just prior to release of the stretched skin after two minutes of stress relaxation;
D. Peak flux following release of the skin extension;
E. Mean Post Extension Flux.

RESULTS

There were 36 examinations performed in all. 28 in venous limbs and 8 normal limbs. There were 2 men and 6 women in the normal group. There were 10 men and 18 women in the venous group. Mean flux values at the various waypoints throughout the extensometer cycle are shown in Table 1. Groups were compared using the Mann - Whitney U test on a computer statistical package (Table 2). The force required to produce the fall in LD signal was used to calculate the fall in flux per unit of force. The median fall in flux per unit force was greatest in the ulcerated group at 5.8 AU., followed by the liposclerotic (3.4 AU), oedematous (2.7 AU) and lastly control group (0.9 AU).

Table 1. Mean Flux Values

Group	Baseline Flux	Minimum During	Maximum During
Control	18.6	8.3	16.8
Ulcer	78.9	30.2	83.1
Oedema	30.9	8.1	36.4
Liposclerotic	82.1	9.9	38.6
No ulcer	46.9	9.2	36.6

Group	Max. Post Extension	Mean % Post Extension	Maximum % Post Extension
Control	82.9	307.6	536
Ulcer	82.6	20.2	115
Oedema	110.5	134.5	296.9
Liposclerotic	138.1	38.8	96
No ulcer	79.7	104.6	234

In summary, it has been shown that laser Doppler flux changes with forced skin extension. The response differs significantly between controls and those with venous disease, thus confirming the importance of mechanical factors in leg ulceration [7].

Table 2. Mann-Whitney Comparisons

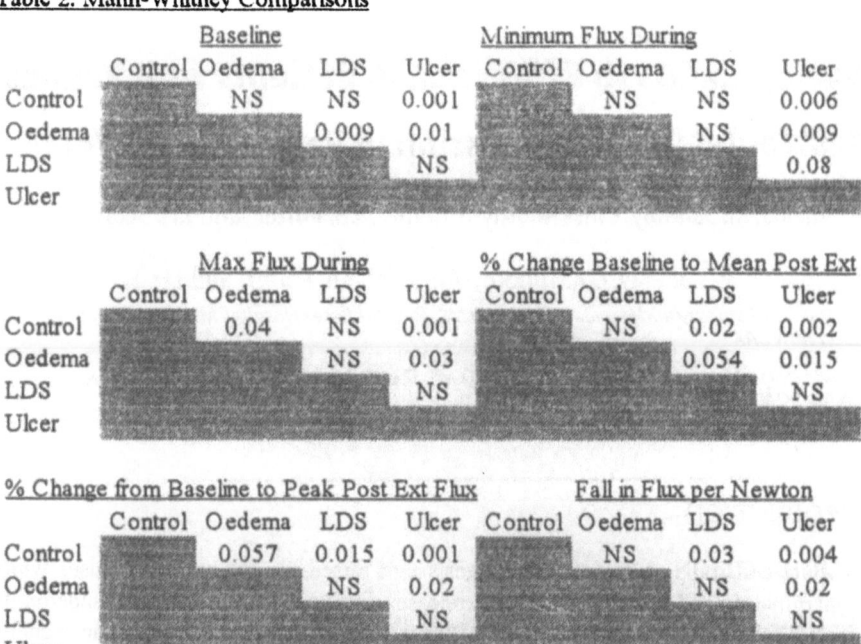

Baseline

	Control	Oedema	LDS	Ulcer
Control		NS	NS	0.001
Oedema			0.009	0.01
LDS				NS
Ulcer				

Minimum Flux During

	Control	Oedema	LDS	Ulcer
Control		NS	NS	0.006
Oedema			NS	0.009
LDS				0.08
Ulcer				

Max Flux During

	Control	Oedema	LDS	Ulcer
Control		0.04	NS	0.001
Oedema			NS	0.03
LDS				NS
Ulcer				

% Change Baseline to Mean Post Ext

	Control	Oedema	LDS	Ulcer
Control		NS	0.02	0.002
Oedema			0.054	0.015
LDS				NS
Ulcer				

% Change from Baseline to Peak Post Ext Flux

	Control	Oedema	LDS	Ulcer
Control		0.057	0.015	0.001
Oedema			NS	0.02
LDS				NS
Ulcer				

Fall in Flux per Newton

	Control	Oedema	LDS	Ulcer
Control		NS	0.03	0.004
Oedema			NS	0.02
LDS				NS
Ulcer				

REFERENCES

1. Cheatle TR, McMullin GM, Farrah J, Coleridge-Smith PD, Scurr JH. Three tests of microcirculatory function in the evaluation of treatment for chronic venous insufficiency. Phlebology. 1990;5:165-72.

2. Henriksen O. Local Sympathetic reflex mechanism in regulation of blood flow in human subcutaneous adipose tissue. Acta Physiol Scand. 1977;Suppl 450:7-48.

3. Mellander S, Oberg B, Odelram H. Vascular adjustments to increased transmural pressure in cat and man with special reference to shifts in capillary fluid transfer. Acta Physiol Scand. 1964;61:34-48.

4. Lansman JB, Hallam TJ, Rink TJ. Single stretch-activated ion channels in vascular endothelial cells as mechanotransducers? Nature. 1987;325:811-3.

5. Shami SK, Scurr JH, Coleridge-Smith PD. The veno-arteriolar reflex in chronic venous insufficiency. Vasa. 1993;22 (3):227-31.

6. Bircher A, De Boer EM, Agner T, Wahlberg JE, Serup J. Guidelines for measurement of cutaneous blood flow by laser Doppler flowmetry. Contact Dermatitis. 1993;28:1-8.

7. Chant ADB. Tissue pressure, posture and venous ulceration. Lancet. 1990;336:1050-1.

Phlebology '95, D. Negus et al. (eds.). Phlebology (1995) Suppl. 1: 108-109

OP/9.3

Neutrophil CD11b Expression in Patients with Venous Disease

D.A. Shields, M. Saharay, C.A. Timothy-Antoine[1], J.B. Porter[1] and J.H. Scurr

Departments of Surgery and Haematology[1], UCLMS, The Middlesex Hospital, Mortimer Street, London W1N 8AA

INTRODUCTION

The white cell trapping hypothesis suggests that raised venous pressure causes white cell margination and activation, with the release of proteolytic enzymes and superoxide radicals, believed responsible for the tissue destruction seen in chronic venous insufficiency [1]. We have demonstrated raised levels of plasma lactoferrin [2] and elastase [3] as markers of neutrophil degranulation in patients with venous disease previously. To investigate this hypothesis further we have looked at the neutrophil adhesion molecule, CD11b/CD18, believed to be the second stage receptor responsible for physiological binding of neutrophils to the vascular endothelium [4]. The current theory of neutrophil adhesion [5] is that L-selectin (CD62L) initially binds to receptors on the capillary endothelium, causing a reduction in flow velocity and allowing second stage receptors to bind. For neutrophils the main second stage receptor is an integrin, CD11b, which consists of a specific α dimer linked to a common ß2 dimer, CD18. This integrin binds to receptors on the vascular endothelium prior to extravascular migration or degranulation. We hypothesised that the raised venous pressure seen in venous disease might correlate with the first stage of binding of neutrophils, in that a low flow state would increase the likelihood of the second stage receptors to bind prior to cell activation. The purpose of this study was to measure CD11b expression in patients with venous disease to see if they had increased neutrophil CD11b expression.

METHODS

Two groups of ten patients with uncomplicated varicose veins and with skin changes of lipodermatosclerosis (LDS) associated with deep venous incompetence (DVI), confirmed on colour duplex imaging, were compared with two groups of age-matched controls who had no history or clinical findings of venous disease. None of these subjects had arterial disease (ankle-brachial index < 0.9), diabetes, any connective tissue disorder, any infection within the previous six weeks, or were on any medication known to alter white cell activity. Blood was taken from an arm vein for full blood count and neutrophil CD11b expression, measured in whole blood using a fluorescent-labelled monoclonal antibody.

RESULTS

	Control	Varicose veins	Control	LDS
Age (years)	42.1 (28-64)	42.6 (25-58)	62.5 (44-81)	65.5 (45-7)
Sex (M:F)	7:3	3:7	5:5	2:8
Neutrophil mean cell fluorescence	1.43 (1.11-2.52)	4.6 (3.12-5.85)	1.53 (1.41-2.44)	1.22 (1.06-1.42)
Difference between medians (95% CI)	2.7 (1.04 to 4.61)		.445 (.02 to 1.32)	
Mann-Whitney U Test	p=.005		p=.028	

Neutrophil CD11b expression given as mean cell fluorescence. Descriptors are medians and interquartile ranges, CI = confidence interval.

CONCLUSIONS

These data show that patients with uncomplicated varicose veins have increased surface expression of neutrophil CD11b, indicating increased neutrophil adhesion, but patients with deep venous insufficiency associated with lipodermatosclerosis have reduced CD11b expression compared to normal controls. It is presumed that neutrophil adhesion prior to activation is important in the initiation of tissue damage in venous hypertension, hence the raised level of CD11b expression in patients with uncomplicated varicose veins. However, neutrophils may not be important in the maintenance of tissue damage once formed, as shown by decreased CD11b levels in patients with DVI and LDS. This suggestion requires further elucidation.

REFERENCES

1.   Coleridge Smith PD, Thomas PRS, Scurr JH, Dormandy JA. Causes of venous ulceration: a new hypothesis. *Br Med J* 1988;**296**:1726-7

2.   Shields DA, Andaz S, Abeysinghe RD, Porter JB, Scurr JH, Coleridge Smith PD. Lactoferrin as a marker of neutrophil activation in venous disease. *Phlebology* 1994;**9(2)**:55-8

3.   Shields DA, Andaz S, Sarin S, Scurr JH, Coleridge Smith PD. Plasma elastase in venous disease. *Br J Surg* 1994:**81**;1496-9

4.   Pardi R, Inverardi L, Bender JR. Regulatory mechanisms in leukocyte adhesion: flexible receptors for sophisticated travelers. *Immunology Today* 1992;**13(6)**:224-30

5.   Yong K, Khwaja A. Leucocyte cellular adhesion molecules. *Blood Reviews* 1990;**4**:211-25

Phlebology '95, D. Negus et al. (eds.). Phlebology (1995) Suppl. 1: 110-112

OP/9.2

Markers of Neutrophil Degranulation in Patients with Venous Disease

D.A. Shields, S.K. Andaz, S. Sarin, R.D. Abeysinghe[1], J.B. Porter[1] and J.H. Scurr

Departments of Surgery and Haematology[1], UCLMS, The Middlesex Hospital, Mortimer Street, London W1N 8AA

INTRODUCTION

The white cell trapping hypothesis suggests that raised venous pressure causes white cell margination and activation, with the release of proteolytic enzymes and superoxide radicals, believed responsible for the tissue destruction seen in chronic venous insufficiency [1]. White cell margination has been shown to occur in the post-capillary venules when blood flow is reduced [2], and Moyses *et al.* [3] and Thomas *et al.* [4] have shown a reduction in the number of white cells leaving a limb on dependency in normal subjects and in patients with venous disease respectively. Increased numbers of white cells have also been demonstrated in the skin of patients with lipodermatosclerosis [5], but there was no direct evidence for the release of products of white cell activation in venous disease. It was, therefore, decided to measure plasma elastase and lactoferrin as markers of neutrophil primary and secondary granule release in patients with venous disease to look for evidence of neutrophil activation.

METHODS

Blood was taken from groups of patients with venous disease and age- and sex-matched normal control subjects with no history or clinical findings of venous disease, and subsequently analysed for plasma lactoferrin or elastase. Venous disease was confirmed on colour Duplex imaging and photoplethysmography. Subjects were excluded if there was any evidence of arterial disease (ankle:brachial index < 0.9), diabetes, connective tissue disorder, recent infection, or if they were on any medication known to alter white cell activity. For lactoferrin four groups of ten patients with uncomplicated varicose veins, varicose veins associated with lipodermatosclerosis (LDS), active venous ulceration (but clinically no infection), and healed ulcers were examined, and for elastase three groups of 15 patients with uncomplicated varicose veins, LDS, and active venous ulceration. Plasma lactoferrin was measured using an ELISA developed by two of the authors [6]; plasma elastase was measured at the Scottish National Blood

Transfusion Centre using a radio-immunoassay [7] which detects both free and bound enzyme, is specific for neutrophil elastase and does not cross-react with platelet or pancreatic elastase. Blood was also taken for a full blood count including neutrophil count.

RESULTS

	Varicose veins	LDS	Active ulcer	Healed ulcer
Patients	669 (441-1063)	422 (327-533)	323 (254-429)	364 (228-511)
Controls	405 (240-509)	223 (115-375)	210 (155-242)	176 (134-203)
Difference between medians (95% CI)	269 (63-602)	199 (49-315)	128 (41-214)	195 (88-344)
Mann-Whitney U Test	p=.016	p=.01	p=.005	p=.0005

Table 1. Plasma lactoferrin in ng/ml. Descriptors are medians and interquartile ranges, CI = confidence interval.

	Varicose veins	LDS	Active ulcer
Patients	25.6 (20.4-39.8)	22.1 (19-33.9)	26 (22.3-39.5)
Controls	18.1 (12.6-30)	17.7 (12.6-21.2)	18.8 (12.4-34.3)
Difference between medians (95% CI)	8 (.8-14)	7.0 (1.4-13.5)	8.5 (1.4-14)
Mann-Whitney U Test	p=.034	p=.007	p=.027

Table 2. Plasma elastase in ng/ml. Descriptors are medians and interquartile ranges, CI = confidence interval.

CONCLUSIONS

These results show that both plasma lactoferrin and elastase are significantly raised in all patients with venous disease compared to their controls and hence systemic neutrophil activation is a feature of venous disease, as suggested by the white cell trapping hypothesis. There was no difference in the white cell count between any of the patient and control groups. Plasma elastase was similar in all patient groups; plasma lactoferrin was highest in the group with varicose veins compared to the other three patient groups, although when corrected for neutrophil count the increase in plasma lactoferrin only reached significance in comparison to the active ulcer group (p=.0052, Mann-Whitney U Test, difference between medians (95% confidence interval) 77 (22-161) ng/ml). It is difficult to understand why the uncomplicated varicose vein group should have a higher plasma lactoferrin than the other, clinically more severe, patient groups, although it may be that chronic stimulation in the more severe disease groups has caused such prolonged degranulation that there is a relative lack of lactoferrin in the neutrophils.

REFERENCES

1. Coleridge Smith PD, Thomas PRS, Scurr JH, Dormandy JA. Causes of venous ulceration: a new hypothesis. *Br Med J* 1988;**296**:1726-7

2. Braide M, Amundson B, Chien S, Bagge U. Quantitative studies of leukocytes on the vascular resistance in a skeletal muscle preparation. *Microvasc Res* 1984;**27**:331-52

3. Moyses C, Cederholm-Williams SA, Michel CC. Haemoconcentration and accumulation of white cells in the feet during venous stasis. *Int J Microcirc Clin Exp* 1987;**5**:311-20

4. Thomas PRS, Nash G, Dormandy JA. White cell accumulation in the dependant leg of patients with venous hypertension - a possible mechanism for trophic skin changes? *Br Med J* 1988;**296**:1693-5

5. Wilkinson LS, Bunker C, Edwards JCW, Scurr JH, Coleridge Smith PD. Leukocytes: Their role in the etiopathogenesis of skin damage in venous disease. *J Vasc Surg* 1993;**17**:669-75

6. Devereux S, Porter JB, Hoyes KP, Abeysinghe RD, Saib R, Linch DC. Secretion of neutrophil secondary granules occurs during granulocyte-macrophage colony stimulating factor induced margination. *Br J Haematol* 1990;**74**:17-23

7. Jackson MH, Collier A, Nicoll JJ, Muir AL, Dawes J, Clarke BF *et al.* Neutrophil count and activation in vascular disease. *Scott Med J* 1992;**37(2)**:41-3

Phlebology '95, D. Negus et al. (eds.). Phlebology (1995) Suppl. 1: 113-115

OP/18.1

Microcirculation and Venous Disease: Effects of Raised Venous Pressure, Walking and Temperature

C. Solomon, A.M. van Rij, J. Walton, T.M. O'Flynn, R.A. Christie and G.B. Hill

Department of Surgery, University of Otago, Dunedin, New Zealand

INTRODUCTION

While the exact pathophysiological mechanisms resulting in the skin changes and ulceration of the skin in the limbs of patients with severe chronic venous insufficiency (CVI) remain unclear, various investigative modalities have increased our understanding of the characteristic microcircultory changes present. In particular, studies using laser Doppler fluximetry (LDF) and transcutaneous oxygen (TcO2) measurements have demonstrated a characteristic "hypoxic hyperaemia"[1]. These measurable differences have been used to support various theories of ulceration and indeed to monitor the response to therapeutic regimens[2,3]. A number of studies have, however, shown conflicting results and various aspects of these measures remain controversial and require further study [4,5]. TcO2 probe temperature, the effects of isolated venous distension, postural change and blood flow changes during walking are addressed in this work.

METHODS

Forty limbs of twenty, age and sex-matched, patients with SVS/ISVS classes of chronic venous insufficiency (0, 1, 2 and 3) were studied. Diabetics and those with significant peripheral vascular disease (ABI < 0.8) were excluded. Clinical examination, air plethysmography and duplex scanning were performed to define the underlying venous disease.

Assessment of the microcirculation

Measurements were conducted in two sessions. Simultaneous monitoring of: TcO2 (mmHg) (probe at 43^{0}C); LDF (ml/min/mg tissue), skin temperature (^{0}C) and transcutaneous carbon dioxide (TcCO2)(mmHg) was performed initially on the right limb (index). TcO2 with a probe at 37^{0}C was measured on the contralateral limb. Probes were placed 5cm above the medial malleolus or within 1.5cm of the ulcer edge in class III limbs. After an initial period of stabilisation (20 minutes), patients were monitored supine, with and without a distended 30cm thigh cuff (50mmHg), standing and then walking on a treadmill for 10 minutes in a controlled temperature environment (21-23^{0}C). The heater of the 43^{0}C TcO2 probe was switched off during a further 2 minute period of walking. All measures were allowed to reach a plateau between each phase. The second session was performed after an interval of 1 hour in a similar way, with the left leg as the index limb.

114

Analysis and statistical tests

Within (paired) and between group (non-paired) differences were assessed for each parameter under different conditions of posture, venous distension and temperature for TcO2. Student's t-test were used to test the significance of differences for normally distributed data, and Mann Whitney U and Wilcoxin tests for non-normally distributed data. Data are presented as means ± standard deviations. The level of declaring statistical significance was set at 0.05.

RESULTS

Transcutaneous oxygen

Levels of TcO2 measured for each group during the study are shown in Figure 1.

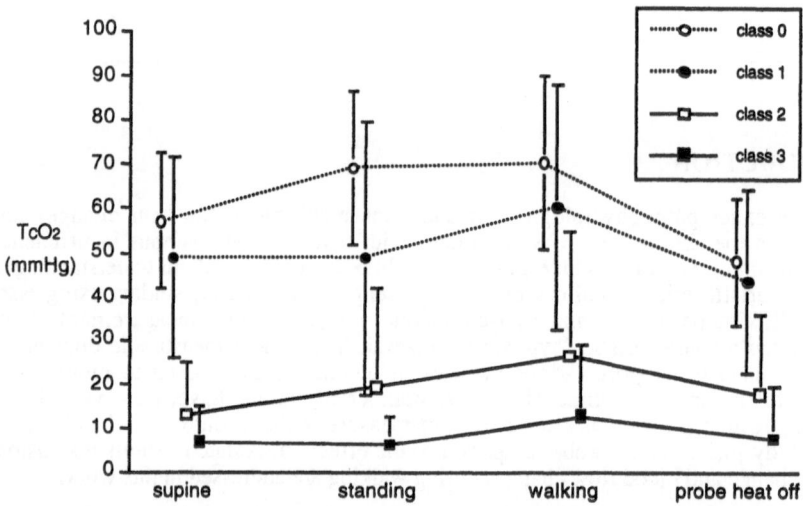

Figure 1: Levels of TcO2 (means ± SD) for each class of CVI at different stages of the study.

Venous distension by thigh cuff inflation did not alter levels of TcO2 (data not shown). TcO2 in Class 1 were not different from normal limbs but Class 2 and 3 limbs had significantly lower levels at all stages. The proportional drop in TcO2 on decrease in probe temperature was similar in all groups. Walking tended to increase TcO2 but this only reached statistical significance in class 1 limbs. At 37° while TcO2 levels were lower in all groups there was still a significant difference between Class 1 and Class 3 limbs.

Laser Doppler

Skin blood flow was significantly higher in classes 2 and 3 and all groups showed a similar proportional lowering of LDF on standing and an increase during walking (Table 1).

Table 1: Ratios of LDF between postural changes and during ambulation.
* denotes significant difference classes 0, 1 vs class 3 p<0.05.

LDF ratio	class 0	class 1	class 2	class 3
standing/supine	0.45 ± 0.17	0.65 ± 0.22	0.55 ± 0.25	0.43 ± 0.35
walking/standing	20.8 ± 6.8	20.0 ±8.8	13.7 ±8.9	9.9 ± 5.9

Transcutaneous CO2

Levels of $TcCO_2$ measured during the study are shown in figure 2.

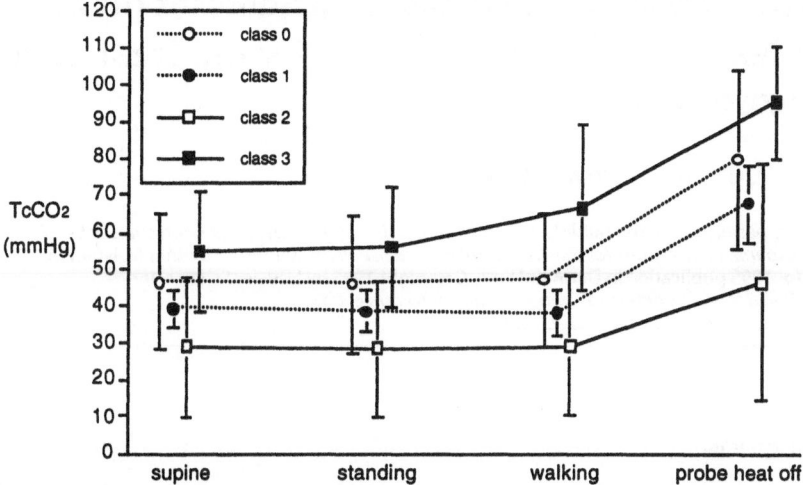

Figure 2: Levels of TcCO2 (means ± SD) for each class of CVI at different stages of the study.

In comparison to class 0 and class 1 limbs, class 3 limbs had significantly higher and class 2 limbs significantly lower levels of CO_2 at all stages of the study. Ambulation did not alter Co2 levels but decreasing probe temperature increased CO_2 significantly in all groups.

CONCLUSIONS AND DISCUSSION

Abnormalities of flow (reflux) in large veins and high venous pressures are particularly evident during postural changes and walking. Microcirculatory changes present in severe venous disease (hypoxic hyperaemia) are, however, present at rest and do not appear to be pronounced by acute changes to posture, ambulation or isolated venous distension. Indeed standing tends to lower skin blood flow in all groups suggesting that the venoarteriolar response is intact, even in ulcerated limbs. While increasing probe temperature enhanced the differences between normal and class 2 and 3 limbs, a relative hypoxia was still present at 37^0. The decrease in transcutaneous oxygen with decreasing probe temperature suggests that liposclerotic skin does have the capacity to vasodilate, and lack of vasodilation does not fully explain the hypoxia that is present. Microcirculatory measures seem useful in separating normal from diseased limbs. It would seem that maximal differences for absolute measures of TcO_2 and LDF occur standing and supine respectively and not during periods of ambulation.

REFERENCES

1. Scurr JH, Coleridge-Smith PD. The microcirculation in venous disease. Angiol 1994; 45:537-41.
2. Christopoulos DC, Nicolaides AN, Belcaro G, Kalodiki E. Venous hypertensive microangiopathy in relation to clinical severity and effect of elastic compression. J Dermatol Surg Oncol 1991; 17:809.
3. Stacey MC, Burnand KG, Layer GT, Pattison M. Transcutaneous oxygen tensions in assessing the treatment of healed venous ulcers. Br J Surg 1990; 77:934-6.
4. Clyne CAC, Ramsden WH, Chant ADB, Webster JHH. Oxygen tension on the skin of the gaiter area of limbs with venous disease. Br. J. Surg. 1985; 72:644-7.
5. Dodd HJ, Gaylarde PM, Sarkany I. Skin oxygen tension in venous insufficiency of the lower leg. J Royal Soc Med 1985; 78:373-6.

Phlebology '95, D. Negus et al. (eds.). Phlebology (1995) Suppl. 1: 116-118

OP/16.4

Identification of Arteriovenous Anastomoses (AVA) by Duplex Ultrasound: Implications for the Treatment of Varicose Veins[1]

A. Kanter[2], M. Gardner[2,3] and M. Issacs[3]

[1] Reprinted by permission of the publisher from *Identification of Arteriovenous Anastomoses by Duplex Ultrasound: Implications for the Treatment of Varicose Veins*, Kanter A, Gardner M, Issacs M, scheduled for 1995 publication in Dermatol Surg. Copyright 1995 by Elsevier Science, Inc.
[2] *In private Practice, Vein Centers of Orange County & Torrance, USA*
[3] *In Private Practice, Vein Care Specialists, USA*

INTRODUCTION

Over the past five decades, various authors have suggested that arteriovenous anastomoses (AVA) may contribute to the etiology of varicose veins [1,2]. Despite angiographic [3], thermographic [4-6], and operative microscopic [5,7] data to support this theory, the exact role AVA may play in the pathogenesis of varicose veins remains controversial [8].

Regardless of their role, pre-treatment anatomical localization of AVA would be of practical significance in order to guide varicose vein treatment, and to avoid direct injection into AVA during sclerotherapy, thereby avoiding potential tissue necrosis from intra-arterial injection.

Duplex ultrasound (DUS) has recently supplanted contrast venography as the gold standard for the diagnosis and treatment of varicose veins [9-12]. Improved technology provides increasingly higher resolution, allowing the accurate identification of smaller vessels than previously possible [10]. Pending resolution of the above controversy, a prospective study was conducted to determine the incidence of AVA identifiable by DUS in patients undergoing pre-treatment evaluation for symptomatic saphenous vein disease due to junctional incompetence.

METHODS

All patients presenting for evaluation of symptomatic varicose veins over a one year period were examined by CW-Doppler, and were admitted to the study if found to have junctional and/or saphenous axial reflux. A total of 510 subjects were derived from three

[1] Reprinted by permission of the publisher from *Identification of Arteriovenous Anastomoses by Duplex Ultrasound: Implications for the Treatment of Varicose Veins*, Kanter A, Gardner M, Issacs M, scheduled for 1995 publication in Dermatol Surg. Copyright 1995 by Elsevier Science, Inc.
[2] In Private Practice, Vein Centers of Orange County & Torrance, USA
[3] In Private Practice, Vein Care Specialists, USA

separate private phlebology practices, utilizing the same ultrasound technician for all cases. DUS examination of all saphenous vein axes was performed with the patient in both the standing and supine positions. One practice used a Diasonics Spectra with color-flow imaging (Center 1, n=252), and the other two centers used a (gray-scale only) Diasonics DRF400 (Centers 2 & 3, n=258); all used a 7.5 mHz probe. (Diasonics, Milpitas, CA).

Whenever an AVA was suspected by visualization of arterial pulsation in close proximity to a saphenous vein or its tributary, a pulsed-doppler was focused at its origin and followed continuously along its length. When a distinct arterial waveform underwent a gradual transition to venous waveform along a visualized continuous vessel, an AVA was determined to be present, and its location noted with respect to position and associated saphenous system.

RESULTS

Twenty-six AVA were found in 19/510 patients (incidence=3.7%), four of whom were found to have multiple AVA; this included 3% (13/439) of females and 8.5% (6/71) of males, with no differences found regarding history of previous vein stripping (10 stripping vs 9 without stripping). Both left (n=12) and right (n=14) sides were affected equally, with most AVA occurring in the greater saphenous system below the knee (Figure 1). All three patients with venous ulcers had AVA present at the base of their ulcers adjacent to the medial malleolus. There was a bimodal peak incidence of AVA with respect to age in the fourth and sixth decades.

Center 1 using a higher resolution ultrasound system detected AVA twice as often as Centers 2 and 3 using lower resolution systems (5.2% [13/252] vs 2.3% [6/258]).

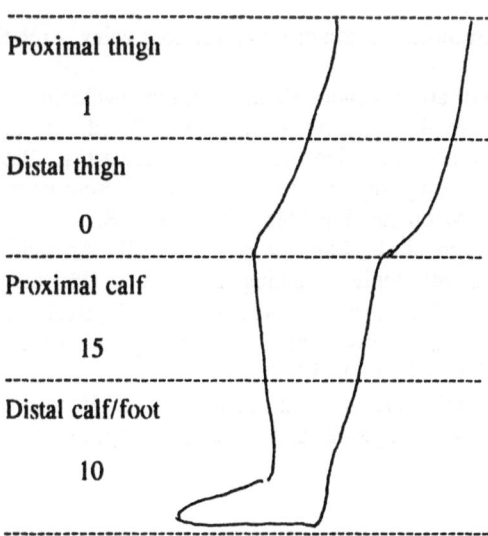

Figure 1. AVA location: 19 in greater saphenous, 6 in lesser saphenous, and one in Vein of Giacomini.

CONCLUSIONS

AVA occurring in association with clinically-apparent saphenous vein disease and identified using duplex ultrasound are most often found in males, below the knee, communicating with the greater saphenous axis, and sometimes at the base of venous leg ulcers. AVA are more likely to be detected using a high-resolution duplex ultrasound system using color-flow imaging.

Because of the anatomical distribution of AVA found in this study, caution is especially recommended when performing sclerotherapy to the saphenous vein axis below the knee. Knowledge of the presence and location of AVA may help to avoid intra-arterial injection during sclerotherapy, and guide the approach to treatment.

REFERENCES

1.   Goldman MP. Sclerotherapy: Treatment of varicose and telangiectatic leg veins. St. Louis; Mosby Year Book, 1991:71-4.
2.   Pratt GH. Arterial varices: a syndrome. Am J Surg 1949;77:456.
3.   Haimovici H, Steinman C, Caplan LH. Role of arteriovenous amastomoses in vascular diseases of the lower extremity. Ann Surg 1966;164:990-1002.
4.   Haeger KHM, Bergman L. Skin temperature of normal and varicose legs and some reflections on the etiology of varicose veins. Angiology 1963;14:473-70.
5.   Schalin L. Arteriovenous communications in varicose veins localized by thermography and identified by operative microscopy. Acta Chir Scand 1981;147:409-420.
6.   Bergqvist D, Bornmyr S. New aspects on thermography as a on-invasive diagnostic method for arteriovenous anastomoses in the extremeties. VASA 1986;15:241-4.
7.   Guis JA. Arteriovenous anastomoses and varicose veins. Arch Surg 1960;81:147-58.
8.   Schalin L. Role of arteriovenous shunting in the development of varicose veins. In: Eklof B, Gjores JE, Thulesius O, Bergqvist D, editors. Controversies in the management of venous disorders. London: Butterworths, 1989:182
9.   Corbett CRR. Meeting Report: annual general meeting of the Venous Forum, Royal Society of Medicine. Phlebology 1993;8:86-8.
10.   Somjen GM, Ziegenbein R, Johnston AH, Royle IP. Anatomical examination of leg telangiectases with duplex scanning. Phlebology 1993;8:940-5.
11.   Depalma RG, Hart NT, Zanin L, Massarin EH. Physical examination, doppler ultrasound, and colour flow duplex scanning: guides to therapy for primary varicose veins. Phlebology 1993;8:7-11.
12.   Valentin LI, Valentin WH, Mercado S, Rosado CJ. Venous reflux localization: comparative study of venography and duplex scanning. Phlebology 1993;8:124-7.

Phlebology '95, D. Negus et al. (eds.). Phlebology (1995) Suppl. 1: 119-121

P050

In Vivo Investigations of the Microcirculation in Venous Ulcers in View of Therapeutic Possibilities

A. Steins, M. Jünger, T. Klyscz, A. Schiek, S. Galler, M.F. Jung and M. Hahn

*Department of Dermatology, University of Tübingen, Liebermeisterstr. 25,
D-72076 Tübingen, Germany*

Chronic venous incompetence is characterized by a stage-dependent microangiopathy. The more advanced the disease is, the more the vascular reserve and the number of cutaneous capillaries decrease (1,2,). These microangiopathic changes are probably among the causes of the trophic disturbances which accompany CVI. The aim of this study is to determine to what extent cutaneous microangiopathy changes or improves during healing. To this end we measured capillary density, transcutaneous oxygen partial pressure and laser Doppler flex at regular intervals during the healing process.

Ten patients took part in the study, 4 women and 6 men with a mean age of $57,3 \pm 9,7$ years and stage III chronic venous incompetence according to Widmer.

The cause of the chronic venous incompetence was a primary varicosis in half of the patient group and post-thrombotic syndrome in the other half.

Two of the patients had a valvular incompetence of the deep veins and eight of the patients one of the long saphenous vein. This was combined in five of the eight with an incompetence of the perforating veins.

A florid ulcer was present in all of the patients. In every case it was a relapse with an average ulcer size of $8,7 \text{ cm}^2 \pm 11.6$ and a mean healing time of 8.6 ± 4.3 weeks.

The patients were examined every 2 weeks at the outset of the study and later every 4 weeks until healing was complete.

Intravital capillary microscopy makes it possible to evaluate capillary density and morphology. The transcutaneous oxygen partial pressure supplies information about how well the capillaries are supplying the affected area („nutritive vascular compartment of the skin") and laser Doppler flux about the capillaries of the deeper vascular plexus („thermoregulative vascular compartment").

The increase in laser Doppler flux (LDF) after the application of heat and arterial occlusion over the initial value (resting value - biological zero value) in the ulcer, ulcer margin area and macroscopically healthy tissue (the proximal calf) was defined as the vascular reserve.

The values measured at the beginning of the study revealed a clear microangiopathy in the venous ulcus cruris and its margin area, but also in tissue that appeared macroscopically healthy.

In addition to a significant reduction in capillary density, even to the point of avascular regions, there are morphological changes in the capillaries themselves: increased dilatation, elongation and arcuate or glomerulate arborization.

This is also reflected in a drop in transcutaneous oxygen partial pressure and vascular reserve, clearly indicating a microcirculatory dysfunction.

Below, the median values measured at the beginning of the study are compared with those measured after the ulcer had completely healed.

	at outset	after healing
capillary density in ulcer (mm^3)	5 (Q_1=1; Q_3=7.5)	21 (Q_1=11; Q_3=23)
capillary density at ulcer margin (mm^3)	34 (Q_1=20; Q_3=36)	42 (Q_1=29; Q_3=55)
tcPO$_2$ ulcer margin area (mmHg)	23 (Q_1=7; Q_3=30)	30 (Q_1=20; Q_3=47)
tcPO$_2$ healthy tissue (mmHg)	36 (Q_1=31; Q_3=45)	42 (Q_1=35; Q_3=50)
LDF heat in ulcer (%)	23.5 (Q_1=6.75; Q_3=60.9)	35.5 (Q_1=13; Q_3=48)
LDF heat at margin (%)	60.5 (Q_1=32.25; Q_3=223)	43.75 (Q_1=19; Q_3=79.5)
LDF heat healthy tissue (%)	86.5 (Q_1=34; Q_3=162)	24 (Q_1=15; Q_3=125)

Capillary density increased markedly during healing, both in the ulcer itself and in the ulcer margin area. In the ulcer margin area a capillary density was measured after the ulcer had healed which corresponded to the retromalleolar values in healthy control individuals (Fig. 1). The transcutaneous oxygen partial pressure measured at the ulcer margin and in macroscopically healthy tissue also rose in both of these areas as the ulcer healed. Laser Doppler flux changed only slightly after the application of heat.

These results lead us to the conclusion that the healing of a venous ulcus cruris is accompanied by an improvement in nutritive cutaneous perfusion, but that the vascular reserve remains limited. With the aid of the examination methods used in our study it is possible to objectify an improvement in microangiopathy after therapeutic intervention.

REFERENCES:
1. Jünger M, Hahn M, Patheiger U, Rahmel B, Rassner G. Morphologische und funktionelle Mikroangiopathie im Ulcus cruris venosum, Thrombose und Thrombosefolgen edited by Wuppermann, T.Richter,H.Konstanz, Schnetzor-Verlag, 1991, p. 121-125.
2. Jünger M, Hahn U, Bort S, Geiger H, May B, Hahn M. Distribution of microangiopathy in chronic venous insufficiency. Int.J.Microcirc.Clin.Exp. (Suppl.) 11: 109,1992.

Phlebology '95, D. Negus et al. (eds.). Phlebology (1995) Suppl. 1: 122-124

PI/3.2

The Microcirculation in Ischemia and Reperfusion

M.D. Menger

Institute for Clinical and Experimental Surgery, University of Saarland, Homburg/Saar, Germany

INTRODUCTION

The microvasculature of striated muscle and skin and, in particular, their endothelial lining cells represent the primary target for injury induced by both ischemia and reperfusion. While the impact of ischemia on the pathogenesis of tissue injury has long been recognized, the particular importance of postischemic reperfusion and reoxygenation to this process has only recently been appreciated.

There are two reperfusion-dependent mechanisms, which have been elucidated to contribute to the aggravation of ischemia-induced tissue injury, either through prolongation of focal ischemia due to perfusion failure of nutritive capillaries (*"no-reflow"*) [1], or through local inflammatory response due to the action of leukocyte-derived cytotoxic mediators, such as reactive oxygen metabolites and proteolytic enzymes (*"reflow-paradox"*) [2,3].

NUTRITIVE PERFUSION DURING ISCHEMIA AND REPERFUSION

Reduction of arteriolar inflow into peripheral tissue (striated muscle and skin) consequently results in the impairment of functional capillary density, ie diminution of the number of red blood cell perfused nutritive capillaries per tissue area, heterogeneity of nutritive perfusion, local tissue hypoxia, and, finally tissue necrosis [4,5,6]. The cause of nutritive perfusion failure may include disturbances of microvascular flow properties and flow conditions of the blood, since isovolemic hemodilution with colloids is effective to improve capillary blood flow and to protect from tissue hypoxia and manifestation of necrosis [7].

In parallel, temporary ischemia followed by reperfusion/reoxygenation is associated with focal nutritive perfusion failure (capillary *"no-reflow"*) and reduction of tissue oxygenation [8]. Although the underlying mechanisms, mediating these microcirculatory disorders, may be different in ischemia/reperfusion when compared with conditions of chronic ischemia, disturbances of microvascular flow properties and flow conditions of the blood seem also to play a major role, because prophylactic isovolemic hemodilution is, similarly, effective to abrogate postischemic capillary perfusion failure [9] and to prevent development of tissue hypoxia [10].

INFLAMMATORY RESPONSE DURING ISCHEMIA AND REPERFUSION

Apart from nutritive perfusion deficits, focal inflammatory response with activation and accumulation of leukocytes within the microvasculature, their adherence to the endothelial lining of postcapillary venules via adhesion receptors, and their release of cytotoxic substances, such as reactive oxygen intermediates, arachidonic acid metabolites and lipid mediators, contribute to the manifestation of tissue injury in ischemia and reperfusion [2,3,11,12]. Indeed, low flow conditions in microvascular networks exert accumulation of leukocytes (PMNs), which interact with endothelial cells of postcapillary venules due to both reduction of wall shear conditions and stimulation of specific adhesion receptors [13,14]. Postischemic reperfusion and reoxygenation further stimulate leukocyte-endothelial cell interactions in venules, ie leukocyte 'rolling' and 'sticking', which, consecutively, lead to disruption of endothelial integrity, increased microvascular permeability and interstitial edema [2].

Adhesion receptors, such as selectins, β_2-integrins and intercellular adhesion molecules, have been suggested to be involved in this inflammatory response. This view is supported by the data of recent in vivo studies, demonstrating that the CD11b/CD18 glycoprotein complex on the surface of leukocytes and the intercellular adhesion molecule-1 (ICAM-1) on the surface of endothelial cells mediate ischemia/reperfusion-induced leukocyte-endothelial cell interaction, because monoclonal antibodies, directed against those two adhesion receptors, effectively attenuate postischemic cellular communications [15].

MICROCIRCULATION IN VENOUS THROMBOSIS

Microcirculatory disorders may also play a pivotal role in venous disease. Intravital microscopic studies have demonstrated the significance of cell to cell interaction between leukocytes, platelets and endothelial cells in the manifestation of arterial and venous thrombosis [16]. However, there is little information on the microcirculation located distal from thrombotic events, probably due to the lack of adequate experimental models, allowing quantitative and repeated microcirculatory analysis over a prolonged period of time.

In order to elucidate the nature and the magnitude of microcirculatory disorders in tissues exposed to venous thrombosis, and to find novel therapeutic regimens to prevent vascular injury and manifestation of necrosis, major efforts should be undertaken to develop experimental models, which allow quantitative and repeated analysis over time periods up to six months. With the use of intravital fluorescence microscopy those models may then provide an approach to study changes of micro-angioarchitecture, microhemodynamics in arterioles, capillaries and venules, as well as cellular and molecular mechnisms involved in postthrombotic disease.

REFERENCES

1. Menger MD, Steiner D, Messmer K. Microvascular ischemia-reperfusion injury in striated muscle: significance of *"no-reflow"*. Am J Physiol 1992;263:H1892-H1900.
2. Menger MD, Pelikan S, Steiner D, Messmer K. Microvascular ischemia-reperfusion injury in striated muscle: significance of *"reflow-paradox"*. Am J Physiol 1992;263:H1901-H1906.

3. Lehr HA, Guhlmann A, Nolte D, Keppler D, Messmer K. Leukotrienes as mediators in ischemia-reperfusion in a microcirculation model in the hamster. J Clin Invest 1991;87:2036-2042.

4. Menger MD, Hammersen F, Barker JH, Feifel G, Messmer K. Tissue pO2 and functional capillary density in chronically ischemic skeletal muscle. Adv Exp Med Biol 1987;222:631-636.

5. Barker JH, Hammersen F, Bondàr I, Galla TJ, Menger MD, Gross W, Messmer K. Direct monitoring of nutritive blood flow in a failing skin flap: The hairless mouse ear skin-flap model. Plast Reconstr Surg 1989;84:303-313.

6. Barker JH, Menger MD, Sack FU, Messmer K. Nutritive Hautdurchblutung bei partieller Ischämie: Wirkung verschiedener vasoaktiver Substanzen. Vasa 1988; 17:37-41.

7. Barker JH, Hammersen F, Galla TJ, Bondàr I, Zeller P, Menger MD, Messmer K. Direct monitoring of capillary perfusion following normovolemic hemodilution in an experimental skin-flap model. Plast Reconstr Surg 1990;86:946-954.

8. Menger MD, Hammersen F, Barker J, Feifel G, Messmer K. Ischemia and reperfusion in skeletal muscle: Experiments with tourniquet ischemia in the awake Syrian golden hamster. Prog Appl Microcirc 1989;13:93-108.

9. Menger MD, Sack FU, Barker JH, Feifel G, Meßmer K. Quantitative analysis of microcirculatory disorders after prolonged ischemia in skeletal muscle: Therapeutic effects of prophylactic isovolemic hemodilution. Res Exp Med 1988;188:151-165.

10. Menger MD, Sack FU, Hammersen F, Messmer K. Tissue oxygenation after prolonged ischemia in skeletal muscle. Therapeutic effects of prophylactic isovolemic hemodilution. Adv Exp Med Biol 1989;248:387-395.

11. Menger MD, Vollmar B, Glasz J, Post S, Messmer K. Microcirculatory manifestations of hepatic ischemia/reperfusion injury. Prog Appl Microcirc 1993;19:106-124.

12. Granger DN, Kubes P. The microcirculation and inflammation: modulation of leukocyte-endothelial cell adhesion. J Leukoc Biol 1994;55:662-675.

13. Granger DN, Benoit JN, Suzuki M, Grisham MB. Leukocyte adherence to venular endothelium during ischemia-reperfusion. Am J Physiol 1989;257:G683-G688.

14. Perry MA, Granger DN. Role of CD11/CD18 in shear rate-dependent leukocyte--endothelial cell interactions in cat mesenteric venules. J Clin Invest 1991;87:1798-1804.

15. Nolte D, Hecht R, Schmid P, Botzlar A, Menger MD, Neumueller C, Sinowatz F, Vestweber D, Messmer K. Role of Mac-1 and ICAM-1 in ischemia-reperfusion injury in a microcirculation model of Balb/C-mice. Am J Physiol 1994; 267:H1320-H1328.

16. Laux V, Seiffge D. Platelet function in the dorsal skin fold chamber of the rat. In Vivo 1993;7:45-52.

17. Menger MD, Lehr HA. Scope and perspectives of intravital microscopy - bridge over from in vitro to in vivo. Immunol Today 1993;14:519-522.

Phlebology '95, D. Negus et al. (eds.). Phlebology (1995) Suppl. 1: 125-127

PI/3.3

The Role of Leucocytes in Venous Ulceration

P.D. Coleridge Smith FRCS

Department of Surgery, University College London Medical School, The Middlesex Hospital, Mortimer Street, London W1N 8AA

The "White Cell Trapping" hypothesis.

The search for the mechanisms of skin damage in venous disease has resulted in investigation of the blood. Moyses *et al* studied the limbs of normal subjects Their subjects sat on a bicycle saddle with the limbs dependent for a period of forty minutes without moving[1]. Blood samples were taken from the long saphenous vein at the ankle. The haematocrit and red cell count increased in parallel. However, the white cell count remained unchanged, despite the increased haematocrit. White cells were being 'lost' from the circulation, which after 40 minutes accounted for a 25% change in the white cell count. Thomas performed a similar study in which he compared patients with normal lower limbs to those of patients with venous disease resulting in lipodermatosclerosis and ulceration[2]. Their patients sat with legs dependent, a less stringent requirement than that of Moyses et al. Blood samples were taken from the long saphenous vein at the ankle. After 60 minutes patients with venous disease were 'trapping' 30% of the white cells and control subjects were trapping 7%. The white cells were "released" when patients lay supine. This lead me to examine the microcirculation using capillary microscopy. I found that venous hypertension appeared to reduce the number of visible capillary loops in patients with venous disease, but not in control subjects[3], suggesting that capillary damage may occur during venous hypertension.

The 'White Cell Trapping' hypothesis proposes that white cell activation follows interaction of leucocytes with endothelium, resulting in release of proteolytic enzymes, superoxide radicals and chemotactic substances. All classes of white cells appear to become trapped so a wide range of phenomena is possible. Monocytes might become activated releasing the cytokines interleukin 1 (IL-1) and tumour necrosis factor alpha (TNF_α)[4]. These may cause endothelial cell activation, in which the endothelium permits the passage of much larger molecules than would normally be the case[5].

Histological Studies

Patients undergoing surgery for venous disease were studied in my laboratory, some with uncomplicated varicose veins, some with liposclerotic skin changes and others with healed venous ulcers. Skin biopsies were taken from the liposclerotic area and histological slides made. The number of white blood cells visible in the upper 0.5mm of the skin in each section was estimated. Patients with normal skin had a low number of white blood cells visible (4 /sq. mm) . There were eight times as many in patients with liposclerotic skin and 40 times as many in patients with healed venous ulcers. We have subsequently undertaken an immunohistological study to determine the types of white cell present in this infiltrate. The majority of cells are macrophages with a T-lymphocyte component, but no excess of neutrophils compared with control sections taken from normal limbs. This infiltrate is a reflection of a chronic inflammatory process, and suggests that an investigation of the cell products of these leucocytes might indicate the mechanisms involved in venous ulceration. We have also been able to identify IL-1 as an inflammatory mediator in this process using immunohistochemical methods[6].

Neutrophil elastase as an indicator of neutrophil activation.

Plasma elastase and lactoferrin were measured as markers of neutrophil degranulation in three groups of 15 patients with uncomplicated varicose veins, lipodermatosclerosis (LDS) and venous ulceration and compared with the values obtained from those in age- and sex-matched control subjects. Blood was taken from an arm vein in all patients and control subjects for full blood count including neutrophil count, erythrocyte sedimentation rate and plasma neutrophil elastase, measured using a radio-immunoassay. Lactoferrin was measured by an enzyme-linked immunosorbant assay (ELISA). Higher levels of elastase and lactoferrin were found in all patient groups compared to their controls. There was no difference in the neutrophil count between the patient and control groups. This provides evidence of increased neutrophil degranulation in patients with venous disease. It is likely that the cause for this is neutrophil activation within the lower limb caused by venous hypertension.

Does short-term venous hypertension cause neutrophil activation?

Venous hypertension should cause neutrophil activation within one hour if the "white cell trapping" hypotheses is tenable. Lactoferrin levels were measured in a group of 15 healthy normal volunteers with no clinical evidence of venous disease using two models of venous hypertension. Blood was taken for lactoferrin assay from the right and left foot and the right arm. A blood pressure cuff was inflated to 80 mm Hg around the right leg for 30 mins and further blood samples taken. There was a significant rise in lactoferrin in the right leg but not in the left leg or arm. During the second part of the experiment volunteers stood without moving for 30 minutes, raising the venous pressure at the ankle. Lactoferrin levels increased significantly in all three limbs. These data confirm that short term venous hypertension results in neutrophil degranulation, in subjects with no evidence of venous disease.

I propose that in patients with venous valvular damage, repeated exposure of the lower limb to neutrophil activation may initiate the trophic skin changes seen in chromic venous insufficiency. This initially leads to the development of the classical histological and capillary microscope appearances seen in liposclerotic skin. Eventually catastrophic failure of the microcirculation, perhaps combined with massive inflammatory cell activation, leads to ulceration.

References

1   Moyses C, Cederholm-Williams SA, Michel CC. Haemoconcentration and the accumulation of white cells in the feet during venous stasis. Int J Microcirc: Clin Exp 1987; 5: 311-320.

2   Thomas PRS, Nash GB and Dormandy JA. White cell accumulation in the dependent legs of patients with venous hypertension: a possible mechanism for trophic changes in the skin. Br Med J 1988; 296: 1693-5.

3   Scott HJ, McMullin GM, Coleridge Smith PD, Scurr JH. Venous ulceration and the role of the white blood cell. J Med Sci & Tech 1990; 14:184-7.

4   Adams DO and Hamilton TA. The cell biology of macrophage activation. Ann Rev Immunol 1984; 2: 283-318.

5   Pober JS. Cytokine-mediated activation of vascular endothelium. Am J Pathol 1988; 133: 426-33.

6   Wilkinson LS, Bunker C, Edwards CW, Scurr JH, Coleridge Smith PD: Leucocytes: Their role in the etiopathogenesis of skin damage in venous disease. J Vasc Surg 1993; 17:669-75

P040

SKIN MICROCIRCULATORY CHANGES DURING CHRONIC VENOUS INSUFFICIENCY: A MODERN STUDY BY MEANS OF LASER-DOPPLER, VIDEOCAPILLAROSCOPY AND TcPO2.

F.Binaghi, P.F.Fronteddu, F.Cannas, F.Mariani, F.Pitzus

Institute of Internal Medicine, Center of Medical Angiology - University of Cagliari (Italy)

Objective: During chronic venous insufficiency (CVI) morpho-functional microcirculatory abnormalities, due to venous hypertension and chronic stasis, develop.The aim of this study was to evaluate the alterations in cutaneous microcirculation in patients with CVI, classified in different stages according to Widmer.

Design: To evaluate microcirculatory changes during CVI at different stages by three modern methods used simultaneously.

Patients: 60 patients with CVI, subdivided in 15 patients for every Widmer's group.

Measurements: by means of laser-doppler (LD), videocapillaroscopy (VC) and transcutaneous oxygen pressure (TcPO2).

Results: In stage 0 only LD parameters shows some changes, particularly a decrease of the effectiveness of venoarteriolar reflex (VAR). The first morphologic alterations found by VC appear in stage I: dilatation of capillaries and pericapillary halo, while TcPO2 is still normal. In fact the PO2 decreases only from the stage II when skin microcirculatory alteration produces a damage of perfusion which is implicated in the pathophysiology of trophic disorders.

Conclusion: It can be affirmed that LD allowed to point out the hemodinamic microcirculatory alteration from the early stage of CVI, while VC and TcPO2 are useful from the following stages.

P048

MICROVASCULAR FAILURE IN THE CVI AND EFFECT OF THERAPY

G M Andreozzi, R Martini, S Signorelli, L Dr Pino, M Barresi, G Failla, G C Busacca, G Pennisi, A Leone

Chair of Angiology - Internal Medicine Dept. "A Francaviglia" - University of Catania - Italy

Postphlebitic syndrome (PPhS) or untreated Varicose Disease (VD) leads to Chronic Venous Insufficiency (CVI). Both begin by damage to the macrocirculation, the deep veins in the first and the superficial in the second. The early stages are almost asymptomatic: the healthy venous system supplies the default of the ill ones. The orthostatic venous pressure (PVos) increases in the ill system but the physiological reduction of the orthodynamic venous pressure (PVos) is still effective ($\geq$60%). The calf muscle pump (CMP. ΔR reflex light pleth.) is still efficient, mostly in PPhS than in VD. The capillaroscopy (capill) shows poor alterations only (small venulo-capillary network) and the transcutaneous gas-analysis has normal values.

The symptoms (orthostatic weight, dyschromia, hypodermitis) appear in the following stages, when the venous hypertension (VH) involves the supplementary system. PVod decreases (37% VD, 20% PPhsS), the cmp becomes inefficient (ΔR(120mV), capill shows the halo-formations, tcpO2 decreases (20-30mmHg), tcpCO2 increases (38-42mmHg). The laser-Doppler (LD), still with functional sufferences at the first stages (VAR 30% Microangiopathy Ind 23%), confirms the microcirculatory damage (MI 11.3%) in the advanced stages. The LFW's decrease and the HFW's increase (reduction/loss of the local autoregulation). This microhemodynamic and metabolic failure induces an involvement in the cellular systems, with reduction of the endothelial protective factors and increase of the macrophagic cytokines. These factors activate the lipoperoxidation and microvascular thrombosis leading to hypodermitis and ulcer finally. The CVI treatment need the reduction of VH, the improvement of the cmp efficacy and of the local autoregulation with reduction of the leucocytic and thrombophilic activation.

P182
RECURRENCE OF VARICOSE VEINS IN RELATION TO THEIR PREOPERATIVE EXTENT

Roztočil K., Přerovský I., Bergmann P.
Institute of Clinical and Experimental
Medicine, Prague, Czech Republic

Objective: To establish whether recurrences of varicose veins following surgery are related to their preoperative extent. **Design:** Clinical follow-up of patients indicated for surgery of varicose veins. **Patients:** 158 legs in 88 patients with primary varicose veins. **Measurements:** Measuring the length of contours of varicose veins before and at 6 months, 1.5, 2.5 and 3.5 years following surgery. **Results:** Mean length of varicose veins in the whole group was 280 cm preoperatively and 2 cm at 6 months, 29 cm at 1.5 years, 44 cm at 2.5 years and 56 cm at 3.5 years following surgery. The more extensive the varices preoperatively, the higher the rates of recurrence, and the more extensive the recurrences were. **Conclusions:** Predispositions responsible for development of varicose veins and present already before surgery remain perhaps the main factor determining their recurrences.

P101
BLOOD CELL VELOCITY AND RELATIVE MICROHAEMATOCRIT IN CHRONIC VENOUS INSUFFICIENCY.

CASSIANI D., BARTOLO M.Jr, CARIOTI B., CARLIZZA A., ALLEGRA C.

DEPT, OF ANGIOLOGY, SAN GIOVANNI HOSPITAL, ROME, ITALY.

Aim of our research was to evaluate red blood cells velocity (rCBV) and relative microhaematocrit (relHct) in patients affected with chronic venous insufficiency (CVI).

28 patients (10 males, 18 females, mean age 56±7), 14 with mild and 14 with severe CVI were investigated by means of dynamic capillaroscopy (Capiflow),in lateral supine position, at the medial ankle.. This method allows to measure rCBV by the dual window and cross correlation technique. Relative microhaematocrit is measured using the black level of the computer as reference and positioning the measurement window on the capillary loop. The computer calculates the difference in density between the window on the loop and the black level of the computer and provides the percentage difference, which is referred to as relative microhaematocrit.
Results:

IVC	CBV (mm/sec)	CT (%)
II stage	0.29±0.16	47.33±7.71
III stage	0.21±0.10	40.16±6.56

*p<0.05

A relevant difference in relHct between the two groups was registered while Cbv didn't change significantly.

Data concerning relHct can be accounted for by a greater optical density, which the computer registers on the capillary loop, since the increase in velocity, even if not significant, causes a quicker transit of red cells with a decrease in gaps. Consequently the window positioned over the capillary registers a higher percentage of black.

P089

EFFECT OF COMPRESSION STOCKINGS ON NEUTROPHIL PRIMING IN CHRONIC VENOUS DISEASE

Salaman R‡§, Hallett M§ Lane I§, Harding K‡
Wound Healing Research Unit‡ and Department of Surgery,§
University of Wales College of Medicine, Cardiff, U.K.

Objective: To determine the effect of compression therapy on neutrophil priming in chronic venous disease.

Design: Ex vivo clinical study of the effect of compression therapy on neutrophil priming.

Patients: 11 affected legs of patients with deep venous incompetence proven by colour duplex ultrasonography.

Measurements: Neutrophils were isolated from arm and leg venous blood samples following 30 minutes of leg dependancy and 5 minutes re-elevation. Samples were repeated following application of a class 2 compression stocking. Priming was assessed by formylated met-leu-phe stimulation of luminol dependant chemiluminesence. Results are expressed in medians (inter quartile range) p is calculated using Wilcoxon paired Z test.

Results: Leg / arm ratio of neutrophil response decreased significantly from 1.31 (0.9-1.7) during leg dependancy to 0.79 (0.62-1.28) following re-elevation (p<0.05). This decrease was abolished after application of a compression stocking (p=0.7). The initial decrease in leg / arm ratio was due to increased priming in the arm sample following re-elevation. Compression therapy significantly reduced priming from all samples (p<0.003). Greatest effect was seen in the arm sample following leg re-elevation (p<0.01) accounting for the normalisation of the leg / arm ratio. Priming was reduced in samples taken from the legs though this did not achieve significance.

Conclusions: Compression therapy may exert its effect by reducing neutrophil priming during leg dependancy.

P090

CHANGES IN LASER DOPPLER FLUX IN NORMAL AND LIPODERMATOSCLEROTIC SKIN DURING COMPRESSION THERAPY

Salaman R‡§, Lane I§, Harding K‡.
Wound Healing Research Unit‡ and Department of Vascular Surgery,§ University of Wales College of Medicine, Cardiff, U.K.

Objective: To measure blood flow in normal and lipodermatosclerotic (LDS) skin during alterations in posture and after the application of a class two compression stocking.

Design: In vivo clinical study.

Patients: 8 patients with 13 legs affected by deep venous incompetence proven by colour duplex ultrasonography.

Measurements: Laser Doppler flux (LDF), concentration of moving blood cells (CMBC) and blood cell velocity (BCV) were recorded using a Periflux 4001 laser Doppler unit connected to a portable computer. Readings were taken during leg elevation and dependancy. A class 2 compression stocking was applied and both sets of readings repeated.

Results: LDS had a higher flux than normal skin during both elevation (p<0.002, paired Wilcoxon) and dependancy(p<0.002), due to significant increases in both CMBC and BCV. Normal and LDS skin showed similar postural reflexes, dependancy causing significant reductions in LDF, with and without compression stockings (p<0.002 in all cases). Compression therapy increased flux in normal (p<0.02) and LDS (p<0.02) skin during leg elevation. This was due to an increase in CMBC (normal p<0.02, LDS p<0.05). BCV was increased in normal skin (p<0.03) though not in LDS (p=0.4). Patterns were similar during dependancy though failed to reach significance.

Conclusions: Compression therapy increases LDF in normal and LDS skin. In LDS this is due to changes in CMBC and not BCV. This effect is decreased by limb dependancy.

Anatomy/Pathology

Clinical Research
Macrocirculatory Pathology

Phlebology '95, D. Negus et al. (eds.). Phlebology (1995) Suppl. 1: 132-134

P096

Patterns of Venous Reflux in the Popliteal Fossa

P. Cariati and G. Lucertini

Cattedra di Chirurgia Vascolare, Università degli Studi, Genova, Italy

INTRODUCTION

Popliteal fossa is a complex area, containing the popliteal vein, the lesser saphenous vein, the Giacomini vein, the gastrocnemius veins and the communicating vein of the popliteal fossa, that can present venous reflux in many sites.

Precise localization of vein valvular incompetence is essential to diagnose exactly and to treat adequately chronic venous insufficiency with venous reflux in the popliteal fossa.

The purpose of this study is the proposal of a classification for venous reflux in the popliteal fossa.

MATERIAL AND METHODS

A retrospective review of a consecutive series of patients with chronic venous insufficiency was carried out from January 1990 to December 1994.

There were 50 patients (60 lower limbs) presenting venous reflux in the popliteal fossa, 19 males and 31 females, ages ranging from 25 to 74 years (mean 47.8).

The assessment was performed by clinical examination, duplex scanning and venography.

Clinical examination showed that the 60 lower limbs presented varicose veins in the popliteal area and/or in the calf.

Investigation of the deep venous system, the gastrocnemius veins, the greater saphenous vein, the lesser saphenous vein and the principal communicating veins was performed using duplex scanning.

Varicography was performed in all cases, ascending venography in 3 and descending venography in 2.

RESULTS

Based on duplex scanning findings and venographic pictures, the incompetent

veins in the popliteal fossa were identified:
- the lesser saphenous vein (LSV) in 31 cases (51.6%);
- the gastrocnemius veins (GV) in 4 cases (6.7%);
- the communicating vein of the popliteal fossa (CV) in 3 cases (5.0%);
- the lesser saphenous vein and gastrocnemius veins in 11 cases (18.3%);
- the lesser saphenous and proximal popliteal (PV) veins in 4 cases (6.7%);
- the lesser saphenous and deep (DV) veins in 3 cases (5.0%).

DISCUSSION

Venous reflux in the popliteal fossa has been widely studied in the last few years [1,2], but only Somjen and coworkers [3] have proposed a classification of venous reflux in the popliteal fossa.

The following clinical and pathophysiologic considerations are essential guidelines of a classification for venous reflux in the popliteal fossa.

It is acknowledged that the severity of chronic venous insufficiency is related to the state of the distal deep veins (the infrapopliteal veins and the popliteal vein). Generally lower limbs with reflux in the distal deep veins show more severe venous disease than those with reflux in the proximal deep veins or in the superficial veins. The incidence of lipodermatosclerosis and ulcer is higher in the former than in the latter [4-9].

Hemodynamic parameters, such as venous pressure and venous refilling time, are often normal or slightly altered in lower limbs with competent distal deep veins. On the contrary, valvular incompetence of the distal deep veins nearly always causes severe changes of these parameters [4,7,10-12].

Therefore four patterns of venous reflux in the popliteal fossa can be identified:
- Type 1: reflux in one or more veins draining into the popliteal vein (eg, valvular incompetence of the lesser saphenous vein);
- Type 2: reflux in one or more veins draining into the popliteal vein and in the proximal popliteal vein;
- Type 3: reflux in one or more veins draining into the popliteal vein, in the proximal popliteal vein and in the superficial femoral vein;
- Type 4: reflux in one or more veins draining into the popliteal vein and in the distal deep veins.

The severity of cases with venous reflux in the popliteal fossa can be easily identyfied using the above classification, for example Types 1 to 3 are generally less severe than Type 4.

This classification is also an useful aid for prognostic evaluation. Types 1 to 3 have a low probability of ulceration, while Type 4 has a high incidence of ulceration.

The planning and results of therapy are also closely related to the classification Types. Lower limbs presenting venous reflux in the popliteal fossa with competent distal deep veins (as in Types 1 to 3) can undergo ligation of incompetent vein(s) draining into the popliteal vein (ie, the lesser saphenous vein, the gastrocnemius veins and the communicating vein of the popliteal fossa) with stripping, if this is indicated (for the lesser saphenous vein only). These cases can present altered hemodynamic parameters, that are

134

normalized after treatment [13]. Lower limbs presenting venous reflux in the popliteal fossa with incompetent distal deep veins (as in Type 4) can undergo a procedure (such as venous valve transplant) for recovery of venous valve function at the popliteal level. In these cases altered hemodynamic parameters will not be normalized after therapy.

The proposed classification offers the possibility to group cases with the same pathophysiologic conditions, allowing a suitable assessment of each case with venous reflux in the popliteal fossa.

REFERENCES

1. Sugrue M, Stanley S, Grouden M, et al. Can pre-operative duplex scanning replace pre-operative short saphenous venography as an aid to localizing the sapheno-popliteal junction? Phlébologie 1988;4:722-5.

2. Vasdekis SN, Clarke GH, Hobbs JT, et al. Evaluation of non-invasive and invasive methods in the assessment of short saphenous vein termination. Br J Surg 1989;76:929-32.

3. Somjen GM, Royle J, Fell G, et al. Venous reflux patterns in the popliteal fossa. J Cardiovasc Surg 1992;33:85-91.

4. Shull KC, Nicolaides AN, Fernandes é Fernandes J, et al. Significance of popliteal reflux in relation to ambulatory venous pressure and ulceration. Arch Surg 1979;114:1304-6.

5. Strandness DE Jr, Langlois Y, Cramer M, et al. Long-term sequelae of acute venous thrombosis. JAMA 1983;250:1289-92.

6. Moore JM, Himmel PD, Sumner DS. Distribution of venous valvular incompetence in patients with the postphlebitic syndrome. J Vasc Surg 1986;3:49-57.

7. Rosfors S, Lamke L-O, Nordstrom E, et al. Severity and location of venous valvular insufficiency. The importance of distal valve function. Acta Chir Scand 1990; 156:689-94.

8. van Bemmelen PS, Bedford G, Beach K, et al. Status of the valves in the superficial and deep venous system in chronic venous disease. Surgery 1991;109:730-4.

9. Hanrahan LM, Araki CT, Rodriguez AA, et al. Distribution of valvular incompetence in patients with venous stasis ulceration. J Vasc Surg 1991;13:805-12.

10. Pearce WH, Ricco J-B, Queral LA, et al. Hemodynamic assessment of venous problems. Surgery 1983;93:715-21.

11. Lindner DJ, Edwards JM, Phinney ES, et al: Long-term hemodynamic and clinical sequelae of lower extremity deep vein thrombosis. J Vasc Surg 1986;4:436-42.

12. Gooley NA, Sumner DS. Relationship of venous reflux to the site of venous valvular incompetence. Implications for venous reconstructive surgery. J Vasc Surg 1988;7:50-9.

13. Christopoulos D, Nicolaides AN, Galloway JMD, et al. Objective noninvasive evaluation of venous surgical results. J Vasc Surg 1988;8:683-7.

Phlebology '95, D. Negus et al. (eds.). Phlebology (1995) Suppl. 1: 135-137

P108

Atypical Varicose Veins: Investigation of the Sources and Pathways of Venous Reflux

A. Viacava

Cattedra di Chirurgia Vascolare, Università degli Studi di Genova, Genova, Italy

INTRODUCTION

In the majority of cases, primary varicose veins are due to incompetence of long saphenous vein, short saphenous vein and medial thigh or lower leg communicating veins.

However, other varicosities, that are less frequently encountered, depend on different sites of venous incompetence. These are commonly called "atypical" and are generally located on the lateral or posterior aspect of the leg or in the pudendal and perineal area [1,2,3,4].

Owing to the variability and complexity of the anatomical connections between superficial and deep veins and their pathophysiological involvment, the various possible sources and pathways of atypical venous reflux are still not thoroughly known.

Their accurate investigation is, therefore, an important point both for cultural and practical purposes.

MATERIAL AND METHODS

From a consecutive series of 360 varicose patients, 36 patients (36 legs with atypical varicose veins) were accurately examined using clinical tests, ultrasounds and varicography.

It was determined that in all 36 patients varicosities didn't depend on conventional sites of incompetence and the various sources and pathways of venous reflux were searched.

Using the clinical assessment as a guide, the Doppler examination (both c.w. and duplex) was employed, first to follow the superficial varicose trunks and the reflux in them, and then to detect the site where they joined a perforating vein and, when possible, up to the connection with the deep venous system.

Varicography was carried out by injecting low osmolar contrast medium into the varicose vein with the patient positioned on a tilting fluoroscopy table equiped with image intensification [5].

The progression of the contrast medium clearly showed the course of the vein, its subfascial connections and the outlet into the deep venous axes.

Films were taken to document the most significant images.

RESULTS

The examined varicose veins were found to be connected with the deep venous system at various different levels and particularly at the following sites (Table. 1):

-pelvic veins (P) (4 cases), laterally through gluteal veins in one case and medially through the obturator vein in three cases and in one of them with an interesting abnormal connection of this vein with the external iliac vein (Fig. 1).

-common femoral vein (CF) (4 cases) and deep femoral vein (DF) (21 cases) through lateral communicating veins of the thigh;

-popliteal vein (Po) (3 cases) through the lateral communicating vein (Thiery vein) [6] in the popliteal fossa;

-gastrocnemius and soleal veins (GS) (3 cases) and peroneal veins (Pe) (1 case) through postero-lateral communicating veins of the calf.

Table 1. Atypical varicose veins: sources of venous reflux

Site	N° of cases	%
P	4	11.2
CF	4	11.2
DF	21	58.3
Po	3	8.3
GS	3	8.3
Pe	1	2.7
Total	36	100

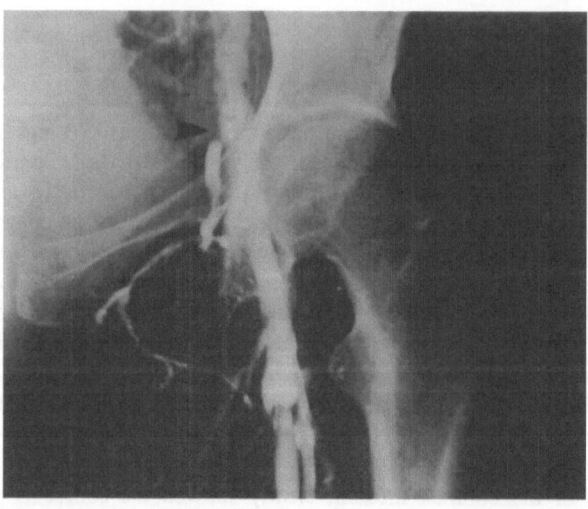

Fig 1. Varicography of atypical varicose veins of the medial aspect of the thigh tributaries of the obturator vein that has, in this case, an abnormal outlet into the external iliac vein.

CONCLUSIONS

On the basis of our experience we can generally conclude that primary varicose veins may be also caused by various sources and pathways of venous reflux different from the long saphenous vein, the short saphenous vein and the medial communicating veins.

The rate of atypical varicose veins in our series was 10%, the most frequent being those connected with the deep femoral vein (58%) . In other series, although not homogenous with ours, the percentage was 25% [3] and 19% [4].

In all cases of atypical varicose veins an accurate assessment of the anatomical and pathophysiological connections is required. For this purpose, we consider, in agreement with other Authors [2,3,7,8], that varicography is the most complete and reliable diagnostic technique. In fact, it gives us the most detailed and, at the same time, panoramic images necessary for correct diagnosis and treatment, expecially in cases with varicose recurrence and/or venous abnormalities.

The study finally suggests that a more profound knowledge of atypical varicose veins allows us to better approach the problem in both theoretical and practical aspects.

REFERENCES

1. Lea Thomas M., Chan O.. Lateral thigh varicose veins: a phlebographic study.Br. J. Radiol. 1988; 61: 372-3.
2. Loveday E.J., Lea Thomas M.. The distribution of recurrent varicose veins: a phlebographic study. Clin. Radiol. 1991; 43: 47-51.
3. Urigo F., Pischedda A., Pinna L., Rovasio S:S:, Maiore M., Canalis G.C.. Ruolo della flebografia nello studio delle varici recidive degli arti inferiori .Radiol. Med. 1993; 85:764-72.
4. Pinzani A., Paoli G., Spreafico G., Gongolo A.. La varicografia nel bilancio preoperatorio delle varici primitive. Radiol. Med. 1989; 77: 504-11.
5. Lea Thomas M.. Ascending and descending phlebography. In Bergan J.J., Kistner R.L. eds.. Atlas of venous surgery. Philadelphia: WB Saunders, 1992; 95-104
6. Thiery L.. Anatomie chirurgicalle de la fosse poplitèe. Phlébologie. 1986; 39:57-66
7. Lea Thomas M., Keeling F:P.. Varicography in the management of recurrent varicose veins. Angiology. 1986; 37:570-5.
8. Ruckley C.V.. Varicography for recurrent varicose veins. In Greenhalgh R.M. ed. Vascular imaging for surgeons. London. Saunders ; 1995: 423-32.

Phlebology '95, D. Negus et al. (eds.). Phlebology (1995) Suppl. 1: 138-140

P079

Comparative Morphometric Study of the Therapeutic Effect of Compression and Saphenectomy Carried out on Cutaneous Biopsies in Patients Affected by Peripheral Chronic Venous Insufficiency

C. Ruggieri[1], S. Massi[1], A.M. Peccatori[1], G. Botta[1], R. Lio[2], C. Miracco[2] and S. Mancini[1]

[1] III Department of General Surgery and Pabisch Center of Phlebology, [2] Institute of Pathological Anatomy, Faculty of Medicine and Surgery, University of Siena, Italy

INTRODUCTION

The alteration of the capillary circulation of the lower limbs in patients with chronic venous insufficiency, studied on biopsy-punch in optic microscopy and the complication caused in the cutaneous district, undergo some significant variation after a 3 months elastic-compression treatment and surgical saphenectomy.

This experience of us wants to analyse quantitative differences on cutaneous biopsies in patients affected by chronic venous insufficiency and treated with elastic-compression therapy for 3 months in a first group, joining also a surgical saphenectomy in a second group.

MATERIALS AND METHODS

We examined 40 punch biopsies of the lower limbs of patients with chronic venous insufficiency.

Patients were divided in 2 groups: the first group (10 patients) has been treated only with compression therapy by means of elastic bandage of 2nd class compression (30-40 mmHg), in the second group (10 patients) the surgical therapy has preceded the compression therapy with the same intensity of the first.

Biopsies were fixed in 10% formalin, paraffin included and processed for the following colourings: Hematoxiylin-eosin, PAS, (to underline possible mucopolysaccharides and basement membranes) Weigert Van Gieson-Mallory (elastic fibers, connective tissue), Perls (haemosiderin deposits).

The section were observed at optic-microscopy and processed on semiautomatic image analyzer (Ibas-Kontron).

For each biopsy 5 microscopical fields were analyzed (250X) using a Weibel grid (Weibel 1979) fixing in this way the percentage occupied area by the following parameters: elastic fiber volume fraction; collagen fiber volume fraction; vascular surface volume fraction; volume fraction occupied by interstitial oedema. The perivascular and interstitial lympho-histiocytary inflammatory infiltration and eventual haemosiderin deposits have been valuated in a semi-quantitative way.

RESULTS

Semiquantitative analysis: utilising specific couloratings no haemosidering deposits have been examined in the cases we treated. The inflammatory infiltration (lymphocyties, histocyties) prevalently perivascular in all cases we examined was of a little importance and it didn't have any modification further to different therapeutical approaches.

Statistic-analysis: results obtained from the percentage count, utilizing the Weibel grid, have been analysed by means of no parametric Wilcoxon test.

Quantitative-analysis: in all examined parameters, the percentage of connective and elastic fibers do not show any statistically significative differences before and after 2 treatments neither in the comparison between the 2 different treatments. Vice verse in both treatments we observe a significant reduction of the dermic oedema. Capillary do not undergo any significant modification after the elastic-compression, while they increase in a statistically significant way if the elastic-compression is joined to saphenectomy.

DISCUSSION

In our experience, after 3 months treatment, both therapeutical procedures are followed by a reduction of the oedema in the derma, while elastic and connective fibers do not seem to have any rilevant modification.

Saphenectomy seems not to have an importance of oedema and, for this, we believe that the only elastic-compression for at least 3 months could reach the same result itself.

Saphenectomy seems not to have an important role in the reduction of oedema and, for this, we believe that the only compression for at the least 3 months could reach the same result itself.

The surgical saphenectomy has an important role when joined to the elastic-compression therapy why it permits a new vascularisation of capillary in the derma.

This statistically significant increasing could express an improovement of the cutaneous trophism taken away clinically in patients subject of our research.

SUMMARY

After having analyzed, in a previous study, the quantitative differences of the micro-vascular tissue space modification on 10 cutaneous biopsies in patients affected by peripheral chronic venous insufficiency after 30 days elastic-compression without surgery treatment, and seen the absence of statistic significant results, the Authors have thought it right to carry on this research bringing some modification to the study.

Further 40 drawings (punch-biopsy) on the lower limbs on patients affected by chronic venous insufficiency have been examined in optic-microscopy.

Byobsies-punch, in both groups have been made before and after the therapeutic treatment.

Patients have been subdivided in 2 groups, the first one has been treated with compression-therapy only, while in the second the surgery therapy has preceded the compression therapy.

Biopsies were fixed in 10% formalin and processed for the routinary colouring, for the colourating of elastic-fibers, of the connective tissue and for the haemosiderin deposits.

The sections were observed by means of a semiautomatic image analyzer and the parameters used are the following: 1) elastic fiber volume fraction, 2) collagen fiber volume fraction, 3) vascular surface volume fraction, 4) volume fraction occupied by interstitial oedema; with semiquantitative method, the perivascular lympho-histiocytary inflammatory infiltrate of the dermis and possible deposits of haemosiderin were evaluated.

All the parameters have been statistically studied before and after the therapeutic treatment carried on for 3 months in both groups.

REFERENCES

1. Curri S.B.. Le microangiopatie. II Ed. Idb Ed, Milano 1986.
2. Curri S.B, Annoni F. Modificazioni del microcircolo cutaneo da elastocompressione nella insufficienza venosa cronica. Centro documentazione W. Pabisch.
2. Merlen J.F., Coget J.M.. Rétentissement micro-vasculo-tissulaire de l'insuffisance veineuse circulatoire. Colle de Bagnoles de l'Orne, mai 1977 (collection R. e C.)
4. Merlen J.F., Coget J.M.. Influences de la compression sur la microcirculation. Société francaise de Phlebologie 1985, 13:12.
5. Merlen J.F., Coget J.M.. Action et reaction des tissus à la compression. Phlebologie 1981, 34:365-373.
6. Lio R., Miracco C., Botta G., Ruggieri C., Massi S., Mancini S.. Quantitative study on cutaneous biopsies of the lower limbs in patients with chronic venous ifailure before and after compression therapy. Phlebologie 92. Eds P. Raymond Martinbeau, R. Prescott, M. Zummo. John Libbey Eurotext, Paris 1992, 85-87.

Phlebology '95, D. Negus et al. (eds.). Phlebology (1995) Suppl. 1:141-143

P257

Teneurs Plasmatique et Veineuse en Composés Réactifs à l'acide Thiobarbiturique (Lipoperoxydes) en Acide Ascorbique et en Fer Total chez les Sujets sains ou Variqueux

P. Joanny[1], P. Barthèlemy[2], J. Steinberg[3], A. Zamora[4], M.de Champvallins[5], G. Pillion[5], M. Portugal[6] and A.M. Pauli[6]

[1] Laboratoire de Physiopathologie Respiratoire Intégrée et Cellulaire - URA 1630 - Faculté de Médecine Nord, Boulevard P. Dramard, 13916 Marseille Cedex 20, [2] Service de Chirurgie Vasculaire - Hôpital Nord, Chemin des Bourrely, 13915 Marseille Cedex 20, [3] Laboratoire de Neurophysiologie et Neurochimie, Faculté de Médecine, 27, Bd J. Moulin 13385 Marseille Cedex 05, [4] Laboratoire des Interactions Cellulaires Neuroendocriniennes - UMR 9941 - Faculté de Médecine Nord, Boulevard P. Dramard, 13916 Marseille Cedex 20, [5] IRIS, place des Pléïades, 92415 Courbevoie, [6] Laboratoire de Chimie Analytique, Faculté de Pharmacie, 27, Bd Jean Moulin, 13385 Marseille Cedex 05

INTRODUCTION

Des travaux épidémiologiques ont montré que la maladie variqueuse pouvait en partie être liée à des facteurs diététiques (1). L'hypothèse d'une implication des formes radicalaires intermédiaires délétères de l'oxygène a été formulée dans la pathogénie de l'insuffisance veineuse (2,7). Ces derniers auteurs ont, en effet, montré l'existence d'une corrélation positive entre le pouvoir anti-lipoperoxydant tissulaire veineux et la teneur en alpha-tocophérol dans la veine normale ou variqueuse. Il était donc intéressant d'élargir ces travaux à l'étude comparative à la fois plasmatique et pariétale veineuse chez les sujets sains et variqueux :

- des teneurs respectives en lipoperoxydes (évalués en tant que substances réactives avec l'acide 2-thiobarbiturique (SR-ATB) marqueurs biologiques reconnus de l'atteinte tissulaire par les radicaux libres oxygénés (RLO) ;

- de la teneur en acide ascorbique réduit (AAR), connu, pour exercer un double rôle protecteur : anti-lipoperoxydant (reconnu comme le plus puissant au niveau plasmatique), anti-radicalaire ainsi que de régénération de l'alpha-tocophérol oxydé;

- de la teneur en fer total (FT), métal de transition, qui est doué d'un effet pro-peroxydant.

MATERIEL ET METHODES

Des segments de veines saphènes internes témoins étaient prélevés et stockés à -70°C (carboglace) lors de l'intervention chirurgicale: soit pour pontage coronarien ou pour greffe veineuse fémoro-distale (âge : $66,70 \pm 2,66$ ans). Des segments de veines saphènes internes variqueuses étaient prélevés lors de saphénectomies (âge : $50,75 \pm 2,69$ ans). Ces échantillons veineux étaient soumis, ainsi que les plasmas autologues héparinés, après 10 minutes à + 4°C, à l'analyse biochimique. La teneur en SR-ATB était mesurée par une microméthode spectrofluorimétrique (3), la teneur en AAR par HPLC (4) et la teneur en

FT par méthode spectrophotométrique (ferrozine) après dialyse automatisée en flux continu (AAT II, Technicon) (5 et 6).

RESULTATS

Les teneurs autologues plasmatique et veineuse en SR-ATB, en AAR et en FT sont données dans le tableau suivant :

Conditions expérimentales	PLASMA			TISSU VEINEUX		
	SR-ATB ng/ml	AAR µg/ml	FT µg/ml	SR-ATB µg/g	AAR µg/g	FT µg/g
Sujets témoins	272,9 ± 10,9 (20) NS	12,4 ± 1,21 (20) NS	0,91 ± 0,05 (20) NS	2,97 ± 0,23 (20) p<0,025	165,3 ± 12,1 (20) p<0,001	51,4 ± 2,9 (20) p<0,001
Sujets variqueux	268,7 ± 21,3 (20)	10,4 ± 1,5 (20)	1,02 ± 0,03 (20)	3,42 ± 0,22 (20)	58,2 ± 4,3 (20)	20,5 ± 2,4 (20)

Valeurs moyennes suivies de l'erreur-standard de la moyenne et du nombre d'expériences (entre parenthèses) ; analyse statistique, test "U de Mann-Whitney

Plasma : pour les teneurs en SR-ATB, en AAR et en FT, il n'existe aucune différence significative entre les sujets témoins et les sujets variqueux. Chez ces derniers, (résultats non montrés), il n'existe aucune différence significative entre ces paramètres biochimiques étudiés respectivement au niveau du membre supérieur et du membre inférieur, la même étude n'a pas pu être réalisée pour des raisons éthiques chez les sujets témoins.

Tissu veineux : il existe une tendance non significative d'élévation de l'hydratation veineuse (+ 14 % exprimé en kg H_2O/kg de matière sèche) dans la veine variqueuse. Les valeurs expérimentales font ressortir une élévation de 15 % (p<0,025) en SR-ATB chez les sujets variqueux, concomitante d'une importante baisse à la fois de la teneur en AAR (-64 % ; p<0,001) et en FT (- 60 %; p<0,001).

DISCUSSION

Il convient de souligner d'emblée l'absence d'altération plasmatique des teneurs en SR-ATB, AAR et FT chez les sujets variqueux, contrastant avec celles que nous observons au niveau pariétal veineux, fait déjà souligné (2) pour la constance de la capacité anti-lipoperoxydante plasmatique et notamment en alpha-tocophérol, chez les sujets témoins et les sujets variqueux, l'alpha-tocophérol ne représentant en fait qu'une faible fraction de celle-ci. L'élévation que nous observons dans la veine variqueuse des taux de SR-ATB confirme clairement la baisse (45 %) de la capacité anti-lipoperoxydante qui a été décrite sur l'homogénat de veine variqueuse par rapport à celui de la veine saine (7). Un argument expérimental supplémentaire en faveur de l'hypothèse d'une genèse radicalaire de la maladie variqueuse réside dans l'importante baisse concomitante en AAR que nous constatons sur la veine variqueuse ; cette baisse en AAR pourrait, ici, vraisemblablement, s'expliquer par une utilisation nettement augmentée de cette molécule correspondant à un mécanisme de défense tissulaire contre la formation excessive de RLO délétères. L'AAR possédant, par ailleurs, un rôle reconnu de réduction de l'alpha-tocophérol oxydé. Ceci pourrait, en partie, rendre compte de la baisse observée en alpha-tocophérol réduit dans la veine variqueuse (2,7), la consommation de ce dernier étant elle-même augmentée au cours du stress oxydatif.

L'interprétation de la baisse en FT du tissu variqueux, que nous observons est difficile dans l'état actuel de nos travaux, d'autant plus que dans la cellule le stockage du fer est intra-cavitaire au sein des molécules de ferritine et cela par l'intermédiaire d'une protéine de transfert spécifique : la transferrine; il n'est pas exclu que la teneur tissulaire de ces protéines puisse être modifiée dans la veine variqueuse. Néanmoins, à titre d'hypothèse de travail, nous pensons que cette baisse de teneur en fer pourrait soit correspondre à un mécanisme tissulaire de défense, soit résulter de l'élévation, connue, de la proportion de tissus collagéniques de type I et III dans le tissu variqueux (8). L'acide hyaluronique est dépolymérisé dans la veine variqueuse, phénomène qui peut s'expliquer par la lipoperoxydation et/ou la formation de RLO (8).

Cependant, il n'est, actuellement, pas possible de préciser si l'atteinte radicalaire observée au stade des varices constituées représente une cause ou une conséquence de la maladie variqueuse.

REMERCIEMENTS

Nous remercions le Docteur Carol DEBY pour ses commentaires et suggestions au cours de ce travail expérimental. (Laboratoire de Biochimie de l'oxygène - Université de Liège)

REFERENCES

1 - Malhotra S.L. An epimiological study of varicose veins in indian workers from the south and north of India. Int. J. Epidemiol. 1982;1:117-8.

2 - Deby C., Hariton C., Pincemail J., Coget J. Decreased tocopherol concentration of varicose veins is associated with a decrease in antilipoperoxidant activity without similar changes in plasma. Phlebology 1989;4:113-21.

3 - Noda Y., Mc Geer P.L., Mc Geer E.G. Lipid peroxide distribution in brain and the effect of hyperbaric oxygen. J. Neurochem. 1983;40:1329-332

4 - Kmetec V. Simultaneous determination of acetylsalicytic, salicylic, ascorbic and deshydroascrobic acid by H.P.L.C. J. Pharm. Biomed. Anal. 1992;10(10-12):1073-076.

5 - Giovanello T.J., Dibenedetto G. Palmer D.N. Peter T.J. Fully automated method for the determination of serum iron and total iron-binding capacity. In : Automation in Analytical chemistry Technicon Symposium 1967. Vol. 1, White Plains, N.Y. Mediated Inc. 1968;185-88.

6 - Stookey L.L. Ferrozine : a new spectrophotometric reagent of iron. Anal. Chem. 1970;42: 779-81.

7 - Deby C., Hariton C., Pincemail J. In : Congrés International "Les Rencontres de l'Angiologie" 12-14 avril 1991, 6

8 - Niebes P. Biochemical studies of varicosis. In : Monographs on Standardization of Cardioangeiological Methods 1977;4:233-39.

Phlebology '95, D. Negus et al. (eds.). Phlebology (1995) Suppl. 1: 144-147
P010

Plasma and Vein Wall Levels of Lipoperoxides, Ascorbic Acid and Total Iron in Healthy Subjects and in those with Varicosities

P. Joanny[1], P. Barthèlemy[2], J. Steinberg[3], A. Zamora[4], M.de Champvallins[5], G. Pillion[5], M. Portugal[6] and A.M. Pauli[6]

[1] Laboratoire de Physiopathologie Respiratoire Intégrée et Cellulaire - URA 1630 - Faculté de Médecin Nord, Boulevard P. Dramard, 13916 Marseille Cedex 20, [2] Service de Chirurgie Vasculaire - Hôpita Nord, Chemin des Bourrely, 13915 Marseille Cedex 20, [3] Laboratoire de Neurophysiologie e Neurochimie, Faculté de Médecine, 27, Bd J. Moulin 13385 Marseille Cedex 05, [4] Laboratoire de Interactions Cellulaires Neuroendocriniennes - UMR 9941 - Faculté de Médecine Nord, Boulevard P Dramard, 13916 Marseille Cedex 20, [5] IRIS, place des Pléïades, 92415 Courbevoie, [6] Laboratoire d Chimie Analytique, Faculté de Pharmacie, 27, Bd Jean Moulin, 13385 Marseille Cedex 05

SUMMARY

The authors report on a study of the levels of lipoperoxides, reduced ascorbic acid and total iron in the plasma and in vein walls in normal subjects and in those with varicosities. The plasma levels were not significantly different in between the 2 groups. There was a non-significant elevation in the level of lipoperoxides in the vein wall in subjects with varicosis together with a significant lowering of the other two factors. It is not clear whether this is cause or effect.

TENEURS PLASMATIQUE ET VEINEUSE EN COMPOSES REACTIFS A L'ACIDE THIOBARBITURIQUE (LIPOPEROXYDES), EN ACIDE ASCORBIQUE ET EN FER TOTAL CHEZ LES SUJETS SAINS OU VARIQUEUX

P. Joanny[1], P. Barthélemy[2], J. Steinberg[3], A. Zamora[4], M. De Champvallins[5], G. Pillion[5], M. Portugal[6], A.M. Pauli[6].

[1] Laboratoire de Physiopathologie Respiratoire Intégrée et Cellulaire - URA 1630 - Faculté de Médecine Nord, Boulevard P. Dramard, 13916 Marseille Cedex 20, [2] Service de Chirurgie Vasculaire - Hôpital Nord, Chemin des Bourrely, 13915 Marseille Cedex 20, [3] Laboratoire de Neurophysiologie et Neurochimie, Faculté de Médecine, 27, Bd J. Moulin 13385 Marseille Cedex 05, [4] Laboratoire des Interactions Cellulaires Neuroendocriniennes - UMR 9941 - Faculté de Médecine Nord, Boulevard P. Dramard, 13916 Marseille Cedex 20, [5] IRIS, place des Pléïades, 92415 Courbevoie, [6] Laboratoire de Chimie Analytique, Faculté de Pharmacie, 27, Bd Jean Moulin, 13385 Marseille Cedex 05.

INTRODUCTION

Des travaux épidémiologiques ont montré que la maladie variqueuse pouvait en partie être liée à des facteurs diététiques (1). L'hypothèse d'une implication des formes radicalaires intermédiaires délétères de l'oxygène a été formulée dans la pathogénie de l'insuffisance veineuse (2,7). Ces derniers auteurs ont, en effet, montré l'existence d'une corrélation positive entre le pouvoir anti-lipoperoxydant tissulaire veineux et la teneur en alpha-tocophérol dans la veine normale ou variqueuse. Il était donc intéressant d'élargir ces travaux à l'étude comparative à la fois plasmatique et pariétale veineuse chez les sujets sains et variqueux :

- des teneurs respectives en lipoperoxydes (évalués en tant que substances réactives avec l'acide 2-thiobarbiturique (SR-ATB) marqueurs biologiques reconnus de l'atteinte tissulaire par les radicaux libres oxygénés (RLO) ;

- de la teneur en acide ascorbique réduit (AAR), connu, pour exercer un double rôle protecteur : anti-lipoperoxydant (reconnu comme le plus puissant au niveau plasmatique), anti-radicalaire ainsi que de régénération de l'alpha-tocophérol oxydé;

- de la teneur en fer total (FT), métal de transition, qui est doué d'un effet pro-peroxydant.

MATERIEL ET METHODES

Des segments de veines saphènes internes témoins étaient prélevés et stockés à -70°C (carboglace) lors de l'intervention chirurgicale: soit pour pontage coronarien ou pour greffe veineuse fémoro-distale (âge : $66,70 \pm 2,66$ ans). Des segments de veines saphènes internes variqueuses étaient prélevés lors de saphénectomies (âge : $50,75 \pm 2,69$ ans). Ces échantillons veineux étaient soumis, ainsi que les plasmas autologues héparinés, après 10 minutes à + 4°C, à l'analyse biochimique. La teneur en SR-ATB était mesurée par une microméthode spectrofluorimétrique (3), la teneur en AAR par HPLC (4) et la teneur en

FT par méthode spectrophotométrique (ferrozine) après dialyse automatisée en flux continu (AAT II, Technicon) (5 et 6).

RESULTATS

Les teneurs autologues plasmatique et veineuse en SR-ATB, en AAR et en FT sont données dans le tableau suivant :

Conditions expérimentales	PLASMA			TISSU VEINEUX		
	SR-ATB ng/ml	AAR µg/ml	FT µg/ml	SR-ATB µg/g	AAR µg/g	FT µg/g
Sujets témoins	272,9 ± 10,9 (20) NS	12,4 ± 1,21 (20) NS	0,91 ± 0,05 (20) NS	2,97 ± 0,23 (20) p<0,025	165,3 ± 12,1 (20) p<0,001	51,4 ± 2,9 (20) p<0,001
Sujets variqueux	268,7 ± 21,3 (20)	10,4 ± 1,5. (20)	1,02 ± 0,03 (20)	3,42 ± 0,22 (20)	58,2 ± 4,3 (20)	20,5 ± 2,4 (20)

Valeurs moyennes suivies de l'erreur-standard de la moyenne et du nombre d'expériences (entre parenthèses) ; analyse statistique, test "U de Mann-Whitney

Plasma : pour les teneurs en SR-ATB, en AAR et en FT, il n'existe aucune différence significative entre les sujets témoins et les sujets variqueux. Chez ces derniers, (résultats non montrés), il n'existe aucune différence significative entre ces paramètres biochimiques étudiés respectivement au niveau du membre supérieur et du membre inférieur, la même étude n'a pas pu être réalisée pour des raisons éthiques chez les sujets témoins.

Tissu veineux : il existe une tendance non significative d'élévation de l'hydratation veineuse (+ 14 % exprimé en kg H_2O/kg de matière sèche) dans la veine variqueuse. Les valeurs expérimentales font ressortir une élévation de 15 % (p<0,025) en SR-ATB chez les sujets variqueux, concomitante d'une importante baisse à la fois de la teneur en AAR (-64 % ; p<0,001) et en FT (- 60 %; p<0,001).

DISCUSSION

Il convient de souligner d'emblée l'absence d'altération plasmatique des teneurs en SR-ATB, AAR et FT chez les sujets variqueux, contrastant avec celles que nous observons au niveau pariétal veineux, fait déjà souligné (2) pour la constance de la capacité anti-lipoperoxydante plasmatique et notamment en alpha-tocophérol, chez les sujets témoins et les sujets variqueux, l'alpha-tocophérol ne représentant en fait qu'une faible fraction de celle-ci. L'élévation que nous observons dans la veine variqueuse des taux de SR-ATB confirme clairement la baisse (45 %) de la capacité anti-lipoperoxydante qui a été décrite sur l'homogénat de veine variqueuse par rapport à celui de la veine saine (7). Un argument expérimental supplémentaire en faveur de l'hypothèse d'une genèse radicalaire de la maladie variqueuse réside dans l'importante baisse concomitante en AAR que nous constatons sur la veine variqueuse ; cette baisse en AAR pourrait, ici, vraisemblablement, s'expliquer par une utilisation nettement augmentée de cette molécule correspondant à un mécanisme de défense tissulaire contre la formation excessive de RLO délétères. L'AAR possédant, par ailleurs, un rôle reconnu de réduction de l'alpha-tocophérol oxydé. Ceci pourrait, en partie, rendre compte de la baisse observée en alpha-tocophérol réduit dans la veine variqueuse (2,7), la consommation de ce dernier étant elle-même augmentée au cours du stress oxydatif.

L'interprétation de la baisse en FT du tissu variqueux, que nous observons est difficile dans l'état actuel de nos travaux, d'autant plus que dans la cellule le stockage du fer est intra-cavitaire au sein des molécules de ferritine et cela par l'intermédiaire d'une protéine de transfert spécifique : la transferrine; il n'est pas exclu que la teneur tissulaire de ces protéines puisse être modifiée dans la veine variqueuse. Néanmoins, à titre d'hypothèse de travail, nous pensons que cette baisse de teneur en fer pourrait soit correspondre à un mécanisme tissulaire de défense, soit résulter de l'élévation, connue, de la proportion de tissus collagéniques de type I et III dans le tissu variqueux (8). L'acide hyaluronique est dépolymérisé dans la veine variqueuse, phénomène qui peut s'expliquer par la lipoperoxydation et/ou la formation de RLO (8).

Cependant, il n'est, actuellement, pas possible de préciser si l'atteinte radicalaire observée au stade des varices constituées représente une cause ou une conséquence de la maladie variqueuse.

REMERCIEMENTS

Nous remercions le Docteur Carol DEBY pour ses commentaires et suggestions au cours de ce travail expérimental. (Laboratoire de Biochimie de l'oxygène - Université de Liège)

REFERENCES

1 - Malhotra S.L. An epimiological study of varicose veins in indian workers from the south and north of India. Int. J. Epidemiol. 1982;1:117-8.

2 - Deby C., Hariton C., Pincemail J., Coget J. Decreased tocopherol concentration of varicose veins is associated with a decrease in antilipoperoxidant activity without similar changes in plasma. Phlebology 1989;4:113-21.

3 - Noda Y., Mc Geer P.L., Mc Geer E.G. Lipid peroxide distribution in brain and the effect of hyperbaric oxygen. J. Neurochem. 1983;40:1329-332

4 - Kmetec V. Simultaneous determination of acetylsalicytic, salicylic, ascorbic and deshydroascrobic acid by H.P.L.C. J. Pharm. Biomed. Anal. 1992;10(10-12):1073-076.

5 - Giovanello T.J., Dibenedetto G. Palmer D.N. Peter T.J. Fully automated method for the determination of serum iron and total iron-binding capacity. In : Automation in Analytical chemistry Technicon Symposium 1967. Vol. 1, White Plains, N.Y. Mediated Inc. 1968;185-88.

6 - Stookey L.L. Ferrozine : a new spectrophotometric reagent of iron. Anal. Chem. 1970;42: 779-81.

7 - Deby C., Hariton C., Pincemail J. In : Congrés International "Les Rencontres de l'Angiologie" 12-14 avril 1991, 6

8 - Niebes P. Biochemical studies of varicosis. In : Monographs on Standardization of Cardioangeiological Methods 1977;4:233-39.

Phlebology '95, D. Negus et al. (eds.). Phlebology (1995) Suppl. 1: 148-150

P013

The Role of 'Vasa Vanorum' in the Pathogenesis of Valvular Incompetence

S.B. Curri[1], A. Visintin and M. Casagrande[2]

[1] Centre of Molecular Biology, Milan, Italy
[2] Casa di Cura Sanatorio Trestino, Trieste, Italy

INTRODUCTION

Little is known on the microcirculatory supply of the venous wall of the lower limbs (1). In previous studies (2,3) severe changes of the 'vasa venorum' (Vv) have been shown in cases of varicosity of the saphenous vein with valvular incompetence. The protocol of the present study is: a) to examine other varicose veins for a comparison of the morpho-histochemical aspects, b) to obtain quantitative data on the dilation of the newformed capillaries proliferating in the varicose vein wall, c) to investigate differences between the pathological changes of the microvessels in the adventitia, media and intima of varicose veins, d) to study the blood flow regulation devices of the 'vasa venorum'.

MATERIALS AND METHODS

A total of 132 samples taken during surgical operations for varicose veins have been examined (great Saph. V, and their collaterals at different points, 'Crosse', mall Saph. V., Cockett's and Boyd's perforators, accessory and antero-lateral Saph.V., Pudendal, Hypogastric and Leonardo's Veins). The histological sections were stained with Haematoxyline-Eosine, P.A.S., Gomori for reticular fibres and Weigert for elastic fibres and examined at 800-1000 x. To measure the lumen diameter of the Vv capillaries a modified (4) Capiflow equipment acc. to Bollinger and Fagrell (5) was used (Computerised Video-Mircorangiology, 4)

Structure and Pathological Changes of the 'Vasa Venorum'

The Vv consist of: (1) small arteries provided by 2-8 layers of smooth muscle cells (s.m.c), (2) precapillary arterioles with only one layer of spirally oriented s.m.c., (3) endarterial blocking devices. These are of two types: cushions with a large base formed by 10-20 layers of obliquely or vertically oriented s.m.c. which form a wall protuberance in the lumen (Type A); pedunculated, polypoid, roundish or oval devices which seems to float in the blood stream (Type B). The volume and velocity of the blood flow directed to the Vv capillaries is regulated by the contraction (opening of the lumen) and respectively by the constriction (closing of the lumen) of the s.m.c. of these blocking devices. Pathological changes of the smallest arteries are narrowing of the media s.m.c., dilation of the lumen, endothelial swelling, massive sclerosis of the adventitial connective tissue (Venous Stasis Histangiopathy, 6). The endarterial blocking devices show an almost complete structural narrowing provoked by a connective fibres newforming and sclerosis, with degenerative changes of the s.m.c. The regulation of blood supply to the vein wall capillaries (Vv) is therefore damaged or abolished (2,3). Furthermore Vv consist of by (4) small vines and by (5) two types of capillaries.

All the venules of the Vv are massively dilated (Max. Diam. of the lumen 97.8+/-64.16nm, Min. Diam. 72.24+/-36.2nm : Fig 1). Venous stasis is therefore diffused even to the Vv efferent system, with decreased blood flow velocity and accumulation of catabolites in the vein wall. The consequence must be an increase in the sclerosis and regressive changes of the s.m.c. of the vein wall. The first type of capillaries are the proper Vv capillaries, functionally even in the preclinical stages of venous stasis. The second type are pathologically newformed capillaries, proliferating in the altered vein wall (Fig. 1). The dilated capillaries (Max. Diam. of the lumen 29.56+/-12.6nm, Min. Diam. 19.27+/-15.4nm) can be observed also in the thickened and dysendothelized intima of varicose veins, up to 40nm from the lumen. Intraparietal microhaemorrhages are a frequent finding. The most dilated are the adventitial capillaries (32+/-8.3nm) followed by the media capillaries (28.7+/-10.6nm) and by the intima capillaries (21.4+/-10.2): intima vs media P = 0.0001, adventitia vs. media P = 00001, adventitia vs. media P=0.01. The abnormal, anarchic proliferation of the newformed Vv capillaries is a constant picture of the varicose disease and can be found in all the veins of the lower limbs.

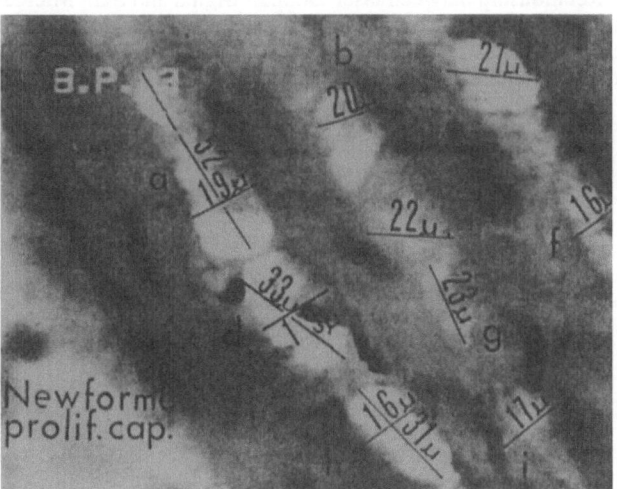

Fig 1. Case nr.13, B.P., f., 40y, Varicosis of great saphenous vein (cross). Chain-like disposed proliferating capillaries in the innerst layers of the media. a-b-c-d-e-f = nuclea of degenerated smooth muscle cells. Haematoxyline-Eosine, 1000 x.

CONCLUSIONS

There are no agreements on the pathogenesis of valvular incompetence: different theories, which include so called 'primitive' valvular incompetence, perforating vein incompetence, arterio-venous anastomoses, a 'primitive weakness' of the venous wall, biochemical factors such as hypoxia of the vein wall with reduction of the oxidative metabolism, have been proposed. It is surprising that so little attention has been given to endothelial damage in varicose veins and to the relationships between endothelial damage and Vasa venorum. These have been 'forget by Phlebologists' (7). Our findings show that the Vasa venora and their pathological changes play a great role in the development of the varicose disease. The functional incompetence of the blocking devices may be the first act of the progressive weakness of the venous wall, followed by the abnormal, anarchic proliferation of the Vv capillaries, similarly to the capillary newforming in diabetic rhetinopathy. It's greatly probable that the hypoxia, the first metabolic wall disorders and finally the 'tonus loss' my find their genesis in the microcirculatory failure of the 'Vasa venorum'.

REFERENCES

1. Fegan W G, The cause of varicose veins, Eur Congr int Un Phlebology, Budapest Sept 6-10, Abstr Book p 17.
2. Curri S B, Significato delle alterazioni anatomo-funzionali dei vasa venarum nella patogenesi del danno parietale venoso e della malattia varicosa, Flebolinfologia 1991; 2/2:23-42.
3. Curri S B, Rapporti tra endotelio, parete venosa e vasa vasorum, in: P Pola (Ed), Proc VI Congr naz Flebologia, Rome Nov 8-11 1989, Monduzzi Publ Co, Bologna 1989: 505-511
4. Curri S B, Allegra C, A Modified CapiFlow Technic for Microangiological and Histochemical Purposes, 17th Eur Conf Microcirculation, London July 6-10, 1992.
5. Bollinger A, Fagrell B, Clinical Capiillaroscopy, a Guide to its Use in Clinical Reasearch and Practice, Hogrefe & Huber Publ Co, Toronto 1990.
6. Vurr S B, Relationship between laser-Doppler Signal and Skin Microangiotectonic in Chronic Venous Stasis, Eur Congr int Un Phlebology, Budapest Sept 6-10, 1993, Abst Book 140.
7. Bollinger A, Personal Communication, London July 6-10, 1992.

Phlebology '95, D. Negus et al. (eds.). Phlebology (1995) Suppl. 1: 151-153

P061

Sonographic Findings of Preserved Saphenous Veins in Primary Varicose Veins Patients

M. Mo[1], R. Adachi[2], Y. Ichikawa[2], T. Kosuge[1], K. Imoto[1], J. Kondo[1] and A. Matsumoto[1]

[1] First Department of Surgery, Yokohama City University School of Medicine, Yokohama, Japan
[2] Department of Cardiovascular Surgery, Yokohama Minami Kyousai Hospital, Yokohama, Japan

Introduction

High ligation of the saphenous vein without stripping (HL) and selective stripping of upper part of the saphenous veins (SS) have been performed to reduce surgical invasiveness of conventional varicose veins surgery and to preserve possible future graft materials for arterial reconstruction[1,2,3]. Aim of this study is to evaluate morphology and competence of preserved saphenous veins by sonography and Doppler flowmetry or Duplex scanning.

Patients and Methods

Twenty-nine legs of 28 patients with primary varicose veins were studied. Diagnosis of primary varicose veins was confirmed by physical examination, sonography and ascending venography. HL was performed in 21 legs whose saphenous veins were incompetent but without aneurysmal formation. Ligation of long saphenous vein (LSV) around knee was also performed. SS was performed in 8 legs. Only competent lower parts of saphenous veins were preserved based on Doppler or duplex findings [2] (Fig. 1). We evaluated 10 of incompetent LSV, 7 of competent LSV, 11 of short saphenous vein (SSV), 1 of competent SSV. Diameter of saphenous vein was measured at thigh, knee and ankle for LSV and knee and ankle for SSV. At the same time, competence of the vein was evaluated by Doppler flowmetry or Duplex scan with compressing and releasing distal parts of the veins. The vein was judged as incompetent when duration of reflux was longer than 0.5 seconds (Fig. 2)

Fig. 1 Patients and Methods

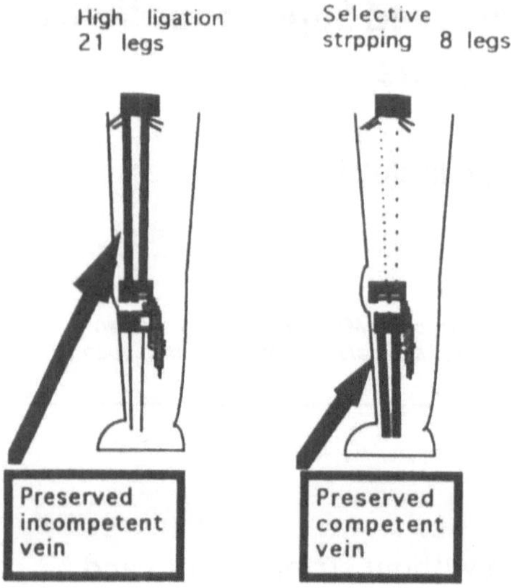

High ligation
21 legs

Selective
strpping 8 legs

Preserved
incompetent
vein

Preserved
competent
vein

Fig. 2 Doppler and Duplex wave form

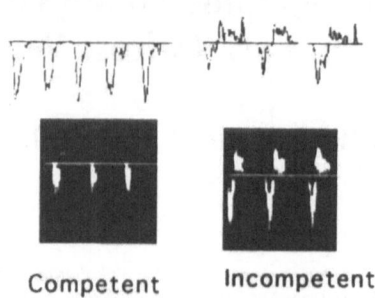

Competent   Incompetent

Results

The most of saphenous veins were patent except two LSV with partial occlusion. Diameter of incompetent LSV at thigh reduced to 4.5 mm from 9.0 mm (p<0.01), and diameter of incompetent SSV at knee reduced to 3.1 mm from 8.2 mm (p<0.01) after HL (Fig 3). Diameter of competent LSV at the knee was not changed significantly after SS. Five of 10 incompetent LSV and 8 of 11 incompetent SSV became competent after the HL. Two of 7 competent LSV and one competent SSV became incompetent after SS (Fig 4).

Morphology and function of preserved saphenous veins normalize in the most of cases with HL and SS. These preserved grafts might be used for future graft materials. Although close observation for recurrence of varicose veins is required in the cases with remained incompetent saphenous veins.

References
1. Jacobsen BH. The value of different forms of treatment for varicose veins. Br J Surg 1979;66:182-5

2. Koyano K, Sakaguchi S. Selective stripping operation based on Doppler ultrasonic findings for primary varicose veins of the lower extremities. Surgery 1988; 103:615-20
3. Rutherford RB, Sawyer JD, Jones DN. The fate of residual saphenous vein after partial removal or ligation. J Vasc Surg 1990;12:422-8

Fig. 4 Competence of LSV

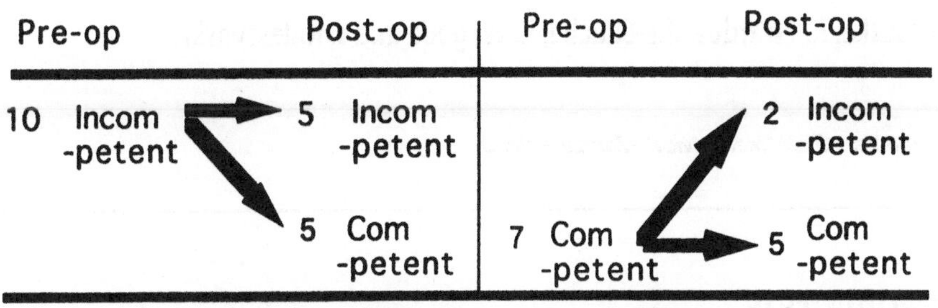

Pre-op	Post-op	Pre-op	Post-op
10 Incom-petent	5 Incom-petent		2 Incom-petent
	5 Com-petent	7 Com-petent	5 Com-petent

Fig. 5 Competence of SSV

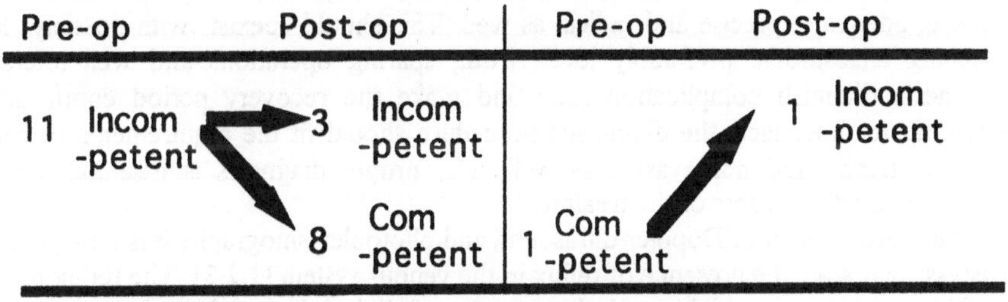

Pre-op	Post-op	Pre-op	Post-op
11 Incom-petent	3 Incom-petent		1 Incom-petent
	8 Com-petent	1 Com-petent	

Fig. 3 Change of LSV and SSV Diameter

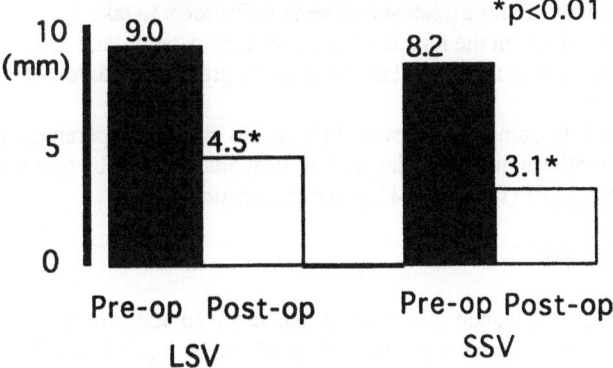

*p<0.01

10 (mm)

9.0   4.5*   8.2   3.1*

Pre-op Post-op   Pre-op Post-op
   LSV   SSV

Phlebology '95, D. Negus et al. (eds.). Phlebology (1995) Suppl. 1: 154-156

P099

Comparative Analysis of Venous Haemodynamic Parameters in Patients Undergoing Surgery due to Long Saphenous Vein Insufficiency

D. Czaczka, K. Twardowska-Saucha, Z. Wygoda and J. Kuleszynski

Varicose Vein Clinic "Medservice" - Zabrze, Poland

INTRODUCTION

Nowadays the surgical methods tend to be less invasive, functionally and cosmetically effective, adequate, precise and radical as well. This should coexist with possibly low burdening anaesthesia, preferably local. Both, sparing operations and well tolerable anaesthesia diminish complication ratio and make the recovery period comfortable. According to these facts the diagnostic procedure should fit the requirements of being adequate, handy and noninvasive as well. The proper diagnosis is essential for the accuracy and effectiveness of the treatment.

With the development of Doppler ultrasound and photopletysmography it is now possible to assess accurately the presence of reflux in the venous system [1,2,3]. The reflux has been shown to correlate highly with the presence of clinical disease. It is also possible to gain many useful information concerning the haemodynamic parameter disturbances which could have an effect on patient's condition. These simple techniques may permit the majority of physicians involved in the treatment of venous disorders to take advantage of the information given on the functional status of their patients, and help them to choose the proper treatment method. They are also of a great use in follow-up especially in outpatient clinic.

The purpose of the paper is to compare the lower limb venous circulation parameters before and at least 6 months after limited stripping and/or local phlebectomy by means of digital photopletysmography (D-PPG) and CW-Doppler examinations.

MATERIAL AND METHODS

The comparative study on haemodynamic parameters of patients who underwent surgery due to long saphenous vein insufficiency was undertaken. 63 patients (72 lower limbs) with LSV insufficiency ranging from 2nd to 4th degree according to Hach's, were surgically treated in Varicose Vein Clinic "Medservice" - Zabrze in one day procedure between November 93' and July 94'. 13 limbs presented 2nd degree , 35 - 3rd degree and 24 - 4th degree of insufficiency. 17 (19%) from the primary number of 89 legs were excluded from the study due to presence so called "non improvable" chronic venous insufficiency in preliminary D-PPG tourniquet test. This was recognised if difference among values of To before and after applying of the tourniquet at the site above and below the knee (i.e. in three measurements) differed one from another no more than 10% [2,4].

The interventions on day care condition concerned limited or complete stripping of the LSV, sometimes combined with: selective dissection of large, insufficient perforators, excisions of certain large tributaries (for example accessory saphenous vein), excision of large varicose plexes, ligation or stripping of the short saphenous vein (SSV), removal of saphenous or non-saphenous varices with atypical refluxes.

Taking detailed patient's history, careful physical, CW-Doppler and D-PPG examinations preceded every operation.

The anaesthesia was obtained by femoral block and infiltration of the local anaesthetics. All the interventions were followed by compression treatment in the period of 4 weeks.

Using continuous wave Doppler probe, emitted 4-8 MHz ultrasound we were able to obtain the information concerning:

1. Localisation and topography of: junctions, main superficial and deep trunks, superficial tributaries, perforating veins, even small thrombi in superficial and some bigger in deep venous system.
2. Refluxes, both - superficial and deep
3. Concomitant arterial insufficiency by means of ankle/arm index

The accuracy was ranging from 60% in case of deep veins to almost 95% in case of superficial venous system [5]. D-PPG combined with the tourniquet test determine whether vein function is improvable or not, that means whether it is possible to predict the improvement of venous system functions after elimination of varicose veins.

We observed two parameters mainly - venous refilling time (To) and venous pump work (Fo), and how they vary during tourniquet test.

The presence and localisation of the reflux in CW-Doppler was examined before and at least 6 months after surgery. Haemodynamic parameters - To and Fo were evaluated with a help of D-PPG in the same terms. These parameters gave us the answer to the questions concerning muscle pump functions and helped us to establish the indications for further sclerotheraphy or compression therapy.

Symptoms such as pain, leg heaviness and oedema were also noted down.

RESULTS

Fig. 1

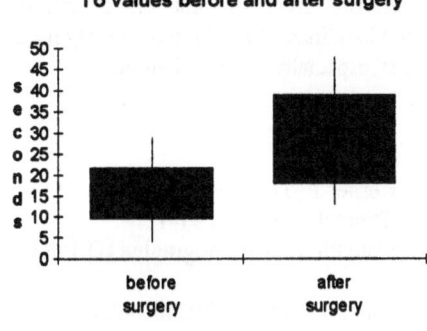

To values before and after surgery

In preoperative diagnostic To values ranged from 4s to 29s (mean:15.6s), while after operation To ranged from 13s to 48s (mean: 28.4s). Standard deviation: 10.6s. In 6 cases prolongation of To < 10%, in 15 cases from 10% to 30% and in 51 limbs more than 30% in comparision with the preoperative maesurement values were observed.

Fig. 2

Fo values before and after surgery

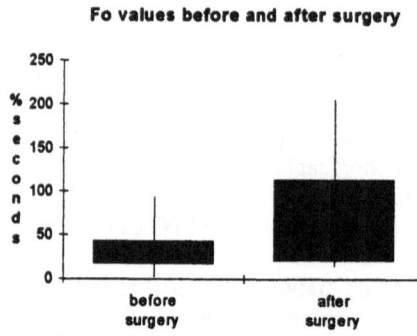

Before surgical treatment Fo values ranged from 2%s. to 94%s (mean: 29.8%s) while after surgery Fo ranged from 15%s to 206%s (mean: 66.9%s).
Standard deviation: 46.6%s.
In 1 limb Fo decreased; in 7 cases the increase <10%, in 16 cases from 10% to 30% and finally in 48 cases bigger than 30% versus the preliminary values were noticed.

Significant decrease or remission of pain were observed in 59 cases, the decrease or remission of oedema in 54 cases. Significant improvement of To (p<0.0001) and Fo (p<0.0023 by means of t-test) were observed. This was accompanied by distinct remission of the symptoms.

CONCLUSIONS

CW-Doppler and D-PPG together with careful physical examination are of a great value in follow-up the effectiveness of surgical treatment.
They reveal significant improvement of haemodynamic parameter values and reduction of reflux after adequate surgery in the majority of cases.
They are also relatively cheap and practically without any adverse effects but demand experience and attentiveness in interpretation.
  Almost 93% of our patients were classified to the surgery, compression sclerotherapy or conservative treatment basing on findings gained by means of these methods. Only 7% demanded more invasive diagnostic procedures to establish the diagnose and treatment plan. This was mainly due to coexistence of other severe diseases.
  We want to emphasise that CW-Doppler and D-PPG are only screening tests with the sensivity ranging from about 60% to 80% and specificity from 75% to 90%.
It should be emphasised that in D-PPG measurements it is important to differentiate the group of patients (about 20%) with so-called "non improvable" chronic venous insufficiency by means of the tourniquet test [4]. In this type of venous insufficiency the active treatment is almost without influence on measurable haemodynamic parameters. That is why we can not predict clinical disease and the effects of treatment exactly in every case. However, because of their simplicity and handiness they could be widely used to assess venous function and topography in patients, especially in outpatient unit.

REFERENCES

1. Blazek V., Schmitt H.J., Schultz-Ehrenburg U., Kerner J. "Digitale Photopletysmographie für die Beivenendiagnostik." Phlebol. u. Proctol. 1989; 18: 91-7
2. Kerner J., Schultz-Ehrenburg U., Blazek V. "Digitale Photopletysmographie (D-PPG) Phlebol. u. Proctol. 1989; 18; 98-103.
3. Grondin L. "Practical aspects of Doppler evaluation and D-PPG" Phlebol. '92; 642.
4. Streićek J. "Sclerotheraphy and digital photopletysmography." Phlebol. '92; 772.
5. Browse N.L. et al. *Book*: "Diseases of the Veins" London; Arnold E., 1988: 80-84.

Phlebology '95, D. Negus et al. (eds.). Phlebology (1995) Suppl. 1: 157-159

OP/21.2

Long Saphenous Vein Reflux in the Groin, the Incidence of "True" Saphenofemoral Incompetence

G.M. Somjen, J. Donlan and A.H. Johnston

The Vascular Centre and Vascular Surgery Unit, Mornington Peninsula Hospital, Frankston, Victoria, Australia

INTRODUCTION

Saphenofemoral incompetence is a frequent clinical diagnosis in patients with large saphenous varicosities.

Earlier studies on mostly large varicose veins, using either continuous wave Doppler or duplex ultrasound as a diagnostic modality, suggested that in the majority of cases the source of long saphenous vein incompetence was reflux from the deep venous system [1,2,3]. Using continuous wave Doppler in nearly 90% of patients with typical saphenous varicosities reflux was found at the saphenofemoral junction [3]. This finding was suggestive of saphenofemoral incompetence. Duplex ultrasound examination arrived at similar conclusions regarding the frequency of venous reflux from the deep veins [4].

More recently it has been shown that "true" saphenofemoral reflux may not be as common as it was suggested in the past [5,6].

The aim of this study was to further clarify the reflux patterns at the level of the saphenofemoral junction in patients with venous incompetence in the proximal long saphenous vein.

PATIENTS AND METHODS

Extremities with varicose veins were investigated. All legs were examined with the duplex scanner and 79 consecutive legs, with reflux in the proximal long saphenous vein were selected for further analysis. In all cases incompetence was present at or within 1 cm of the saphenofemoral junction.

Duplex scanning was performed with the patient standing, and cephalad venous flow was induced in the long saphenous vein by manual calf muscle compression. Manual compression was executed in a way that strong centripetal flow was detectable both in the long saphenous vein and in the common femoral vein on augmentation. Reflux was detected on release of calf compression. Reflux was considered to be present if reverse flow lasted longer than 0.5 second. Colour flow imaging was performed first, then reflux was tested by pulsed Doppler interrogation at selected levels in the vicinity of the saphenofemoral junction (Figure 1).

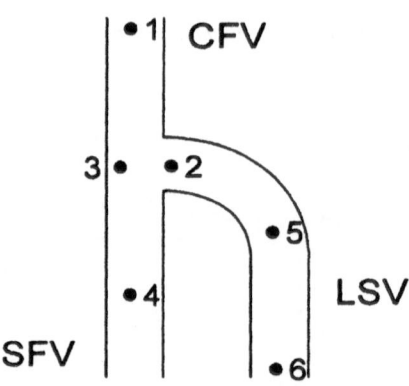

Figure 1. Doppler sampling sites in the femoral and long saphenous veins.
(CFV: common femoral vein; SFV: superficial femoral vein; LSV: long saphenous vein).

The venous flow and reflux on augmentation were documented by recording the Doppler wave forms. Duplex scanning was performed with a high frequency imaging transducer coupled to a real time colour duplex scanner (ATL Ultramark 9 HDI Scanner, Advanced Technology Laboratories, Bothell, Washington, L10-5 MHz linear array transducer).

RESULTS

Reflux pathways at the level of the saphenofemoral junction were examined in 79 legs, all of which displayed venous incompetence in the long saphenous vein.

In 33 instances (41.7%) reflux was detectable both in the long saphenous vein and the proximal common femoral vein, indicating "true" saphenofemoral incompetence. The femoral vein below the saphenofemoral junction was found competent in these cases.

In 43 extremities (54.4%) reflux was limited to the long saphenous vein only and the first saphenous valve remained competent. The most frequent source of reflux in legs without true saphenofemoral reflux appeared to be the venous network draining the lower abdominal wall (eg. superficial epigastric, superficial circumflex iliac and superficial external pudendal veins). These relatively large veins which joined the long saphenous vein just distal to its first valve were readily demonstrated by the duplex ultrasound.

In the remaining 3 legs (3.7%) extensive deep and superficial venous incompetence was the ultrasound diagnosis; reflux was present in the femoral vein above and below the junction with the incompetent long saphenous vein.

CONCLUSION

This study revealed that "true" saphenofemoral incompetence existed in less than half of the legs when reflux was present in the proximal long saphenous vein at the level of the saphenofemoral junction. Our findings further emphasise that "trans-fascial escape" (reflux from the deep veins) may not be a precondition of extensive long saphenous vein incompetence and related varicose veins.

REFERENCES

1. de Groot WP. Treatment of varicose veins: Modern concepts and methods. J Dermatol Surg Oncol 1989;15:191-98.

2. Schultz-Ehrenburg U, Hubner HJ. Reflux diagnosis with doppler ultrasound. Findings in angiology and phlebology 1989;35:1-47.

3. Goren G, Yellin AE. Primary varicose veins: topographic and haemodynamic correlations. J Cardiovasc Surg 1990;31:672-77.

4. Quigley FG, Raptis S, Cashman M and Faris B. Duplex ultrasound mapping of sites of deep to superficial incompetence in primary varicose veins. Aust NZJ Surg 1992;62:276-78.

5. van Bemmelen PS, Sumner DS. Laboratory evaluation of varicose veins In: Bergan JJ, Goldman MP, eds. Varicose veins and telangiectasias, diagnosis and treatment. St Louis, Missouri: Quality Medical Publishing Inc, 1993: 73-84

6. Somjen GM. Anatomy of the superficial venous system. J Dermatol Surg Oncol 1995;21:1-11

Phlebology '95, D. Negus et al. (eds.). Phlebology (1995) Suppl. 1: 160-163

P023

Classification des sur 607 Echo Marquages: Aspect Evolutif de La Maladie Variqueuse

D. Creton

Espace Chirurgial Ambroise Paré, rue Ambroise Paré 54100 F. Nancy

Classification of Varices Following Doppler Mapping in 607 Patients: Natural History of Varicosis

SUMMARY

The author describes a survey of 607 pre-operative mappings of varices by echo-Doppler: in only 48% of patients was stripping of the long saphenous vein justified; in 27% it was entirely normal and in 25% it was normal in parts.

CLASSIFICATION DES VARICES SUR 607 ECHO MARQUAGES : ASPECT EVOLUTIF DE LA MALADIE VARIQUEUSE

D. CRETON

Espace Chirurgical Ambroise Paré, rue Ambroise Paré 54100 F. NANCY

INTRODUCTION

La pratique du marquage préopératoire nous montre qu'il existe une grande variété de types de varices des membres. Cette variété augmente avec la précision des examens hémodynamiques préopératoires. Il nous a semblé intéressant de classer ces différentes variétés de varices et de les comparer aux tranches d'âge des patients pour essayer de comprendre l'évolution de la maladie variqueuse.

MATERIEL ET METHODE.

En 1994 nous avons étudié 607 cartographies préopératoires consécutives réalisées avec un écho doppler ESAOTE AU 530 sonde 7,5 Mz 10 Mz pour des patients porteurs de varices essentielles . Durant cette période 740 interventions de varices ont été réalisées, dont 133 interventions pour récidives variqueuses, exclues de cette étude.

RESULTAT

Nous avons défini plusieurs types anatomiques de varices correspondant à plusieurs types hémodynamiques de disfonctionnement.
- 272 reflux intéressant toute la saphène interne de la crosse à la région malléolaire soit 45 % notés SIL .
- 87 reflux intéressant la saphène interne proximale de la crosse jusqu'à la bifurcation jambière haute, région des perforantes de Boyd, respectant les saphènes jambières soit 14 % notés SIC.
- 47 reflux intéressant la saphène externe soit 7,7 % notés SE.
- 20 reflux associés intéressant la saphène externe et la saphène interne en tout ou partie soit 3,3 %.
- 5 reflux sur une veine perforante fosse poplitée isolée soit 0,8 % notés VFP.
- 23 reflux isolés sur une branche saphène essentiellement saphène interne avec une fonction saphène interne normale soit 3,7 % notés BSI .
- 64 varices isolées alimentées par des perforantes, par des veines génito crurales ou sous cutanées abdominales ne concernant pas la saphène interne soit 10 % notés Vi.
- 21 reflux isolés sur une branche de la crosse saphène interne 3,4 % notés C/2 (20 sur une branche antéro externe et une sur une branche postéro interne) reflux respectant toujours la continence de la valve préostiale saphène et parfois la valve ostiale saphène.
- 29 reflux courts intéressant une demi saphène haute proximale de la crosse jusqu'à mi cuisse et se déversant sur une branche saphène souvent antéro externe soit 4,7 % notés S/2 H.

162

- 21 reflux courts intéressant une demi saphène basse distale alimentée par une perforante de cuisse ou par des veines honteuses externes respectant la crosse et la terminaison saphène soit 3,4 % notés S/2 B.
- 13 reflux courts intéressant une demi saphène moyenne respectant la partie terminale saphène et la partie distale jambière soit 2,1 % notés S/2 M.

Si l'âge des patients à la date de l'intervention ne correspond pas à la date d'apparition des varices, c'est un mode de comparaison efficace pour apprécier l'âge d'apparition de ces différentes sortes de varices. La courbe en pourcentage cumulé des patients, en fonction des tranches d'âge au jour de l'intervention dans chaque catégorie de varices, montre que les varices ne concernant pas la saphène interne en totalité sont opérées chez des patients plus jeunes alors que les reflux complets saphène interne concernent des patients opérés plus tard. Dans la tranche d'âge 30-40 ans on retrouve :
- 25 % des reflux saphène interne long SIL
- 30 % des reflux saphène interne court SIC
- 33 % des reflux partiels saphène S/2
- 37 % des reflux partiels sur la crosse C/2
- 48 % des reflux sur une veine isolée Vi
- 58 % de reflux sur une branche saphène BS
Environ moitié de la pathologie non saphène est déjà opéré entre 30 et 40 ans pour seulement 1/3 de la pathologie saphène interne.

- 50 % des reflux d'une branche saphène BS sont opérés à 33 ans
- 50 % des reflux d'une veine isolée Vi sont opérés à 36 ans
- 50 % des reflux partiels de la crosse C/2 sont opérés à 40,5 ans
- 50 % des reflux partiels saphène S/2 sont opérés à 42 ans
- 50 % des reflux courts saphène SIC sont opérés à 42,5 ans
- 50 % des reflux complets saphène SIL sont opérés à 44 ans

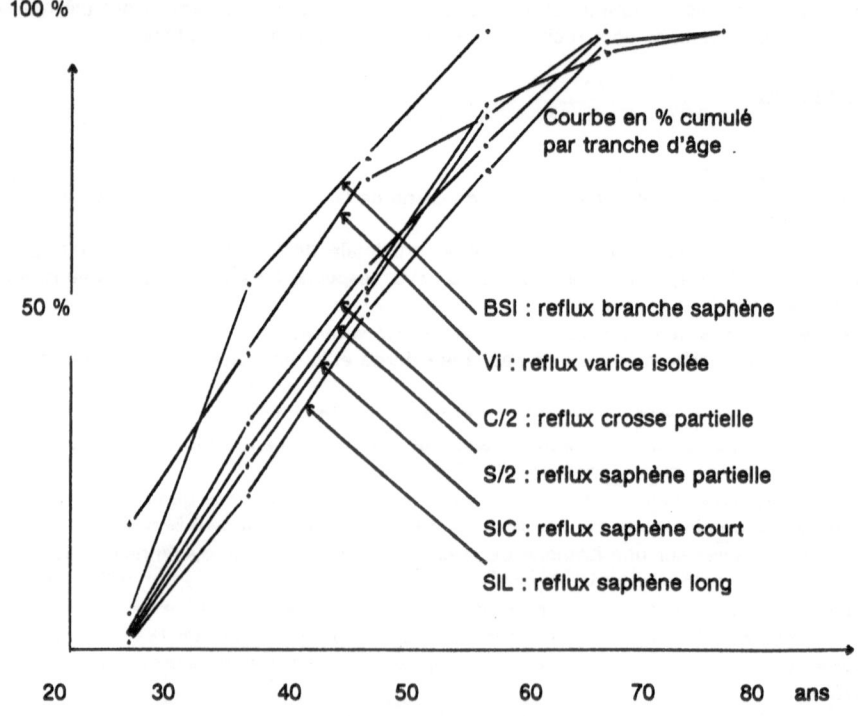

COMMENTAIRE

- Cette étude montre que la saphène interne présente un reflux complet uniquement dans 48 % des cas ; dans 27 % des cas elle n'est pas du tout concernée par la maladie variqueuse , et dans 25 % des cas elle ne reflue que sur une partie.
Cela signifie qu'une intervention conservatrice saphène est possible presque une fois sur deux. Sur les 150 cas de reflux partiels saphène interne : les 20 derniers cm de saphène proximale sont sains dans 22 % des cas , dans 20 % des cas la saphène est saine entre le 20 derniers centimètres et la partie haute de jambe et dans 86 % des cas elle est saine au niveau jambier.

- L'étude des pourcentages cumulés montre très nettement que les varices opérées les plus tôt sont celles qui n'intéressent pas la saphène interne. La maladie variqueuse semble débuter par une pathologie des branches saphène qui sont opérées en moyenne 10 ans avant la maladie saphène complète. Tab I.
Le reflux des branches saphène semble apparaître avant le reflux du tronc saphène, lui-même. Cette hypothèse corrobore un phénomène souvent rencontré : lorsqu'il existe un reflux sur une branche saphène, la saphène d'aval a souvent un diamètre supérieur à la saphène d'amont de quelques millimètres [1] et elle présente souvent un reflux long de faible vitesse ; la supression de la branche saphène refluante normalise le diamètre de la saphène d'aval et fait disparaître le reflux. Cela justifie l'intérêt de traiter tôt ces modes de début de la maladie variqueuse pour préserver la fonction saphène interne . L'idée qu'un reflux dans une branche puisse détériorer la continence du tronc correspondant a été développée par Somjen [2] au niveau de la jonction saphéno poplitée.
La pathologie variqueuse isolée semble apparaître environ 7 ans avant la pathologie saphène, et parmi les patients présentant une maladie saphène, sur une période d'environ 5 ans on retrouve la même graduation année par année, avec au début le reflux partiel sur la crosse, puis le reflux partiel saphène , puis le reflux court et enfin le reflux complet sur la saphène correspondant au stade ultime de la maladie variqueuse touchant tout l'arbre saphène, tronc et branches.

CONCLUSION

L'écho marquage préopératoire est fondamental pour différencier plusieurs catégories de varicoses.
Le stripping saphène interne long ne se justifie que dans 48 % des cas. Dans 27 % des cas la saphène est entièrement normale et dans 25 % des cas elle est saine en partie.
Les varices isolées et surtout les reflux isolés sur une branche saphène respectant la fonction saphène semblent apparaître 10 ans avant le reflux tronculaire saphène, ce qui laisse penser que la pathologie de l'arbre saphènien débute par une pathologie des branches avant de toucher le tronc. L'ablation de ces branches saphène pourrait être un traitement préventif de la pathologie du tronc saphène.

REFERENCES

1 VIDAL - MICHEL JP., BOURREL Y. , ENSALLEM J., BONERANDI JI. Aspect chirurgical des crosses saphènes internes modérément incontinentes par " effet siphon" chez le patient variqueux. Phlébologie 1993; 46 : 143-147

2 SOMJEN GM., ROYLE JP., FELLE G., ROBERTS AK., HOARE MC., TONG Y. Venous reflux patterns in the popliteal fossa. J Cardiovasc Surg 1992 ; 33 : 85-91

Phlebology '95, D. Negus et al. (eds.). Phlebology (1995) Suppl. 1: 164-168

PI/4.13

Hypothèses Etiologiques des Récidives Variqueuses Saphène Interne: Etude Anatomique Sur 211 Cas

D. Creton

Espace Chirurgial Ambroise Paré, rue Ambroise Paré 54100 F. Nancy

Hypotheses on the Aetiology of Recurrences of Varices of the Long Saphenous System: Anatomical Study of 211 Patients

SUMMARY

The author reports on an echo-Doppler study in 211 patients with post-operative recurrent varicose veins with reflux at the groin. The frequency of a residual long saphenous incompetence in recurrent varicose veins is stressed.

HYPOTHESES ETIOLOGIQUES DES RECIDIVES VARIQUEUSES SAPHENE INTERNE : ETUDE ANATOMIQUE SUR 211 CAS.

D. CRETON

Espace Chirurgical Ambroise Paré, rue Ambroise Paré, 54100   F. NANCY

INTRODUCTION
Le traitement chirurgical du reflux saphène interne s'accompagne d'un pourcentage important de récidives évalué à 20 % 25 % avec des extrêmes allant de 7 % [1] à 65 %[2]. Leur traitement chirurgical représente environ 20 % [3] des interventions de varices . L'étude écho doppler préopératoire et l'étude anatomique peropératoire de nos réinterventions inguinales pour récidives saphène interne nous a permis d'émettre une hypothèse hémodynamique de la récidive saphène interne.

MATERIEL ET METHODE
Entre 1992 et 1994 sur 2149 exérèses variqueuses réalisées sur 1889 patients, nous avons opéré 419 récidives variqueuses, (19,4 % de notre activité). Parmi celles-ci, nous avons uniquement étudié les récidives en territoire saphène interne comportant une reprise de l'ancienne crossectomie pour reflux inguinal : soit 211 cas.
Nous avons étudié le type anatomique d'alimentation du **reflux haut résiduel** et son importance hémodynamique.
1°) jonction fémoro variqueuse importante , notée C+.
Parmi les reflux résiduels issus d'un ancien moignon de crossectomie nous avons appelé par ordre décroissant en fonction de la taille de la jonction et de l'importance du reflux
C+ 1 les crosses saphènes internes intactes
C+ 2 les jonctions par une branche unique résiduelle de l'ancienne crossectomie .
C+ 3 les jonctions par une branche bifurquée résiduelle de l'ancienne crossectomie .
C+ 4 les jonctions par une branche trifurquée résiduelle
C+ 5 les jonctions par une branche quadrifurquée résiduelle.
C+ 6 les reperméabilisations de la crosse sur mauvaise ligature de la crosse.
C+ 9 les reflux par des perforantes de cuisse situés à moins de 15 cm du moignon.
C+ 10 les reflux par une volumineuse branche juxta inguinale (double crosse).
2°) jonction fémoro variqueuse mineure, néoangiogénèse sur le moignon fémoral noté :
C0 7 pour l'angiogénèse inguinale sur le moignon fémoral.
C0 8 pour le reflux alimenté par les veines honteuses externes
C0 11 pour le reflux alimenté par des veines sous cutanées abdominales,

Nous avons étudié le type de **varice résiduelle à la cuisse** .
1°) saphène résiduelle refluante d'un diamètre supérieur à 3 mm et d'une longueur supérieure à 15 cm (S+). Nous avons noté.
S+ 1 quand elle se raccorde directement avec la source de reflux sur la veine fémorale.
S+ 2 quand il existe une néojonction saphéno fémorale de 2 à 10 cm.
S+ 3 quand elle est isolée plus basse sans communication visible avec la voie profonde,
S+ 4 quand elle est alimentée par une veine honteuse externe ,
S+ 5 quand elle est alimentée par une perforante de cuisse.
2°) varice diffuse non systématisée notée S0.

L'importance de la récidive, appréciée globalement sur sa tolérance clinique est jugée par le délai en année , entre la première intervention et la réintervention . Afin de comprendre le rôle pathologique des différents critères anatomiques de la récidive et leur importance hémodynamique, nous avons étudié les corrélations entre ces différents facteurs.

RESULTAT

L'origine du reflux   Tab I
- L'exploration de l'ancienne crossectomie a permis de retrouver 137 reflux importants : 35 crosses intactes (C+ 1) dont 19 en continuité avec la saphène interne. 52 % sont des reflux sur crosse intacte ou sur volumineuse branche unique résiduelle, les autres sont des reflux retrouvés sur la branche basse de bifurcation du tronc résiduel.Un peu moins de la moitié de ces reflux communiquent avec une saphène résiduelle.
- 15 perforantes hautes de cuisse (C+ 9) (C+ 10) sont responsables du reflux dont une volumineuse (C+10) située à 5 cm en-dessous du moignon de crossectomie, correspond à l'abouchement d'une deuxième saphène méconnue lors de la première intervention.
- 47 fois un reflux mineur est constaté : néoangiogénèse ou néojonction fémoro variqueuse microscopique (C0 7). Aucun moignon de ce type n'est en communication avec la saphène résiduelle.
- 12 fois un reflux mineur est issu de varices génito crurales (C0 8)
- 1   reflux mineur (C0 11) vient de branches sous cutanées abdominales.

Le type de saphène résiduelle   Tab I
- Dans 93 cas nous avons retrouvé des varices diffuses (S0) sans tronc saphène au niveau de la cuisse : dans 55 % des cas elles sont associées à un reflux majeur sur jonction fémoro variqueuse résiduelle (C+) et dans 31 % elles sont associées à un reflux mineur (C0 7) de type néojonction fémoro variqueuse microscopique.
- Dans 118 cas , la saphène résiduelle a un diamètre de 3 à 15 mm et une longueur de 15 à 70 cm. 60 saphènes (S+ 1) communiquent avec la fémorale par l'intermédiaire d'une branche résiduelle de la crosse.
- 16 saphènes (S+ 2) communiquent avec la veine fémorale par l'intermédiaire de macro vaisseaux tortueux de 2 à 10 cm assurant la jonction entre la saphène et la veine fémorale de deux façons :
8 fois par un reflux majeur sur une branche résiduelle (C+)
8 fois par un reflux mineur de type angiogénétique (C0 7)
- 25 saphènes (S+ 3) souvent plus basses sur la cuisse ne communiquent avec aucune source de reflux. Aucun contact n'est visible entre la saphène résiduelle et la veine fémorale au niveau de l'ancienne crossectomie.16 s'accompagnent d'un reflux majeur (C+) et 9 d'un reflux mineur (C0 7).
- 10 saphènes sont alimentées par des varices honteuses externes génitocrurales ,
- 7 sont alimentées par une perforante haute de cuisse.

Corrélation entre type de reflux et type de saphène   Tab II
Les associations retrouvées sont par ordre de fréquence .
- reflux résiduel + saphène résiduelle C+ / S+ 85 (43 %)
- reflux résiduel + varice diffuse C + / S 0 52 (26%)
- absence de reflux inguinal + varice diffuse C0 / S0 34 (17 %)
- absence de reflux inguinal + saphène résiduelle C 0 / S+ 26 (13 %)
Préférentiellement   C+ sera associé à S+ (62 %) et   C0 à S0 (57 %).

Corrélation entre le type de récidive et le délai de réintervention   Tab III Tab IV
L'intervalle entre l'intervention et la récidive varie de 1 à 41 ans, l'analyse des courbes en fréquence cumulées des patients réopérés montre que celles-ci sont semblables pour les différents paramètres de la récidive avec un décalage de temps.
50% des patients présentant un type C0 S+ sont réopérés 9 ans après et 75% 15ans après.
50%des patients présentant un type C+ S+ sont réopérés 10 ans aprèset 75%20 ans après.
50% des patients présentant un typeC0 S0 sont réopérés 11ans après et 75% 19ans après.

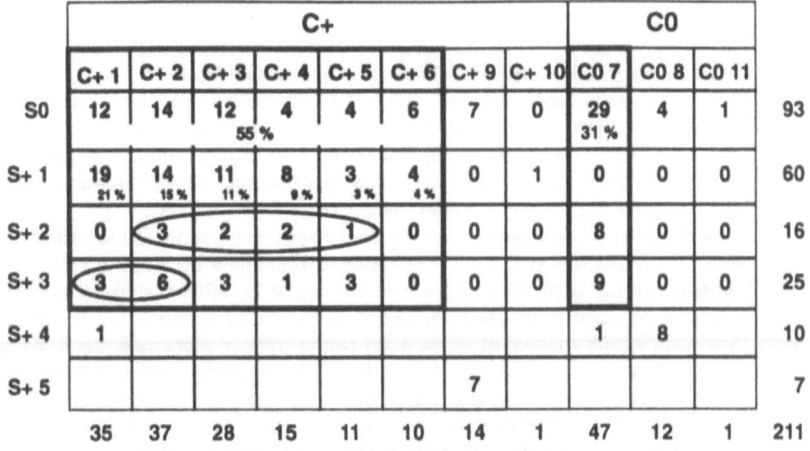

	C+								CO			
	C+ 1	C+ 2	C+ 3	C+ 4	C+ 5	C+ 6	C+ 9	C+ 10	CO 7	CO 8	CO 11	
S0	12	14	12 55 %	4	4	6	7	0	29 31 %	4	1	93
S+ 1	19 21 %	14 15 %	11 11 %	8 9 %	3 3 %	4 4 %	0	1	0	0	0	60
S+ 2	0	3	2	2	1	0	0	0	8	0	0	16
S+ 3	3	6	3	1	3	0	0	0	9	0	0	25
S+ 4	1								1	8		10
S+ 5							7					7
	35	37	28	15	11	10	14	1	47	12	1	211

**Tab I - Corrélation entre le type de reflux inguinal C
et le type de saphène résiduelle S**

	S+	S0	
C+	85	52	137
CO	26	34	60
	111	86	

	S+	S0	
C+	62 %	38 %	100 %
CO	43 %	57 %	100 %

	S+	S0
C+	76 %	60 %
CO	24 %	40 %
	100 %	100 %

**Tab II - Corrélation entre le type de reflux inguinal
résiduel et le type de saphène résiduelle**

C+ = C+ 1 2 3 4 5 6 10
CO = CO 7 8 11
S+ = S+ 1 2 3 4

Type de récidive	% cumulé 50 %	75 %	Type de récidive	% cumulé 50 %	75 %
C+	12	20	CO	10	17
C+ S+	10	20	CO S+	9	15
C+ S0	13	19	CO S0	11	19
S+	10	20	S0	13	19
C+ S+	10	20	C+ S0	13	19
CO S+	9	15	CO S0	11	19
Années			Années		

**Tab III - Intervalle entre la première
intervention et la réintervention
pour récidive**

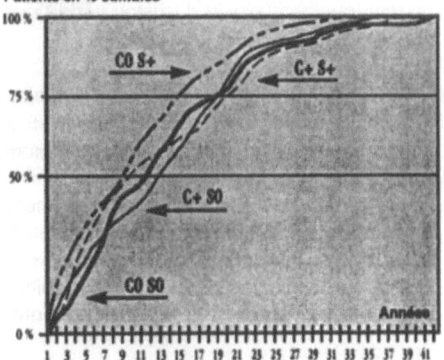

**Tab IV - Intervalle entre la première
intervention et la réintervention
pour récidive**

50% des patients présentant un type C+ S0 sont réopérés 13ans après et 75% 19ans après

DISCUSSION

De cette étude se dégage l'idée qu'il existe deux types de récidives . 1 °) des récidives de type hémodynamique où le reflux inguinal résiduel et la saphène résiduelle acceptrice du reflux sont intriqués. 2°) Des récidives de type angiogénétique où les reflux sont mineurs. En effet , la persistance d'un reflux inguinal important s'accompagne d'une saphène acceptrice du reflux dans 62 % des cas, et la persistance d'une saphène résiduelle refluante s'accompagne d'un reflux inguinal majeur dans 76 % des cas.

Dans les récidives de type hémodynamique les deux hypothèses sont : ou le reflux résiduel inguinal développe , entretien et détériore la saphène résiduelle ou c'est le reflux dans la saphène résiduelle qui par aspiration aggrave et développe le reflux résiduel inguinal .

Parmi les 41 saphènes résiduelles sans connexion directe avec le reflux (S+ 2, S+ 3) 58 % sont associées à un reflux majeur et 41 % à un reflux mineur alors que dans les seules néojonctions fémoro saphènes macroscopiques (S+ 2) la proportion est la même . D'autre part dans les reflux résiduels majeurs il y a deux fois moins de néojonctions fémoro saphènes que de saphènes isolées 6 % pour 12 %, ce qui montre que les phénomènes d'angiogénèse sont indépendants des phénomènes de pression. Si la saphène a été enlevée complètement, la récidive est alimentée dans plus de la moitié des cas par un reflux inguinal important 55%, et dans 31% des cas par un reflux mineur (C0 7) Tab I. Dans les saphènes résiduelles branchées directement sur le reflux inguinal, les reflux résiduels majeurs sont plus fréquents. 21 % sont en connexion avec des crosses intactes, et 4 % avec des branches de faible débit. Les saphènes résiduelles développant une néojonction fémoro saphène macroscopique (S+ 2) sont en connexion avec des reflux inguinaux moyens , alors que les saphènes résiduelles isolées (S+ 3) sont associées à des reflux majeurs importants , ce qui montre que l'angiogénèse est un phénomène qui se développe en dehors des effets de pression.

Si l'étude de l'intervalle entre les deux interventions ne traduit pas le délai d'apparition de la récidive , cet intervalle est un élément objectif de comparaison des différents types de récidives. Il faut , par ailleurs, tenir compte de la moins bonne tolérance clinique de certaines formes de récidives.Si, sur un reflux résiduel inguinal on laisse en plus une saphène , on raccourcit le délai de reprise de 3 ans . Si malgré une bonne crossectomie on laisse une saphène on raccourcit la reprise de 2 ans, cela souligne le rôle important de la saphène résiduelle dans la récidive. Par contre, curieusement, laisser un reflux inguinal en plus d'une saphène résiduelle rallonge plutôt le délai de réintervention, et si sur une saphénectomie complète on réalise une crossectomie complète le délai raccoucit de 2 ans. Cette discordance s'explique par le fait que ces deux derniers types de récidives sans saphène avec reflux mineur sont de type angiogénétique et certainement moins bien tolérés cliniquement par la diffusion variqueuse inesthétique et souvent douloureuse.Tab IV

CONCLUSION

Cette étude permet de montrer le rôle majeur de la saphène interne résiduelle dans la pathogénie de la récidive, rôle important où sont intriqués les phénomènes de réservoirs accepteurs du reflux sus jacent et des phénomènes d'aspiration ascendante de reflux. Ces récidives de type hémodynamique sont souvent issues d'une faute chirurgicale, elles sont étonnament bien tolérées dans le temps et logiquement de traitement chirurgical.

L'angiogénèse ou néojonction fémoro saphène microscopique ne semble pas influencée par les effets de pression, c'est un phénomène de cicatrisation indépendant des pressions de reflux, ces types de récidives sont moins bien tolérées cliniquement et le traitement chirurgical devrait s'effectuer dans des conditions maximum d'atraumatisme.

REFERENCES

1 RILVINS S. The surgical cure of primary varicose vein. Br. J. Surg. 1975 ; 62 : 913-17

2 ROYLE JP. Recurrent varicose vein . World J. Surg . 1986 ; 10 : 944-53

3 DAVIES GC. The Lothian surgical audit . Medical Audit News 1991; 1 : 26-7

Phlebology '95, D. Negus et al. (eds.). Phlebology (1995) Suppl. 1: 169-171

OP/18.6

Telangiectasias: Incidence, Classification, and Relationship with the Superficial and Deep Venous Systems: a Double-Blind Study

Pauline Raymond-Martimbeau, MD[1] and J.L. Dupuis, MD[2]

[1] The Dallas Non-Invasive Vascular Laboratory, Dallas, Texas, USA
[2] St. Joseph Hospital, La Malbaie, Quebec, Canada

INTRODUCTION

Telangiectasias are a difficult clinical entity, from both an etiologic and a diagnostic standpoint. Recently, noninvasive diagnostic imaging modalities, particularly B-scan duplex ultrasonography with or without color, have allowed researchers to gain better insight into telangiectasias, including their underlying reflux points and degrees of reflux. In an earlier study, it was shown that apparently idiopathic telangiectasias connect with underlying venous systems, and a system was proposed for classifying these lesions [1]. The present study was undertaken to expand upon preliminary findings with respect to the incidence and classification of telangiectasias, as well as their relationship with the superficial and deep venous systems.

PATIENTS, MATERIALS, AND METHODS

This double-blind study involved 525 consecutive patients, including 506 women and 19 men, with a mean age of 50.2 years (range, 18 to 80 years). All patients were seen between January 1993 and January 1995. Four hundred fifty-two patients were Caucasian, 52 were Hispanic, and 21 were African American. All cases involved a diagnosis of telangiectasia without underlying incompetence of saphenous veins and/or perforators (diameter >2 mm). No patient had a history of previous sclerotherapy. A total of 884 telangiectatic zones were studied. The patients had the following symptoms and signs in the affected extremity: a burning sensation (416/525; 80%), heaviness (319/525; 61%), throbbing (246/525; 47%), fatigue (238/525; 45%), pain (72/525; 14%), edema (44/525; 8%), and skin changes (23/525; 4%). Medications included hormones (254/525; 48%), diuretics (24/525; 5%), antihypertensive drugs (18/525; 3%), antiinflammatory agents (14/525; 2.6%), antidiabetic drugs (4/525; 0.8%), and thyroid agents (4/525; 0.8%).

All patients were examined in the standing and reclining positions. The telangiectasia zones were evaluated by means of inspection, palpation, percussion (Schwartz maneuver), Doppler ultrasound (with a 5-MHz probe), photoplethysmography, duplex ultrasound (gray-scale, with a 7.5-MHz probe), and color Doppler ultrasound as needed. The degree of reflux (mild, moderate, or severe) was determined at each site. The telangiectasia zone was then classified according to a modified version of the previously reported Dupuis/Raymond-Martimbeau classification [2], based on origin of the point of reflux. This scheme includes 5 patterns: in <u>Pattern I</u>, clustered purple telangiectasias join with a venulectasia and the deep venous system; in <u>Pattern II</u>, clustered purple telangiectasias join with a venulectasia and the saphenous system; in <u>Pattern III</u>, clustered purple or red telangiectasias link 2 venulectasias, 1 afferent and 1 efferent, each joining a reticular vein; in <u>Pattern IV</u>, clustered purple or red telangiectasias link to a single venulectasia, which joins a reticular vein; and in <u>Pattern V</u>, red telangiectasias are apparently isolated, without connection to venulectasias and reticular veins. Reflux is severe in Patterns I and II, moderate in Pattern III, mild in Pattern IV, and nonexistent in Pattern V.

All telangiectasias underwent sclerotherapy, performed with the patient lying down. Each pattern was injected with either 1) iodine sodium iodide (ISI) diluted with 25% dextrose combined with 10% sodium chloride (D-HS)* or 2) sodium tetradecyl sulfate (STS)** diluted with 0.9% sodium chloride, according to a protocol described elsewhere [1]. If more than 1 lesion was present in a single patient, the lesions were treated sequentially, starting with Pattern I and ending with Pattern V. Each injection was made at the point of reflux. A selective dressing was then applied at the injection site. For Patterns I and II, multiple cotton balls were applied with 3-inch nonallergenic tape and worn for 24 hours; for Patterns III to V, a cotton ball was applied with 1/2-inch nonallergenic tape and worn for 2 hours. For day wear (duration, 1 to 2 weeks), a graduated compressive stocking, exerting <20 mmHg of pressure, was prescribed in asymptomatic patients with Pattern III to V lesions. A similar stocking, exerting >20 mmHg of pressure, was prescribed for symptomatic patients and those with Pattern I or II lesions. Immediate ambulation was prescribed, and patients were advised to engage in low-impact exercise.

Postsclerotherapeutic assessment was performed to elucidate venous and perivenous tissue changes. Complications were noted. If occlusion of the point of reflux was not induced by the initial injection, patients underwent further treatment, according to a predetermined injection protocol, at 7- to 30-day intervals. After the completion of treatment, all patients underwent follow-up assessment at 3, 6, 12, and 24 months.

---

*Pharmacy preparations
**Wyeth Laboratories, Philadelphia, Pennsylvania, USA

RESULTS

Presclerotherapeutic assessment revealed the following telangiectatic patterns: Pattern I, 78 cases (8.8%); Pattern II, 111 cases (12.6%); Pattern III, 233 cases (26.4%); Pattern IV, 396 cases (44.8%); and Pattern V, 66 cases (7.4%). Compared with the overall study population, whose mean age was 50.2 years, patients with Pattern I and II lesions were older, and those with Pattern III to V lesions were younger.

Postsclerotherapeutic assessment with duplex ultrasound, revealed successful occlusion of the point of reflux and spontaneous disappearance of the telangiectatic zones after a single treatment in 64 (82%) of the Pattern-I cases, 96 (86.5%) of the Pattern-II cases, 205 (88.2%) of the Pattern-III cases, and 348 (87.9%) of the Pattern-IV cases. In the remaining cases, successful occlusion was achieved with 1 or 2 additional treatments.

Within 2 years after sclerotherapy, recurrent disease was seen at 32 (3.6%) of the 884 sites.

DISCUSSION

For many decades, telangiectasias without underlying venous incompetence appeared to be idiopathic. This double-blind study shows that, in a significant number of such cases, hypertension originating in the deep venous and/or saphenous system produces precisely located points of reflux that are detectable subcutaneously and manifested by telangiectasias visible on the skin surface. It also shows that, by eradicating these specific points of reflux, one can control the disease and stop its progression, with minimal side effects. Because Patterns I and II tend to occur in patients older than 50 years of age who are taking concomitant medications, these patterns may be related to drug intake, particularly hormone replacement therapy, or the aging process.

Further studies should be undertaken to evaluate the relationship between telangiectatic patterns and other variables such as skin complexion and texture, eye color, hair color, and lifestyle. Studies should also be performed with other sclerosants.

CONCLUSION

Telangiectasias connect with the deep and saphenous venous systems in a significant percentage of cases. Far from being simple cosmetic problems, they can produce signs and symptoms of reflux from underlying venous systems in the absence of apparent incompetence.

REFERENCE

1. Raymond-Martimbeau P. Sclerotherapy of superficial varicose veins. Can J Dermatol 1994; 6:602-11.
2. Raymond-Martimbeau P, Dupuis JL. Telangiectasias. Proceedings, North American Society of Phlebology Meeting, Maui, Hawaii, February 1994.

Phlebology '95, D. Negus et al. (eds.). Phlebology (1995) Suppl. 1: 172-174

OP/21.5

Artificial Computer Neural Networks for the Assessment of the Results of Venous Calf Air Plethysmography

C. Solomon[1], N.K. Kassabov[2], M. Bailey[2], S.F. Greig[1] and A.M. van Rij[1]

[1] Department of Surgery and [2] Department of Information Science,
University of Otago, Dunedin, New Zealand

INTRODUCTION

The high cost of venous ulceration in both social and economic terms has highlighted the need for accurate diagnosis and early appropriate treatment or preferably detection of the at-risk limb and ulcer prevention. Until fairly recent times diagnostic procedures have tended to be either invasive (direct venous pressure measurement or venography) or limited as in the case of photoplethysmography. Duplex scanning has markedly influenced the accuracy of the anatomical location and description of venous pathology, while air plethysmography (APG) now provides accurate measures not only of venous filling but also of calf muscle pump function and venous outflow. The extent to which APG correlates with direct AVP recordings is controversial. There is a large overlap of variables between classes of chronic venous insufficiency (CVI) which limits the usefulness of single variables for either assessing the risk of venous ulceration or for measuring the adequacy of treatment. The question remains whether the many variables assessed by APG provides any real advantage over single variables such as venous filling time, which can be assessed by simpler and quicker methods.

Artificial neural networks are powerful computer tools that can detect important interactions of a given set of variables which effect outcome and which are not obvious on usual statistical testing [1]. Model neurones were first used in the early 1940's and since the 1980's have found their way into many applications in biology, psychology, industry and more recently in medicine and vascular laboratory assessments [2]. This study reports our experience of applying neural network techniques to the analysis of the results of 1,500 APG assessments for lower limb venous disease.

METHODS

Artificial neural networks

Artificial neural networks are comprised of a number of input variables (nodes) and a number of predetermined output options (nodes). Neural networks can be created with or without a hidden layer (Figure 1). Networks without a hidden layer do not take into account interaction of variables and produce a result equivalent to multiple logistic regression.

Configuring and training the networks

Two sets of neural networks (with and without hidden layers) were trained, on a subgroup of data records (1,200), to recognise APG variable patterns on the basis of class of CVI. Input variables were venous filling index (VFI) venous filling time (VFT) residual volume

(RV) residual fraction (RF) ejection volume (EV) ejection fraction (EF) VFT/RF ratio and patient age. The first set were trained to recognise and classify limbs according to three outputs: normal limbs (class 0), limbs with uncomplicated varicose veins (class 1) and limbs with skin changes or ulceration (classes 2 and 3). The second set were trained only to distinguish between normal limbs (class 0) and those with skin changes (classes 2 and 3).

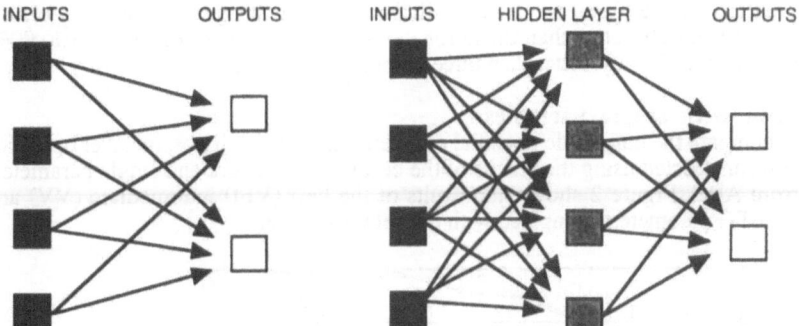

Figure 1: Schematic representation of artificial neural networks with and without hidden layers

Testing the trained networks
The trained networks were subsequently tested on the remaining set of records not used for training. The networks classify each test limb by a probability score of belonging to each output node. Probabilities between 0.4 and 0.6 were considered ambiguous. Based on the number of correct classifications, the sensitivity, specificity, positive predicative value and negative predicative value for ulcerative disease of each network was assessed. Receiver operating characteristic curves were drawn to compare the usefulness of using a neural network with a hidden layer (combined variables assessed by APG with their associated interactions) to the use of single variables.

RESULTS

Three output node networks
These networks were trained to recognise and classify limbs into three groups. Normal limbs (class 0), uncomplicated varicose veins (class 1) and limbs with skin changes or ulceration (class 2 or 3). The results of testing networks without and with a hidden layer are shown in tables 1 and 2 respectively.

Table 1: Results of testing a three output node neural network without a hidden layer.
Bold indicates correct classification.

Network classification	Actual classification		
	normal (class 0)	varicose veins (class 1)	skin changes (class 2 or 3)
normal	**65%**	15%	10%
varicose	20%	**40%**	15%
skin changes	5%	40%	**70%**
ambiguous	10%	5%	5%

Table 2: Results of testing a three output node neural network with a hidden layer.
Bold indicates correct classification.

Network classification	Actual classification		
	normal (class 0)	varicose veins (class 1)	skin changes (class 2 or 3)
normal	**75%**	50%	10%
varicose	15%	**20%**	20%
skin changes	5%	25%	**70%**
ambiguous	5%	5%	0%

Two output node networks

Ideally, a network capable of determining ulcer risk would assess the probability of a limb belonging to a skin change or a non-skin change group. Therefore a set of simpler networks, with and without hidden layers were trained to recognise normal and ulcerated limbs only. Results of testing these networks were significantly better than the three output node sets. The sensitivity (89%), specificity (85%), false positive rate (14.5%) and positive predictive value (86%) for detecting limbs with skin changes or ulceration (class 2 and 3 disease) were all better than those for the non-hidden layer network (88%, 80%, 20%, 79%) and significantly better than three output node network.

Comparison with single variables

In order to compare the diagnostic potential of determining at-risk limbs, receiver operating curves were constructed using the results of the best neural network and single parameters derived from APG. Figure 2 shows the results of the best (VFI) intermediate (VV) and worst (EV) APG parameters compared with the network score.

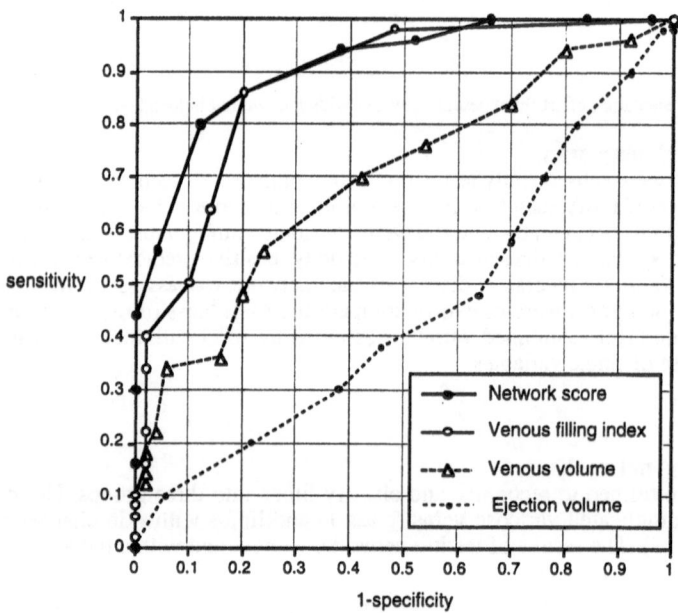

Figure 2: ROC curve for best (VFI) intermediate (VV) and worst (EV) APG parameters and network score in differentiating normal limbs (class 0) from limbs with skin changes (class 2 and 3).

Artificial neural networks were adequately trained to accurately classify the severity of venous haemodynamic abnormalities. Combining multiple variables did not increase the predictive power over simple measures of venous filling. Training the networks to recognise limbs according to clinical classification may have limited their potential to distinguish sub-groups of limbs such as those with primary calf muscle pump failure. The next step would be to apply self organising neural networks[3], which have the ability to cluster similar data without predefined restrictions. This approach may have both diagnostic implications and may also improve our understanding of the ulcerative process.

REFERENCES

1. Davalo E, Naïm P. Neural Networks, London: MacMillan Press,1991.
2. Kohonen T. Self organisation and Associative Memory, Berlin :Springer-Verlag, 1988.
3. Baykal N, Reggia JA, Yalabik N, Erkmen A. Interpretation of Doppler blood flow velocity waveforms using neural networks. Proc Annu Symp Comput Appl Med Care 1994; 1994:865-9.

Phlebology '95, D. Negus et al. (eds.). Phlebology (1995) Suppl. 1: 175-177

OP/3.5

Degree of Failure of Elastic Laminae in Superficial Veins Correlates with Severity of Venous Insufficiency

C. Solomon, G.T. Jones, A. Moaveni and A.M. van Rij

Department of Surgery, University of Otago, Dunedin, New Zealand

INTRODUCTION

The importance of valvular incompetence, venous reflux and high ambulatory venous pressure in the pathogenesis of venous ulceration has been well documented [1]. Why only a proportion of patients with significant reflux go onto develop severe disease is not clear and why similar findings on duplex examination are associated with varying levels of physiological dysfunction [2] has not been well explained. The hypothesis of this study was that following the onset of reflux through incompetent valves there is variable damage to the vein wall with loss of elasticity. This change would shorten the venous filling time and back pressure profile in the microcirculation. This preliminary study examines the histological differences in the walls of superficial veins of patients with varying degrees of chronic venous insufficiency due to superficial vein incompetence alone.

METHODS

Patient Selection and Tissue Sampling

Samples of superficial varicose veins from below the knee were obtained from patients undergoing superficial vein surgery as part of their routine clinical management. In total, 23 diseased limbs from 20 patients were sampled (9 class 1, 4 class 2, 10 class 3). 22 of these 23 limbs had received full physiological (using standard air plethysmography (APG)) and duplex ultrasound evaluation prior to surgery. Measurements included: 2-second outflow ratio, venous filling time, and the venous filling index. The dissection, removal and handling of the venous tissue was performed gently to minimise stretching and damage to the venous wall. Two additional cases served as controls (class 0); one from a below-knee amputation (secondary to peripheral vascular disease), and the other from superficial venous dissection for coronary artery bypass grafting.

Tissue Preparation

Two methods were employed in the preparation and fixation of the venous tissue. In 12 of the 23 samples, cross-sections of the veins were fixed in fresh 10% phosphate buffered formalin, embedded in paraffin, sections cut (5μm thick) using a standard microtome and stained with Verhoeff's elastic tissue stain and Curtis' modified van Gieson counterstain. The remaining 11 samples were dissected open and longitudinally pinned out, taking care not to overstretch the veins. Tissue was fixed in 2.5% gluteraldehyde, post-fixed in 1% osmium tetroxide, and embedded in epon resin. Semi-thin sections were cut and stained with Toluidine Blue. Both techniques stain elastic tissue (fibres and laminae) but formalin fixation may contract tissue and possibly lead to an underestimation of elastic tissue which is avoided by pinning out.

Histological Examination

Specimens were examined blind, by conventional light microscopy. Images for analysis were captured onto a computer image analysis system via a high resolution video camera attached to the light microscope. 6-8 captures of the intimal-medial border were taken at pre-designated points of reference at 66 times magnification. The degree of elastic tissue failure and intimal thickening in each specimen was assessed by three measurements:
(1) Percentage of elastic tissue intact (%ET), representing the fraction of the total intimal-medial border which contained elastic tissue along it; (2) Average fragment length (AFL in µm), the average length of the elastic tissue fragments present and (3) Intimal thickening, graded semi-quantitatively from 0-4 (0 having no intimal thickening, 1 with 2-4 smooth muscle cell layers, 2 with 5-9, 3 with 10-14 and 4 with 15 or more layers). Although in most cases the intimal thickening was relatively homogeneous, in those with less uniform intimal thickening the area with maximal thickening was used for assessment. Measurements were made using the public domain NIH Image program, version 1.57 (written by Wayne Rasband at the U.S. National Institute of Health and available from the Internet by anonymous ftp from zippy.nimh.nih.gov or on floppy disk from NTIS, 5285 Port Royal Rd., Springfield, VA 22161, part number PB93-504868).

Statistics

Analysis was performed using SPSS 4.0 software for Macintosh. Between-group comparisons were by one-way analysis of variance with Duncans multiple range test, for parametric measures, and Wilcoxin rank sum test for non-parametric measures. Correlation of histological measures with physiological variables, age, and duration of disease were by calculation of Spearmans correlation coefficients. Statistical significance was set at $p<0.05$. Data quoted are means $\pm$ 1 standard deviation.

RESULTS AND DISCUSSION

No difference was found for the measurements of percentage of elastic tissue intact (%ET) and average fragment length (AFL) between the two methods of preparation for veins of the same CVI class. It was therefore concluded that the two methods of preparation made no difference to the elastic tissue measurements and all data was duly pooled.

Histologically there was a distinct difference between the class 1 and 3 specimens. Class 1 specimens had generally intact elastic laminae, usually only slight intimal thickening and well organised layered medial smooth muscle cells. Class 3 specimens, however, were usually grossly devoid of intact elastica and often had a severely thinned medial layer with overlying intimal thickening. These qualitative observations are substantiated by the measurements made.

Age of class 1 patients was on average 10 years younger than class 2 and 3 patients but this group was made up of a younger and older subgroup (35 vs 75 years) and correspondingly different durations of varicosities (11.5 vs 46.0 years). Interestingly the younger group had less intact elastic tissue.

Overall, class 1 veins had a higher percentage of elastic tissue (54 ± 19) than those with class 2 (24 ± 11) and class 3 (23 ± 22) disease (Fig. 1.). The controls although few in number had higher levels of percentage intact elastic tissue (58 and 88). There was no significant correlation between this measurement and the duration of disease or the age of the patients.

Fragment lengths were quite variable but were greatest in class 1 (46.0 ± 34.3) and least in class 3 (11.2 ± 10.9) $p < 0.05$.

There was no statistically significant difference in the degree of intimal thickening between the classes of CVI nor was the degree of intimal thickening significantly correlated with the age of the patients. There was however a correlation between intimal thickening and AFL ($r = -0.48$).

A statistically significant correlation was found between both measures of elastic tissue failure (AFL & %ET) and 2 second venous out flow ratio as measured by APG prior to surgery. Similar, but not statistically significant, trends were found with VFT and residual volume. The failure of venous elastic tissue components could be expected to result in such physiological changes via altered vessel wall elasticity.

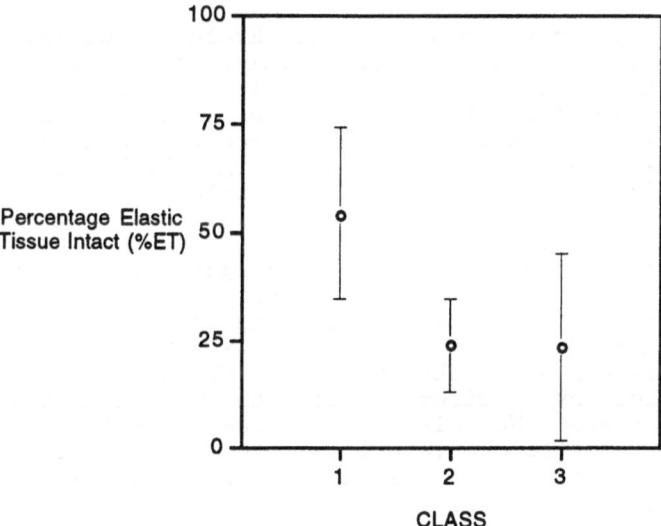

Figure 1. Plot showing Percentage Elastic Tissue Intact (mean ± SD) by class of CVI.

These findings support the hypothesis that there are differences with severity of CVI in the elastic tissue components of superficial veins. This may represent an important step in the pathogenesis of venous ulcerations.

REFERENCES

1. Nicolaides AN, Hussein MK, Szendro G,.Christopoulos D,.Vasdekis S, Clarke H. The relation of venous ulceration with ambulatory venous pressure measurements. J Vasc Surg 1993; 17: 414 - 9.
2. van Rij AM, Solomon C,.Christie R. Anatomic and physiologic characteristics of venous ulceration. J Vasc Surg 1994; 20: 759-6.

Phlebology '95, D. Negus et al. (eds.). Phlebology (1995) Suppl. 1: 178-179

P297

Clinico-Evolutive Classification of Chronic Venous Insufficiency and the Therapeutic Correlation in Each of its Stages

E.A. Enrici[1] and H.S. Caldevilla[2]

[1] Phelobology & Lymphology Department, Post-Graduate School of Health Sciences, Argentine Catholic University, Buenos Aires, Argentina
[2] Argentine College of venous & Lymphatic Surgery, Buenos Aires, Argentina

The authors have developed a Clinico-Evolutive Classification based on personal experience, clinical observation and invasive and non-invasive exploration. Its importance is supported by the fact that, being evolutive, it indicates the different stages of the disease which necessitate selective therapeutic methods, both medical and surgical. For clarity's sake, the following stages are being detailed:
STAGE 1
Recent Posthrombotic Syndrome,
described by the recently deceased distinguished Argentine angiologist Prof. Dr. Samuel Rascovan:
It extends from acute venous thrombosis up to rechanneling. The leg presents ædema, congestion, pain and hinders walking. Vascular laboratory shows obstruction in the deep venous system with sufficiency in direct perforants and a normal superficial venous system. Non-elastic compression during first 3 months, and an elastic one later, ease up blood deviation through the muscular vascular system, protecting the superficial and perforant systems.
STAGE 2
Compensatory Hypertrophy of the Mulsculo-Venous Pump:
The onset is at the rechannelling of the deep venous system. Muscular hypertrophy produces increase in the calf and leg due to the higher compensatory activity demanded by the deep venous hypertension. There are no congestion or ædema as in the previous stage. Deep venous urgery may be contemplated in accordance with clinical surrender, deep venous residual hypertension and valvular recovery period, the aim being to brake the course of the disease in its early stage and attain higher operative recovery, above all if the patient is young.
STAGE 3
Compensatory Hypertrophy of the Musculo-Venous Pump and the Surge of Secondary Varicosities:
This usually happens when the deep valvulo-parietal damage is severe, and is more pronouced when rechannelling incomplete.

A superficial venous plathology is developed by means of the
"blow-out" mechanism due to an insufficient perforant. Treatment should increase the musculo-venous pump´s compensatory
development by means of programmed exercise. If this is not
attained, direct venous surgery on the deep venous system and
ligation of all the insufficient perforants and secondary varices should be performed. Action on varicose veins only is
not recommended as it would immediately develop what the authors call "acute inflammatory hypodermitis" as its superficial buffing would be cancelled, leaving the transversal
backflow mechanism intact. Pain, ædema and local inflammation
promptly ensue, generating hypodermitis, solution of which is
by subfascial surgery to the perforant system.

STAGE 4

Advanced Chronic Venous Insufficiency Syndrome:
This stage, which involves all ætiologies, signals the moment
when progress of the disease decompensates the Buffer Venous
Circuit by means of the aformentioned mechanisms. Clinical
characteristic is indurative hypodermitis in the leg. Also,
ædemas and/or ulcers may appear, an epiphenomenon that com-
plicates the development of the disease. Vascular laboratory
and invasive exploration show direct perforant system insufficiency with decompensation of the Buffer Venous Circuit. In
more severe, usually posthrombotic, cases a global perforant
insufficiency of the leg may be diagnosed. Cockett´s and Lin-
ton´s operations respectively, with the modifications advocated by the authors, are the proper choice here. The deep venous hypertension should be evaluated according to studies
carried out, and possible surgery to this system should be
considered once perforant surgery has been completed and the
veno-musculo-articulation pump totally recovered.

STAGE 5

Phlebo-Arthrosic Syndrome:
Here hypodermitis totally surrounds the leg and, depthwise,
engages tendons and the tibio-astragallous articulation, preventing the musculo-venous pump from operating. Thus a state
of deep venous hypertension is created. Clinically there is a
pronounced atrophy of the musculo-venous pump, and large ulcers. Treatment is medical and, through kinetic treatment,
aims at recovering pump and joint movement. If reverting the
disease to stage 4 is attained, surgery to the different affected systems will be indicated in accordance with the
aforementioned comments.

STAGE 6

Posthrombotic Lymphædena:
This is the most severe of all stages and the hardest to
treat. Tissue sclerosis, which progressively damages arterial
and lymphatic capillaries, added to repeated infections, produces a chronic inflammatory condition which at leg level
destroys the delicate lymphatic system. The ædema which
originally presented high flow and low proteic content, becomes highly proteic. This favours the proliferation of
fibroelements, thus producing a vicious circle that nourishes
and aggravates the disease. Surgery attains no success here;
medical and local & general kinetic therapy is recommended,
always accompanied by external compression.

P289

VENOMUSCULAR PUMP IN PERIPHERAL VEINS: A TRIAL WITH INTRAVENOUS LIDOCAINE ADMINISTRATION

Watanabe R, Yui N

1st Div Dept Int Med, Wakayama Red Cross Hospital, Japan

Objective: To demonstrate venomuscular pump in peripheral veins using intravenous lidocaine administration.
Design: A case-control study.
Patients: Five controls (group 1) and 10 patients with cardiovascular disease (group 2) were studied.
Measurements: In group 1, 10mg of lidocaine was injected in the cubital vein with the measurements of cubital and central venous pressure (Pv, Pcv). In group 2, 5 and 10 mg of lidocaine were injected in the cubital vein with the measurement of Pv.
Results: In group 1, after 10 mg of lidocaine injection Pv increased from 8.6 ± 1.4cm H_2O to 10.5 ± 1.4 cm H_2O (M±SEM), and Pcv was 7.8 ± 1.6cm H_2O to 7.7 ± 1.4 cm H_2O before and after drug injection. The difference in means of Pv before and after injection was significant (p<0.01). In group 2, Pv changed after 5 mg lidocaine from 13.8 ± 1.9 cam H_2O to 12.3 ± 2.8 cm H_2O, and after 10mg injection Pv changed from 14.2 ± 2.1 cm H_2O to 16.3 ± 5 cm H_2O. The differences in these means were not significant. In 3 of 10 patients Pv dropped zero level after 5 or 10mg drug injection. On the contrary, in one patient Pv rose 17 and 30 cm H_2O after 5 and 10 mg drug injection.
Conclusions: Change of Pv after lidocaine injection could be explained by the drug induced dysfunction of venous pumping mechanism in peripheral veins.

P291

VENOUS PAIN OF LOWER LIMBS IN DAILY PHLEBOLOGY CLINICAL AND THERAPEUTIC STUDY

Tartour J, Lenica D, Gallo R

Consultation de Phlébologie (service de cardiologie)
Centre Hospitalier général Ballanger 93602 Aulnay Sous Bois France

Objective: To evaluate importance of pain as motive of consultation in our patients, and to assess effectiveness of treatment on pain
Design: Retrospective review of a consecutive series
Patients: 100 patients coming to our medical consultation of phlebology
Intervention: After elimination of no-phlebological pain, patients had treatment associating surgery or/and sclerotherapy + vaso-activ drugs and elastic compression
Measurement: Clinical examination to assess effectiveness of treatment of venous pain at 1, 2 and 6 months follow-up after first consultation.
Results: In 78 percent of patients treatment was effective on venous pain. In other 22 percent of patients treatment was no-effective (chronic venous insufficiency was associated with other cause of pain, or no good observance of treatment and of hygiene and dietetic advice).
Conclusions: Venous pain is a very important motive of consultation in daily phlebology. Clinical examination has to be very perfect at the first consultation. Treatment was effective on venous pain in 78 percent of cases.

PI/4.4

DUPLEX SCAN ASPECT OF RECURRENT VARICOSES VEINS AFTER SURGICAL TREATMENT OF THE LONG SAPHENOUS VEIN

F. VIN, F. CHLEIR

HOPITAL NOTRE DAME DE BON SECOURS
PARIS

Many studies shown recurrent varicose veins after surgical treatment.
One hundred patients after ligation-division and stripping of the long saphenous vein between 1 and 10 years were checked.
After a clinical examination all the patients were investigated by duplex scan examination to locate the differents leaking points and to have a map of varicose veins.
We found sometimes a reflux from perforating veins at the thigh and in the medial aspect of the leg. But the most important location of the reflux were the groin.
We describe all the type of communication between the deep and the superficial system and the percent of each one.

P293

CHRONIC VENOUS INSUFFICIENCY
WITHOUT SURGICAL POSSIBILITIES
STUDY WITH AMINAFTONE (CAPILAREMA)

P. Albino, S. Luz, S. Carvalho, A. Farrajota

VASCULAR SURGERY UNIT-H.ST.MARTA-LISBOA-PORTUGAL.

Objective: Review the benefit of Aminaftone (Capilarema) in patients with chronic venous insufficiency showing occlusive syndromes or deep venous insufficiency with scarce chances of venous revascularization/reconstruction.

Design: Clinical consecutive Trial

Patients: 20 consecutive patients (22 legs).

Measurements: The authors study 20 consecutive patients (22 legs) with Duplex Scan, Oclusion Plestimography, Photopletismography and in case of doubts with venography (ascendent/descendent) to characterize the venous situation.36% of cases have active ulcer.
They were submited during a four weeks period to a protocol based on prophylaxis measures, local treatment to the ulcers, and aminaftone as basic venotropic (225 mg/dia).After they were submeted to a new Oclusion Pletismography and another Photoplestismography.

Results: All patients benefit with the treatment. That have a good tolerance the patients with ulcers have limited the extension of this lesion. They don't have any significant improved in Oclusion Pletismography or Photopletismography.

Conclusion: There are evidence that this protocol with aminaftone benefit the patients with advance venous insufficiency but there aren't evidence of hemodynamic improved.

P020

EXTENSION AND HEMODYNAMIC SEQUELAE OF INFERIOR V. CAVA (IVC) OBSTRUCTIONS IN PATIENTS WITH ECHINOCOCCUS MULTILOCULARIS (EM)

U. Hoffmann, A. Fleiner-Hoffmann, K. Hagspiel, W. Schöpke, M. Fried, A. Bollinger.
Department of Internal Medicine, Divisions of Angiology and Gastroenterology and Department of Radiology, University Hospital Zurich, Switzerland

Objective: To study the extension of IVC obstructions in the region of the liver hilus due to EM and their hemodynamic consequences on venous return of the lower limbs.

Design: Prospective, uncontrolled investigation.

Patients: In an ongoing study 6 of 13 patients with known compression or thrombosis of IVC were investigated to date.

Measurements: CT and/or MRI scan of the abdomen and pelvis to assess extension. Colour-coded duplexsonography (CD) of the lower limb veins to evaluate venous return and valvular function. Dynamic plethysmography with treadmill exercise to analyse venous return under maximum work load (4 out of 6 patients).

Results: In 2 patients thrombosis of IVC extended from the confluence of the liver veins to the renal veins. One patient exhibited tight stenosis of the IVC at the liver hilus. In one patient thrombosis of IVC reached distally to the renal veins and in another the iliac and femoropopliteal veins on the right side were involved. With CD in 5 patients a phasic flow on both legs was found in the iliacofemoral veins. Two patients exhibited femoropopliteal valvular insufficiency presenting with chronic venous insufficiency stage II and III. Dynamic plethysmography showed normal findings in the 4 patients investigated. None of them compained about venous claudication.

Conclusion: Preliminary results indicate that compression of the VCI due to EM of the liver results in venous thrombosis with variable extension. Hemodynamic sequelae for venous return from the lower limbs are only moderate.

P021

SINGLE SITE SKIN EXTENSIBILITY IN VENOUS DISEASED AND CONTROL LOWER LIMBS

Shutt AM, Dodds SR, Cowan AR, Chant ADB.

Dept. Vascular Surgery, Royal South Hants Hospital, Southampton, UK.

Objective: To determine differences at a single site in the mechanical characteristics of skin in venous diseased and normal subjects using parameters derived from a double exponential curve fit for the stress relaxation of skin.

Design: In vivo comparison using a uniaxial skin extensometer of medial gaiter skin in two planes. Non linear least squares fit analysis by an iterative numerical method.

Patients: 25 patients with proven deep venous insufficiency later subdivided into ulcerated (VU), liposclerotic (VS) or oedematous (VO) groups, and 7 controls (Cont).

Measurements: Maximal force required to achieve a 30% forced extension of skin over ten seconds. The stress relaxation curve for skin was followed for two minutes and data taken from these curves for the numerical modelling. Each curve could be expressed by a series of 4 parameters labeled A to D. Comparisons between venous subgroups and control groups were then made.

Results: Mann Whitney comparisons: Female group maximal forces Control vs. VU $p < 0.04$, VU vs. VS $p < 0.02$. Parameter A; VS vs. VO $p < 0.03$, Cont vs. VU $p < 0.05$. Parameter B; Cont vs. VU $p < 0.02$. Parameter C; Cont vs. VS $p < 0.008$. Parameter D; Cont vs. VS $p < 0.002$. Similar differences found in male group.

Conclusions: There are wide variations in values derived for stress relaxation curves. Liposclerotic skin is generally more resistant to a 30% forced extension. In contrast skin in close proximity to ulcers stretches more easily.

OP/20.5

OEDEMA OF THE LIMBS WITH SIGNS OF VENOUS IN-VOLVEMENT IN ONCOLOGICAL PATIENTS: DIAGNOSTIC AND PROGNOSTIC VALUE

Pirovano C, Zanolla R, Balzarini A, Martino G.

Physical Therapy and Rehabilitation Department, National Cancer Institute, Milan, Italy

Objective: To evaluate the diagnostic and prognostic meaning of the characteristics of the limbs oedema in oncological patients with suspected local relapse, compared with investigational data.

Design: Retrospective review of a consecutive series.

Patients: 80 women operated on for breast cancer with homolateral arm oedema and 34 patients (13 women and 21 men), affected by different neoplasms and with lower limbs oedema.

Intervention: Assessment of 1) dimensions and 2) characteristics of the oedema (firmness, cyanosis, pitting, skin and nail dystrophy) 3) pain, range of motion, sensitive troubles 4) time of temporal increase.

Measurements: In all the cases, in addition to physical examination and to the pointing out of significant symptoms, were performed Echocolordoppler (ECD) of the involved limbs and Magnetic Resonance (MR) of their proximal regions.

Results: In all the cases the signs and symptoms suggested venous involvement and possible local neoplastic relapse. ECD revealed deep venous thrombosis and/or extrinsic compression of vascular structures in only 8% of the cases. MR showed progressive disease in 28% of the patients. Clinical or investigational signs of relapse became evident, in all the cases, in few weeks or months.

Conclusions: The characteristics of the oedema, associated with symptoms, are very important for the early diagnosis of progressive neoplastic disease. ECD may fail in showing slight venous involvement and MR may be able to disclose local relapse only late.

P028

SUPERFICIAL FEMORAL VEIN LEIOMYOSARCOMA
Diagnostic and therapeutic management

L CASTELLANI; R MARTINEZ; D GARCES; S ROUCHET; A de MURET; G CALAIS; P ROSSET

Service di Chirurgie Cardio-Vasculaire CHR TROUSSEAU

Vascular leiomyosarcomas are rare tumours of the soft tissues. The great majority of cases occurred in the inferior vena cava.

Only a few cases of peripheral leiomyosarcomas have been reported.

At this time neither limited margins' resection, nor the benefits of post-operative radiotherapy and chemotherapy are well established.

To try to analyse progress in diagnostic and therapeutic management; we report the case of a woman of 65 years of age who underwent surgical treatment for a leiomyosarcoma of the superficial femoral vein at our institution by an entire resection of a ten centimetre tumour with vascular reconstruction. The histological examination confirmed the vascular nature of the fusiform leiomyosarcomatous cell proliferation (stage II).

Phlebography, computed tomography and magnetic resonance imaging can support the preliminary approach but only the pathology examination of the tumour could definitely establish the diagnosis.

The disease prognosis is dominated by the local tumour recurrence and the high metastatic risks that could be reduced by post-operative radiotherapy and chemotherapy.

It is very difficult to draw conclusions about the management of this infrequent and rare tumour, but the experience of this case and opinions of various authors suggest that adjuvant therapy protocol still needs some working on and that the optimal therapeutic modality of leiomyosarcoma of the superficial femoral vein remains surgical.

OP/21.6

FOOT VOLUMETRY, QUANTITATIVE DUPLEX SCANNING AND CLINICAL STAGING, PRELIMINARY RESULTS

Lassvik C, Toulesius O
Department of Clinical Physiology, University Hospital, Linköping, Sweden

Background: Some patients with large varicose veins and reflux have few subjective symptoms of skin changes whereas others with seemingly minor reflux are very distressed due to pain, swelling and seven ulcers. The reason for this may be manyfold; location and amount of reflux, duration of venous insufficiency, type of activities such as sitting, standing, walking and possible risk factors.

Objective: To relate findings of detailed mapping and quantification of leg vein insufficiency in the superficial and deep systems, with symptoms and signs.

Design: Patients referred for leg vein mapping are investigated. A detailed list of clinical data is obtained, including verification of objective signs (staging of the disease).

Measurements: Foot volumetry with determination of expelled volume and refilling flow before and after exclusion of superficial reflux. Quantitative Duplex colour flow scanning with determination of mean and maximal reflux, valve closure time in deep and superficial systems.

Results and Conclusions: In this ongoing study it is our experience that there is an obvious discrepancy between degree of venous reflux and severity of presenting clinical impairment. The obtained objective parameters, however, constitute useful guidelines for decision to active treatment and as prognostic indications.

Physiology

KENDALL *INTERNATIONAL*

T.E.D.   ResT.E.D.   SCD   TSCD   Lastosheer

As the leading manufacturer of
clinically proven anti-embolism hosiery worldwide,
Kendall's brand names,
T.E.D. and **ResT.E.D.**,
will be well known to many clinicians.

The sequential graduated compression devices
for the prevention of DVT (**SCD**)
and treatment of chronic venous insufficiency (**TSCD**)
are more recent introductions.

These devices can be used alone,
or in combination with
T.E.D., or **Lastosheer** Compression Stockings,
for enhanced performance.
Both systems are clinically proven to be effective.

For more information on any of these products, please contact:

Kendall Healthcare Products - Europe
2 Elmwood, Chineham Business Park, Crockford Lane, Basingstoke, Hants RG24 8WG
TEL: 01256-708880   FAX: 01256-708071

Phlebology '95, D. Negus et al. (eds.). Phlebology (1995) Suppl. 1: 187-188

P003

Males with a Standing Profession: Diurnal Volume Changes of the Lower Legs

R.M.A. Krijnen[1], E.M. de Boer[1], H.J. Adèr[2] and D.P. Bruynzeel[1]

Departments of[1] Dermatology and[2] Epidemiology, Free University Hospital, Amsterdam, The Netherlands

INTRODUCTION

Oedema is one of the first clinical signs of venous insufficiency. Clinically, swelling of the lower legs is present, which gives rise to a tired and heavy feeling in the legs. Slight oedema is difficult to register. In this study an optical leg volume meter was used. We recently described the (intra-individual) reproducibility of measurements obtained with an adapted optical leg volume meter. In our laboratory the intra-individual reproducibility was calculated at 0.46% [1]. According to the literature, these are accurate measurements [2]. The aim of the present study was to investigate intra-individual volume changes of the lower legs in the course of a working day in persons with a standing position at work. Healthy employees were compared with employees having venous insufficiency. The presence of subjective complaints of the legs in relation to diurnal volume changes was investigated.

MATERIALS AND METHODS

A field trial was performed in 13 factories, mainly in the meat and shoe industry. 278 workers (mean age 38 years) were examined for the presence of CVI (chronic venous insufficiency) by physical examination, Doppler ultrasound investigation and light reflection rheography.
Major CVI was defined as the presence of skin changes (Widmer II-III) or signs of deep venous insufficiency or stem varicosis.
Minor CVI was defined as beginning corona paraplantaris phlebectatica, ankle flare or side branch varicosis without dermal changes.
Individuals with only intracutaneous varices were considered not to have CVI.
Subjects were asked for the presence of complaints of the legs, such as a tired and heavy feeling or pain in the legs. The volume of the lower legs was measured with an optical leg volume meter at the beginning and the end of 2 working days.

188

RESULTS

A volume increase was found in 70% of the legs.

No signs of CVI were found in 71% of the workers. The mean diurnal volume change in this group was +1,6%.

Minor CVI was present in 11% of the workers. The mean diurnal volume change in this group was +2,6%.

Major CVI was present in 18% of the workers. The mean diurnal volume change in this group was +3,6%.

These differences are statistically significant.

In persons with CVI, a high volume increase of the legs was associated with subjective complaints of the legs, as well for a tired feeling as for pain. In healthy persons this was not the case.

CONCLUSIONS

A volume increase of the lower leg in the course of the working day is common (70%). In workers with CVI a higher mean volume change was found. In workers with CVI, but not in the healthy, a higher volume increase was associated with subjective complaints of the legs.

ACKNOWLEDGMENTS

This study was made possible by a grant of the Praeventiefonds.

REFERENCES

1. Hebeda CL, de Boer EM, Verburgh CA, Krijnen RMA, Nieboer C, Bezemer PD. Lower limb volume measurements: standardization and reproducibility of an adapted volometer. Phlebology 1993;8:162-6.

2. Fischbach JU, Monter B, Göltner E. Messungen von Armödemen durch optoelektronische Volumetrie. Phlebol u Proktol 1986;15:184-189.

Phlebology '95, D. Negus et al. (eds.). Phlebology (1995) Suppl. 1: 189-192

P033

Alpha$_{2a}$-Adrenoceptors in Microsomes from Human Saphenous Vein

R. Hanf, G. Le Filliatre, F. Mardon, P. Luce and M. Finet

Laboratoire INNOTHERA, 10 Avenue Paul Vaillant Couturier, BP35, 94111 Arcueil Cedex

INTRODUCTION:

The sympathetic control of vascular tone involves different types and subtypes of adrenoceptors. If both alpha 1 and alpha 2 adrenoceptors induce venous and arterial constriction, α_2-adrenoceptor activation is the main process at the venous level. Based on the distinct affinities of a series of drugs, four different α_2-adrenoceptors have been pharmacologically characterized: α_{2A}, α_{2B}, α_{2C}, α_{2D}. In human, three genes (namely: α_2C2 α_2C4 α_2C10), localized on chromosomes 2, 4 and 10 encode for respectively α_{2B}, α_{2C} and α_{2A}-like subtypes.

Little is known on the subtype of α_2-adrenoceptor expressed at the venous level. Smith et al. (1992) have shown that contractile response of the human saphenous vein to noradrenaline was partially antagonized by the α_{2B} selective antagonist: prazosin. In order to characterize the pharmacological profile of the α_2-adrenoceptor(s) present in human saphenous vein microsomes, equilibrium binding experiments have been performed using 3H RX821002 as specific but not selective ligand of α_2-adrenoceptors. Results were compared to those obtained on rat kidney microsomes which were described to contain homogeneous population of α_{2B}-adrenoceptors.

METHODS

Membrane isolation procedures:
Human saphenous veins were obtained from 48±4 year-old women undergoing surgical removal of varicose veins. Microsomes have been obtained according to Dacquet et al. (1988). Microsomes were resuspended (1-2 mg/ml) in the binding medium: 5 mM MgCl2; 100 mM TRIS/HCl (pH=7,4). Aliquots were stored at -80°C until use.

Rat kidney microsomes were obtained according to Borton et al. (1991).

190

Radioligand binding studies:

All experiments were performed in a final volume of 1 ml binding buffer. Equilibrium binding has achieved during 60 minutes incubation at 20°C. This was done using 100 μl aliquots of membrane suspension in the presence of increasing concentrations of 3H RX821002. Membrane bound radioligand was separated from free by rapid filtration through GF/C Wathman filters using a Skatron cell harvester. Non-specific binding was determined in the presence of 100 μM noradrenaline. The Kd and Bmax values were calculated from computer assisted analysis using EBDA-LIGAND program. Competition studies were carried out in the presence of 5 nM 3H RX 821002. Inhibition constants (Ki) were calculated according to the equation of Cheng and Prussoff.

RESULTS and DISCUSSION

Saturation curve of 3H RX821002 binding and the corresponding linear Scatchard analysis showed that 3H RX 821002 binds to an apparently homogeneous population of $\alpha 2$ adrenoceptors in human saphenous vein microsomes. This specific binding was likely related to smooth muscle cell receptors since endothelium and adventice have been removed before membrane isolation procedure.

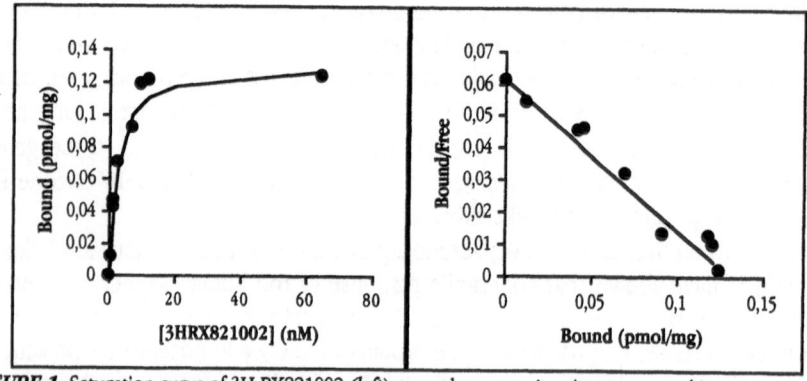

FIGURE 1: Saturation curve of 3H RX821002 *(left)* on saphenous vein microsomes and its corresponding Scatchard analysis *(right)*. This experiment is representative of four on four distinct microsomal fractions. Mean values for Kd and Bmax were 4,07±2,06 nM and 0,120±0,020 pmol/mg.

Competition studies using selective and non selective $\alpha 2$ receptor ligands are summarized in Table 1. These experiments show that the $\alpha 2$ adrenoceptors present in human saphenous vein membrane vesicles share the fundamental characteristics of the $\alpha 2A$ subtype: high affinity for oxymetazoline, low affinity for prazosin. This is exemplified by the marked differences in potencies of these two selective drugs between saphenous vein and rat kidney, an $\alpha 2B$ containing tissue. Furthermore, IC50 values of chlorpromazine and corynanthine ($\alpha 2B$ prefering ligands) were very high.

		Human saphenous vein		Rat Kidney	
Ligand	selectivity	IC50 (nM)	nH	IC50 (nM)	nH
Prazosin	α_{2B}	**41028±19000**	**0.48**	**558±481**	**0.629**
Oxymetazoline	α_{2A}	**112±28**	**0.65**	**391±123**	**0.585**
Yohimbine	NS	21±5	1.1	80±30	0.791
Idazoxan	NS	194±109	0.83	139±109	0.584
Phentolamine	NS	122±27	0.62	31±18	0,795
Corynantine	$\pm\alpha_{2B}$	5272±2046	0.56	ND	ND
Chlorpromazine	α_{2B}	3451±464	0.61	ND	ND
Naphazoline	$\pm\alpha_{2A}$	289±53	0.74	ND	ND

TABLE 1: Calculated IC50 and nH values of the drugs which have been tested on human saphenous vein and rat kidney microsomes. NS = non-selective drug. ND= not determined

pKi values derived from competition experiments on human saphenous vein membranes were compared to those found by Devedjian et al. on COS7 cells encoding for the three human genes: $\alpha_{2}C2$, $\alpha_{2}C4$, $\alpha_{2}C10$ (Figure 2). An excellent correlation has been revealed whith $\alpha_{2}C10$ human gene product (slope=1,1; R=0,944) while only poor correlations have been found with $\alpha_{2}C2$ or $\alpha_{2}C4$ human gene products.

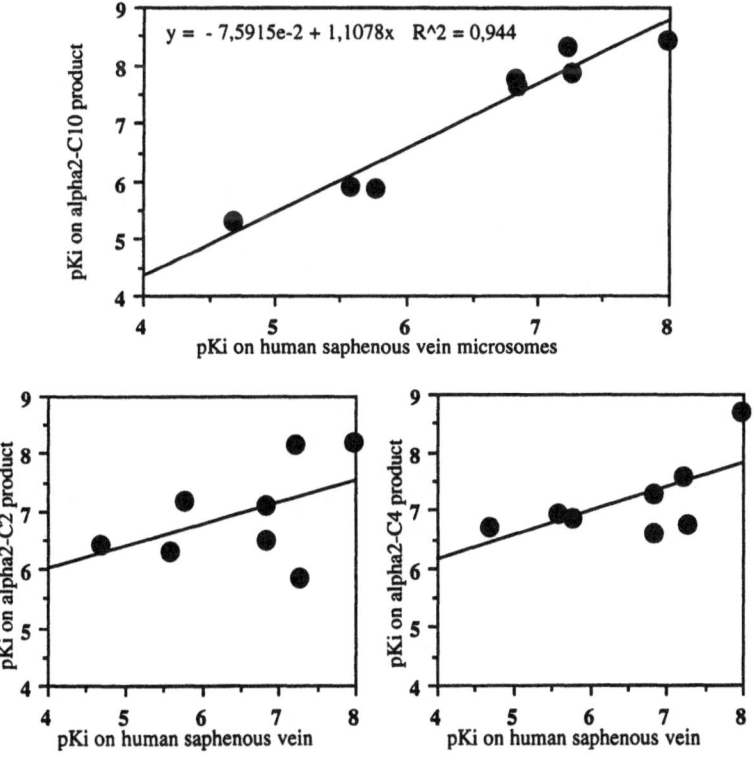

FIGURE 2: Correlations between pKi measured in this study and those found by Devedjian et al. (1994) on microsomes isolated from COS 7 cells transfected with a2C10, a2C2 or a2C4.

Therefore, the α2A pharmacology described in Table 1 probably reflects the α2C10 human gene expression in smooth muscle cells from saphenous vein. However, further mRNA analysis are necessary to confirm this hypothesis.

In order to assess the physiological relevance of this biochemical study, in vitro contractile experiments are actually performed using rings of human varicose saphenous vein. As previously described by Smith et al. (1992), preliminary results show that 1 μM prazosin induces a rightward shift in the noradrenaline response curve. However, this might be due to the copresence of both α1 (highly sensitive to prazosin) and α2 adrenoceptors and not necessarily demonstrates the presence of the α2B subtype.

REFERENCES: Borton et al. (1991) J. Auton. Pharmacol. 11:247-253. Dacquet et al. (1988) Biochem. Biophys. Res. Com. 152:1165-1172. Devedjian et al. (1994) Eur. J. Pharmacol. 252:43-49. Smith et al. (1992) Br. J. Pharmacol. 106:447-451

Phlebology '95, D. Negus et al. (eds.). Phlebology (1995) Suppl. 1: 193-195

P034

Histamine Receptor Subtypes Involved in Macromolecular Permeability and Microvessel Vasomotricity Regulation: Quantitative Analysis using an Intravital Fluorescence Microscopy Model

G. Gimeno[1], P. Carpentier[2] and M. Finet[1]

[1] *Pharmacology Department INNOTHERA Laboratories, 10 av PV Couturier, 94110 ARCUEIL, France*
[2] *Medicine & Vascular Biology Laboratory, Joseph Fourier University, 38043 GRENOBLE Cedex, France*

INTRODUCTION

Many authors postulated that increased macromolecular permeability was associated with venodilatation. The aim of this study was to determine the involvement of histamine receptors on microvessel vasomotricity and macromolecular permeability using an intravital videomicroscopy model.

MATERIAL AND METHODS

Experiments were realised in hamster dorsal skin fold preparation (1) which gave simultaneous access to many parameters such as macromolecular permeability (local on observed tissues and general vascular on observed microvessels) and vasomotricity (arteriolar and venular) on unanaesthetized animals. Animals were mounted under a microscopic stage and vessels were differenciated with a fluorescent marker (FITC Dextran 150,000) administered via a catheter inserted in the jugular vein. General inflammatory reaction was induced by intravenously-injected histamine (1mg/kg). Different H1 and H2 receptors antagonists were intravenously-injected 15' prior to histamine.

Experiments were performed under a black and white CCD video camera (HPR 610) connected on top of a microscope (Ergolux LEICA) and were recorded (Mitsubishi BV 2000-E) on video tapes to be analyzed off-line. The images (256 grey levels and 511x511 pixels) are representing a small part of the tissue containing microvessels (arterioles, venules, postcapillary venules and capillaries).

Video records were used to determine the macromolecular permeability by establishing a curve quantifying the number of pixel in each grey level (0 = black ; 255 = white). Two peaks can be observed on that curve : the left one centred in the low grey levels was representing the extravascular fluorescence and the right one, centred in higher grey levels, was corresponding to the intravascular fluorescence (2, 3). Peaks were moving (on grey level scale) during histamine-induced hyperpermeability state. The number of grey levels between the center of each peak (before histamine injection) was corresponding to the maximal permeability.

Vasomotricity was calculated by measuring arteriolar and venular diameter each 5 minutes. Values were expressed in percent of diameter before the injection and area under curve were compared. A ratio (treated/solvent) >1 determined a vasodilatation and a ratio <1 showed a vasoconstriction.
Statistical significance was calculated with Student's t-test. The level of significance was taken at $P < 0.05$.

RESULTS AND DISCUSSION

Local macromolecular permeability : histamine (1mg/kg)-induced maximal effect was 12.53±8.03 (percent of calculated maximal permeability). H1 receptor antagonists (1mg/kg) : triprolidine or mepyramine completely blocked this response : 1.78±2.17 and 3.84±2.45. The H2 receptor antagonist ranitidine (1 mg/kg) was ineffective against histamine response : 11.24±5.4. However, another H2 receptor antagonist cimetidine (1mg/kg) strongly blocked this histamine effect : 2.64±3.44.
General vascular macromolecular permeability reached 56.60±6.35 (percent of calculated maximal permeability) in presence of histamine (1mg/kg). H1 receptor antagonists triprolidine or mepyramine inhibited the general permeability with respectively 2.42±1.6, 5.30±2.62. H2 receptor antagonists ranitidine or cimetidine (1 mg/kg) partially inhibited this general effect with respectively 14.59±7.3 and 22.58±2.77.

Effects of H1 antagonists (triprolidine, pyrilamine) or H2 antagonists (ranitidine, cimetidine) on general vascular permeability and local permeability

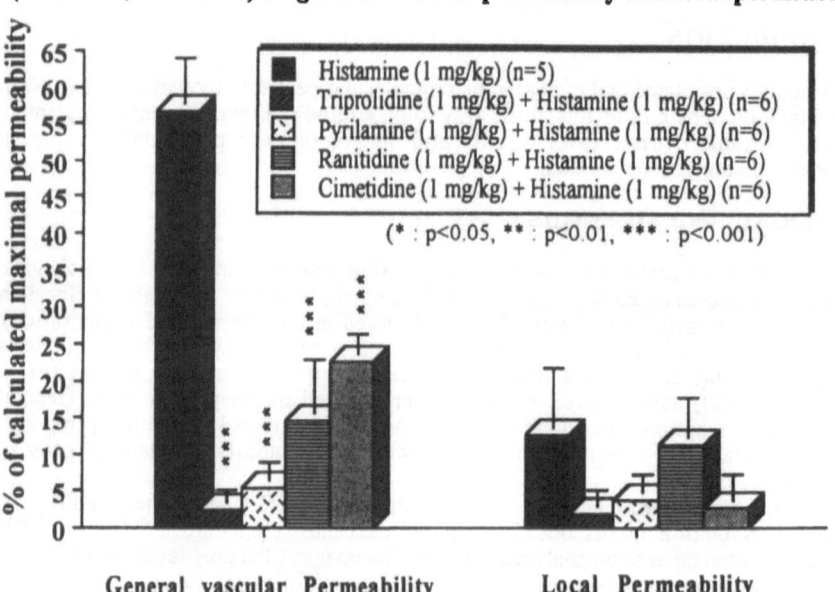

Venular diameter measurements on the same animals showed a dilatation (vs 0.9 % saline) in histamine conditions. H1 antagonists did not inhibit the venodilatation but H2 antagonists prevented this effect.

Arteriolar diameter measurements gave prominence to a constriction in presence of histamine. This effect seemed to be inhibited by both H1 and H2 antagonists. However these measurements were highly confused by arteriolar vasomotion observed during experiments.

Effects of H1 antagonists (triprolidine or pyrilamine) or H2 antagonists (ranitidine or cimetidine) on venular and arteriolar diameters

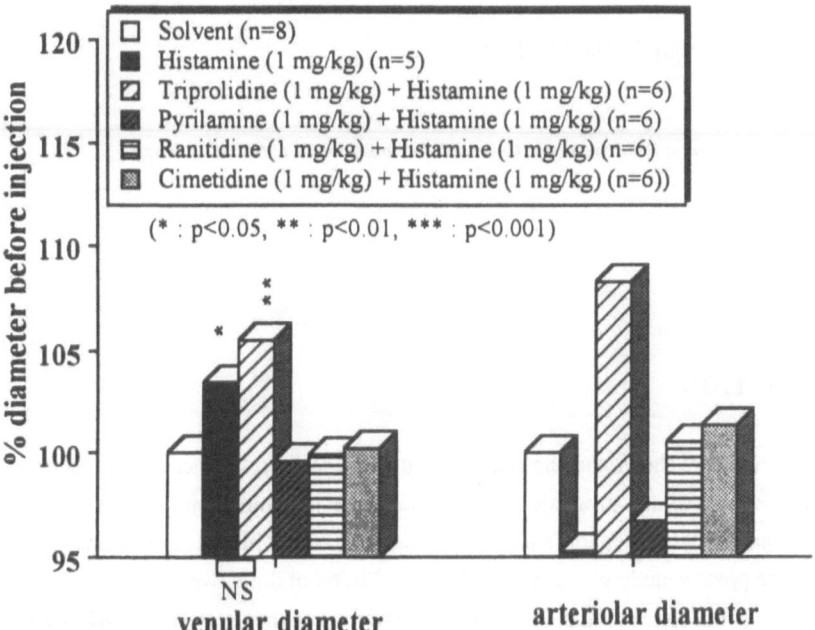

CONCLUSION

H1 and H2 histamine receptor subtypes are involved in macromolecular permeability but H1 contribution is highly superior than H2. Venodilatation observed (in presence of histamine) in the same experiments seems to be produced by H2 activation. Further experiments could define if H1 involvement in permeability was potentiated by H2-induced venodilatation.

REFERENCES

1- B. ENDRICH, K. ASAICHI, A. GÖTZ, K. MESSMER. Technical report : a new chamber technique for microvascular studies on unanesthetized hamsters.Res. Exp. Med. 1980; 177: 125 - 134.
2- G. GIMENO, P. CARPENTIER, M. FINET. A new technique using intravital videomicroscopy for macromolecular permeability measurement. 18° European Congress of Microcirculation (ROME September 1994)
3- G. GIMENO, P. CARPENTIER, M. FINET. Nouvelle approche basée sur la videomicroscopie pour mesurer la permeabilité vasculaire aux macromolécules 29° Congrès du Collège Français de Pathologie Vasculaire (PARIS Mars 1995)

Phlebology '95, D. Negus et al. (eds.). Phlebology (1995) Suppl. 1: 196-198

Effect of Pregnancy on the Effects of L-Name and Indomethacin on Arterial and Venous Pressures in Anaesthetized Intact and Pithed Rats

F. Le Marquer-Domagala and M. Finet

Innothéra, Department of Pharmacology, 10 avenue PV Couturier, 94111 Arcueil, France

INTRODUCTION

Various studies have shown that arterial pressure and vascular resistance are decreased in pregnant compared to non-pregnant rats [1]. Variations of sympathetic tone and/or NO production, and/or cyclooxygenase pathway could be involved in these alterations. Thus, the aim of the present study was to assess the contribution of these two factors on arterial pressure and venous tone in female rats. We compared a treatment with L-NAME (a NO-synthase inhibitor), in presence of indomethacin (inhibitor of cycloxygenase) or its vehicle, in both virgin and pregnant anaesthetized intact or pithed rats.

METHODS

Measurements were made on 2 month-old virgin and pregnant (17 days of gestation) Wistar rats, anaesthetized with an intraperitoneal injection of sodium pentobarbital (Sanofi; 60mg/kg). Biotrol cannulas were inserted into the left carotid, the vena cava and the right femoral vein for respectively the measurement of mean arterial pressure (MAP), central venous pressure (CVP) and the injection of products. Total venous tone was determined by measuring pressures after inflating a Fogarty cannula (Baxter, 2F), inserted in the righ atrium via the right external jugular vein [2], which allowed the calculation of the mean circulatory filling pressure (MCFP). An intratracheal cannula was inserted to allow an artificial ventilation at 76 breaths.min-1, with a volume of 2,5ml room air. Anaesthetized virgin and pregnant rats were divided into 2 groups, group I:

intact rats, group II: pithed rats [3], then rapidly put under artificial respiratory with a rodent ventilator (Harvard). Bolus injection of L-NAME (Sigma Chemical Co; 30mg/kg) were made on both groups, 15 min. after a continuous perfusion of indomethacin (Sigma Chemical Co; 5mg/kg/min) or its vehicle (trizma 0,2M; Sigma Chemical Co). The three vascular parameters were measured during 1 hour after injection, and compared to control values measured just before injection.

RESULTS

In intact rats, MAP and MCFP were significantly lower in pregnant compared to virgin rats. After pithing, values were not significantly different. Perfusion of indomethacin did not modify those pressures. Basal values are shown in Table 1.

Table 1. Basal values of mean arterial pressure (MAP) and mean circulatory filling pressure (MCFP) (mm Hg)

		MAP	MCFP
Intact	Virgin n=8	135±4	6,6±0,3
	Pregnant n=8	115±6	5,5±0,2
		-15% p<0,05	-17% p<0,01
Pithed	Virgin n=8	49±3	4,6±0,3
	Pregnant n=8	49±1	5,0±0,2
		NS	NS

In intact rats treated by trizma, L-NAME-induced rises in MAP (expressed as % of initial MAP) were not significantly different in the pregnant compared to the virgin group, but significantly lower when the same comparison was made in pithed animals (-54%; p<0,01). Comparatively to intact rats, L-NAME-induced rises in MAP is higher in pithed rats in both virgin and pregnant rats. The rise in MCFP observed with L-NAME was significant only in pithed rats, and was significantly lower in the pregnant compared to the virgin group (-51%; p<0,05). Results are shown in Table 2.

In anaesthetized intact rats, indomethacin significantly decreased the effect of L-NAME in virgin (-26%,p<0,05), without effect in pregnant rats. After pithing, indomethacin compared to trizma, decreased the difference of effect of L-NAME on MAP between virgin and pregnant rats, but those difference remained significant (Table 3). In presence of indomethacin, we did not observe any difference on MCFP between virgin and pregnant rats (Table 3).

Table 2. Maximum effect of a bolus injection of L-NAME (30 mg/kg) on MAP and MCFP in presence of trizma. Results are expressed as percentage of initial values

		MAP		MCFP	
Intact	Virgin	+27±2			
	Pregnant	+29±1	*NS*		
Pithed	Virgin	+127±13		+33±6	
	Pregnant	+57±5	*-54%,p<0,01*	+16±4	*-51%,p 0,05*

Table 3. Maximum effect of a bolus injection of L-NAME (30 mg/kg) on MAP and MCFP in presence of indomethacin (5mg/kg/min) or its vehicle (trizma), in pithed virgin and pregnant rats. Results are expressed as percentage of initial values

		MAP		MCFP	
		+Trizma	+Indometh	+Trizma	+Indometh
Pithed	Virgin	+127±13	+120±12	+33±6	+26±4
	Pregnant	+57±5	+79±7	+16±4	+20±5
		-54%,p<0,01	*-34%,p<0,01*	*-51%,p<0,05*	*NS*

DISCUSSION

These results suggest that between anaesthetized intact, virgin and pregnant rats, there is no difference in endogenous NO contractile effect, but a low pressure tone induces a different increase of this effect between the two groups.

The pressor effect of L-NAME on MAP is dependent on the vascular tone and modulated by a sympathetic reflex [4], but this effect of L-NAME is more dependent of the sympathetic reflex in virgin, and is partially cyclooxygenase-dependent in pregnant rats.

REFERENCES

1.   Ueland K, Novy MJ, Peterson EN, Metcalfe J, Maternal cardiovascular dynamics. Am J Obstet Gynecol 1969;105:856-864.

2.   Yamamoto J, Trippodo NC, Frohlich ED, Total vascular pressure-volume relationship in the conscious rat. Am J Physiol 1980;238:H823-H828.

3.   Shipley RE, Tilden JH, A pithed rat preparation suitable for assaying pressor substances. Proc Soc Exp Biol Med 1947;64:453-455.

4.   Le Marquer-Domagala F, Finet M, Freslon J-L, Effects of L-NAME and indomethacin on arterial and venous pressures in anaesthetized intact and pithed spontaneously hypertensive rats (SHRs). 52° réunion de l'association française des pharmacologistes 1994;A44:99.

Phlebology '95, D. Negus et al. (eds.). Phlebology (1995) Suppl. 1: 199-201

P005

In Vitro Determination of the Rabbit Saphenous Vein Compliance

F. Le Marquer-Domagala, P. Luce, F. Mardon, C. Rouvreau and M. Finet

Innothéra, Department of Pharmacology, 10 avenue PV Couturier, 94111 Arcueil, France

INTRODUCTION

The venous system plays an important function in the regulation of the cardiac output, and compliance modification is one of the regulatory factors. The aim of this study was to determine the venous compliance on in vitro conditions, and to measure the effect of noradrenaline (NA), the physiological venotonic neuromediator.

METHODS

Saphenous veins of male rabbit (2,5kg) were cut in segments of 1 cm length, mounted on one side on a catheter, connected to a syringe which allowed perfusion of modified Krebs solution, while the other side was ligatured. They were then immersed in a 20-ml bath (37°C) containing a modified Krebs solution, aerated by a gas mixture (95%O_2 and 5%CO_2). With perfusion flow, inside vessel volume increased so as pressure, which was measured with a Gould transducer connected to the vein, allowing to establish the compliance curve. Two rates of perfusion have been tested: 0,15ml/h and 0,375ml/h, and at the higher rate of perfusion, cumulative doses of NA (10-7 to 10-5M) were injected in the bath, and an other pressure-volume curve was studied.

Compliance is defined as the change in volume (dV) resulting from a change in pressure (dP), and is represented by the slope of the pressure-volume relationship.

$$(C) = (dV) / (dP)$$

RESULTS

Pressure-volume curve was studied from 0 to 100 mm Hg, which corresponds to a maximum volume inside the vein of 2700µl. This curve was obtained in 7min. We observed three portions of the curve with some differences between veins in function of their length, but generally, we obtained two linear portions between 0 to 10 mm Hg and above 30 mm Hg, and a non-linear one between 10 to 30 mm Hg (Fig.1.). In the linear portion, the slope of the curve was calculated to determine the compliance of controls.

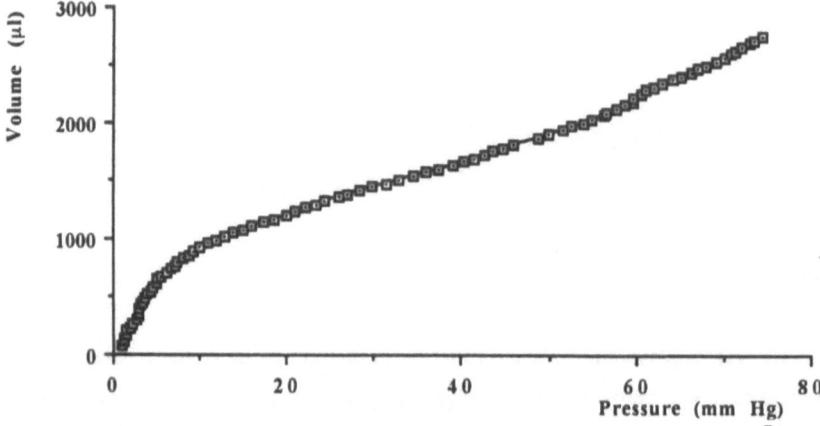

Fig.1. Pressure-volume curve of a saphenous vein.

We have studied the pressure-volume curve of the vein at physiological pressures. In this condition and compared to the control, NA induced a significant diminution of the compliance, the value being respectively 0,69; 0,16 and 0,08µl/mm Hg, at 10-6 and 10-5M (Fig.2.).

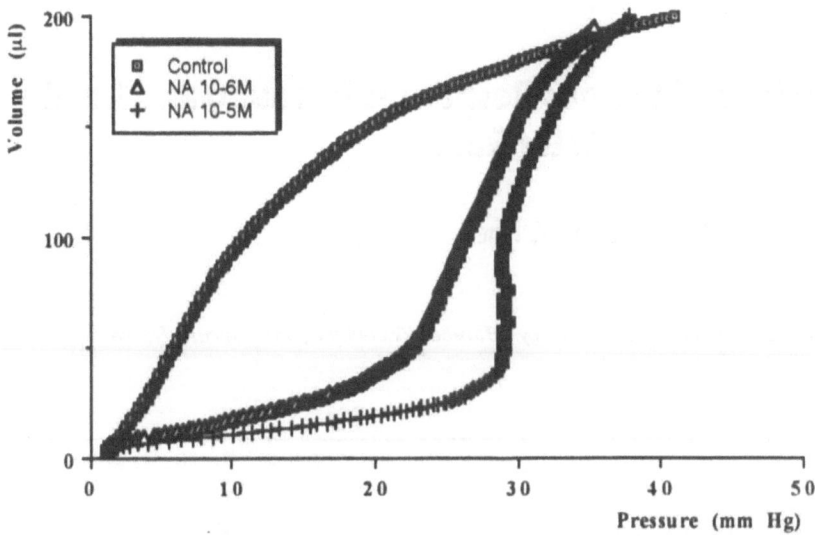

Fig.2. Effect of NA (10-6 and 10-5M) on the pressure-volume curve of a saphenous vein

DISCUSSION

This in vitro model allows the direct measure of venous compliance, represented by the slope of the pressure-volume curve of the vein at low pressure levels. This compliance is significantly reduced by the venoconstriction induced by noradrenaline administrated in the bath.

Phlebology '95, D. Negus et al. (eds.). Phlebology (1995) Suppl. 1: 202-204

P006

Arterial and Venous Responses in Young and Aged Anaesthetized Wistar Rats

F. Le Marquer-Domagala and M. Finet

Innothéra, Department of Pharmacology, 10 avenue PV Couturier, 94111 Arcueil, France

INTRODUCTION

In human, aging is associated with a variety of alterations in cardiovascular function, including a decreased cardiac output and an augmented blood pressure [1]. The purpose of this study was to test the influence of age in anaesthetized rats, on arterial and venous parameters.

METHODS

Measurements were made on 3, 12 and 20 month-old Wistar rats, anaesthetized with an intraperitoneal injection of sodium pentobarbital (Sanofi; 60mg/kg). Biotrol cannulas were inserted into the left carotid, the vena cava and the right femoral vein for respectively the measurement of mean arterial pressure (MAP), central venous pressure (CVP) and the injection of products. Total venous tone was determined by measuring pressures after inflating a Fogarty cannula (Baxter, 2F), inserted in the right atrium via the right external jugular vein [2], which allowed the calculation of the mean circulatory filling pressure (MCFP). An intratracheal cannula was inserted to allow an artificial ventilation at 76 breaths.min-1, with a volume of 2,5ml room air. Rats were rapidly put under artificial respiratory with a rodent ventilator (Harvard). Pressure responses to noradrenaline (NA, 1µg/kg/min=ED 30) were measured, and in a second time, MAP and CVP were measured for the study of the dose-response curve of acetylcholine (ACh: 0,1 to 100 µg/kg), so as the duration of ACh vasodilatation, indicated by time from maximum of effect to time when pressure had restored to 50% of maximum pressure change (T1/2; s). In a third time, bolus injection of L-NAME (Sigma Chemical Co; 30mg/kg) were made and vascular parameters measured during 1 hour.

RESULTS

Aging did not modify MAP, but we observed a significant increase of CVP at 12 and 20 months, so as a significant increase of MCFP. Basal values are shown in Table 1.

Perfusion of NA induced a significant higher increase of MAP in the 12 and 20 months old rats compared to the 3 months old rats (respectively +98% and +72% of augmentation). The effect of NA on CVP was not modified with age, but we observed a significant higher increase in MCFP in 12 and 20 months old rats (respectively +80% and +48% of augmentation). The effects of NA are summarized in Table 2.

Table 1. Basal values of mean arterial pressure (MAP), central venous pressure (CVP) and mean circulatory filling pressure (MCFP) in anaesthetized 3, 12 and 20 month-old Wistar rats.

(mm Hg)	3 Months	12 Months		20 Months	
MAP	145±3	145±5	NS	143±4	NS
CVP	2,9±0,2	3,8±0,3	+29%,p<0,05	3,7±0,3	+26%,p<0,01
MCFP	7,9±0,3	8,5±0,2	NS	8,9±0,4	+14%,p<0,001

Table 2. Effect of perfusion of NA (1µg/kg/min) in anaesthetized 3, 12 and 20 months old wistar rats, on mean arterial pressure (MAP), central venous pressure (CVP) and mean circulatory filling pressure (MCFP) (expressed in difference of pressure with basal values).

(mm Hg)	MAP	CVP	MCFP
3 months	7,4±2,9	0,1±0,1	1,7±0,3
12 months	23,3±4,6 p<0,05	0,1±0,2 NS	3,3±0,5 p<0,05
20 months	18,9±3,6 p<0,05	0,3±0,2 NS	2,7±0,4 p<0,05

Bolus injection of ACh induced a dose-dependent diminution in MAP, which was not modified with age, but at the higher concentration, a significant diminution of the T1/2 in the 12 and 20 months old rats, compared to the 2 months old rats was observed (respectively -25% and -35% of inhibition) (Fig.1.).

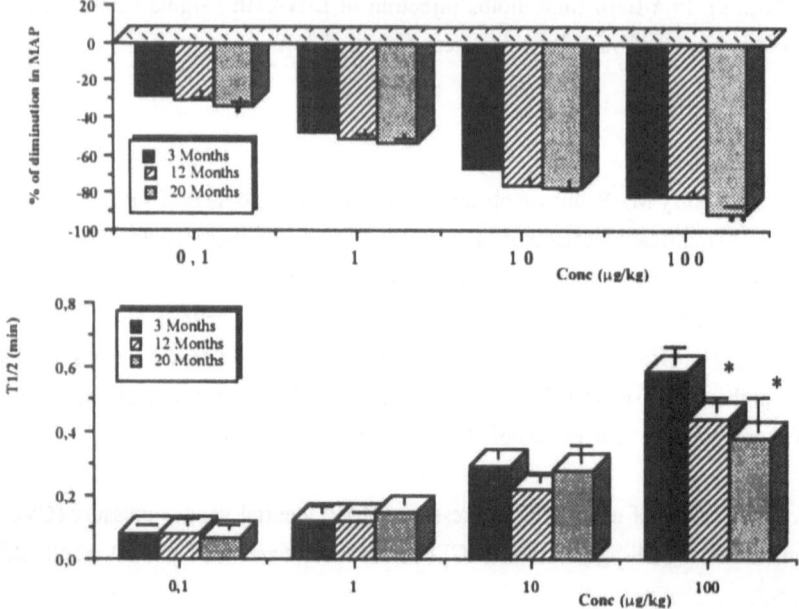

Fig.1. Dose-response curve to ACh on MAP (expressed as % of diminution of MAP), and on T1/2 (expressed in times:min.), for the 3, 12 and 20 month-old rats.

Bolus injection of L-NAME (30 mg/kg) induced a maximum increase at 5 min. in MAP, which was lower comparatively to 3 months old rats, but not significantly, in the 12 and 20 months old rats (respectively +40%, +26% and +27% of increase).

DISCUSSION

In the anaesthetized old rats, no change in MAP was observed with age, but there was an increase in CVP and MCFP, resulting in a diminution of the compliance, which is the contrary to changes reported in the venous pathology. Moreover, we observed with age, in arterial as in venous bed, an increased vasoconstrictor and a reduced vasodilator responsiveness to respectively NA and ACh. However, we did not obtain direct evidence to suggest that NO production is affected by aging.

REFERENCES

1.   Docherty JR, Cardiovascular responses in ageing: a review. Pharmacol Rev 1990;42(2):103-125.
2.   Yamamoto J, Trippodo NC, Frohlich ED, Total vascular pressure-volume relationship in the conscious rat. Am J Physiol 1980;238:H823-H828.

Phlebology '95, D. Negus et al. (eds.). Phlebology (1995) Suppl. 1: 205-208

P035

Effects of Venous Stasis on Blood Viscosity, Micro-Circulatory Dynamic, Cutaneous Oxygen Pressure

C. Le Devehat and T. Khodabandehlou

Unité de Recherches d'Hémorhéologie Clinique, Pavillon Jules Renault, Centre Hospitalier, 58033 NEVERS, France

INTRODUCTION

Numerous works have pointed-out the hemorheologic disorders which occur in patients with chronic venous insufficiency *"CVI"* (1,2,3). It was also demonstrated that the cutaneous blood flux and vasomotion are disturbed during this pathology (4,5). Disturbances caused by the disease occur mainly in stasis areas (1) and are accompanied by a tissue hypoxia. The hemorheologic and microcirculatory abnormalities are involved in a vicious circle (3) and may be more amplified when the stasis is extended within both the time and the space. This work aims to evaluate the hemorheologic and microcirculatory effects of venous stasis in patients with *CVI*. A procedure to provoke a venous stasis was first tested in healthy subjects and thereafter applied on patients with *CVI*.

PATIENTS AND METHODS

24 patients suffering from a *CVI* and 11 healthy subjects with no pathology and medication participated in this study. The age (mean +- SD) of the patients was 61 +- 8 years and of the controls 36 +- 16 years. The male to female ratio in the patients was 5 : 19 and in the controls 8 : 3. They were studied in supine position. Microcirculatory measurements consisted of continuous record of cutaneous blood flux *"BF"* as well as vasomotion by laser doppler fluxmeter (PF4 Perimed) and also the transcutaneous oxygen pressure (TcPO2) by Oxymonitor SM 361 (Hellige) with arterialization at 44°C. Both measurements were performed on the dorsum of the foot before, during and after a provoked venous stasis *"PVS"*. The venous stasis was provoked by a cuff inflated at 100 mmHg during 15 minutes around the knee.

Hemorheologic measurements were performed on blood samples drawn from a foot vein before and at the end of the *PVS*. We measured red blood cell *"RBC"* aggregation by the Sefam erythro-aggregometer (6) based on the analysis of changes in backscattered light through the blood suspension in a Couette flow. The derived parameters are *RBC* aggregation time *"AT"* expressed in $(mPa)^{-1}$, *RBC* aggregation index *"AI"* and *RBC* disaggregation shear stress *"DSS"* expressed in (mPa.) Relative blood viscosity was determined by a Couette viscometer (LS 30 Contraves) at shear rates of 128.5 s-1 and 1.285 s-1. *RBC* deformability was evaluated by the Hanss hemorheometer (7) based on the filtration of a red cell suspension through nucleopore membrane with pore diameter of 5 μm. Results are expressed as a rigidity index *"RI"*. All these measurements were performed at a temperature of 37°C. We also determined plasma viscosity *"PV"* by a capillary viscometer (KSPV4 Myrenne), plasma fibrinogen level *"FGN"* by a thrombin clotting time technique and albumin level *"ALB"* by a colorimetric technique. The changes of all the parameters induced by the *PVS* were analysed by Student "t" test for paired data. Comparisons between groups were assessed by Student "t" test for unpaired data.

RESULTS AND DISCUSSION
1. Hemorheologic parameters (Tables 1-2)
Before stasis, only hematocrit and albumin were significantly lower in patients. Plasma fibrinogen and viscosity were similar in patients and controls. Erythrocyte aggregation, evaluated at a standard hematocrit of 40% was within normal ranges in patients. Erythrocyte deformability as appreciated by viscometric measurements at high shear rate and filtration technique, was not different in patients. However, we could detecte hemorheologic disturbances in *CVI* patients when applicating the *PVC*. Indeed hematocrit, plasma fibrinogen and viscosity were found to be increased and explain the erythrocyte hyperaggregation as shown by a decrease in *AT* and also increases in *AI* and *DSS* as a result of *PVS*. In contrast to patients, in the subjects all parameters excepted the *DSS* remained unchanged following the *PVS*. From these findings, one can be drawn to conclude that patients with *CVI* even in the absence of hemorheologic abnormalities at rest, might be predisposed to develop them in stasis situations. Future studies should aim to determine whether the *PVS*-induced change would depend on the severity and duration of venous insufficiency and whether it can be reestablished by therapeutic measures.

		BEFORE STASIS		END OF STASIS
Hematocrit (%)	Controls	42.3 ± 3.40	NS	42.3 ± 5.10
		$p < 0.05$		NS
	patients	39.2 ± 3.20	$p < 0.001$	40.8 ± 3.90
Fibrinogen (g / l)	Controls	2.68 ± 0.20	NS	2.70 ± 0.20
		NS		NS
	patients	2.71 ± 0.44	$p < 0.001$	3.10 ± 0.56
Albumin (g / l)	Controls	45.6 ± 5.10	NS	48.0 ± 4.90
		$p < 0.01$		NS
	patients	42.8 ± 2.30	$p < 0.01$	44.9 ± 3.20
Plasma viscosity (mPa.s)	Controls	1.30 ± 0.03	NS	1.31 ± 0.06
		NS		NS
	patients	1.27 ± 0.06	$p < 0.01$	1.32 ± 0.05

Table 1 : Hematocrit, Fibrinogen, Albumin and plasma viscosity before and at the end of the stasis in 11 healthy subjects and in 24 patients with CVI

		BEFORE STASIS			END OF STASIS	
Aggregation time (mPa)-1	Controls	21.2 ± 3.50	NS	NS	20.7 4.80	
		NS			p < 0.02	
	patients	19.8 ± 4.10	p < 0.001		16.1 ± 3.70	
Erythrocyte aggregation index	Controls	31.8 ± 2.30	NS	NS	32.7 ± 3.90	
		NS			p < 0.02	
	patients	33.3 ± 3.80	p < 0.001		37.6 ± 4.30	
RBC disaggregation Shear stress (mPa)	Controls	300 ± 32	p < 0.05		321 ± 37	NS
		NS			NS	
	patients	303 ± 46	p < 0.01		343 ± 63	
Relative viscosity of blood at 128.5 s-1	Controls	3.50 ± 0.20	NS		3.60 ± 0.20	NS
		NS			NS	
	patients	3.40 ± 0.20	NS		3.40 ± 0.20	
Relative viscosity of blood at 1.285 s-1	Controls	8.30 ± 0.70	p < 0.01		9.40 ± 0.80	NS
		NS			NS	
	patients	9.40 ± 1.70	p < 0.001		11.0 ± 2.30	
Red blood cell Rigidity index	Controls	11.2 ± 0.80	NS		10.3 ± 1.36	NS
		NS			NS	
	patients	10.6 ± 1.80	NS		10.7 ± 1.60	

Table 2 : Hemorheologic parameters before and at the end of the provoked venous stasis in 11 healthy subjects and in 24 patients with chronic venous insufficiency

		BEFORE STASIS	END OF STASIS	AFTER STASIS
LASER DOPPLER BLOOD FLUX "BF"(PU)	Controls	12.6 ± 6.6	2.2 ± 1.7	29.6 ± 11
		NS	NS	p < 0.01
	Patients	8.4 ± 6.9	1.9 ± 1.2	17.3 ± 9.0
VASOMOTION FREQUENCY (Cycles/minute)	Controls	3.7 ± 1.7	1.0 ± 0.5	7.4 ± 3.3
		p < 0.001	NS	p < 0.001
	Patients	1.9 ± 1.0	0.8 ± 0.9	4.1 ± 1.9
TcPO2 (mm Hg)	Controls	70.2 ± 7.70	42.0 ± 13.0	75.5 ± 14.0
		p < 0.01	NS	p < 0.001
	Patients	55.3 ± 16.0	36.0 ± 19.0	61.4 ± 19.7

Table 3 : Microcirculatory parameters before stasis, at the end of stasis and after stasis in 11 healthy subjects and in 24 patients with CVI.

2. Microcirculatory parameters (Table 3)

When compared to the controls before stasis, patients showed a small but insignificant difference in blood flux while other authors (8,9,10) showed higher resting flux. However, it should be noted that these authors studied the patients presenting liposclerosis or venous ulcerations or those with venous hypertension. Vasomotion frequency and TcPO2 were already significantly lower in patients. The lower TcPO2 values we found are in accordance with those reported eleswhere (11,12). Differences we already noticed before stasis become more pronounced after stasis. Our data indicate that there is a difference in the dynamic of microcirculation which would explain the abnormal cutaneous oxygenation.

REFERENCES

1. LE DEVEHAT C., BOISSEAU M., BERTRAND A. Signification physiopathologique des paramètres microrhéologiques dans l'insuffisance veineuse. Hémorhéologie et agrégation érythrocytaire EM Intern VOL 2, 1989, 175-180.
2. LE DEVEHAT C., VIMEUX M., BONDOUX G., BERTRAND A. Hemorheological factors in the pathophysiology of venous disease. Clin. Hemorh. Vol 9, 1989, 237-246.
3. DORMANDY JA., NASH G. Importance of red cell aggregation in venous pathology. Clin. Hemorh. Vol 7, 1987, 119-122.
4. SHAMI SK., CHEATLE TR., SCURR JH. Vasodilatory capacity in venous disease. Phlebology 92, J. Libbey Eurotext 1992, 94-96.
5. SHAMI SK., CHITTENDEN SJ., SCURR JH. How is blood flow altered in venous disease. Phlebology 92, J. Libbey Eurotext 1992, 80-90.
6. DONNER H., SIADAT M., STOLTZ JF. Erythrocyte aggregation : approach by light scattering determination. Biorh.25, 1988, 367-375.
7. HANSS M. Erythrocyte filtrability measurement by the initial flow rate method. Biorh. 20, 1983, 199-211.
8. LEU AJ., YANAR A., GEIGER M., FRANZECK UK., BOLLINGER A. Microangiopathy in chronic venous insufficiency before and after sclerotherapy and compression treatment. Results of a one year fellow up study. Phlebology 8, 1993, 99-106.
9. BELCARO G., CHRISTOPOULOS D., NICOLAIDES A. Skin flow and swelling in post phlebitic limbs. Vasa 18, 1989, 136-139.
10. BALCARO G., GRIGG M., RULO A., NICOLAIDES A. Blood flow in the perimalleolar skin in relation to posture in patients with venous hypertension. Ann. Vas. Surg. 3 (1), 1989, 5-7.
11. FAGRELL B. Microcirculatory disturbances. The final cause for venous leg ulcers ? Vasa 11, 1982, 101.
12. FRANZECK U., BOLLINGER A., HUCH R., HUCH A. Transcutaneous oxygen tension and capillary morphologic characteristics and density in patients whith chronic venous incompetence. Circul. 70, 1984, 806.

Phlebology '95, D. Negus et al. (eds.). Phlebology (1995) Suppl. 1: 209-213

PI/5.3

La Pompe Veineuse du Mollet: Etude des Pressions Veineuses et Musculaires

P. Barthèlemy, Y. Alimi and C. Juhan

Service de Chirurgie Vasculaire, Hôpital Nord, Marseille, France

The Calf Pump: Study of the Venous and Vascular Pressures

SUMMARY

The authors have studied the simultaneous pressures within the long saphenous vein, the popliteal vein and the three fascial compartments of the leg by means of micro-catheters. Values in normal subjects in various positions and during a Valsalva manoeuvre are reported.

LA POMPE VEINEUSE DU MOLLET: ETUDE DES PRESSIONS VEINEUSES ET MUSCULAIRES

P. Barthèlemy, Y. Alimi, C. Juhan

Service de Chirurgie Vasculaire, Hôpital Nord, Marseille, France

INTRODUCTION

L'analyse des différents éléments permettant le retour du sang veineux des membres inférieurs vers le coeur, souligne le rôle essentiel joué par la pompe musculaire du mollet en orthostatisme et lors de la marche (2, 6). De nombreuses études épidémiologiques, y compris celles que nous avons réalisées (3), ont montrées la grande fréquence de la pathologie veineuse superficielle et/ou profonde, chez la femme. L'origine de ces affections peut se trouver dans un dysfonctionnement de la pompe musculaire du mollet. Les éléments de la littérature ne permettent pas de définir avec précision le mode de fonctionnement normal de cette pompe. C'est pour cela que nous avons réalisé, afin d'établir les valeurs normales lors d'un certain nombre de situations courantes (décubitus, orthostatisme, station accroupie, marche), chez des sujets sains et volontaires, la mesure simultanée des pressions intramusculaires, intraveineuses profonde et superficielle, grâce à l'introduction et à l'enregistrement de cinq micro-cathéters placés dans chacune des trois loges de jambes, dans la veine poplitée et dans la veine saphène interne. Les modifications des pressions veineuses et l'étude du retentissement musculaire ont été étudiées lors de manoeuvres de Valsalva qui agissent directement sur la colonne de sang.

SUJETS ET METHODES

SUJETS

Etant donné la nette prédominance féminine de la pathologie veineuse des membres inférieurs et afin d'obtenir un groupe homogène, l'étude a porté sur 9 jeunes femmes nullipares de 18 à 25 ans, ayant donné leur consentement éclairé écrit, indemne de toute pathologie veineuse, lymphatique ou artérielle des membres inférieurs.

METHODES

Au terme d'un échodoppler couleur réalisé systématiquement, sur le sujet debout, pour controler l'absence de pathologie veineuseet repérer la veine poplitée et la veine saphène interne à la cheville. Une anesthésie locale a été réalisée avant toute ponction. La veine poplitée (VP) a été ponctionnée chez le sujet, debout, un cathéter y a été introduit. Un autre cathéter a été placé dans la veine saphène interne, juste au-dessus de la malléole interne.

La prise de pression musculaire a utilisé la technique de perfusion microcapillaire décrite par Stif J. et coll. (5). Après l'anesthésie locale un micro-cathéter intramusculaire a été implanté dans chacune des trois loges de jambe : loge postérieure profonde (LPP), loge postérieure superficielle (LPS) et loge antéro-externe (LAE). Un enregistrement simultané des cinq chaînes de pressions a été réalisé lors d'une série de situations étudiées selon un protocole standard. Les différentes mesures ont été effectuées selon les caractéristiques suivantes. La manoeuvre de Valsalva a été contrôlée sur colonne de mercure (dépression de 40mmHg), la dorsiflexion maximale du pied a consisté en une pression exercée d'environ 15kg, la flexion plantaire maximale du pied a consisté en une pression exercée d'environ 40kg, les manoeuvres effectuées debout ont été réalisées sur un membre en charge et la marche a été réalisée sur un tapis roulant défilant à 3km/h, avec une pente de 10%.

I- Valeurs normales des pressions musculaires et veineuses lors des différentes situations (Pressions moyennes et erreurs standards en mmHg).

	L.P.P.	L.P.S.	L.A.E.	V.S.I.	V.P.
DECUBITUS					
Repos	12,5+2,7	8,0+2,4	12,0+5,7	16,3+3,3	14,6+3,9
Valsalva	14,7+3,4	8,7+2,8	9,4+2,9	17,4+3,7	30,1+4,3
DorsFlex.Pied	31,8+5,7	14,8+6,7	68,7+13,7	24,6+6,4	18,2+4,4
FlexPlant.Pied	52,3+12,2	16,8+9,3	22,6+8,3	18,2+3,1	17,7+4,2
Flex.c/t	34,7+4,9	26,4+4,2	34,2+4,4	27,3+4,0	52,2+15,6
ASSIS					
Repos	-1,0+4,4	0,5+5,1	2,8+2,5	54,2+7,8	23,9+3,5
Valsalva	1,0+6,0	0,2+6,0	3,1+3,4	62,5+11,3	41,6+4,9
DorsFlex.Pied	16,1+7,2	7,8+7,7	54,9+12,6	74,9+13,3	25,7+3,5
FlexPlant.Pied	33,4+11,6	4,3+6,9	17,7+8,1	62,4+8,6	26,2+3,4
Flex.c/t	48,6+5,4	42,8+9,5	51,8+7,1	83,7+11,3	51,7+18,9
DEBOUT					
Repos	16,1+8,3	7,3+9,7	14,8+6,5	74,7+11,5	39,2+5,6
Valsalva	33,5+9,3	7,9+8,9	23,3+6,0	90,9+10,4	65,1+5,6
DorsFlex.Pied	39,1+13,3	9,8+10,8	82,3+13,8	87,4+11,7	42,3+4,3
FlexPlant.Pied	76,8+17,4	11,1+10,4	20,0+11,5	79,6+12,6	41,5+4,1
Flex.c/t	62,8+10,4	35,6+7,0	64,6+6,8	103,0+6,1	57,0+14,5
ACCROUPIE					
Repos	64,2+16,6	66,7+23,9	70,1+17,2	62,7+19,0	56,7+14,9
Valsalva	74,7+14,3	55,4+24,5	68,9+14,5	74,4+19,1	67,4+14,4
MARCHE					
Pied Sol	61,9+14,4	18,6+9,8	51,3+7,6	51,0+7,3 5	56,9+5,8
Pied en l'Air	32,3+20,2	9,1+10,9	43,3+7,8	38,6+8,2	35,0+10,6

II- Etude des variations de pressions musculaires et veineuses lors des différentes situations

Au repos, une élévation significative des pressions dans la VSI et dans la VP est notée lors de la position assise, lors de la position debout, et lors de la position accroupie. Les pressions moyennes dans les trois loges musculaires n'augmentent significativement que dans la position accroupie. La manoeuvre de Valsalva est responsable de variations significatives des pressions veineuses et musculaires uniquement lors du décubitus dans la VSI et lors de la position debout dans la VSI et la VP. La dorsiflexion du pied entraîne une augmentation très significative de la pression dans la LAE en décubitus, en position assise et en position debout. L'élévation des pressions dans la LPP est moins importante en décubitus, en position assise et position debout, alors que l'augmentation des pressions dans la LPS n'est pas significative quelle que soit la position. Cette augmentation contrastée des pressions selon la loge musculaire étudiée s'associe à une valeur élevée du gradient de pression dans la VSI, en décubitus, en position assise et en station debout. L'élévation moyenne de la pression dans la VP est uniquement significative en décubitus. La flexion plantaire du pied est associée à une hausse significative des pressions uniquement dans la LPP, en décubitus, en position assise et en station debout. Les pressions veineuses n'augmentent pas significativement quelle que soit la position. La flexion de la cuisse sur le thorax est responsable d'une augmentation significative des pressions musculaires dans les trois loges musculaires et dans les trois positions étudiées (décubitus, assis, debout) à l'exception de la LPS en position debout. L'élévation des pressions veineuses est significative dans la VSI et dans la VP, uniquement en décubitus et en station debout mais pas en position assise. Au cours de la marche les valeurs de pressions musculaires et veineuses sont significativement plus élevées dans la position pied au sol par rapport à la position pied en l'air.

DISCUSSION

L'étude des pressions dans le système veineux profond et superficiel des membres inférieurs (1,2,6), met en évidence la complexité des gradients de pression veineux et le rôle essentiel joué par la pompe musculaire du mollet. Les pressions veineuses superficielle et profonde s'égalisent lors du décubitus, grâce au bon fonctionnement du réseau des veines perforantes. Cette étude permet d'émettre des hypothèses de fonctionnement physiologique de la pompe musculaire du mollet: l'augmentation significative des pressions veineuses liée à l'action de la pression hydrostatique, n'est pas transmise dans les loges musculaires, du fait de l'action protectrice possible des valvules veineuses. A l'inverse, l'augmentation première des pressions dans les loges musculaires nécessaire à la statique du corps, est directement transmise dans les secteurs veineux superficiel et profond. La manoeuvre de Valsalva bloque le retour veineux. Ainsi, lors du décubitus, la pression augmente significativement uniquement dans la VP, et lors de la station debout, les pressions s'élèvent significativement dans la VSI et dans la VP. Les pressions dans les loges musculaires n'augmentent pas en décubitus mais s'élèvent significativement dans la LPP et dans la LAE en station debout. L'action protectrice des valvules veineuses, évoquée précédemment, semble dépasser lors de l'hyperpression réalisée debout. En position assise et accroupie, il semble que l'hyperpression ne puisse être transmise à la colonne de sang veineux. Pour agir dans le même temps sur la VSI et sur la VP, l'obstacle à la transmission de l'hyperpression doit se situer au niveau de l'arcade crurale et peut être une compression de la veine fémorale commune par l'arcade crurale. L'étude séparée de ces différents mouvements permet une meilleure compréhension de l'action de chacune des loges musculaires. Chaque mouvement entraîne une très significative augmentation de la pression dans une seule loge musculaire et est associé à une légère élévation des pressions veineuses. L'action normale de la pompe musculaire lors des mouvements de dorsiflexion-flexion plantaire du pied, n'entraîne pas d'élévation significative de la pression dans la VP ; ceci peut être interprété comme une vidange veineuse satisfaisante et est corrélé avec les résultats de Raju et coll. (4). Par contre, l'activité musculaire est responsable de variations significatives des pressions dans le système veineux superficiel. Il existe lors de la flexion de la cuisse sur le thorax une élévation significative des pressions dans les trois loges musculaires, pratiquement dans toutes les positions. Parallèlement, les pressions veineuses augmentent, de façon significative, en décubitus et en station debout, mais pas en position assise. Ceci semble confirmer le rôle d'obstacle que joue l'arcade crurale sur la veine fémorale commune et qui entraîne donc une hyperpression dans les systèmes veineux superficiel et profond lorsque le sujet passe d'une position sans angulation (décubitus ou station debout) à une position angulée, lorsqu'il plie la cuisse sur le thorax. Par contre, cet obstacle existant déjà en position assise, il ne peut être responsable une nouvelle fois d'une augmentation significative des pressions veineuses. Après 5 minutes de marche, la pression dans les deux loges musculaires postérieures est significativement plus élevée chaque fois que le pied sur le sol comparée à la situation où le pied quitte le sol. La pompe musculaire du mollet qui doit évacuer l'augmentation d'afflux de sang issu du système artériel, agit donc bien en deux temps : une période systolique, lorsque le pied est en appui, avec élévation significative des pressions veineuses et une période diastolique, lorsque le pied quitte le sol, avec pression veineuse basse. Nous mettons en évidence une augmentation significative des pressions dans la VP comme dans la VSI durant la phase systolique, traduisant une action concomitante des loges musculaires sur le système veineux profond comme sur le système superficiel. Une élévation significative de pression dans la LPS, survient électivement durant la marche. Ainsi, l'action de la LPS semble être préférentiellement recrutée. La LPS n'intervient pas sur les pressions veineuse, lors du maintien d'un tonus statique mais lors d'un mouvement dynamique.

BIBLIOGRAPHIE

1. Arnoldi CC. Physiology and pathophysiology of the venous pump of the calf (Chp 2). In : « Controversies in the management of venous disorders » Eklöf B, Gjöres JE, Thulesius O, Bergqvist D, Eds., London Butterworth, 1989;6-23.

2. Browse NR, Burnand KG, Lea Thomas M. Physiology and functional anatomy (Chp3). In : « Diseases of the veins pathology, diagnosis and treatment ». Edward Arnold Eds, London, 1988; 53-69.

3. Juhan C, Barthèlemy P, Alimi Y, Morati N, Lelong B, Dominguez M, Flori A. Etude de la prévalence de l'incontinence des veines jumelles par échodoppler couleur (modifications de la stratégie thérapeutique). Bull Acad Natle Méd 1993;177:233-41.

4. Raju S, Fredericks R, Lischman P, Neglen P, Morano J Observations on the calf venous pump mechanism: determinants of postexercise pressure. J Vasc Surg 1993;17:459-69.

5. Styf JR, Körner LM. Microcapillary infusion technique for measurement of intramuscular pressure during exercice. Clin Orthop 1986; 207: 253-62.

6. Sumner DS. Hemodynamics and patholphysiology of venous disease. In : « Vascular Surgery » Rutherfors R D, 3rd Ed. SAUNDERS, Philadelphia, 1989.

Phlebology '95, D. Negus et al. (eds.). Phlebology (1995) Suppl. 1: 214-216

P111

Estrogen and Progesterone Receptors in Normal Saphenous Vein in Male and Female

A. Mashiah[1], H. Ben-Hur[2], H.H. Thole[3] and S.S. Rose[4]

[1] Department of Vascular Surgery and [2] Obstetrics and Gynecology, Kaplan Hospital, Rehovot, Israel
[3] Max-Planck-Institute für Experimentalle Endokrinologie, Hannover, Germany
[4] University Hospital of South Manchester, Manchester, England

INTRODUCTION

The involvement of hormones in the etiology of varicose veins is well known. During pregnancy, when levels of estrogen are elevated, big and painful varicose veins develop in the lower limbs. They regress after birth when sex-hormone levels decline [1,2]. The increased frequency of varicose veins in females is another manifestation of the hormonal participation in the underlying causes of varicose veins.

In order to study the mechanism of this hormonal effect, we have investigated the presence of estrogen and progesterone receptors in the walls of normal veins, and compared their expression in males and females.

METHODS

Saphenous veins samples were obtained from a group of patients, 20 males and 10 females, 60-80 years of age, undergoing coronary artery bypass surgery. Vein samples were frozen instantly in isopentane cooled by liquid nitrogen and stored at -70°C.

Immunohistochemical detection of estrogen and progesterone receptors was accomplished by indirect immunostaining of vein samples. Cryostat sections, 7 μm thick, were fixed in Zamboni fixative solution, washed with phosphate buffered saline (2 x 5 min.) and in tenfold diluted normal goat serum (20 min.) to reduce non-specific binding. Consequently the slides were incubated overnight (at 4°C) with monoclonal anti estrogen-receptor antibodies (13H$_2$ from the Max-Planck Institute) [3-5] or anti progesterone-receptor antibodies (Enco scientific Services, Newcastle UK), or with normal mouse IgG1 (Zymed Labs, USA). After washing, Protein A coupled to horseradish peroxidase (Strep A-B reagent Universal Kit, Diagnostic Products, Los Angeles USA) was applied to the sections (30 min.). Following additional washes immunoperoxidase staining with diaminobenzidine was performed. After dehydration and mounting slides were examined by light microscopy and nuclear stain intensity was scored on a scale of 1 (low) to 4 (high) by three independent observers.

RESULTS

Ninety-seven vein samples, 32 from females and 65 from males were checked, and estrogen and progesterone receptors were identified and estimated visually. In the large majority of the veins such hormone receptors were found (Figures 1-3). They were localized in the nuclear area of the intima and media in both female and male saphenous veins. In the adventitia the estrogen and the progesterone receptors were found only in the nuclei of the vasa vasorum. The density of receptors scored in males was high (score of 3-4) for both receptor types. In one man (out of twenty) no progesterone receptors were noted. In one woman (out of ten) no estrogen receptors were found but progesterone receptors were present.

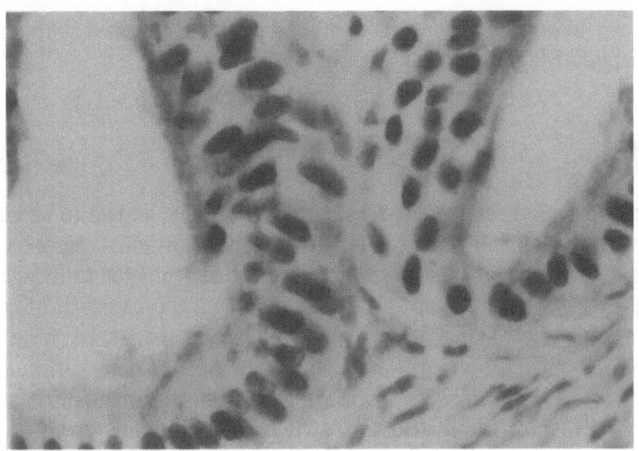

Figure 1: Endometrial gland stained with anti estrogen-receptor. Magnification x 600. (Control slide)

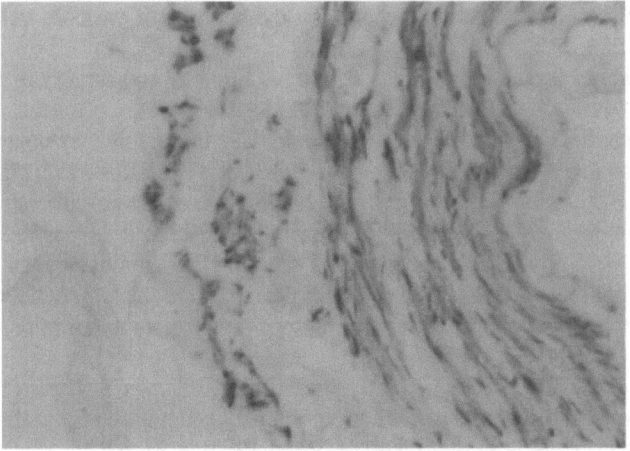

Figure 2: Female saphenous vein stained with anti estrogen-receptor. Notice the endothel, muscularis and vasa vasorum. Magnification x 150.

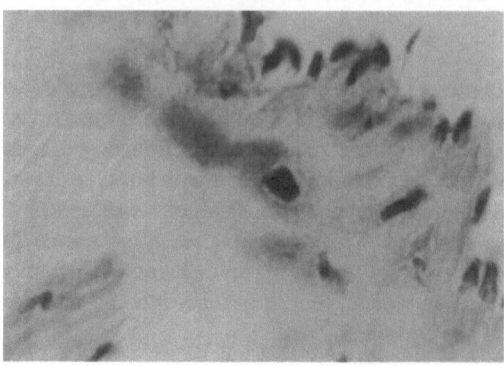

Figure 3: Nuclei in normal saphenous vein wall positively stained with anti estrogen-receptor. Magnification x 600

DISCUSSION

Estrogen receptors have been previously detected in organs known to be controlled by sex steroids, such as uterus or bones [6,7]. However the hormonal effect on veins is not established, and no mechanism has been proposed for such alleged influence, especially in the case of varicose veins. This is the first report on the presence of sex hormone receptors in vein walls. Their presence and apparent abundance were surprising, especially when found in older patients, and with no apparent difference between females and males. All patients had blood estradiol levels normal for their age.

The prevalence and distribution of these sex-hormone receptors in different age groups and in particular in varicose veins is presently under investigation.

REFERENCES

1. Rose S, Ahmed A. Some thoughts on the etiology of varicose veins. Cardiovas Surg 1986;27:534.
2. Mashiah A, Sternfeld M, Reina A, Hod I. Etiology and surgery of varicose veins of the lower limb. Harefua 1991;121:191-192.
3. Sierralta WD, Thole HH. Immunogold labelling of the cytoplasmic estradiol receptor in resting porcine endometrium. Cell Tissue Res 1992;270:1-6.
4. Thole HH, Assignment of the ligand binding site of the porcine estradiol receptor to the N-terminal 17 kDa part of domain E. FEBS Lett 1993;320:92-96.
5. Jacob F, Tony HP, Schneider D, Thole HH. Immunological detection of oestradiol receptor protein in cell lines derived from the lymphatic system and the haematopoietic system: variability of specific hormone binding *in vitro*. J Endocrinol 1992;134:397-404.
6. Ben-Hur H, Mor, Insler V, Blickstein I, Amir-Zaltsman Y, Kohen F. Assessment of estrogen receptor distribution in human endometrium by direct immuno-fluorescence. Acta Obstet Gynecol Scand 1995;74:97-102.
7. Ben-Hur H, Mor G, Blickstein I, Likhman I, Kohen F, Dgani R, Insler V, Ornoy A. Localization of estrogen receptors in long bones and vertebrae of human fetus. Calcified Tissue 1993;53:91-96.

Phlebology '95, D. Negus et al. (eds.). Phlebology (1995) Suppl. 1: 217-219

P083

Mechanical Properties of Lower Limb Skin. A Natural Compression Stocking

A.M Shutt, S.R. Dodds, A.R. Cowan and A.D.B. Chant

Department of Vascular Surgery, Royal South Hants Hospital, Southampton, UK

INTRODUCTION

Skin has a complex microstructure, composed of multiple fibre networks surrounded by interstitial fluid. The networks are covered by a layer of partially keratinized epithelium. In engineering terms skin has highly unusual properties. It is a viscoelastic anisotropic non linear inhomogenous time dependent material which demonstrates the properties of hysteresis, stress relaxation and creep [1]. Viscoelasticity is a combination of elastic characteristics, (a material is elastic if it returns to its predeformed shape on removal of the deforming force), and viscous flow, seen when a fluid is stressed. Anisotropy reflects the variable deformation that can occur as a result of an applied force and varies with the direction in which the force is applied. Stress relaxation is seen when deformed skin is held at constant length, the force required to maintain it at that length diminishes with time. Creep is the increase in length seen when subjected to constant load. Hysteresis is seen when skin has a load applied to it and then removed. The loading and unloading curves do not overlap and an hysteresis loop can be seen. Skin has fascinated a number of earlier workers. Some mechanical properties of skin have been examined in venous disease previously [2]. However, some characteristics were modelled using a single decay exponential term, which due to the viscoelastic behaviour of skin, is not likely to be accurate. We investigated the peak forces required to perform a 30% extension of skin at multiple sites on the lower limb and also followed the stress relaxation of skin at these sites. Stress relaxation was modelled with a double exponential decay term. We were particularly interested in the skin characteristics over the gaiter area.

METHOD

A normal male subject was examined in a temperature and humidity stable environment with a uniaxial skin extensometer (Dermatronics Ltd. UK). A 30% extension of non preconditioned skin was performed over ten seconds at a number of sites on the lower limb. Measurements were made parallel to the long bone axis and perpendicular to it. Peak force at the end of the 30% extension was measured. The stress relaxation curve was recorded for two minutes. Values from each of these curves were taken and analysed by a non linear least squares fit by an iterative numerical method. The skin relaxation is described by a biexponential decay equation of the form:

$$Force\ (t)\ =\ \left(\frac{Fmax\ -\ Fres}{2}\right) \cdot \left(e^{-ct}\ +\ e^{-dt}\right)\ +\ Fres$$

where Fmax is the peak force, and Fres the residual force, C and D are the exponential terms applied. A stress relaxation curve can be seen in Figure 1.

Figure 1.

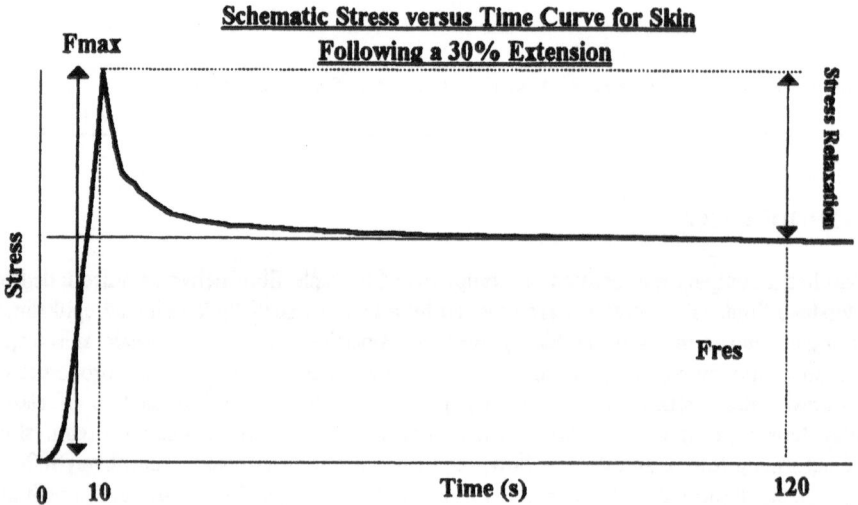

RESULTS

Table 1. Peak Forces (N) at Different Sites on the Lower Leg of a Normal Male Subject

	Perpendicular to Long Bone Axis	
Position on Leg	Medial	Lateral
Upper Thigh	1.6	6.1
Above Knee	2.9	5.1
Below Knee	2.0	10.4
Gaiter	3.9	25.7

Note how easily the skin on the medial side of the leg in the perpendicular direction is extended. Conversely, laterally, a greater force is required. Both medially and laterally the gaiter area requires greater forces to extend it.

DISCUSSION

The complete results indicate, that in general, the skin nearer the ankle requires a greater force to undertake the 30% extension, i.e. it is a stiffer structure and more resistant to stretch. This probably reflects an adaptation to increased forces in the area. It does in effect form a natural compression stocking.

REFERENCES

1. Tregear RT. Physical functions of skin. London: Academic Press, 1966:

2. Mourad MM, Edwards C, Marks R. Skin extensibility in the gravitational syndrome. Bioeng Skin. 1989;4:199-215.

Phlebology '95, D. Negus et al. (eds.). Phlebology (1995) Suppl. 1: 220-221

OP/6.8

CD11b/CD18 as a Marker of Neutrophil Adhesion in Experimental Venous Hypertension

D.A. Shields, S.K. Andaz, C.A. Timothy-Antoine[1], J.H. Scurr and J.B. Porter[1]

Departments of Surgery and Haematology[1], UCLMS, The Middlesex Hospital, Mortimer Street, London W1N 8AA

INTRODUCTION

The white cell trapping hypothesis suggests that raised venous pressure causes white cell margination and activation, with the release of proteolytic enzymes and superoxide radicals, believed responsible for the tissue destruction seen in chronic venous insufficiency [1]. To investigate this we have demonstrated raised levels of plasma lactoferrin [2] and elastase [3] as markers of neutrophil degranulation in patients with venous disease, and also have shown raised levels of lactoferrin in normal volunteers exposed to short term experimental venous hypertension [4]. To investigate this hypothesis further, in this study we have measured neutrophil CD11b, an integrin expressed on the surface of neutrophils during adhesion to vascular endothelium, as a marker of neutrophil adhesion during short-term experimental venous hypertension. This integrin is believed to be the second stage receptor responsible for physiological binding of neutrophils to the vascular endothelium prior to extravascular migration or degranulation [5]. The number of CD11b/CD18 molecules expressed on the cell surface is dependent on the state of cell activation, and neutrophils are able to upregulate rapidly surface integrins from intracellular stores. This upregulation occurs within minutes, and leads in turn to increased binding of the adhesion receptors. We thought that the raised venous pressure seen in venous disease might correlate with the first stage of binding of neutrophils, in that a low flow state would increase the likelihood of the second stage receptors, the integrins, to bind prior to cell activation. Accordingly, we decided to measure neutrophil CD11b levels in a group of normal volunteers exposed to a 30 minute period of venous hypertension.

METHODS

A group of 20 volunteers (M:F 17:3, mean age 30, range 20-40) with no previous history or clinical evidence of arterial or venous disease was investigated. All the volunteers had no other disease, were on no medication known to alter white cell activity, and were free from any recent bacterial or viral infection. A cannula was placed in a dorsal foot vein of each volunteer, who lay supine for 30 minutes. Blood was then taken for full blood count and neutrophil CD11b. The volunteer then stood supported for 30 minutes to raise venous pressure, after which a second blood sample was taken. They then returned to the supine position for 10 minutes, when a third sample was taken. This last sample was taken on the assumption that many neutrophils would become adherent to the vascular endothelium during the period of venous hypertension and so would not be available for measurement. A ten minute period with no hypertension was thought adequate to allow sufficient downregulation of adhesion to occur for the neutrophils to return to the systemic circulation, associated with the increased shear forces on the neutrophils caused by the increase in venous blood flow. The samples were assayed for CD11b using a fluorescent-labelled monoclonal antibody in whole blood, and a full blood count and neutrophil count were also performed.

RESULTS

	Basal	Standing	Supine
Neutrophil mean cell fluorescence	119.4 (63.5-145.7)	107.8 (77.1-160.7)	142.2 (108.3-216.7)
Difference between medians (95% CI)	-3.05 (-25.1 to 18.95)		41.5 (17.5 to 75.3)
Wilcoxon Test	p=.8		p=.005

Neutrophil CD11b expression given as mean cell fluorescence. Descriptors are medians and interquartile ranges, CI = confidence interval.

There was a fall in the white cell:red cell ratio after standing (median 1.21 to 1.17) followed by a rise on return to the supine position (median 1.35, p=.001, Wilcoxon) as white cells re-entered the systemic circulation.

CONCLUSIONS

There is a reduction in the number of white cells leaving a normal limb during short term venous hypertension, associated with a small fall in CD11b expression. This is followed by a rise in the number of white cells released into the circulation on return to dependency, with a significant increase in neutrophil CD11b. This suggests that there is increased binding of neutrophils to vascular endothelium during short term venous hypertension, as would be expected from the white cell trapping hypothesis.

REFERENCES

1. Coleridge Smith PD, Thomas PRS, Scurr JH, Dormandy JA. Causes of venous ulceration: a new hypothesis. *Br Med J* 1988;**296**:1726-7

2. Shields DA, Andaz S, Abeysinghe RD, Porter JB, Scurr JH, Coleridge Smith PD. Lactoferrin as a marker of neutrophil activation in venous disease. *Phlebology* 1994;**9(2)**:55-8

Phlebology '95, D. Negus et al. (eds.). Phlebology (1995) Suppl. 1: 222-224

P046

Lactoferrin Release in Experimental Venous Hypertension

D.A. Shields, S.K. Andaz, R.D. Abeysinghe[1], J.H. Scurr and J.B. Porter[1]

Departments of Surgery and Haematology[1], UCLMS, The Middlesex Hospital, Mortimer Street, London W1N 8AA

INTRODUCTION

The white cell trapping hypothesis suggests that raised venous pressure causes white cell margination and activation, with the release of proteolytic enzymes and superoxide radicals, believed responsible for the tissue destruction seen in chronic venous insufficiency [1]. White cell margination has been shown to occur in the post-capillary venules when blood flow is reduced [2], but there is little evidence to show white cell activation under these circumstances. To investigate this, lactoferrin, a secondary granule content of neutrophils, was measured as a marker for neutrophil degranulation during experimental venous hypertension in normal volunteers.

METHODS

A group of 15 normal subjects (M:F 11:4, mean age 31) was investigated during two models of venous hypertension. None of the volunteers had any history or clinical findings of venous or arterial disease, no other systemic disease, were on no medication known to alter white cell activity, and had no recent infection. Venous blood was taken from cannulae in both feet and the right arm for full blood count, neutrophil count and plasma lactoferrin, measured using an ELISA [3], during application of a pneumatic tourniquet to 80 mmHg for 30 minutes to the right leg while supine, five minutes after release of the tourniquet, and then during a 30 minute period of standing.

RESULTS

	Tourniquet		Standing	
	Pre	Post	Pre	Post
Lactoferrin (ng/10^6 neutrophils)	59 (39.6-91.8)	62.5 (49.3-93)	59.2 (41.9-89.6)	61.4 (39.7-99.5)
Difference between medians (95% CI)	6.92 (.002-16.2)		5.73 (1.2-10)	
Wilcoxon	p=.002		p=.001	

Lactoferrin levels expressed as median (interquartile range). Descriptors are medians and interquartile ranges. CI = confidence interval.

The total white cell count, neutrophil count and percentage neutrophil count were also recorded throughout the experiment. There was a significant increase in the white cell count in the right leg during application of the tourniquet and on standing, and in the left leg and arm following release of the tourniquet and on standing. The white cell:red cell ratio was also calculated to look for white cell trapping (indicated by a fall in the WBC:RBC ratio). During the first part of the experiment, the WBC:RBC ratio showed a non-significant fall in the tourniquet leg and a non-significant increase in the other two limbs. During the second part of the experiment there was a non-significant fall in both lower limbs. Previous workers have shown a significant fall in the WBC:RBC ratio in the leg in patients with chronic venous insufficiency [4] and a non-significant fall in normal controls [5], and our work is in agreement with this finding.

CONCLUSIONS

The results show a significant rise in plasma lactoferrin compared to baseline values in two models of venous hypertension, demonstrating that short-term venous hypertension causes neutrophil activation in normal subjects. These findings provide the first direct evidence of neutrophil activation caused by venous hypertension in the lower limb in normal subjects, in keeping with the white cell trapping hypothesis.

REFERENCES

1. Coleridge Smith PD, Thomas PRS, Scurr JH, Dormandy JA. Causes of venous ulceration: a new hypothesis. *Br Med J* 1988;**296**:1726-7

2. Braide M, Amundson B, Chien S, Bagge U. Quantitative studies of leukocytes on the vascular resistance in a skeletal muscle preparation. *Microvasc Res* 1984;**27**:331-52

3. Devereux S, Porter JB, Hoyes KP, Abeysinghe RD, Saib R, Linch DC. Secretion of neutrophil secondary granules occurs during granulocyte-macrophage colony stimulating factor induced margination. *Br J Haematol* 1990;**74**:17-23

4. Thomas PRS, Nash GB, Dormandy JA. White cell accumulation in the dependent legs of patients with venous hypertension: a possible mechanism for trophic changes in the skin. *Br Med J* 1988;**296**:1693-5

5. Moyses C, Cederholm-Williams SA, Michel CC. Haemoconcentration and accumulation of white cells in the feet during venous stasis. *Int J Microcirc Exp* 1987;**5**:311-20

3. Shields DA, Andaz S, Sarin S, Scurr JH, Coleridge Smith PD. Plasma elastase in venous disease. *Br J Surg* 1994:**81**;1496-9

4. Shields DA, Andaz S, Abeysinghe RD, Porter JB, Scurr JH, Coleridge Smith PD. Neutrophil activation in experimental venous hypertension. *Phlebology* 1994;**9(3)**:119-24

5. Pardi R, Inverardi L, Bender JR. Regulatory mechanisms in leukocyte adhesion: flexible receptors for sophisticated travelers. *Immunology Today* 1992;**13(6)**:224-30

P092
HAEMODYNAMICS OF INCOMPETENT PERFORATOR VEINS IN PRIMARY VARICOSE VEINS.

M. Cappelli, R. Molino Lova, S. Ermini, A. Turchi, and G. Bono.
Phlebologic Surgery, 46 Datini Street, Florence, Italy.

Objective: To investigate perforator veins behavior in the upstanding patient at muscle compression-relaxation test. Perforator veins are divided into 2 groups: "terminal", i.e. below which the saphenous vein is competent, and "non terminal", i.e. below which the saphenous vein is still incompetent.
Design: Retrospective review of a consecutive series.
Patients: 68 consecutive patients suffering from primary varicose veins.
Measurements: Venous flow direction at pulsed Doppler, once the perforator vein has been identified by Echography.
Results: During muscle relaxation "non terminal" perforator veins always presented with reflux, while "terminal" perforator veins never presented with reflux, mostly showing suction from superficial to deep veins.
Conclusions: 1) It is possible to transform a "non terminal" into a "terminal" perforator vein either by surgical interruption of the saphenous vein 1 inch below the incompetent perforator vein, for therapeutic purpose, or by external compression, for diagnostic purpose. 2) Perforator vein incompetence itself does not always mean reflux.

P093
"PRIVATE CIRCULATIONS": AN HAEMODYNAMIC APPROACH TO VARICOSE VEINS.

M. Cappelli, R. Molino Lova, S. Ermini, A. Turchi and G. Bono.
Phlebologic Surgery, 46 Datini Street, Florence, Italy.

Objective: To classify haemodynamic patterns of venous circulation in primary varicose veins.
Design: Retrospective review of consecutive series.
Patients: 600 consecutive patients.
Measurements: Echo-Doppler haemodynamic "map".
Results: 2.2% of varicose veins were related to "Deflux", i.e. retrograde flow without escape point. 97.8% of varicose veins were related to "Reflux", i.e. retrograde flow with escape point. According to Franceschi and Bailly anatomical classification of venous-venous shunts these last cases have been diveded in: Type 1 Shunt (35.4%), Type 2 Shunt (2.2%), Type 3 Shunt (48.5%), Type 4 Shunt (5.2%), Type R3 Shunt (6.5%). (Preliminary report).
Conclusions: The study provides haemodynamic guidelines to understand circulatory patterns of varicose veins, essential requirement to be fulfilled for a proper haemodynamic surgical treatment.

P037

EXERCISE INDUCED PURPURA AND DECREASED VENOUS DISEASE

J C J M Varaart[1], M Prins[1], A H M Vermeulen[2], H A M Newmann[1]

Department of Dermatology[1] and Pathology[2], Academisch Zeikenhuis Maastrict, The Netherlands

Objective: Several studies have been performed to determine the effects of duration exercise on the vascular system and microcirculation. However, few investigators have studied the influence of extremes of exercise on their subjects.

Design: Open clinical study

Patients: 54 lower extremities from 54 long-distance walkers (37 males, 15 females, sex in 2 not registered) were evaluated within one hour following an 80 kilometre walk. The mean age was 42 ± 11 years (age range 22-71 years).

Measurements: Extremities were evaluated by light reflex rheography LRR (Laumann 1000 Rheography[R] Selb, Germany). From 10 legs with purpura skin biopsies were taken for routine and immunofluorescence investigations.

Results: The mean value of the VRT found in the studied 52 LRR curves was 9.5 ± 5.6 seconds (range 1.0-26 seconds) which was significant (p<0.001) different from normal values. No statistical significant difference of VRT was found between the left and right leg or between men and women. All skin specimens showed signs of a leucocytoclastic vasculitis with deposition of c3 and IgC in some.

Conclusions: This study clearly shows that endurance exercise can induce a decompensation of the venous muscle pump function as well as a leucocytoclastic vasculitis.

OP/18.2

EFFECTS OF COMPRESSION ON THE MICROCIRCULATION OF THE HEEL

A Abu-Own, K Sommerville, J H Scurr and P D Coleridge Smith

Department of Surgery, Middlesex Hospital, London, UK

The heel is a common site for the development of pressure ulceration. LASER Doppler fluxmetry was used to assess the effects of compression on the skin microcirculation of the heel in 10 patients at risk of developing pressure ulceration and 10 age-matched healthy subjects. An acrylic indenter with a slot to accommodate a low profile LASER Doppler probe was used to apply compression to the heel region. A pressure sensor was used to measure the applied compression. The resting LASER Doppler flux was measured with the subject lying supine. Compression forces were then applied in increments and the corresponding interface pressure (IP) and LASER Doppler flux (LDF) recorded. The IP and LDF were also measured from the heel while the subject was lying on a low air-loss bed and then on a conventional hospital (NHS) bed.

RESULTS

IP (mmHg)	0	30	50	35(Airbed)	60(NHS)
LDF(AU):					
Controls*	271(153-379)	170(60-290)	20(15-50	47(15-115)	15(10-18)
Patients*	124(68-247)	65(36-130)	25(18-70)	40(29-44)	10(5-23)

AU = arbitrary units; * = median (interquartile range)

The resting LDF is lower in the patient group compared to the control group (p < 0.05). Compression of the heel caused a progressive decrease in LDF in both groups. Compression greater than 50 mmHg as well as lying on an NHS bed (IP = 60 mmHg) reduced the LDF signal to a minimal value (biological zero). On a low air loss bed (IP=35 mmHg) the median LDF was 17% of the resting value in the control group and 32% in the patient group. The results indicate that the heel microcirculation is vulnerable to compression. A low air loss bed maintained the IP sufficiently low to prevent complete cessation of the heel microcirculation.

P100

ARTERIO-VENOUS TRANSIT TIME A METHOD TO MEASURE VENOUS
VELOCITY

Rai D, Chokshi V,
Dept of Surgery and Radiology,Interfaith Medical Center,
555Prospect place, Brooklyn,N.Y.11238,U.S.A.

Objective:To understand the motion of venous blood in
patients having generalised incompetency of venous valves
(IVV)of lower extremity deep and superficial venous system.
Design:Study is conducted in 10 patients and control
group of 2 normal individuals.Patients with IVV underwent
Vein Valve Transplantation to below knee popliteal vein.
Measurement:20Mci of technitium 99 in human serum albumin
is injected into femoral artery of diseased leg.The data
are aquired by gamma camera,stored and processed by
computer. The time activity curve is generated by drawing
region of interest on iliac vein of same side.In each
patient pre & post operative gamma varient graph obtained.
tl-is the transit time to reach peak activity.This
depends on the velocity of arterial and venous blood.
Results:The mean preop tl is 160 sec.The mean postop tl
is 72sec.Mean fall of tl following surgery is 88sec.
(p=0.0431).Mean tl in control group is 28sec.Given the
same arterial ankle index,blood pressure,cardiac rate &
body temperature,pre & post operative change in tl is due
to change in venous velocity.None of these patients show
significant changes in venous ambulatory pressure(VAP)
Conclusion:Chronic venous insufficiency 2ry to IVV- in
these patients the study of venous velocity gives a better
understanding of venous haemodynamics and a better
parameter for diagnosis and follow up in pre and post
operative state,than the study of VAP.

The Investigation of Venous Disorders

Aethoxysklerol®

worldwide №1 of sclerosing agents
№1 des sclérosants dans le monde

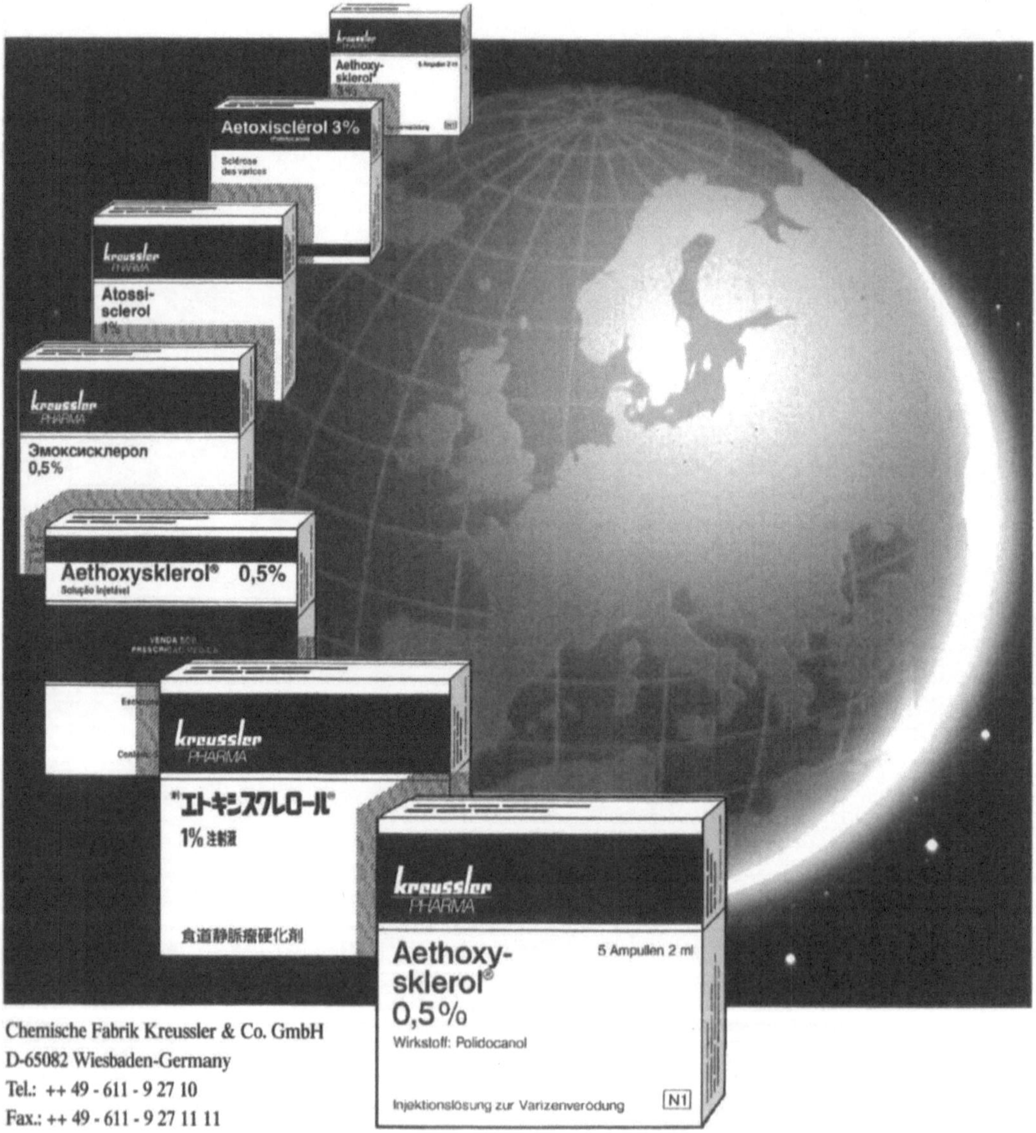

The Investigation of Venous Disorders

Non Invasive Investigations

The Investigation of Venous Disorders

Non Invasive Investigations
Duplex Scanning

Phlebology '95, D. Negus et al. (eds.). Phlebology (1995) Suppl. 1: 233-235

P059

The Incidence of Atypical Refluxes in Varicose Veins: Instrumental Pre-Operative Study

R. Pepe[1], E. Marchitelli[1], R. Gloria[1], M. Guazzaroni[2], M. Pocek[2], A. Romagnoli[2] and G. Simonetti[2]

[1] Department of Angiology, S. Eugenio Hospital of Rome, Italy
[2] Department of Radiology, S. Eugenio Hospital, Tor Vergata University of Rome, Italy

INTRODUCTION

Normal sapheno-femoral and sapheno-popliteal crosses can be found in evaluating so-called primary varicose veins. In these cases superficial refluxes can be related only to primary deep venous insufficiency or to incompetence of typical or atypical perforating veins.These ones have to be considered the main cause of some postoperative relapses.[1,2,3]. For these reasons the role of a complete and carefoul preoperative evaluation has to be underlined.
Duplex scanning with colour flow imaging has recently emerged as the newest technological advancement in the non invasive evaluation of vascular diseases and has recently been applied to the study of venous diseases else. The direct visualization of deep or superficial refluxes expecially in unusual sites, and evaluation of quantitative / functional prevalence became possible[4,5,6].
Actually, the third generation of instruments offering a high resolution can be able to detect the most of venous refluxes, although discendent venography still has to be considered the gold standard in diagnostic[7,8].
The aim of this study is to evaluate the cause of venous refluxes in so-called primary varicose veins and compare these findings with the presence of typical or atypical ones.

MATERIAL AND METHODS

110 patients (157 legs, mean age = 52.3) with so-called primary varicose veins underwent preoperative evaluation from January to June 1994. The study have been performed by a standard echo-color-doppler imaging (Apogex CX Interspect) and all superficial and deep veins of the thigh and leg were explored; particulary, the common site of perforators (Hunter, Boyd, Dodd, Cocckett, etc.) and the uncommon ones, clinically detected. In case of unclear source of refluxes, a dynamic retrotrade phlebography was performed (Telec Siemens Siregraph) through common femoral vein (4-5 Fr catheters, non osmotic contrast Iopamiro 300, 20 cc, 5 ml /sec).

RESULTS

By ultrasound evaluation were detected one or two typical crossing refluxes in 31 legs. In some of these patients, a deep venous incontinence involving superficial femoral vein, was also correctly identified. In these cases the two combined refluxes were prevalent, while combined perforators insufficiency was found in 15 legs.

In 79 legs no crossing refluxes were identified, however, typical or atypical perforating veins insufficiency and/or primary deep veins incontinence, causing segmental saphenous vein varicosities were detected.

In 14 legs, where duplex-scan did not show the refluxes sources, pelvic or femoral veins refluxes were correctly identified by phlebography (tab 1).

Table 1.

N ° LEGS = 157	%	REFLUXES SITES
64	40,7	7 saph. fem. cross 8 saph. fem. cross and superficial fem. vein. 5 saph. popl. cross 11 saph. popl. cross and saph. popl.
79	50	26 thighs perforators 14 legs perforators 19 thighs and legs p. 16 superficial fem. vein and thighs perforators. 4 deep fem. vein and thigs perforators
14	8, 9	3 epigastric vein 4 hypogastric vein 3 epi. and hypo. vein 4 pelvic and deep fem. vein and thighs perf.

CONCLUSIONS

1) In our series, diagnostic evaluation carried out by ultrasound imaging techniques seems to point out the poor frequency of crossing refluxes in saphenuos varicose veins.

2) A correct preoperative protocol that includes a carefoul investigation of deep and perforator veins refluxes in order to reduce postoperative relapses that are mostly due to these causes.

3) Deep venous system incontinence is frequently associated with crossing or perforators refluxes

4) Pelvic veins are often the source of undetected ultrasound refluxes.

5) Although echo-color-doppler imaging has high sensibility and specificity, retrograde

dynamic phlebography still remains the gold standard in unclear cases.

REFERENCES

1.Gottlob R. May R. Venous valve. London New York, J. Libbey Ed. 1986.
2.Van Cleef J.F. Griton Ph. Ribreau C. Lemaire R. Venous valves and tributary veins. Phlebology 1991,6,223.
3.Sarin S. Scurr J.H. Coleridge Smith P.D. Should we strip the long saphenous vein? A randomized controlled trial. In:Phlébologie 92.Eds P.Raymond-Martimbeau, R.P.Prescott, Zummo M.. J.Libbey Eurotext,Paris 1992.
4.Rosfors S. Bygdeman S. Nordstrom E. Assessment of deep venous incompetence: a prospective study comparing duplex scanning with discending phlebography. Angiology-The J.Vasc.Dis. 4:463,1990.
5.Nicolaides A.N. Quantification of venous reflux by means of duplex scanning. J. Vasc. Surg. 10: 670,1990.
6.Vasdekis S.N. Hobbs J.T. Nicolaides A.N. Colour flow imaging in the assessment of chronic venous insufficiency. In: Advances in vascular Pathology,vol.2 Medica,Amsterdam,1989 Excepta.
7.Morano J.U. Raju S. Chronic venous insufficiency: assessment with descending venography. Radiology 174;441,1990.
8.Almgren B.Eriksson J. Bylund H. Lorelius L.E.: Phlebographic evaluation of non thrombotic deep venous incompetence : new anatomic and functional aspect. J. Vasc. Surg. 11 : 389, 1990.

Phlebology '95, D. Negus et al. (eds.). Phlebology (1995) Suppl. 1: 236-237

PI/5.6

Detection of Venous Reflux in Patients with Chronic Leg Ulcer: Comparison of Hand-Held Doppler Performed by a Nurse Specialist with Colour Duplex Ultrasound Performed by a Radiologist

A.W. Bradbury, P.L. Allan[1], J. Brittenden, A.A. Milne, E.A. Nelson and C.V. Ruckley

Vascular Surgery Unit, University Department of Surgery and [1] Department of Radiology, Royal Infirmary of Edinburgh, Lauriston Place, Edinburgh EH3 9YW Scotland

INTRODUCTION

Although nurse specialists are playing an increasingly important role at the interface between hospital and community care of patients with chronic venous ulcers it is unclear which tasks can be delegated. Traditionally, nurses have been restricted to applying dressings and compression therapy prescribed by doctors. In future the nurse specialist may become more involved in diagnosis and initiation of treatment. Before this can happen it needs to be demonstrated that the nurse specialist can correctly assess the patient and recognise when specialist medical input is required. In order to achieve this goal the nurse specialist must first be shown capable of excluding other causes of ulceration (most importantly arterial impairment and malignancy) and of positively diagnosing the presence of venous disease i.e. reflux.

In order to address the latter question we have prospectively compared hand-held Doppler examination, performed by a nurse specialist (EAN), with colour flow duplex assessment, performed by a Consultant Radiologist (PLA), in the assessment of venous reflux in 152 patients presenting with leg ulceration.

PATIENTS AND METHODS

A prospective study of 152 patients presenting to a leg ulcer clinic. Venous reflux was assessed by a nurse specialist using hand-held Doppler. Colour flow duplex ultrasonography of the deep and superficial venous systems of both legs was performed by a single Consultant Radiologist (PLA) using an Acuson 128 machine (Acuson, Mountain View, California, USA) and either a 5-MHz or 7.5-MHz transducer. Patients were examined standing and in a 30° head up tilt. Venous reflux was elicited using manual calf compression and Valsalva manoeuvre and was considered pathological if reverse flow exceeded 0.5 secs [1].

RESULTS

The mean age of the patients was 69 (range 22-89) years and the male to female ratio was 55:97. The sensitivity, specificity, positive and negative predictive value of hand-held Doppler compared with duplex were as follows: termination of long saphenous vein (81, 33, 85, 30%), long saphenous vein at mid-thigh (84, 33, 88, 27%), short saphenous vein termination (70, 37, 76, 30%), common femoral vein (37, 77, 86, 24%), popliteal vein above knee (54, 74, 83, 41%) and popliteal vein below knee (54, 71, 79, 38%).

CONCLUSIONS

The present study compares hand-held Doppler, performed by a nurse specialist , with duplex examination, performed by a Consultant Radiologist, in the assessment of venous reflux in patients with chronic leg ulceration. The results indicate that in the assessment of superficial reflux within the long and short saphenous veins hand-held Doppler possesses a high sensitivity and positive predictive value but low specificity and negative predictive value. The proportion of un-informative scans where the vein in question could not be adequately visualised was low and similar with respect to both investigative modalities.

By contrast, in the assessment of deep venous reflux, especially at the groin level, hand-held Doppler possessed a low sensitivity but high specificity. This converse with respect to the superficial system is interesting as it has been suggested that duplex scanning is oversensitive in the detection of venous reflux and may be detecting reflux within the deep system that is not clinically significant. With respect to the popliteal fossa, the relatively poor accuracy in assessment of short saphenous and popliteal reflux is perhaps to be expected as the anatomy of the short saphenous vein is highly variable, making it difficult to be certain as to which vessel is being examined by the hand-held Doppler.

In summary the present study demonstrates that hand-held Doppler ultrasound assessment of superficial venous reflux by a nurse specialist is sufficiently accurate in the assessment of superficial reflux that it may be used in the diagnosis and assessment of leg ulcers in the community. This information, together with the exclusion of significant arterial disease by means of Doppler generated ankle:brachial pressure indices (ABPI), allows the nurse specialist to diagnose accurately the great majority of leg ulcers. By contrast, hand-held Doppler assessment of reflux within the deep venous system is unreliable.

For this reason, in cases of doubt or where ulcers are atypical and/or intractable, there should be no hesitation in referring the patient to hospital for a more thorough evaluation that would including duplex ultrasonography.

REFERENCE

1. Sarin S, Sommerville K, Farrah J, Scurr JH, Coleridge Smith PD. Duplex ultrasonography for assessment of venous valvular function of the lower limb. Br J Surg 1994; 81:1591-1595.

Phlebology '95, D. Negus et al. (eds.). Phlebology (1995) Suppl. 1: 238-240

P060

Colour Doppler Sonography in Diagnosis of Patients with Chronic Oedema of Low Extremities

M. Simka, K. Ziaja, T. Drazkiewicz, M. Blaszczynski, T. Urbanek and M. Szewczyk

1st Department of General and Vascular Surgery, Silesian School of Medicine in Katowice, Poland

Introduction

Changes in venous system of low extremity are common causes of limb oedema. There can be: obstruction of deep veins, valvular insufficiency, or combination of pathologies mentioned above. The management of chronic venous insufficiency (surgery, phlebotropic agents, compression stockings) depends on kind of pathology in venous system of the extremity [2, 3].

Methods

In 61 patients (11 males, 50 females, aged 17-65 years, mean age 38 years) with chronic low extremity oedema with suspected venous origin sonographic investigations of 69 extremities were performed. Patients were examined in supine position. Measurements were made with pulsed-wave Doppler colour-flow system with 5 MHz probe (Prisma, Diasonics). To induce retrograde flow in femoral vein Valsalva manoeuvre was used, and manual compression of the thigh was performed to induce popliteal vein reflux. Reflux time greater than 0.5 s was regarded as significant [1, 4, 6].

Sonographic findings

	femoral vein	popliteal vein
Occlusion of vein	2	6
Insufficient valves	21	44
Unchanged vein	46	19

Deep venous system of the extremity:

Occlusion	− 3
Occlusion + insufficient valves	− 4
Insufficient valves	− 54
Unchanged deep venous system	− 8
Total	− 69

When sonographic findings were doubtful (5 cases - 7 %) - phlebography was performed [5]. There were no good candidates for surgical treatment. In patients without changes in venous system investigation of lymphatic system was recomended.

Conclusion

In our opinion colour doppler sonography is excellent diagnostic procedure sufficient to establish diagnoses in most cases and allows to avoid phlebography which is associated with risk of phlebitis and irradiation of the patient.

References:

1. Araki CT, Back TL, Padberg FT, Thompson PN, Duran WN, Hobson RW. Refinements in the ultrasonic detection of popliteal vein reflux. J Vasc Surg 1993, 18: 742-8.
2. Cheatle TR, Scurr JH, Coleridge Smith PD. Drug treatment of chronic venous insufficiency and venous ulceration: a review. J Roy Soc Med 1991, 84: 354-8.
3. Hopkins NF, Wolfe JH. Deep venous insufficiency and occlusion. BMJ 1992, 304: 107-10.
4. Iafrati MD, Welch H, O'Donnell TF, Belkin M, Umphrey S, McLaughlin R. Correlation of venous noninvasive tests with the Society for Vascular Surgery / International Society for Cardiovascular Surgery clinical classification of chronic venous insufficiency. J Vasc Surg 1994, 19: 1001-7.
5. Masuda EM, Kistner RL. Prospective comparison of duplex scanning and descending venography in the assessment of venous insufficiency. Am J Surg 1992, 164: 254-9.
6. Weingarten MS, Branas CC, Czeredarczuk M, Schmidt JD, Wolfarth CC. Distribution and quantification of venous reflux in lower extremity chronic venous stasis disease with duplex scanning. J Vasc Surg 1993, 18: 753-9.

Phlebology '95, D. Negus et al. (eds.). Phlebology (1995) Suppl. 1: 241-242

V/2.8

Echo Color Flow in Varicose Veins: Its Usefulness

G.P. Avruscio[1], F. Battocchio[2], L. De Santis[2], M. Baldan[2], D. Celi[2], O. Terranova[2] and G.P. Signorini[1]

1 Servizio di Angiologia, 2 Cattedra di Chirurgia Generale e Divisione di Chirurgia Geriatrica, Università di Padova, Italy

INTRODUCTION

Several methods, either direct or indirect, are available to study lower limbs varicose disease [1-5]. Clinical examination, always necessary, cannot sometimes define the status of the disease and the integrity of deep venous system.

Aim of this study is to demonstrate the suitability of echo doppler in the evaluation of this disease as direct, non invasive method. It facilitates the surgeon both in the preoperative mapping and in the therapeutic decisions. Other advantages depending on this method are the avoiding of invasive tests and the decreasing of recurrences.

METHODS

During the year 1994 613 patients suffering from varicose disease underwent our observation (475 females and 138 males with age ranging from 19 to 82 years). All the patients had been submitted to clinical examination and echo doppler color flow; 155 patients had also been evaluated with ascending phlebography.

Echo doppler color flow and phlebography were performed by different specialists with the purpose of removing the doubts raised by clinical examination.

Echo doppler color flow succeeded in demonstrating 196 cases of isolated insufficiency of the greater saphenous vein, 95 cases of isolated insufficiency of the lesser saphenous vein, 97 cases with combined insufficiency of greater and lesser saphenous veins concerning the same limb, 80 cases of varicophlebitis, 53 cases of recurrent varicose disease, 92 cases of bilateral varicose veins.

Suggestion to perform phlebography depended on the suspect of lesser varicose vein insufficiency or recurrent varicose disease.

Diagnostic reliability of these two instrumental methods was then confirmed by surgical evaluation.

RESULTS

Clinical examination was little reliable in case of anomalous confluence of lesser saphenous vein, recurrent varicose disease, research of site and number of incontinent perforating veins, demonstration of extension of varicophlebitic process. It could discover valvular incontinence in 85% of cases.

Preoperative mapping of varicose veins with echo doppler color flow had always been confirmed as correct by surgical exploration.

Phlebography was reliable in evaluating anatomical variations and in the study of post surgical recurrencies. Otherwhile it was less reliable in evaluating incontinent perforating vessels. In 4 cases this test was not performed due to contrast medium intollerance; in 7 patients there had been low importance side effects and in no patient serious side effects.

DISCUSSION

In the past phlebography was considered the gold standard in the evaluation of varicose disease and mainly for anatomical variations and recurrencies.

The invasiveness, the high costs, the necessity of preliminary assesment for the use of a contrast medium and the risk of side effects reduce nowadays its indications favouring the reserch of new instrumental methods. In this view echo doppler appears to be a cheaper, direct method, easily performable and repeatable by experts , able to give informations on veins morphology and fluxes dynamics.

In our experience echo doppler has been reliable in the evaluation of varicose disease and useful in therapeutic decisions making, setting itself as a valid choice as compared with phlebography.

The best results in the evaluation of patients suffering from varicose disease have been however obtained through a strict cooperation between the surgical and medical equipes.

REFERENCES

1. De Santis L, Avruscio GP, Cafiero F, Battocchio F, Rugna A, Lupia M et al. Diagnosi, monitoraggio e terapia delle varicoflebiti. Minerva Angiologica 1994; 19 (Suppl 1): 119-121.
2. Rabe E, Fratila A, Kluess H, Kreysel HW. Color duplex investigations of insufficient perforating veins at the lower leg. Phlebologie 1992; Eds John Libbey Eurotext, Paris; 562.
3. Avruscio GP, Battocchio F, De Santis L, Benedetti L, Verlato F, Salmistraro G et al. Evaluation of clinical examination, echo-doppler color flow (angiodynography) and varicography in the assesment of the short saphenous vein. Phlebologie 1992; Eds John Libbey Eurotext, Paris; 598-600.
4. Battocchio F, Avruscio GP, De Santis L, Benedetti L, Vigo M, Signorini GP et al. Studio comparativo: clinico, strumentale e radiologico della piccola safena. Flebolinfologia 1991; 2: 3-5.
5. De Santis L, Battocchio F, Avruscio GP, Signorini GP, Benedetti L, Infanti L et al. Diagnosi a confronto nella chirurgia della piccola safena: ecodoppler color flow e varicografia. Acta Chir Ital 1992; 48: 1040-1044.

Phlebology '95, D. Negus et al. (eds.). Phlebology (1995) Suppl. 1: 243-244

P069

The Significance of Diagnostic Ultrasound in Postthrombophlebitic Syndrome

R.M. Grigoryan[1], G.I. Kuntsevich[2], I.V. Shutikhina[2] and S.V. Sapelkin[1]

[1] Department of Vascular Surgery and [2] Department of Ultrasound Diagnostic, Vishnevsky Institute of Surgery, Moscow, Russia

INTRODUCTION:

Among different methods of diagnosis of postthrombophlebitic syndrome (PTPS) of lower extremities, ultrasound plays an important role. For a long time radiological investigations were considered the "gold standard" in the evaluation of venous reflux and postthrombotic syndromes of the deep venous system (1,2). The reliability of non-invasive diagnosis has increased substantially and the role of phlebography has diminished since duplex scanning appeared. The purpose of our research was to compare the results of clinical, ultrasound and radiological contrast methods of diagnosis.

METHODS:

The cohort comprised 52 patients with both non-complicated and complicated forms of PTPS who were divided in two groups depending on the severity of chronic venous insufficiency (CVI). (Reporting Standards in Venous Disease by the Ad Hoc Committee (3)): Group A (46 patients) - class 1-2 with mild or moderate CVI, Group B (6 patients) - class 3 with severe CVI.

Clinical, functional, ultrasound (Doppler, duplex scanning) and radiological investigations were performed on patients preoperatively. DS was performed with Diasonics (DRF-1000 or Acuson 128XP (linear transducer 5 and 7 MHz). The inferior vena cava and iliac veins were also investigated. Diagnosis of PTPS was based on identification of a heterogeneous mass in the vein lumen, incompetent veins and stenosis of veins. Ascending phlebography was performed according to standard methodolgy with and without a tourniquet above the ankle.

RESULTS:

33 patients in Group A had post-thrombotic changes detected by phlebography, located in the upper third of the femoral vein. In 4 cases the postthrombotic changes were confimed by DS. However, the interpretation of one of the results by using only DS was insufficient because the altered segment of the femoral vein was beyong the reach of ultrasound. In one case a small filling defect on phlebography in the area of sapheno-femoral junction after thrombosis was not diagnosed. These alterations followed thrombosis of calf veins.

In 15 cases valvular incompetence of posterior tibial veins was diagnosed by DS, and in 11 cases incompetence of calf perforators.

In 4 cases in Group B (6 patients) multifocal changes of deep veins was diagnosed. In 2 of 4 cases the diagnosis was established only by phlebography. DS in the cases did not permit evaluation of the femoral vein in Hunter's canal and the calf veins.

Accuracy of Doppler according to the results of our research was 54%.

CONCLUSIONS:

Ultrasound DS together with clinical studies of patients with PTPS gave reliable information in 92% of cases, sufficient to determine the correct treatment. The importance of DS in such pathology is much greater compared with ultrasound Doppler. It is considered reasonable to perform DS and phlebography when thrmbotic processes are localised in the regions inaccessible to ultrasound and multifocal alterations, especially in the cases of paitents with CVI 2-3 class. DS is also preferable in studies of the dynamics in PTPS.

Duplex scanning allows us to determine the extent and type of surgical intevention that should be carried out in patients with chronic venous insufficiency.

REFERENCES:

1. Neglen P, Raju S. A comparison between descending phlebography and duplex Doppler investigaion in the evaluation of reflux in chronic venous insufficiency. A challenge to phlebography as the "gold standard". J Vasc Surg 1992; 16; 867-93.
2. Rosfors S, Bygderman S, Nordstrom E. Assessment of deep venous incompetence; a prospective study comparing duplex scanning with descending phlebography. Angiology 1990; 41: 463-8.
3. Porter JM, Rutherford RB, Clagett GP, et al. Ad Hoc Committee. Reporting standards in venous disese. J Vasc Surg 1988; 8:172-81.
4. Baxter GM et al. Colour Doppler ultrasound in deep venous thrombosis. A comparison with venography. Clin Radiology 1990; 1:31-6.

Phlebology '95, D. Negus et al. (eds.). Phlebology (1995) Suppl. 1: 245-247

P070

Duplex-Scanning of the Direction of the Saphenopopliteal Junction

E. Rabe, G. Gallenkemper, H. Kluess and H.-W. Kreysel

Department of Dermatology, University Bonn, Germany

Introduction

The sapheno-popliteal junction (SPJ) between the short saphenous vein (SSV) and the popliteal vein (PV) is a region of frequent anatomical variations (3). Color duplex (CD) has been demonstrated to be a reliable method in the examination of the SPJ (1,2,3).
In our investigation we wanted to show if there is any difference between the SPJ anatomy in healthy subjects and in patients with an insufficiency of the short or long saphenous vein (LSV).
During the examination the patient was in the standing position with the knee of the examined leg slightly bended.
We looked for the level of the SPJ and for the direction from which the SSV joined the PV using a circular grading system with $360°$ at the dorsal position.
A SPJ between $340°$ and $380°$ was definend as dorsal and between $200°$ and $339°$ as lateral.
Insufficiency of a vein was defined when the reflux-time, demonstrated by CD, was longer than 0,5 sec.

Patients and method

We examined the SPJ in 210 legs by color-duplex (Philips, Quantum, 7,5 and 5,0 MHz)in a longitudinal and a transverse view. The mean age of the patients was 57 years (19-89).
The results in the different groups were compared by t-test for unconnected samples.

Results

In 178 of the 210 legs (84,8 %) the SSV joined the PV between the knee-fold and 3 cm above. In 15 legs (7,1 %) we found a higher and in 6 legs (2,9 %) a lower junction. In 11 cases (5,2 %) no SPJ could be demonstrated (tab. 1)

Junction	number of legs 210	%
kneefold and 3 cm above	178	84,8
higher	15	7,1
lower	6	2,9
no	11	5,2
together	210	100,0

Tab. 1   Localisation of SPJ

Only the 178 legs with the junction in the popliteal fossa were used in the following results (tab. 2).
In 78 legs (37,1 %) no pathological reflux could be demonstrated.
In 44 legs (24,7 %) an insufficient SSV was present. In 65 legs (36,5 %) the LSV was insufficient and in 47 legs (26,4 %) the popliteal vein showed reflux

Insufficiency	legs	dorsal SPJ (%)	lateral SPJ (%
SSV (all)	44	22	78
LSV (all)	65	45	55
LSV (isolatet)	36	67	33
PV	47	-	-
no insufficiency	78	72	28
SSV + LSV	6	33	67
SSV + LSV + PV	12	0	100

Tab. 2   number of insufficient veins

The higher number of legs in table 2 results from the combination of insufficient veins. In 36 legs we found an isolated LSV-insufficiency and in 11 legs an isolated SSV-insufficiency.
In 27 of the 44 legs (61,3 %) with an insufficient SPJ also the PV showed reflux.
In the healthy legs 72 % of the SPJ was found dorsal and only 28 % "lateral". In contrary in legs with an insufficient SSV 78 % of the SPJ was "lateral" and only 22 % "dorsal". In the 36 legs with an isolated LSV-insufficiency 67 % had a dorsal and 33 % a lateral junction.
There was a statistically significant difference between the localisation of the SPJ in healthy legs and in legs with an SSV-insufficiency with $p < 0,01$.

Discussion

We could show, that in the sufficient SPJ the SSV joins the PV from the dorsal direction in most of the cases (72 %). In contrary the insufficient SSV joins the PV laterally in 78 % of the cases. In these legs we could also demonstrate reflux in the PV in 61 %. This correlates with the results of Somjen in 1992 who found reflux in the PV in about half of his patients with SSV-insufficiency (5). Hauser and Brunner found popliteal reflux in more than 80 % of legs with SSV-insufficiency (4). It seems, that the PV is involved in the process of degeneration in patients with SSV-insufficiency. In the developement of SPJ-insufficiency the PV may rotate and the junction gets in a lateral position in most of the cases.

References

1. Bork-Wölwer L, Wuppermann Th, Verbesserung der nichtinvasiven Diagnostik der V. saphena magna- und der Vena saphena parva-Insuffizienz durch die Duplex-Sonographie. Vasa 1991; 20:343-7
2. Brizzio E, Desimone J, L'action de l'insufficience veineuse a l'étage poplité sur la crosse de la saphene externe et les veines jumelles. Phlébologie 1992; 45: 291-6
3. Engel AF, Davies G, Keeman JN, von Dorp TA, Colour flow imaging of the normal short saphenous vein. Eur J Vasc Surg 1994; 8:179-81
4. Hauser M, Brunner U, Neue pathophysiologische und funktionelle Gesichtspunkte zur Insuffizienz der Vena saphena parva. Vorläufige Mitteilung. Vasa 1993; 22:338-41
5. Somjen GM, Royle JP, Fell G, Roberts AK, Hoare MC, Tong Y, Venous reflux patterns in the popliteal fossa. J Cardiovasc Surg 1992; 33:85-91

Phlebology '95, D. Negus et al. (eds.). Phlebology (1995) Suppl. 1: 248-249

P071

Duplex-Scanning of the Depth of the Saphenofemoral Junction

E. Rabe, G. Gallenkemper, H. Kluess and H.-W. Kreysel

Department of Dermatology, University Bonn, Germany

Introduction

The sapheno-femoral junction (SFJ) between the long saphenous (LSV) and the femoral vein (FV) can be situated more or less deep under the skin surface. Color-duplex has been demonstrated to be a reliable method to localize the insufficient SFJ (1,2).

In our investigation we wanted to evaluate the variability of the depth of the SFJ. This knowledge plays an important role during the operation of the insufficient LSV.

Patients and method

We examined 152 legs in 76 patients. 48 where women, 28 men. The mean age of the patients was 53,9 years (16-87). The mean weight was 73,6 kg (47-122) and the mean height was 170,2 cm (154-187).

We devided the patients in three groups: normal weight, underweight and overweight, according to Broca-index.

The examinations were done by color-duplex sonography (Philips Quantum, 7,5 and 5,0 MHz) in a longitudinal and a transverse view. The distance between the skin surface and the proximal angle of the SFJ, skin-junction-distance (SJD), was calculated in mm.

The results in the different groups where compared by t-test for unconnected samples.

Results

The results are demonstrated in table 1.

The mean SJD in all legs was 21,5 mm, 21,3 in women and 21,8 in men. The shortest SJD in our collective was 9,0 and the longest 48,0 mm. The lowest values were 10,4 mm in the normalweight, 9,0 mm in the underweight and 14,2 mm in the overweight group. The highest values where 37,0, 37,0 and 48,0 mm in the three collectives.

	age (years)	height (cm)	weight (kg)	SJD (mm)
all legs	53	170	74	21,5
women	53	166	70	21,3
men	53	178	80	21,8
normal-weight (nw)	55,9	169	69	19,4
women nw	54,6	166	66	18,5
men nw	59	177	76	21,7
under-weight (uw)	41,9	173	61	16,7
women uw	36,2	169	56	16,3
men uw	47,6	177	67	17,1
over-weight (ow)	56,8	170	90	27,8
women ow	59,3	165	85	28,7
men ow	52,7	179	97	26,4

Tab. 1   mean values

There was a statistical significant difference in the SJD values with p < 0,001 between normalweight and overweight as well as between orverweight and underweight patients. The difference between normals and the unterweight group was slightly significant with p = 0,02.

Discussion

Our results show, that there is a great variability of the depth of the sapheno-femoral junction. In the examined 152 legs the SJD ranged from 9 to 48 mm.
Although there is a statistical significant difference between the three weight groups the single values are widely overlapping.
The preoperative duplex-examination provides the surgeon not only with information about the sufficiency or insufficiency of the venous system (2) but also with an exact information about the depth of the SFJ for the planing of the operation. This information can not be given by phlebography.

References

1. Bork-Wölwer L, Wuppermann Th, Verbesserung der nichtinvasiven Diagnostik der V. saphena magna- und der Vena saphena parva-Insuffizienz durch die Duplex-Sonographie. Vasa 1991; 20:343-7
2. Payne SP, London NJ, Jagger C, Newland CJ, Barrie WW, Bell PR, Clinical significance of venous reflux detected by duplex scanning. Br J Surg 1994; 81:39-41

Phlebology '95, D. Negus et al. (eds.). Phlebology (1995) Suppl. 1: 250-252

P073

Preoperative Duplex Ultrasonography of the Internal Mammary Artery and Saphenous Vein in Young Persons and Patients with Atherosclerosis

J. Iwinski, T. Petelenz, P. Weiss and A. Gruszka

III-rd Clinic of Cardiology, Silesian Academy of Medicine, Katowice, Poland

INTRODUCTION

In patients who undergo coronary artery bypass surgery, the internal mammary artery (IMA) is considered to be the bypass conduit of choice. Superior long-term patency rates have been reported for IMA grafts compared with saphenous vein grafts. At the end of the first 10 postoperative years, 69 % to 83 % of IMA grafts versus 41 % to 63 % of vein grafts were still patent.[1-4] The greater long -term durability of the IMA graft has never been fully explained, although numerous studies have revealed the histologic differences between these two types of conduit material as well as the differences in their responses to stress[5-8]. The purpose of this study was to determine :
1) normal reference values for artery and vein wall properties and blood flow data
2) influence of ageing, gander for IMA and SV proprieties
3) validate responsiveness to nitroglycerine (NTG).

METHODS

Patients populations. The study population consisted of patients 100 (who underwent duplex mapping between August 1992 - & December 1994) ranging in age from 18 to 79 years, mean 49 +/- 17. There were 50 young subjects (25 men /M/ , 25 woman /W/) mean age 29 +/- 3,1 years and 50 patients with a atherosclerosis [O] without symptoms of venous or arterial disease. 25 M and 25 W / mean age 59 +/- 6,7 years /

Technique. Imaging was performed by using Sonos 1500 /Hewlett Packard USA/ equipped with a 4,5 MHz and 7,5 MHz transducers. The study was done with subjects in a supine position an identical manner in air conditioned room with room temperature of about 25 - 26 $^{\circ}$ C.

Artery studies. Five cardiac cycles were averaged for measurement. Transthoracic imaging was performed through the third intercostal space. IMA diameter and intimomedial thickness were measured in diastole /minimal diameter /and in systole/ maximal diameter/. The intimomedial thickness was measured according to the method of Pignoli at all [9]. Peterson's elastic modulus [10] (Ep) was calculated as $(\Delta P/\Delta D) \times D$, where D-diastolic IMA diameter, ΔD-systolic diameter-diastolic diameter ,ΔD =systolic /SBD/- diastolic blood pressure/ DBP/ Stiffnes index (β) is calculated as β= ln (SBP/DBP) /ΔD/D.

Venous studies. The greater saphenous veins in each leg on all patients were sequentially examined. The internal diameter of the vein (d) , thickness (th) , flow volume (Qv) was measured above the ankle 5 cm superior to the medial malleolus. The artery and the venous studies were repeated after 30 min during intravenous infusions of NTG (25 µg/min.)

RESULTS

IMA th averaged 0,62 +/- 0,13 mm in both M and W.Y had a significantly
 smaller th 0,51 +/- 0,11mm /p<0.05/. There were no differences in B between M and W in Y,but B and E increased, specially in W in O /p<0,05/.The D of IMA did not differ significantly among the Y for M and W.In O D was significantly lower 2,7 +/- 0,4 mm vs 3,3 +/- 0,2mm, especially in W 2,6 +/- 0,3mm, vs 2,9 +/- 0,6 mm.Age was highly positively correlated with th /r=0,78, p< 0,01/,d /r=0,71, p<0,05/,E /0,79, p < 0,05/, B /r=0,81, p<0,05/. NTG caused vasodilatation of IMA and rise of Q in both groups but significantly in the O D 2,7 +/- 0,4mm, vs 3,1 +/- 0,5mm post NTG, Q 69 ml/min vs 79ml/min post NTG.The Doppler spectrum of velocity was thriphasic in all groups.V max was higher but non significantly in the O.SV th averaged 0,42 +/- 0.09mm in both M and W Y had a significantly smaller th, 0,32 +/- 0,1mm /p<0.05/.The D of SV did not differ significantly among Y for M and W.In O D of SV was significantly higher 3,0 +/- 0,3 mm vs 2,8 +/-0,1mm.Age positively correlated only with th /r=0,76, p< 0.01/.NTG caused vasodilatation of SV and rise of Qv in Y and O but significantly in O /3,0 +/- 0,13 vs 4,1 +/- 0,2 mm, Qv 97,2 +/- 12ml/min vs 101 +/- 18ml/min post NTG.

CONCLUSIONS

Ageing significantly increases Th,E,B of IMA and th of SV.W have significantly higher B,E lower D only in IMA of older group. The intravenous infusion of NTG induced vasodilatation both IMA and SV but significantly higher of SV only in O group.These results may suggest some form of endothelial dysfunction, greater in venous system.

REFERENCES.

1). Grover F.L., Johnson R.R., Marshall Gr., Hammermeister K.E.. Impact of Mammary Grafts or Coronary Bypass Operative Mortality and Morbidity. Ann Thorac Surg 1994 ; 57:559-69
2). Joyce F.S., Mc Carthy P.M., Taylor P.C., Cosgrove III D.M., Lytle B.W. Cardiac Reoperation in Patients with Bilateral Internal Thoracic Artery Grafts. Ann Thorac Surg 1994; 58:80-5
3). Van Der Meer J., Hillege H.L., Van Gilst W.H. de la Rivire A.B., Dunselman P.H., Fidler V., Kootstra G.J., Mulder B.J.M., Pfisterer M., Lie K.I. A Comparison of Internal Mammary Artery and Saphenous Vein Grafts After Coronary Artery Bypass Surgery. Circulation 1994; 90:2367-2374
4). Green G.E., Cameron A., Goyal A., Wong S., Schwanede J. Five - Year Follow - up of Microsurgical   Multiple Internal Thoracic Artery Grafts .Ann .Thorac Surg 1994; 58:74-9
5). Shimshak T.M., Giorgi L.V., Johnson W.L., Mc Conahay D.R., Rutheford B.D., Ligon R., Hartzler G.O. Application of Percutaneous Transluminal Coronary Angioplasty to the Internal Mammary Artery Graft J Am Coll Cardiol 1988 ;12:1205-14
6). Nquyen H.C., Grossi E.A., Le Boutiller M., Steinberg B.M., Rifkin D.B., Baumann G.F.Colvin S.B.,Galloway A.C.Mammary Artery Versus Saphenous Vein

252

Graft:Assessment of Basic Fibroblast Growth Factor Receptors.Ann Thorac.
Surg.1994;58 :308 -11.
7)Canver Ch.C.,Zwolak R.M.Preoperative Evaluation of the Right Internal Thoracic
Artery for Coronary Surgery.Ann.Thorac.Surg.1994;57:440-3.
8)Fujiwara T.,Kajiya F.,Kanazawa S.,Matsuoka S.,Wada Y.,Hiramatsu O.,Kagiyama
M.,Ogasawara Y.,Tsujioka K.,Katsumura T.Comparison of Blood-Flow Velocity Wave
forms in Different Coronary Artery Bypass Grafts.Circulation 1988;78:1210-1217.
9)Pignoli P.,Tremoli E.,Poli A.,Oreste P.,Paoletti R.Intimal plus medial thickness of the
arterial wall a direct measurement with ultrasound imaging.Circulation 1986; 74:1399-
406.
10)Peterson LN.,Jensen RE.,Parnell R.Mechanical properties of arteries in
vivo.Circ.Res.1960; 8:622-39.

Phlebology '95, D. Negus et al. (eds.). Phlebology (1995) Suppl. 1: 253-256

P074

Venous Reflux in Patients with Skin Changes Associated with Chronic Venous Insufficiency

J. Iwinski, T. Petelenz, A. Gruszka and P. Weiss

III-rd Clinic of Cardiology, Silesian Academy of Medicine, Katowice, Poland

INTRODUCTION

Chronic venous insufficiency /CVI/ leads to skin changes including lipodermatosclerosis, atrophy blauche and ulceration.The underlying causes of CVI are incompetence of superficial, deep and perforating veins. /PV/ .Previously,venographic study was needed to determine the pattern of obstruction and insufficiency. With increasing frequency duplex ultrasonography are being used in place of venography for this determination.The addition of colour flow imaging to duplex scanning provides visualisation of reflecting, partially obstructed or completely obstructed venous segments[1..4].Quantification of reflux times can then be obtained with spectral analysis[5].The purpose of this study was to assess the distribution and extent of venous reflux in patients with skin changes associated with chronic venous insufficiency.

METHODS

One hundred sixty seven patients with involvement of 288 limbs were examined between 1992 and 1994. There were 98 female /52,7%/ and 69 male /47,3 %/ patients with a mean age of 59,6 years /range 23 to 75/.Limbs were examined with an Sonos 1500 /Hewlett-Packard USA/ with colour -flow imaging facility using a 4,5MHz pulsed Doopler probe. The sapheno-femoral junction,the common femoral vein /CFV/,and the origin of the deep and superficial femoral veins / DFV,SFV/ were examined with the patients in the standing position. The femoropopliteal segment and the axial calf veins / posterior and anterior tibial, peroneal and gastrocnemial veins /GV/ were examined with the patient in the sitting position facing the examiner and with the foot resting on a stool. To investigate incompetent PV of the calf,active ankle dorsiflexion and passive calf compression above the perforator were used. The long and short saphenous veins /SV/ were examined along their whole length with intermittent calf compression and sudden release. Foot compression was used when examining their distal ends. The superficial veins were examined distal to probe with the same procedure like SV. The reflux was documented by recording the Doppler wave form.Reflux was taken to be present in a vein only if reverse flow of > 1,0s. duration was present.Reflux of shorter duration was considered physiologic,that is the brief distal flow was recorded just before the closure of the valves.In the superficial veins incompetence was taken to be present if reflux was demonstrated > 50% of the length of the vessel. The CFV and popliteal veins were considered incompetent if reflux was visualised below their respective junctions with the long SV,and the short SV and GV.The deep calf veins were taken to be incompetent if reflux was detected in any segment along their length, except at junctions with incompetent PV. PV were considered incompetent if flow was from deep to superficial during active dorsiflection of the ankle followed by superficial to deep flow during relaxation of the calf muscle or if flow was from deep to superficial on passive compression of the calf.

RESULTS

The deep,superficial and both systems insufficiency was 21,7%, 29,0%, 47,2% respectively.Isolated PV incompetence occurred in 2,1% of extremities.Deep venous incompetence /DVI/ was associated with superficial vein reflux in 53% limbs.In the absence of DVI, superficial vein insufficiency /SVI/ was often associated with PV and GV incompetence. See table 1,2,3.

CONCLUSION

Skin changes are more common associated with DVI than with SVI.However SVI can be associated with ulceration,specially in presence distal reflux.Perforator incompetence frequently accompanies superficial venous reflux in limbs with skin changes.

Location of reflux in 157 legs without skin changes defined by Doppler ultrasonography with colour-flow imaging.

Tab (1)

	Extent of reflux					
	Superficial veins	Deep veins	Gastrocne-mius veins	Perfora-ting veins	Total	%
Superficial incompetence	+	-	-	+	10	
	+	-	+	+	2	
	+	-	-	-	36(48)	22,9
Deep incompetence	+	+	-	+	6	
	+	+	-	+	4	
	+	+	+	-	2	
	-	+	+	+	2	
		+	-	+	6	
		+	-	-	22(41)	14,0
Perforating vein incompetence	-	-	-	+	2(2)	1,3
No reflux					65	41,4

Incidence of skin changes in 109 legs in relation to location of reflux defend by Doppler ultrasonography with colour-flow imaging.

Tab (2)

	Extent of reflux					
	Superficial veins	Deep veins	Gastrocne-mius veins	Perfora-ting veins	Total	%
Superficial incompetence	+	-	-	+	16	
	+	-	+	-	3	
	+	-	+	+	3	
	+	-	-	-	32(54)	49,5

respective junctions with the long SV,and the short SV and GV.The deep calf veins were taken to be incompetent if reflux was detected in any segment along their length, except at junctions with incompetent PV. PV were considered incompetent if flow was from deep to superficial during active dorsiflection of the ankle followed by superficial to deep flow during relaxation of the calf muscle or if flow was from deep to superficial on passive compression of the calf.

RESULTS

The deep,superficial and both systems insufficiency was 21,7%, 29,0%, 47,2% respectively.Isolated PV incompetence occurred in 2,1% of extremities.Deep venous incompetence /DVI/ was associated with superficial vein reflux in 53% limbs.In the absence of DVI, superficial vein insufficiency /SVI/ was often associated with PV and GV incompetence. See table 1,2,3.

CONCLUSION

Skin changes are more common associated with DVI than with SVI.However SVI can be associated with ulceration,specially in presence distal reflux.Perforator incompetence frequently accompanies superficial venous reflux in limbs with skin changes.

Location of reflux in 157 legs without skin changes defined by Doppler ultrasonography with colour-flow imaging.

Tab (1)

	Extent of reflux					
	Superficial veins	Deep veins	Gastrocne-mius veins	Perfora-ting veins	Total	%
Superficial incompetence	+	-	-	+	10	
	+	-	+	+	2	
	+	-	-	-	36(48)	22,9
Deep incompetence	+	+	-	+	6	
	+	+	-	+	4	
	+	+	+	-	2	
	-	+	+	+	2	
		+	-	+	6	
	+	-	-		22(41)	14,0
Perforating vein incompetence	-	-	-	+	2(2)	1,3
No reflux					65	41,4

Incidence of skin changes in 109 legs in relation to location of reflux defend by Doppler ultrasonography with colour-flow imaging.

Tab (2)

	Extent of reflux					
	Superficial veins	Deep veins	Gastrocne-mius veins	Perfora-ting veins	Total	%
Superficial incompetence	+	-	-	+	16	
	+	-	+	-	3	
	+	-	+	+	3	
	+	-	-	-	32(54)	49,5

Deep incompetence	+	+	+	+	7
	+	+	-	+	4
	-	+	+	+	1
	-	+	-	-	37(49)	44,9
Perforating vein incompetence	-	-	-	-		
	-	-	-	-		
	-	-	-	-		
	+	-	-	+	4(4)	3,7
No reflux					2(2)	1,8

Table 2- continued

Incidence of ulceration in 22 legs in relation to location of reflux defined by Doppler ultrasonography with colour - flow imaging

Tab (3)

	Extent of reflux					
	Superficial veins	Deep veins	Gastrocne-mius veins	Perfora-ting veins	Total	%
Superficial incompetence	+	-	-	-	3(3)	13,6
Deep incompetence	-	+	-	-	1	
	-	+	+	+	1	
	-	+	-	+	16(22)	72,7
Perforating vein incompetence	-	-	-	-	-	
No reflux						

REFERENCES

1/Weingarten S.,Branas C.C.,Czeredarczuk BA.,Schmidt J.D.,Wolferth C.C. Distribution and quantification of venous reflux in lower extremity chronic venous stasis disease with duplex scanning.J.Vasc.Surg.1993;18:753-9.

2/Labropoulos N.,Leon M.,Nicolaides L.M.,Ortega F.,Chan P.Venous reflux in patients with previous deep venous thrombosis:Correlation with ulceration and other symptoms. .J.Vasc.Surg.1994;20:20-6.

3/Lambropoulos N.,Leon M.,Nicolaides A.N.,Giannoukas A.D..,Volteas N.Superficial venous insufficiency. Correlation of anatomic extent of reflux with clinical symptoms and signs.J.Vasc.Surg.1994;20:953-8.

4/Lees T.A.,Lambert D.Patterns of venous reflux in limbs with skin changes associated with chronic venous insufficiency.Br.J.Surg.1993;.80;6,.725-728.

5/van Bemmelen PS.,Bedford G.,Beach K.,Strandness D.E.Quantitative segmental evaluation of venous valvular reflux with duplex ultrasound scanning. J.Vasc.Surg. 1989;10:425-3

Phlebology '95, D. Negus et al. (eds.). Phlebology (1995) Suppl. 1: 257-259

OP/21.3

Accuracy of the Colour Coded Doppler Examination and Phlebography in Diagnosis of Vein Diseases

T. Urbanek, Z. Ziaja, M. Simka, M. Szewczyk, T. Drazkiewicz and M. Blaszczynski

I Department of General and Vascular Surgery, Silesian School of Medicine, Katowice, Poland

INTRODUCTION:

There is no sufficiently good and effective method of varicose vein examination. Phlebography - traditional and a very popular diagnostic method is considered as a " golden standard" today [6]. Unfortunately phlebographic method is conected with many complications that seems to be a very dangerous and burdensome examination for the patients (risk of phlebitis, allergy reaction, irradiation)[6]. The aim of our study is the prospective evaluation of the accuracy of colour doppler sonography in comparison with traditional phlebography in the diagnosis of varicose veins.

MATERIAL and METHOD:

This study was done prospectively by four independent physicians ; ultrasonographist (US doppler examination), radiologist (PHLEBOGRAPHY), surgeon (all results were checked intraoperatively). The results were recorded on a previously prepared special standard report. An independent physicians compared the results.

30 patients with varicose veins were preoperatively examined with colour doppler ultrasound. All ultrasound examinations were performed by the same ultrasonographist experienced in vessel ultrasound diagnostics. We used a Prisma -Diasonic (duplex -scanning) ultrasound system, with colour coded doppler, with linear probe 5 -7 MHz. The examination was done in the supine position. We used two tourniquets at a distance of 15 cm on the leg. Between them we examined the patency of the deep vein system, diameter of saphenous veins, deep vein valves competence, location of the incompetent perforating veins . To be able to see veins better, we pressed the surroundings of the probe. We changed the place of the tourniquets lengthwise to the leg. To find incompetent perforating veins we applied pressure to the lower leg or the foot. To visualize an incompetence of saphenous veins we applied pressure to the thigh and after reliving the pressure we observed the reversed flow in

the saphenous vein or saphenous vein junction. In the similar manner we examined the patency of deep veins and the competence of deep vein valves. (using distal or proximal pressure)

After the US doppler examination ascending phlebography was done in the same patients. All results were recorded on the special standard report.

Than all patients were operated upon for varicose veins.All examinations results and intraoperatively findings were compared by an independent physician.

RESULTS:

During operation we verified the results of the previous examination.

Table 1   (results of US Colour-Coded Doppler examination in comparison with phlebography and intraoperative findings)

	US doppler	Phlebo graphy	Operation-conformity
patency of the deep vein system	30 (100%)	30 (100%)	
increased diameter of the saphenous vein	20	20	20 (100%)
increased diameter of the small saphenous vein	16	16	16 (100%)
competence of the saphenous junction	10		10 (100%)
incompetence of the saphenous junction	20		20 (100%)
competence of the small saphenous vein junction	9		9 (100%)
incompetence of the small saphenous vein junction	21		21(100%)
total number of the incompetent perforating veins in all patients	82	98	102 (96%)phlebogrphy (80%)US doppler
deep vein valves incompetence	13		

Sensitivity:	Colour Coded Doppler Ultrasound	-80%
	Phlebography	-96%
Specificity:	Colour Coded Doppler Ultrasound	-100%
	Phlebography	-100%

DISCUSSION:

Using the ultrasound color doppler we can exactly examine the superficial and deep vein system looking for any anatomical or hemodynamical disorders [1,2,3]. The US doppler is a non-invasive method with very high sensitivity and specificty[3,4]. One of the most important advantages is repeatability. The estimation of patency of deep vein system and diameter of increased saphenous vein was 100% successful in comparison with phlebography and the intraoperative findings. The are still problems to visualize all incompetent perforating veins [5] which was possible only in 80%. (We hope that this percentage of visualizated incompetent perforatores will increase with the experience of ultrasonographist). On the other hand it is very easy to show vein valves incompetence using US doppler, which is impossible in standard ascending phlebography [4,6]. We think that US doppler examination should be used for screening and the diagnoses in cases of:
1. Patients with varicose veins after deep veins thrombophlebitis.
2. Patiens with smal varicose veins operated upon for cosmetic reasons.
3. Patients with big varicose veins in the whole leg, to find the cause of it
(e.g incompetent perforating veins) to decrease the operation range.
4 Patients with contraindicationes for phlebography examination

CONCLUSIONS:

Colour - coded doppler sonography performed by an experienced ultrasonographist is the method of choice in the diagnoses of vein diseases. Phlebography should be performed only in cases where doubt exist.

REFERENCES

1. Bagi P., Schroeder T., Silesian H., Lorentzen J. Real time B-mode mapping of the greater saphenous vein. Eur. J. Vasc Surg. 1989;3, 103-105.
2. Bork-Woelwer L. Wuppermann T. Verbesserung der nicht-invasivem Diagnostic der V. saphena magna und V. saphena parva -Insuffizienz durch die Duplex -sonographie Vasa 1991;20 : 343-347
3. Grouden M.C. Stanley S.T. Colgan M.P Kent P. Sheehan S. Moore D.J. Shanik D.G. Triplex imaging of the saphenofemoral junction is the test of choice in patients with primary varicose veins. Venous Forum Annual Meeting, Royal Society of Medicine, London 16.Oct.1992, Phlebology 1993;8,41
4. Kalodiki E. Calahoras L. Nicolaides A. N. Make in easy: duplex examination of the venous system. Phlebology 1993;8,17-21
5. Rabe, Fratila A., Kluess H. , Kreysel H. W. Colour-duplex-investigations of insufficient perforating veins of the lower leg.Phlebology' 92 Montreal 1992, 1, 574.
6. Ziaja K. , Drążkiewicz T., Błaszczyński M., Diagnosis in phlebology, European Congress of the International Union of Phlebology Budapest 6-10 September 1993, 469-472

PI/5.5

COLOR-CODED DUPLEX SONOGRAPHY IN PATIENTS 12 YEARS AFTER ACUTE DEEP VEIN THROMBOSIS

Franzeck UK, Schalch I, Bollinger A,
Dept of Medicine, Angiology Division, University Hospital, Zürich, Switzerland

Objectives: To evaluate patency and valvular incompetence of initially thrombosed veins in patients with DVT 12 years after the acute event by color-coded Duplex sonography and to compare the findings with initial phlebography.

Design: Prospective study of consecutive patients with DVT.

Patients: 39/58 patients of a low risk population 12 years after acute DVT of the lower extremities.

Measurements: All patients had ascending phlebography in the acute stage and color Duplex sonography examinations (Acuson XP 128) of the lower extremity veins were performed 12 years later.

Results: 39/58 patients could be studied 12 years after the acute event. A total of 124 vein segments were evaluated (13 segments with rethromboses). 58.6% of the segments showed complete recanalisation, 19.8% exhibited partial occlusions, and 12.6% were completely occluded. Reflux was present in 51.4 % of the segments.

Conclusions: Deep vein thromboses of the calf resolved almost in all cases (complete recanalisation in 96.6%, with reflux in 58.6%). Thromboses of the thigh were patent in 78%. However, in the presence of excellent collateralisation via the deep femoral vein and a patent common femoral vein thromboses of the superficial femoral and popliteal vein were not recanalized (22% of 41 superficial femoral vein segments).

OP/2.3

CONFRONTATION BETWEEN DUPLEX SCAN AND SURGERY IN THE SHORT SAPHENOUS VEIN STUDY

Lemasle P H, Uhl JF, Lefebvre-Vilardebo M, Tamisier D, Baud J M.

Centre de Chirurgie Esthétique des Varices, 75116 Paris, France.

Objective: As a part of the checking before surgery of the incompetent short saphenous vein (SSV):
1) to evaluate the agreement between the duplex scan (DS) results and the surgical verifications.
2) to research predictive DS signs of the presence of a terminal trunk common to the SSV and sural veins.

Design: Prospective study of a consecutive series.

Patients: 80 patients had a preoperative DS study and varicose surgery within 48 hours. Only the patients with a complete surgical exploration of the sapheno-popliteal junction (SPJ) were included in the protocol of confrontation DS vs surgery.

Measurements: DS looks for the presence of the SPJ, and determine then its height in relation with the bend of the knee, and its implantation side in the popliteal vein.

Results: 1) the agreement between DS and surgery is proved to be very satisfactory for the height of the junction: 89.6% (CI 95%, 80 to 96%), and the implantation side: 87% (CI 95% 77 to 93%). 2) There is a highly significant relation between the presence of a common trunk and
- the implantation height < 3cm: p < 0.0004 (Chi 2)
- the posterior-internal implantation side: p < 0.0001

Conclusions: The duplex scan is the essential but adequate investigation before SSV surgery.

OP/3.4

RELIABILITY OF DUPLEX SCANNING FOR THE MEASUREMENT OF VENOUS INCOMPETENCE IN EPIDEMIOLOGICAL SURVEYS.

Evans C, Leng G, Stonebridge P, Lee A, Allan P, Fowkes F, Department of Public Health Sciences, University of Edinburgh, Edinburgh, Scotland.

Objective: To measure the repeatability of duplex scanning as a method of measurement of venous incompetence, for future use in an epidemiological study of venous disease of the legs.
Design: Repeatability study.
Patients: 21 patients with current or healed venous ulcers who had attended a vascular clinic at the Royal Infirmary of Edinburgh.
Measurements: Duplex scans were performed by two of three observers at visit 1, and by a different pair of observers at visit 2. Duration of reflux was measured at six points on the deep and superficial veins of each leg.
Results: Two of the observers showed no significant inter-observer variability. The other pairs of observers agreed at the majority of segments, but differed at the popliteal vein (p<0.001), and superficial femoral and common femoral veins (p<0.005). Intra-observer variability was generally good. Observer 2 showed no significant variability, although observer 1 differed at the common femoral and superficial femoral veins (p<0.05), and observer 3 differed at the short saphenous vein (p<0.05).
Conclusions: Inter- and intra-observer reproducibility of duplex ultrasound in the measurement of reflux duration was good at certain sites, but appeared to be influenced by position of the vein and by observer training and experience. Thus it is important for an epidemiological study such as the Edinburgh Vein Study that observers are carefully trained and that their performance is monitored regularly.

OP/16.5

PREOPERATIVE SAPHENOUS VEIN MAPPING BY DUPLEX SONOGRAPHY

Papacharalambous G., Psaroudakis A., Christoforidis M. and Panoussis P.
"KAT" Hospitals
2, Nikis str, Kifissia, Athens 145 61, GREECE

The preoperative assessment of the saphenous vein to determine its anatomical configuration and pathological changes is an important advance to operative management especially in patients undergoing arterial reconstuction with autogenous vein graft. The adequacy of the GSV as arterial substitute was evaluated preoperatively in 40 patients by Duplex sonography. In 39 of them, below knee arterial in situ bypass was required for limb salvation, and in 1 case with severe subclavian artery stenosis, an arterial bypass from contralateral subclavian to post-stenotic segment of the subclavian aretery, was placed. Veins were judged adequately based on size, compressibility and absence of stenosis or thrombosis. With this method mapping of the vein was performed, with all major and minor branches. All the veins were explored duing the operation. In 9 cases with unsuitable veins, a venogram was performed. Preoperative mapping was accurate in 97.5%.

In conclusion, it seems that this method improves the operative planning and tends to be a routine preoperative method of greater saphenous vein assessment.

P114

DUPLEX SCANNING OF GREAT SAPHENOUS VEIN IN CHILDREN: CALIBRES AND REFLUXES

Schadeck M, Allaert FA
Ecole Européenne de Phlébologie, Hôpital Notre-Dame de Bon-Secours, Paris, FRANCE

Objective: To observe the evolution of saphenous diameters according to age, size, weight and to localise the levels of the reflux.
Design: Prospective study on a consecutive series.
Patients: 70 children or teenagers between 6 to 18 years old whose mothers have a varicose disease, at a stage I or II of the classification and grading of chronic venous disease in the lower limbs.
Intervention: We point out age, size and weight. The examination of GSV is realised with a Duplex scanning Esaote AU 530 with a sectorial probe of 10-13 Mhz, in standing position.
Measurements: The echotomographic study begins at the sapheno femoral junction in longitudinal and transversal cross section with measurement of the diameter at 30 mm from the ostium. Then it is made at the superior third, the medium third and the inferior third of the thigh.
At each level, we consider the maximal diameter, avoiding the valvular apparatus.
The haemodynamic study is realised at each one of these levels and below the knee.
Results: The importance of the saphenous diameter is not related to age or weight but proportional to the size of the young patient.
The refluxes affect more than 25% of the children.
Conclusions: We can establish a curve of the saphenous diameters according to the size of the patient. The research of the reflux must be made above and below the knee.

P072

CLINICAL AND INSTRUMENTAL DEMONSTRATION OF A 2ND CROSSE EXISTENCE AT 1/3 MIDDLE THIGH

Riba U, Ghilardi F, Vercelli A, Zanchi R.
Centro Studio Nalattie Vascolari - Torino - Italia
Centro Italiano di Diagnostica Medica Ultrasonica Torino - Italia

Objective: Clinical observation and medical literature study evidence the frequent existence of a point placed at about 1/3 middle thigh, as the site of relevant reflux, connected or not, with ostec sapheno-femoral reflux.
Design: Anatomical structure of venous channels and perforans were studied as reflux cause in the site
Patients: 120 patients with clinical reflux evaluation at 1/3 middle thigh were observed under Eco-colour-Doppler test.
Measurements: Anatomical, clinical and instrumental data were correlated to ascertain the structure of this important reflux site.
Results: In observed patients evidence was found of insufficiency in various combinations of Hunter perforans, of the Hunter anastomotical channel with Leonardo vein and the genu-supreme anastomised vein with entero-exterior crural scarf, together forming a single anatomical functional reflux unity.
Conclusions: Insufficiency existing in the site, from haemodynamic viewpoint causes formation of a real second crosse of saphenous vein with varicose syndrome and sapheno-femoral crosse insufficiency (1st crosse).

V/2.11

INTERMITTENT SEQUENTIAL PNEUMATIC COMPRESSION OF THE FOOT AND CALF IN THE DUPLEX DIAGNOSIS OF VENOUS DISEASES AND IN THE PREVENTION OF DEEP VEIN THROMBOSIS.

Vetorello G.F.M.D., Gasbarro V.M.D., Mascoli F.M.D., Donini I.M.D., Pozza E.M.D.
Institute of Clinical Surgery - University of Ferrara C/so Giovecca 203, 44100 ITALY.

Deambulation is a focal point both in the diagnosis of venous diseses and in the prevention of deep vein thrombosis (DVT). At the Institute of Clinical Surgery an Intermittent Sequential Pneumatic Compression System (ISPC) was used to achieve the haemodynamic effects of deambulation without activating the osteo-articular leves.

1) To apply ISPC to a Duplex instrument : the sequence was a 0.5 sec compression of the foot, 0.5 sec compression of the calf and 1 sec decompression. The following pressures wre used in orthostatism: 150mmHg at the calf and 100 mmHg at the foot. The Orthodynamic Venous Reflux Index (OVRI) was checked at the Saphenous-Femoral Junction (S-FJ). Subsequently the Indexes were correlated to the pathology morphology.

2) To identify the average Duplex values during contraction in orthostatism and to reproduce them using an appropriately modified ISPC system in clinostatism in order to prevent DVT.

Experimental:
Two groupsof 40 normal subjects and 250 phlebopathological subjects were studied with Duplex-ISPC. In clinostatism one mixed group of 92 subjects (phlebopathological and normal) was subjected to separate application of the ISPC pumps (foot, then calf, then associated) in order to verify the best combination to achieve the haemodynamic effects of orthodynamism.

Results:
1) Four classes of Reflux were determined at the S-FJ and correlated to type of morphology.

2) Application of the calf pump alone at 100mmHg made it possible to achieve the velocimetric patterns obtained in orthostatism with Duplex-ISPC.

3) The orthodynamic in vivo evaluation with Phlebodynamometry on tapis roulant confirm the data we used, and gives us the possiblity to use the device as DVT prevention.

OP/21.4

PREOPERATIVE IMAGING IN VARICOSE VEINS OF THE LOWER EXTREMITY: COLOUR DUPLEX FLOW IMAGING OR VENOGRAPHY?

Authors: M Baldt, K Böhler, T Zontsich, B Schneider, M Breitenscher, K Turetschek, G H Mostbeck

Purpose: To compare with diagnostic capability of colour duplex flow imaging (CDFI) to ascending (AVG) and descending venography (DVG) in the preoperative assessment of varicose veins of the lower extremity.

Materials and Methods: In this prospective study, both the deep and superficial venous systems of 137 limbs (112 patients) with clinical evidence of severe varicosis were investigated by CDFI and AVG performed by two independent observers. We compared the diagnostic capabilities of both techniques in the assessment of venous superficial and perforating venous system. DVG was also performed in the first 52 limbs to assess the diagnostic capability of CDFI in the diagnosis of deep and superficial venous reflux, which was graded from I to IV.

Results: Variant venous anatomy was diagnosed by AVG and/or CDFI in 21 limbs. Two cases were missed and one case was misinterpreted by AVG. CDFI was inconclusive in another two cases, which were subsequently clarified by AVG and varicography.
Variant varicosis (n=52) was missed in seven cases by venography and in one case by CDFI. CDFI findings in five cases of variant varicosis were inconclusive. AVG was slightly superior to CDFI in the detection of posthrombotic venous changes, but no cause of venous occlusion was missed There was good agreement between CDFI and DVG in the grading of superficial and deep venous reflux (quadratic-weighted kappa [k] 0.75 and 0.79 respectively). Excellent agreement was found between AVG and CDFI in the grading of long (k:0.96) and short (k:0.94) saphenous vein reflux. There were more incompetent perforating veins detected by AVG and varicography than by CDFI, but the latter technique allows direct preoperative marking of the skin, which is beneficial for the surgeon.

Conclusion: These results suggest that CDFI is a valuable preoperative diagnostic imaging tool before venous stripping which is capable of replacing invasive AVG and DVG. Only patients with inconclusive CDFI results (eg. complex variant venous anatomy) should proceed to AVG and varicography.

The Investigation of
Venous Disorders

Non Invasive Investigations
Other Non Invasive Investigations

Phlebology '95, D. Negus et al. (eds.). Phlebology (1995) Suppl. 1: 265-266

PI/5.2

Calibrated Photoplethysmography (C-PPG)

A. Fronek and W.P. Bundens

Departments of Surgery and Bioengineering, Ubiversity of California San Diego, La Jolla, California 92093-0643, USA

INTRODUCTION

Photoplethysmography (PPG) is widely used in the diagnosis of venous disease. Good correlation of refill times with valvular insufficiency has been shown [1]. Lack of a method of calibrating displacement changes, however, represents a significant drawback. Quantification of optical reflectance could offer important information related to the amount of blood displacement during tests performed. The described calibration can quantify blood volume displacement and is not influenced by skin color or thickness.

METHODOLOGY

The light intensity of the light emitting diode (LED) is temporarily reduced by a known amount (e.g. 5%) by inserting a resistor in the circuit providing power to the LED [2]. Changes in optical reflectance induced by changes in blood displacement (e.g. by dorsiflexion) are then related to the initial, calibrated change in optical reflectance.

SUBJECTS

Fifty controls and 17 patients with various types of venous disease have been examined. The probe was attached to the skin about 8 cm above the medial malleolus and the examination started with the subject in the sitting position. After reaching the steady-state, the calibration was performed. The subject was then asked to lie down, with the legs elevated until a new steady-state condition was reached. The subject was then asked to sit up again with the feet touching the floor and with the knee bent at approximately 120°. After reaching a new steady-state, the subject was asked to perform eight dorsiflexions at a rate of approximately 1/sec. and a continuous recording of the light reflectance was obtained.

The following parameters were obtained: blood volume displacement induced by tilting (Tilt Displacement Volume TDV), followed by standardized exercise (dorsiflexion) Exercise Displacement Volume (EDV). In addition, recovery time was recorded. The efficiency of the veno-muscular pump was expressed by EDV/TDV.

RESULTS

	TDV %	EDV %	Recovery Time (sec)	EDV/TDV
Normal	9.1 ± 3.7	3.3 ± 1.5	55 ± 32	0.51 ± 0.23
Varic.Vs.	11.0 ± 8.2	3.3 ± 1.8	2 ± 11	0.36 ± 0.14
Ven.Ulc.	13.0 ± 4.2	1.4 ± 0.8	7.8±5.6	0.12 ± 0.06

± St. Deviation

It can be seen that the efficiency of the veno-muscular pump decreased significantly in both patient groups (differences are statistically significant $p < 0.01$). While the recovery time is a sensitive index of venous insufficiency, the utilization of tilt and exercise induced volume displacements makes it possible to evaluate the overall efficiency of the veno-muscular pump.

CONCLUSION

The described calibrated photoplethysmography permits quantitative evaluation of the venous circulation. It can be used in conjunction with any existing photoplethysmograph.

REFERENCES

1. Miles C, Nicolaides AN. Photoplethysmography: principles and development. In: Nicolaides AN, Yao JST, editors. Investigation of Vascular Disorders. London: Churchill Livingstone, 1981:501.
2. Fronek A. Recent developments in venous photoplethysmography. In: Bernstein EF, editor. Vascular Diagnosis. St. Louis: Mosby, 1993:930-933.

Phlebology '95, D. Negus et al. (eds.). Phlebology (1995) Suppl. 1: 267-269

P052

New Leg Position for Venous Photoplethysmography

H. Ishibashi, T. Ohta, H. Kazui, R. Kato and T. Tsuchioka

Second Department of Surgery, Aichi Medical University, Nagakute, Aichi, Japan

INTRODUCTION

A new position for photoplethysmographic (PPG) assessment of lower leg venous function is described and tested. The aim of this study is the selection of patients for more limited stripping operation [1, 2].

MAREIAL AND METHODS

Forty-five limbs with primary varicose veins in the distribution of the greater saphenous vein (GSV) were studied. Limbs with lesser saphenous vein incompetence or with varices mainly due to incompetent perforators were excluded.

The subjects were asked to sit on a chair with their knees flexed and feet on the floor (Fig. 1A, sitting position) [3]. The 50% venous refill time (VRT/2) after five dorsiflexions of the foot was measured in seconds. After measurement in the sitting position, the subjects were asked to lie supine on a table with their knees flexed and their feet on the floor, keeping their knees 5 to 10 cm higher than the heart (Fig. 1B, supine body position). The PPG measurement was repeated in this position.

Ascending venography was carried out on all limbs. The table was tilted 60 degrees to the horizontal plane (head up), and three rubber tourniquets were applied, just above the malleolus, immediately below the knee joint and on the upper one third of the thigh. First, incompetent perforators of the lower leg were observed, followed

268

by the deep vein system of the lower leg and the thigh. After arrival of enough contrast agent to fill up to the ilio-femoral vein, the three tourniquets were released sequentially from the thigh to the ankle under increased intra-abdominal pressure. X-ray films were taken and valvular competence of the GSV was assessed by how far the contrast agent refluxed down from the femoral vein into the GSV with the Valsalva maneuver.

Grades of reflux in the GSV on venography were stratified as followed: AK-reflux; reflux to a level above the knee, i.e., the GSV incompetence is limited to the thigh, BK-reflux; reflux to a level below the knee, i.e., the GSV incompetence is extended to the lower leg.

Figure 1. PPG measurement positions

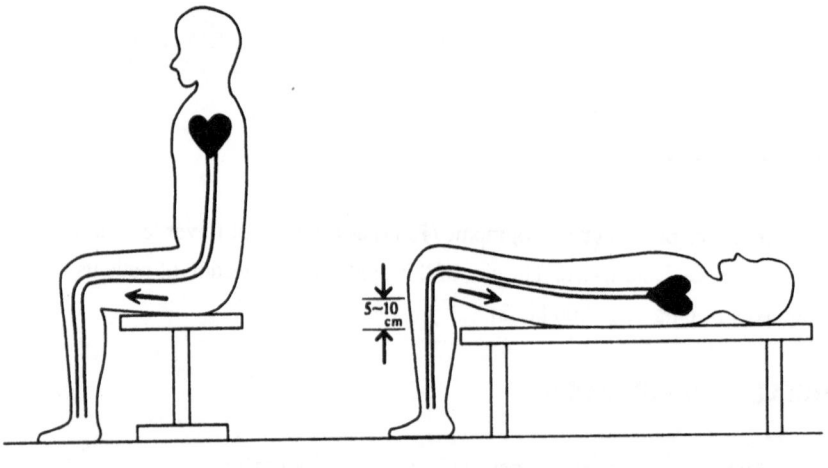

5~10 cm

A. Sitting position   B. Supine body positon

RESULTS

Evaluation of GSV valve function by ascending venography was possible in 41 (92%) limbs. Twenty-five limbs (55%) had an GSV with AK-reflux, and 16 limbs (36%) had an GSV with BK-reflux. In the other 4 limbs, contrast agent ascended into the GSV despite the tourniquets and evaluation of reflux was deferred.

In the 25 AK-reflux limbs, the VRT/2 in the supine body position was significantly longer than in the sitting position (7.6 ± 0.6 sec. vs. 5.0 ± 0.4 sec., p<0.01). On the other hand, in the 16 BK-reflux limbs, the VRT/2 was 4.7 ± 0.6 sec. in the sitting position and 5.5 ± 0.4 sec. in the supine body position--no significant differences. In the supine body position, the VRT/2 of the AK-reflux limbs was significantly longer than the VRT/2 of the BK-reflux limbs (7.6 ± 0.6 sec. vs. 5.5 ± 0.4 sec. p<0.01).

Incompetent perforators of the lower leg were detected by ascending venography. Ten of 45 limbs (22%) had no incompetent perforators. One incompetent perforator was found in 16 limbs (36%), two were found in 8 limbs (18%), three in 9 limbs (20%), and four in 2 limbs (4%). The mean number of incompetent perforators was 1.5 ± 0.2 per limb.

In the 10 limbs with no incompetent perforators, the VRT/2 in the supine body position was significantly longer than the VRT/2 in the sitting position (8.0 ± 1.1 sec. vs. 5.1 ± 0.3 sec., p<0.05). On the other hand, in the 35 limbs with incompetent perforators, the VRT/2 was 4.7 ± 0.4 sec. in the sitting position and 6.2 ± 0.4 sec. in the supine body position. Difference was not significant. In the supine body position, the VRT/2 of the limbs without incompetent perforators was significantly longer than the VRT/2 of the limbs with incompetent perforators (8.0 ± 1.1 sec. vs. 6.2 ± 0.4 sec., p<0.05).

CONCLUSION

PPG measurement in the supine body position gave a better separation of VRT/2's between limbs with a competent GSV in the lower leg and those with an incompetent one, and also between limbs with incompetent perforators and those without them. The supine body position is recommended for assessing lower leg venous function.

REFERENCES

1. Rivlin S. The surgical cure of primary varicose veins. Br J Surg 1975; 62: 913-7.
2. Koyano K, Sakaguchi S. Selective stripping operation based on Doppler ultrasonic findings for primary varicose veins of the lower extremities. Surgery 1988; 103: 615-9.
3. Alam S, Sakurai T, Yano T, Shionoya S, Hirai M. Hemodynamic assessment of chronic venous insufficiency. Jpn J Surg 1991; 21: 154-61.

Phlebology '95, D. Negus et al. (eds.). Phlebology (1995) Suppl. 1: 270-271

P055

The Value of Light Reflection Rheography for the Diagnosis of Varicosis

E.M. de Boer[1], A.J.C. Mackaay[2], A.H.M. Dur[2], P.D. Bezemer[3] and R.M.A. Krijnen[1]

Departments of [1] Dermatology, [2] Vascular Surgery and [3] Epidemiology, Free University Hospital, Amsterdam, The Netherlands

INTRODUCTION

Light reflection rheography (LRR) is often advocated as a practical screening test in the diagnosis of varicose veins. Differentiation between superficial and deep venous insufficiency has been achieved (1,2).

The aim of the present study was to determine the value of light reflection rheography (LRR) as a diagnostic tool in patients with varicose veins and as a screening for varicose veins in healthy workers

MATERIALS AND METHODS

A prospective study was performed in outpatients (group I) and on the working floor (group II).

Group I consisted of all otherwise healthy patients referred to the departments of dermatology or vascular surgery for untreated uncomplicated primary varicose veins. 123 individuals were included, mean age 47 years, 168 legs.

Group II consisted of males included in a field study on the presence of chronic venous insufficiency in workers with a standing position. 374 workers were included, mean age 37 years, 748 legs.

LRR measurements (AV-1000™ Hemodynamics inc.) were taken just above the medial malleolus in the sitting patients. A refilling time of at least 25 seconds after 10 dorsiflexions was considered normal. In case of a shorter refilling time a tourniquet was adjusted just above the probe. A normalisation of the refilling time indicates an insufficient superficial venous system. An unchanged refilling time indicates a deep venous insufficiency.

In group I LRR recordings were compared with colour-coded Duplex investigations as gold standard.

In group II LRR recordings were compared with clinical evaluation in combination with cw-Doppler ultrasound as a gold standard.

RESULTS

Group I: With Duplex investigation superficial varicose veins were proved in 156 legs. LRR showed 33 cases, sensitivity 21%.
Group II: Clinical evaluation with Doppler investigation indicated superficial varicose veins in 189 legs. LRR showed 14 cases, sensitivity 7%.

CONCLUSIONS

LRR appeared of no value in the diagnosis of superficial varicose veins in patients with a history of primary varicosis.
LRR appeared of no value in screening a working population on the presence of varicose veins.

REFERENCES

1. Neuman HAM, Boersma IDS. Light reflection rheography. A non-invasive diagnostic tool for screening for venous disease. J Dermatol Surg Oncol 1992;18:425-30.
2. van Walsum ADP. The value of light reflexion rheography and Doppler ultrasound examination in the diagnosis of varicose veins. Summary of thesis. Phlebology Digest 1993;5:26-9.

Phlebology '95, D. Negus et al. (eds.). Phlebology (1995) Suppl. 1: 272-274

OP/16.6

Computerised Plethysmography Diagnosis of Proximal Deep Vein Thrombosis

Pereira Alves, J. Neves, A. Formiga, L. Mata, M. Neves and Alves Pereria

Hospital St. Antonio Capuchos, Serviço 6, Unidade Vascular, Centro Investigação de Veias, R Garrett 74 s/1 1200 Lisboa, Portugal

INTRODUCTION

Deep vein thrombosis is an elusive and frequently misdiagnosed disorder.

Good clinical practice requires objective confirmation of thrombosis before starting active treatment.

Contrast venography has been regarded as the elected method for this confirmation (1,2). Nevertheless it is invasive, disturbing and with potential risks even with newest contrast media. Interpretation is not always easy.

Doppler ultrasound has shown in several studies (3,4), sensitivity and specificity for detection of proximal venous thrombus.

Colour doppler ultrasound facilitates new data to determine location, extent and age of venous thrombus. Although non-invasive, well tolerated and with no risks, it is time consuming, expensive and extremely operator dependent. An operator is not always available.

Phlebotest, a computerised plethysmography, based on volumetric variations of calf, between calf filling during thigh occlusion and emptying after rapid deflation of occlusion, is an easy quick, non-invasive method, that can be done by a doctor, a technician or a nurse without difficulties.

Different plethysmographic techniques have shown good sensitivity for detection of proximal deep vein thrombosis (5,6).

The objective of our study is to compare the results of colour doppler ultrasound and phlebotest.

METHODS

50 patients with clinical suspicion of proximal deep vein thrombosis have been examined with colour doppler ultrasound and phlebotest by independent clinicians with no knowledge of previous results.

Methodology of sonographic examination starts with the femoral veins, with the patient in supine position. With transverse and longitudinal cuts, the femoral vein is examined from the inguinal region to the adductor canal and up to the common iliac vein. The popliteal vein is scanned with the patient in the prone position and flexion.

We have studied: intraluminal echoes, vein compressibility by simple hand pressure, doppler flow and its respiratory phasicity and augmentation with distal vein compression, vein diameter and valvular status.

Intraluminal echoes and absence of spontaneous or provoked doppler flow are positive signs of thrombosis. Total or partial incompressibility, venous dilatation, valvular paralysis and high flux in collaterals are indirect signs of thrombosis (7).

The methodology of phlebotest is as follows: the patient is in supine position with heels supported in a special frame with an angle of about 90° and maintained in this position. Application of cuffs around the thighs. Strain gauges around larger diameters of the calves. The cuffs are inflated with a special pump to a pressure of 60mm. Maintenance of this pressure until 5 signals are heard, and at the 5th signal quick deflation of the cuffs. Occlusion and sudden deflation provokes changes in the leg volumes that are displayed as curves and numbers that are analysed by a computerised system and printed out, indicating: free flow, obstruction or equivocal.

RESULTS

42 positive patients with colour echodoppler and phlebotest
2 negative patients both with colour echodoppler and phlebotest.
6 equivocal patients with phlebotest showed: 3 partial femoral occlusions, 2 popliteal thrombosis, one negative.
No false positive cases with phlebotest.

CONCLUSIONS

Phlebotest seems to have the same sensitivity as colour doppler ultrasound in detecting proximal deep vein thrombosis.
Equivocal phlebotest results should be confirmed by colour echodoppler.
Phlebotest is less expensive, more easy and rapid and so can be used as a screening method and when good echodoppler operator is not available.

REFERENCES

1. Norman L Browse, Kevin G Burnand and Michael Lea Thomas - Diseases of the Veins - Pathology, Diagnosis and Treatment - Edward Arnold, 1988.
2. C Vaughan Ruckley: Clinical Problems in Vascular Surgery - Robert B Galland and Charles A C Clyne, Edward Arnold, 1994.
3. Lensing A, Prandoni R e al. - Detection of deep vein thrombosis by real time B-mode ultrasonography - N Engl J Med, 1989, 320:342.
4. Becker D M Philbrick J T, Abbitt P L - Real time ultrasonography for the diagnosis of lower extremity deep vein thrombosis - Arch Intern Med, 1989, 149:173.
5. Mullick S C, Wheeler, H B, Songesten G P - Diagnosis of deep venous thrombosis by electrical impedance. Am J Surg 1970, 119:417.
6. Bygdeman S, Ascleberg S, Hindmarsch T - Venous plethysmography in the diagnosis of deep venous thrombosis. Acta Med Scand 1977, 202: 319
7. M Duzat - Ultrasonographie Vasculaire Diagnostic - Le Diagnostic de thrombose veineuse profonde au stage aigu par l'examen doppler, 391: 402, Editions Vigot 1991.

Phlebology '95, D. Negus et al. (eds.). Phlebology (1995) Suppl. 1: 275-276

PI/5.4

Air Plethysmography; A Critical Evaluation of the Technique in Patients with Chronic Venous Insufficiency

L.E.G. Bossuyt and H.A.M. Neumann

Department of Dermatology, Academisch Ziekenhuis Maastricht, The Netherlands

Plethysmography is a valuable non-invasive quantitative technique in clinical studies of the venous blood circulation of the extremity, based upon the measurement of limb volume changes. It gives an insight into the functional aspects of the venous sytem, but provides no information on venous anatomy.

Since the original description of Brodie at the beginning of this century, various methods have been developed for measuring volume changes of the limb. All these methods measured volume changes of only a small segment of the leg, which are not necessarily representative for volume changes of the entire leg.

In 1987, Christopoulos and Nicolaides described a new technique of air plethysmography that measures volume changes of the entire calf. It has been stated that in patients with venous disease, air plethysmography can separate the relative contribution of different abnormalities as insufficient muscle pump fiction, and reflux in the superficial and deep venous system [1].

During the period october 1993 to july 1994, we studied 68 patients with the air plethysmograph (APG-1000^R, ACI Medical).

The indications were mainly patients with an history of deep venous thrombosis, chronic venous insuffiency with suspicion of deep venous insufficiency, and functional complaints of unclear origin.

Most of our patients were also investigated by a battery of other examinations such as Light Reflexion Rheography, doppler ultrasonic examination, duplex scan and/or phlebography. In preticular, we were interested in the correlation between APG and Light Reflexion Rheography, because both are stated to be screening tests for venous disease.

There were 9 failures in our studies, this represents approximately 13 % of our patient group. Most of the failures were due to musculosceletal disorders or orthostatic hypotension. The investigation could indeed only be performed on mobile patients, who can readily stand and perform tiptoe exercises.

The correlation between the refilltime (LRR) and the Venous Filling Index was not significant. A correlation between the severity of the clinical symptoms of venous insufficiency and the calf muscle pump function was demonstrated, confirming the data of Nicolaides [2].

We also checked the formula that has been established empirically by Nicolaides for determining the Ambulatory Venous Pressure (AVP=RVF- 1.5 / 0.98) by studying pressure - volume correlations.We found that individual differences of pressure -volume curves have to be taken into account.

Our conclusions were that APG indeed represents a usefull method for evaluating overall venous function, and muscle pump funtion. As the other methods of plethysmography, it gives information on functional aspects of venous obstruction.

However, we do not have the impression that it does allow accurate differentiation between reflux in the deep venous system, insufficient perforator veins of the lower legs and insufficient calf muscle pump function. There are also some important patient-related limitations.

REFERENCES

1. Christopoulos D, Nicolaides AN, Belcaro G. Air Plethysmography. Phlebology Digest 1992;4:4-10.
2. Christopoulos D, Nicolaides AN, Cook A, Irvine A. Pathogenesis of venous ulceration in relation to the calf muscle pump function. Surgery 1989;106:829-35.

Phlebology '95, D. Negus et al. (eds.). Phlebology (1995) Suppl. 1: 277-279

PI/5.7

A Reliable Non Invasive Method for Venous Compliance Measurement

P. Zamboni, M.G. Marcellino, D. Quaglio, G. Vasquez, A.P. Murgia, L. Pisano and C. Feo

Department of Surgery and Vascular Laboratory, University of Ferrara, Italy

INTRODUCTION.

Venous compliance is a physical index of mechanical wall properties (1-5), which could be useful in clinical practice to assess the progression of chronic venous insufficiency. We carried out a study to assess the clinical feasibility of routine, non-invasive determination of saphenous vein wall compliance, and the reliability of such measure in discriminating between early and advanced stages of varicose vein disease.

METHODS

Twenty-three consecutive patients referred to our vascular laboratory for superficial venous incompetence were classified as being in stage 1, 2 or 3 of venous insufficiency according to the international criteria (6).
Vascular compliance results from the following formula (1):
Compliance ratio $(\% / \text{mm Hg} * 10^{-2}) = 2 (D1-D0)/D0 (P1-P0) * 100$.
Measurements were performed blindly in each patient using two different methods.
Method 1 is a non-invasive procedure that we propose for routine use and method 2 is a partially invasive method used as control.
Method 1. In the formula above, D0 is the diameter of the long saphenous vein measured by duplex scanning after 10 seconds of 30° leg elevation in supine position. The ultrasonic beam was directed toward a rectilinear and non dilated part of the saphenous tract selected.
D1 is the long saphenous vein diameter measured in the same point during quiet standing. P0 is equal to the constant value of 7 mm Hg previously determined. Finally, P1 is the pressure of the hydrostatic column, i.e. the distance between the heart level and the point of saphenous diameter measurement in quiet standing.
Method 2. D0 and D1 were calculated by duplex scanning like in method 1, whereas P0 and P1 were measured using a needle transducer inserted in the long saphenous in the same point of diameter measurement.
Paired t test was used for comparison of the parameters assessed with the two methods. Correlation coefficients were also calculated and tested for significance by standard procedures.
Furthermore, statistical comparison of the distribution of non-invasive compliance values among the three clinical stages of disease was performed using the Kruskal-Wallis test.

RESULTS

The actual mean value of P0 recorded with the invasive method 2 was 6.9 mm Hg
(range: 4-14 mm Hg), showing a close correspondence with the constant value of 7 mm
Hg used in method 1.
The mean value of P1 calculated with method 1 was 70.4 mm Hg (range: 60-91 mm
Hg). The correspondent value calculated invasively with method 2 was 74.4 mm Hg (56-
100 mm Hg).
D0 and D1 measured by duplex scanning were 3.1 mm (1.9-7.3) and 6.4 mm (3.8-8.9),
respectively. Finally, compliance was 3.0% (0.7-6.9%) with method 1 and 3.1% (0.7-
6.6%) with method 2.
The differences in the values of the venous pressures measured in quiet standing with the
two methods showed no statistically significant difference. A high degree of correlation
was also demonstrated ($r = 0.9248$, $P < 0.0001$).
Compliance assessed with the two methods also showed no significant differences and a
highly significant correlation ($r = 0.9887$, $P < 0.0001$).
 The distribution in ordered categories of increasing compliance assessed non-invasively
and worsening disease stage is reported in Table1. Visual inspection of the table suggests
that severity of disease increases as venous compliance worsens. This was confirmed by
the Kruskal-Wallis test ($H = 14.704$, $P = 0.0006$).

Table 1- Distribution of 23 cases of venous insufficiency in categories of increasing
venous compliance (non-invasively assessed).

	Compliance ratio (% / mm Hg$*10^{-2}$)				
	<2	2-3	3-4	>4	Total
Stage 1	0	1	7	3	11
Stage 2	1	7	1	0	9
Stage 3	3	0	0	0	3

DISCUSSION.

An accepted method of vein compliance determination is still lacking. On the other hand,
vein diameters and pressures can be determined echographically and intravenously,
respectively. The accuracy of both measures is widely accepted (7,8). Since venous
diameters and pressures are the only variables needed to calculate venous compliance, we
studied the applicability of a method of assessment of venous compliance that substitutes
the invasive measurement of intravenous pressure with a simple, non-invasive calculation
of the hydrostatic pressure based on the length of the limb under investigation.
The main finding of our study is that non-invasive measures of venous compliance
closely correlate with compliance values obtained from intravenous pressure
determination. In addition, our data confirm that saphenous pressure in quiet standing
corresponds to hydrostatic pressure with great approximation (9).
Using duplex scanning and a simple ruler, saphenous compliance can be calculated in
just a few minutes with no discomfort for the patient. Hence, it has potential possibilities
for routine clinical application.
Finally, our data also show that venous compliance calculated with the method that we
propose could be useful for differentiating between stages of CVI and in the assessment
of saphenous vein wall damage.

REFERENCES

1. Baird RN, Kidson IG, L'Italien GJ, Abbott WM: Dynamic compliance of arterial graft. Am. J Physiol.1977; 233(5): H568-72.
2. Abbott WM, Megerman J, Hasson JE, L'Italien GJ, Warnock DF: Effect of compliance mismatch on vascular graft patency. J.Vasc.Surg. 1987; 5: 376-82.
3. Davies AH, Magee TR, Baird RN, Sheffield E, Horrocks M: Vein compliance: a preoperative indicator of vein morphology and of veins at risk of vascular graft stenosis. Br.J.Surg. 1992; 79: 1019-21.
4. Baird RN, Abbott WM: Elasticity and compliance of canine femoral and jugular vein segments. Am. J.Physiol.1977; 233(1): H15-21.
5. Davies AH, Magee TR, Hayward J, Harris R, Baird RN, Horrocks M: Non-invasive methods of measuring venous compliance. Phlebology 1992; 7: 78-81.
6. Porter JM, Clagett GP, Cranley J: Reporting standards in venous disease. J.Vasc.Surg 1988; 8: 172-81.
7. Talbot SR: B-mode evaluation of peripheral veins. Semin.Ultrasound CT MR 1988; 9: 295-319.
8. Bjordal RI: Pressure patterns in the saphenous system in patients with venous leg ulcers. Acta Chir. Scand. 1971; 137:495-501.
9. Hojensgard IC, Sturup H: Venous pressure in primary and postthrombotic varicose veins. Acta Chir.Scand. 1949; 99: 133-53.

Phlebology '95, D. Negus et al. (eds.). Phlebology (1995) Suppl. 1: 280-282

P095

Invasive and Doppler Pressure Measurements in Venous Diseases

M. Bartolo, P.L. Antignani, A.R. Todini and P.M. Nicosia

Department of Angiology, s. Camillo Hospital, Rome, Italy

INTRODUCTION

The Doppler venous pressure measurement has been used by us since 1977 [1] with interesting results and usefulness not only in the diagnosis of venous diseases but also in the management of therapeutical treatment both in acute and chronic venous pathologies. There is difference between the tensive noninvasive Doppler values, repeatedly proposed by us [2,3,4,5,6] and those quoted by literature and obtained by invasive method [7,8]. A contemporary measurement of venous pressures by means of the two methods has been necessary to validate the Doppler method and to confirm its utility in venous diseases.

METHODS

We measured in orthostatism in 52 subjects (average age 42,8 y.; 29 women and 23 men: 20 healthy cases and 32 phlebopatic subjects with venous insufficiency of various degree) the venous pressure at the ankle by means of Doppler method and, simultaneously, the venous pressure by direct puncturing of the superficial vein at the back of the foot by means of electromanometric method, in orthostatism.

To make the diagnosis, we used only a clinical criterion of selection because the inquiry concerns exclusively the evaluation of the homologies of the two methods of the pressure measurement.

The Doppler method consists of inflation of a sphygmomanometer cuff above the ankle to a pressure estimated to be slightly greater (120 mmHg in orthostatism and 60 mmHg in clinostatism) than the pressure to be measured. The Doppler probe is placed over the long saphenous or posterior tibial veins at the level of the medial malleolus, or over the short saphenous vein at the lateral malleolus. The cuff pressure is dropped and a "blowing noise" is heard when venous pressure is reached.

The invasive venous pressure has been by direct puncturing of a superficial vein at the back of the foot by means of an electromanometric capsule Statham placed at the same height of the cuff used for an indirect measurement (to outweight the hydrostatic gradient between two levels of recording). The signals recorded during the measurements were sent to a polygraph OTE on photographic paper, making the two manometric traces coincide on the same scale.

RESULTS

The average values of orthostatic invasive pressure were similar:
81,70+/-6,62 mmHg in the healthy subjects
79,33+/-7,70 mmHg in the patients;
the average Doppler pressure values were:
67,62+/-13,63 in the healthy cases
92,08+/-11,67 in the phlebopatic ones.
The Doppler pressures in the healthy subjects are lower than invasive ones by 17% while in the phlebopatic ones are higher by 16%.
These results allow us to note that:
- in all the cases and in every functional condition the beginning of the doppler "blowing noise" allows us to recognize, reading the sphygmomanometer, the real venous pressure existing in the zone examined;
- in the healthy subjects the inflation of the cuff does not modify the invasive venous pressure; during the compression of the cuff we can notice the typical Doppler noise that at its beginning coincides with the unlock of the venous flow at the moment when the venous and pneumatic pressure are equal and the two manometric tracings intersect.
A decrease of venous invasive pressure follows and it slowly comes back at the basic orthostatic level. If the tests are repeated, we obtain a progressive and homologous lowering of invasive and Doppler pressures;
- in the phlebopatic subjects during the compression of the cuff there is a hypertension as regards to the basic values. Also in this case, the pressure noticed by the Doppler method always coincides with that existing in the vein; subsequent to the Doppler measurement there is a more limited and more transient tensive decrease than normal or no fall of pressure is noted. The repetition of measurements does not produce a reduction of pressure regards the orthostatic basic level.

DISCUSSION

The comparison between the invasive values and Doppler ones of venous pressure allow us to conclude that:
- The pressure noticed usually on the sphygmomanometer at the moment when the decompression of the cuff allows the venous outflow again is really that existing inside the vein.
- The disparity between two methods mainly consists in the fact that in orthostatism all the authors [7,8] notice in static conditions venous pressures equal both in normal subjects and in phlebopatic ones while with the Doppler method we note a hypertension in phlebopatic subjects and a hypotension in normal ones.
This disparity is not due to a mistake in the procedure but is due to the presence of the cuff which is the only experimental condition that differentiates the invasive test from the Doppler one.
- The venous district below the cuff reacts on the obstacle caused by the hypertension of the air-tube with two different forms: there is hypotension in case of a normal venous involvement, achieving in this way a kind of squeezing of blood from the leg by means of the venous pump of the calf; hypertension in case of spoiled reactivity or reflux or

282

reduced total vascular lumen in case of phlebopathy. These are the reasons for the difference of orthostatic invasive data by non invasive ones.
- So, the venous dynamic response which is induced by the cuff implies physiopathologic implications much finer than the simple invasive test in static conditions, and is like a watershed between the normal and pathologic subjects.

REFERENCES

1) Bartolo M. Phlebodopplertensiometry, a non invasive method for measuring venous pressures. Folia Angiol. 1977; 25:199-205.
2) Bartolo M, Pittorino L, Antignani PL. Rilevazione della pressione venosa con metodica doppler e possibili cause di errore. Atti III congresso nazionale della Società Italiana di Patologia Vascolare, Torino: Minerva Medica, 1981:48-51.
3) Bartolo M, Nicosia PM, Antignani PL et al. Non invasive venous pressures measurements in different venous diseases. Angiology 1983; 34:717-24.
4) Bartolo M, Antignani PL, Di Folca A et al. Mesure de la pression veineuse avec le Doppler: standardisation de la méthode. Phlebologie 1984; 37:103-5.
5) Bartolo M, Antignani PL, Nicosia PM et al. Mesure de la pression veineuse avec le Doppler: données statistiques. Phlebologie 1984; 37:106-9.
6) Bartolo M, Nicosia PM, Antignani PL. Validation of doppler venous pressure measurement. Acta of X World Congress of Union Internationale de Phlebologie, London: John Libbey, 1989:220-224.
7) Bjordal RI. Simultaneous pressure and flow recordings in varicose veins of the lower extremity. Acta Chir Scan 1970; 136:601-5.
8) Hobbs JT. The treatment of venous disorders. Lancaster: MTP Press Ltd.,1977.

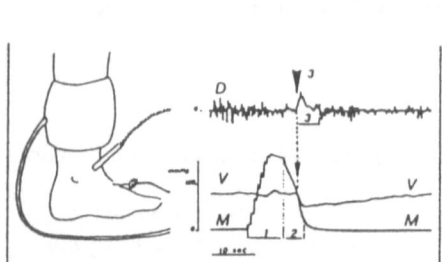 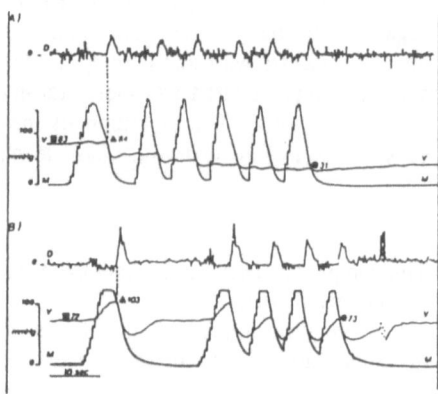

Fig.1: Phlebodopplertensiometry and invasive relieves in normal subject. The two manometric traces (invasive venous pressures and pressures in the cuff) are in the same scale.
D: Doppler trace; V: invasive V.P.; M: pressure in the cuff; 1: insufflation; 2: decompression; 3: Doppler "noise of wind".

Fig.2: Repeated measurements of venous pressure in orthostatism.
A: normal subject: there is a decrease of tensive values.
B: phlebopatic subject: the repeated measurements cause a poor pressure decrease.

■ basal V.P.   D: Doppler trace
▲ initial Doppler relief   V: invasive V.P.
● final Doppler relief   M: cuff pressure

Phlebology '95, D. Negus et al. (eds.). Phlebology (1995) Suppl. 1: 283-285

P062

Photoplethysmography at the Upper Third of the Leg in the Study of Patients with Varicose Veins in the Lower Limbs - A New Technique

Solange S. Meyge Evangelista, MD and Franklin P. Fonseca, MD

Department of Angiology and Vascular Surgery of Hospital das Clinicas,
Universidade Federal de Minas Gerais, Brazil

INTRODUCTION

Among the various noninvasive methods of venous insufficiency assessment, photoplethysmography provides quantitative data of venous refilling, is easy to undertake, and does not demand highly trained personnel. In the past, the venous refilling time (VRT) values obtained with this practice were directly related to the venous pressure as measured at the saphenous vein of the ankle [1].

The purpose of this study was to assess patients with partial reflux at the trunk of the saphenous vein with the use of a new methodology, after observing that in such cases, the standard method produced values within normality range. This new approach, if properly divulged, increases the utilization possibilities of that methodin the assessment of superficial venous insufficiency.

METHODS

Thirty and six limbs of patients with varicose veins in the lower limbs, who had no evidence of deep venous reflux, and some who had evidence of insufficiency associated with perforated veins, were prospectively investigated. Patients were either asymptomatic or had light to moderate evidence of venous insufficiency. Patients were nineteen females and one male, aged 24 to 66 years.

All 36 tests were undertaken with the use of "Medacord PVL Enhanced" of "Medasonics, Inc." with the probe placed at the posterior portion to medial malleolus and fixed with double-sided adhesive tape. Each patient was seated with legs hanging free, and requested to perform five dorsiflexions. Venous refilling time (VRT) was recorded at baseline. All 36 tests were repeated with the probe placed below knee in its medial face, posterior to the internal saphenous vein course. Reflux evidence and location were confirmed by Duplex Scanning analysis.

RESULTS

In the 36 limbs investigated, 36 tests with the probe either fixed at the malleolus or at the upper third of the leg, were performed. Results recorded are as follows:
PROBE AT THE UPPER THIRD OF THE LEG:

- 25 limbs with VRT alteration on photoplethysmography, confirmed by Duplex Scanning.
- 4 limbs with VRT within normality and reflux as verified on Duplex Scanning.
- 3 limbs with changes in VRT, not confirmed on Duplex Scanning.
- 4 limbs without changes on photoplethysmography or Duplex Scanning.

RATES RECORDED ON TEST COMPARISON:
Sensitivity, 86.2%; Specificity, 57.1%; and Accuracy, 69.4%.

PROBE AT MALLEOLUS LEVEL:
- 19 limbs with decreased VRT, confirmed by Duplex Scanning.
- 1 limb with decreased VRT and no reflux as indicated by Duplex Scanning.
- 10 limbs within normality evidence of reflux on Duplex Scanning.
- 6 limbs had no reflux evidence on VRT testing and Duplex Scanning.

RATES RECORDED ON TEST COMPARISON:
Sensitivity, 65.5%; Specificity, 85.7%; and Accuracy, 69.4%.

PROBE ON BOTH PREVIOUS SITES:
- 26 limbs had evidence, on both methods, of venous refilling at the trunk of the saphenous vein.
- 3 limbs with normal VRT values; no reflux on Duplex Scanning.
- 3 limbs with normal VRT values; evidence of reflux on Duplex Scanning.
- 4 limbs without any changes on both methods.

RATES RECORDED ON TEST COMPARISON:
Sensitivity, 89.66%; Specificity, 57.14%; and Accuracy, 83.33%.

Surgical treatment of primary varicose veins of the lower limbs with preservation of the internal saphenous vein. Pre and postoperative evaluation with photoplethysmography. Venous refilling time (seconds).

Fig. 1: Preoperative procedures – August 08, 1994   Fig. 2: Postoperative procedures – 8 months later

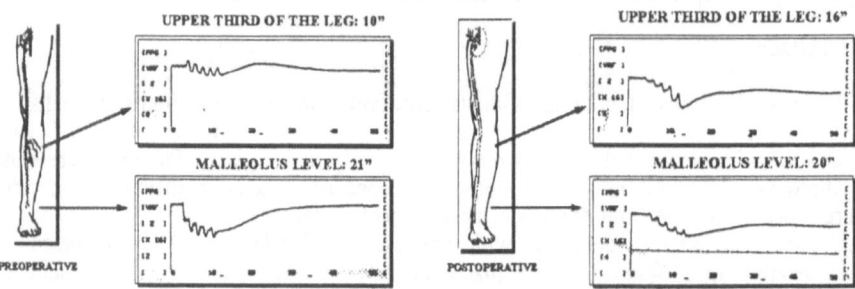

DISCUSSION

The data obtained with probe placed at the upper third of the leg were compatible with reflux evidence, as confirmed by Duplex Scanning, in a greater number of cases than those verified with probe fixed at malleolus. It may, probably, be accounted to the fact that cases of partial refilling at the trunk of saphenous vein till below knee were recorded when the probe was fixed in that particular site, and not evident when probe was fixed at the malleolus. Whenever a reflux was evident in all the trunk of the vein, photoplethysmography data were compatible with those of Duplex Scanning, both below knee and at malleolus.

Whenever a patient had an evidence of refilling at the trunk of saphenous vein only in the leg, caused by a perforating system incompetence, or showed refilling of the short saphenous vein with long saphenous vein, VRT changed at malleolus and remained within normality below knee.

Photoplethysmography reflects a regional change of venous insufficiency [2], with probe positioning being critical in the evaluation of reflux evidence.

All patients investigated showed normal DVS, and the aim was to assess superficial venous insufficiency only.

Photoplethysmography proved to be an efficient tool in the assessment of superficial venous insufficiency [3]. Therefore, it is apparent that both photoplethysmography and Duplex Scanning are necessary instruments for a complete evaluation of venous function.

CONCLUSION

A relationship has been shown associated with the distribution of superficial venous insufficiency and transducer positioning, when tests are undertaken. The methodology of placing the probe at the upper third of the leg increases the sensitivity of the test, allowing for a quantitative assessment based on reflux extent, in the circumstances of refilling not to reach all the trunk of the saphenous vein, which is not feasible with traditional testing.

SUMMARY

The authors present a new methodology for the tests performed with photoplethysmography, on the diagnosis of venous refilling of patients, utilizing, in addition to probe positioning at the distal portion the leg, an evaluation with this same probe positioned at the upper third of the leg in a study of 36 limbs of 20 patients.

When comparing the data obtained with Duplex Scanning – noninvasive Gold Standard Method – we found a greater compatibility of results recorded with photoplethysmography undertaken at the upper third of the leg.

Keywords: Photoplethysmography, Duplex Scanning, varicose veins of the lower limbs.

REFERENCES

1. Abramowitz H, Queral LA, Flinn WR, et al. The use of photoplethysmography in the assessment of venous insufficiency: A comparison to venous pressure measurements. Surgery 1979;86:434-41.
2. Rosfors S. Venous photoplethysmography: Relationship between transducer position and regional distribution of venous insufficiency. J Vasc Surg 1990;11:436-40.
3. Sarin S, Shields DA, Scurr JH, Coleridge Smith PD. Photoplethysmography: A valuable noninvasive tool in the assessment of venous dysfunction? J Vasc Surg 1992;16:154-62.

Phlebology '95, D. Negus et al. (eds.). Phlebology (1995) Suppl. 1: 286-288

P098

Digital Photoplethysmography (D-PPG) in Comparative Analysis of Venous Haemodynamic Changes in Lower Limbs of Patients With and Without Recurrence after Surgical Treatment Modo Babcock

J. Kuleszynski, D. Czaczka, Z. Wygoda and K. Twardowska-Saucha

Varicose Vein Clinic "Medservice", Zabrze, Poland

OBJECTIVE

The aim of this paper is to compare two groups of patients with and without the presence of recurrent varicose veins after surgical treatment modo Babcock performed in the past.

INTRODUCTION

The main causes of recurrent varices are improper surgical procedures and sometimes anatomical abnormalities in veins topography [1,2,3].

The incomplete surgical removal of varicose veins, however remains the primary reason for their recurrence. The flush ligation at the sapheno-femoral junction seems to be rather simple but this is not always so. Inadequate dissection and examination of fossa ovalis leads to several common errors. The most frequent ones are:

-incisions 2-3 cm. below the groin without proper control of sapheno-femoral junction,
-persistence of the large tributary,
-ligation and dissection of the large tributary instead of LSV (for example anterolateral superficial vein),
-incomplete localisation of all sites of reflux or leaving connecting branches between deep and superficial systems,
-sometimes anatomical abnormalities such as double LSV may cause intraoperative difficulties.

Causes mentioned above may produce recurrences especially if the surgeon is not very experienced. Generally it is estimated that in more than 50% of patients early recurrences because of improper diagnose or treatment are present. Late recurrence (after 4-5 years) may sometime result from the natural progression of the disease.

Patients with recurrent varicose veins present wide spectre of complaints concerned with haemodynamic disturbances. In our Varicose Vein Clinic "Medservice" around 24% out of the total number of patients were admitted due to complaints caused by persistence or relapse of the disease after surgery which had been performed in the past. It was often observed during diagnostic procedure that haemodynamic parameters assessed in digital photopletysmography were distinctly worse than in these who underwent an adequate operation of the same type.

MATERIAL, METHODS AND RESULTS

In our Clinic 50 patients with recurrent and 50 patients without recurrent varicose veins were examined. These groups were statistically comparative as far as the age and time after operation are concerned. The age of the patients ranged from 21 to 69 years (mean: 47 yr.) and postoperative time ranged from 4 to 32 years (mean 13 yr.). Both groups underwent the same type of surgery i.e.: ligation of sapheno-femoral junction followed by stripping of LSV modo Babcock. Patients were operated on in several Polish clinics. They attempted our ambulatory because of several subjective complaints and/or cosmetic reasons.

The recurrence of disease was assessed by means of physical examination, simple tests and the presence of the reflux in CW-Doppler. Haemodynamic parameters such as: venous refilling time (To) and venous pump work (Fo) were evaluated with a help of D-PPG [4,5,6].

The first investigated group revealed reflux and incompetence of sapheno-femoral junction.

In some cases reflux was heard in accessory anterior or posterior vein, one of the other large tributaries or in incompetent thigh perforator with positive Valsalva test (atypical reflux).

The second group of patients presented no reflux during CW-Doppler examination of sapheno-femoral junction and through thigh perforators, however some crural perforators were incompetent in some cases. The Valsalva test was negative.

In the group of patients without recurrence of the disease - To ranged from 7s. to 48s. (mean: 22,4s.) and Fo ranged from 5s%. to 142s%. (mean: 48,5s%.).

The typical D-PPG graph is presented at figure 1.

On the other hand in the group of patients with recurrent varicosity - To ranged from 6.s to 28s. (mean 13,4s.) and Fo ranged from 3s%. to 57s% (mean 17,3s%.).

The typical D-PPG graph is presented at figure 2.

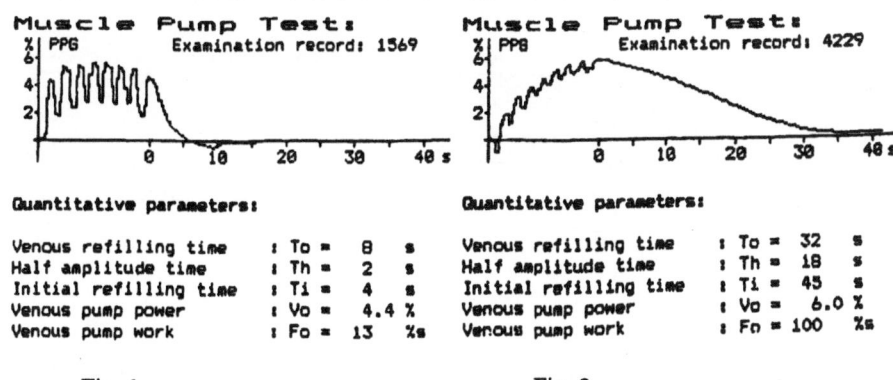

Fig. 1   Fig. 2

Statistical analysis of D-PPG parameters by means of t-test revealed significant differences between these two groups. Patients without recurrence presented longer venous refilling time and bigger venous pump work than patients with recurrent varicose veins (To - p< 0.0001 and Fo - p< 0.0001). These parameters highly correlated (p<0.001) with the subjective complaints such as pain, oedema or leg heaviness.

CONCLUSIONS

Analysis revealed significant differences among examined subjects. In many cases quite good haemodynamic results were observed in patients who underwent surgery performed with attentiveness and accuracy despite of imperfection and obsoleteness of the method.

We intend to stress that the effectiveness of treatment method depends mainly on the experience of the operating team.

Extensive evaluation of the venous disorders and accurate ligation of every site from deep to superficial reflux are crucial in preventing recurrences.

Doppler ultrasound and D-PPG examinations are accurate in deep and superficial venous system assessment both in preliminary and control diagnostics. They make it possible to find out the localisation of the reflux and to estimate the haemodynamic parameters disturbances which could have an effect on the patient's condition. They are also of a great use in the after treatment follow-up.

When necessary, complementary surgery or compression sclerotherapy must be carefully performed according to basic principles.

REFERENCES

1. Juhan C., Haupert S. et al. "Recurrent varicose veins" Phlebology 1990; 5: 201-211.
2. Ruckley C.V., Beggs I. "Recurrent varicose veins: an anatomical analysis" Phlebologie '92; 1187-89.
3. Belardi P., Lucertini G. "Re-exploration of the sapheno-femoral junction in recurrent varicose veins: properties of Li's technique" Phlebologie '92; 1195-97.
4. Kerner J., Schultz-Ehrenburg U., Blazek V. "Digitale Photopletysmographie (D-PPG) Phlebol. u. Proctol. 1989; 18; 98-103.
5. Grondin L. "Practical aspects of Doppler evaluation and D-PPG" Phlebol. '92; 642.
6. Browse N.L. et al. *Book*: "Diseases of the Veins" London; Arnold E., 1988: 80-84.

Phlebology '95, D. Negus et al. (eds.). Phlebology (1995) Suppl. 1: 289-291

P077

Foot Clinical Examinations in the Phlebolymphopatic Patient

E.O. Brizzio[1-2], Alberta Cumaldi[2] and Gustavo Martinez Lacabe[2]

[1] *Vascular Peripheral Laboratory, 965 San Martin St. 1st Floor Buenos Aires, Argentina*
[2] *Argentine School of Phlebology of AMA, 1171 Sante Fe Av, Buenos Aires, Argentina*

INTRODUCTION

The awareness of the fact that there exists a dose relationship among **plantar support alterations**, **walking dynamics** and **phlebopathologies**, makes it necessary for the phlebologist to give the patient an accurate semeiology, for the purpose of diagnosing and correcting static and dynamic pathologies which could affect the etiology and recurrence of the varicose veins.

OBJECTIVE

To provide the phlebologist with a semiologic foot routine, that he can currently use.

METHODS

1) **Interrogation**. 2) **Clinical examination**. 3) **Instrumental examination**.

*1) Interrogation:

    a) Static habits.
    b) Dynamic habits.
    c) Sporting habits.
    d) Walking habits.
    e) Footwear habits.
    f) Orthesis therapy prescribed prior to medical consultation.

*2) Clinical examination: It includes the examination of:
  a) **Dorsal or anterior side**: We examine
the dorsal side and toes,
morphological alterations and distribution
of the toes.

b) **Posterior side**: The alignment of the axial line of the leg with the axial line of the heel forms a physiological angle of 5° (five degrees). Above this value, we refer to valgum foot, below this value or in case of inverted angle we talk about, varus foot.

c) **Internal side**: We examine:

        1) The curvature of the longitudinal arch.
        2) Abnormal protuberance of the tibial malleolus or internal malleolus.
        3) Tuberosity of the head of the astragalus.
        4) Tuberculum of the scaphoid.

d) **External side**: We examine the external margin plantar support.

e) **Position of the foot with respect to the axial line** of the leg (supination, pronation).

f) **Position of the axial line of dorsal side with respect to the axial line** of the heel, normal toe, abducted toe, adducted toe.

g) **Podoscopic plantar examination**:

We make: monopodalic analysis.
    bipodalic analysis:
        comparison of weight

We observe:

    1) Plantar support
       regularity or irregularity.
    2) Plantar impression:
       a) Surface of the heel.
       b) Medial metatarsal
         impression.
       c) Increase or reduction of
         the plantar cavity.
    3) Impression of both toes.

h) **Dissymetries**: Dissimilarity in limb length, particularly noticed when heels are paired.

*3)Instrumental Examination We carry out a:

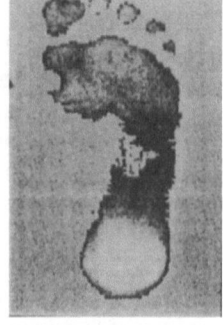

1) **Static instrumental examination**: This examination is performed with the aid of video podoscopy and digitalized videopodoscography. The former consists of the instrumental acquisition of plantar support images. The latter means the digitalization and subsequent loading of such images onto the computer,and the possibility to work with contrast as provided by the computer. The desired goals here consist of obtaining a documentary record of plantar support and manipulating color contrast in order to expand existing information.

2)**Dynamic instrumental examination**: This examination is performed in actual motion whit the shoes the patient normally wears, according to usual habits. For this purpose we carry out computerized podobarometry, which is an instrumental method that permits us to register the progression of plantar support areas within a dynamic walking context.

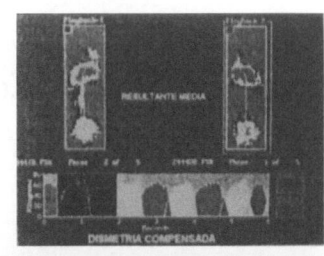

3)Study of the deep plantar venous system with color Echo-Doppler:

This instrumental means permits us the identification of venous axis,their differentiation from surrounding axis, the measurement of the ampoule of the internal plantar vein and finally, an analysis of plantar system permeability.

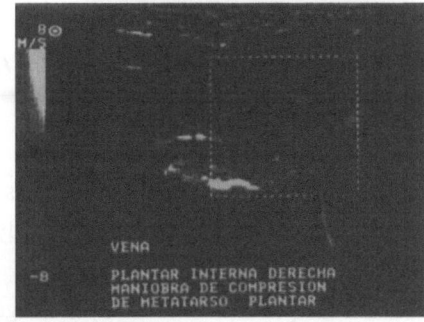

CONCLUSIONS

As a consequence of the application of this methodology of work we can conclude that: 1) Whoever looks for something finds it. 2)The incidence of pathology of toe in phlebolymphopatic patients is very high. 3) Diagnosis and eventual compensatory treatment does improve and optimize the results of phlebological treatments.

REFERENCE:

{1) **Bacci Pier Antonio**. Il plantare a lievitazione per una cura dinámica della patología del piede.
  Centro de Flebología, Universidad de Siena. Italia

{2)**Bacci pier Antonio**. Piede e flebopatíe.
  Centro de Flebología. Universidad de Siena.

{3)**Bonnel F. Claustre**. Le pied, una merveille architectural.

{4)**Chiappara G. Dagnini. M.T. Giuliano. Giacche**.
  Phlebologie et podologie. Phlebologie 1993. 46 No.2 275-286

{5)**Goldcher A**. Podologie. Ed. Naman 1991

{6) **Gillot Ce**. La semelle de Lejars Phlebologie 1993. 46 No 2 173-196

{7)**F.PLas, E.Viel, Y. Blanc** . La marche humaine Ed.Masson. 1989

Phlebology '95, D. Negus et al. (eds.). Phlebology (1995) Suppl. 1: 292-294

P068

Accuracy of Venous Compliance Measurements with Air Plethysmography (APG)

M. Chauveau

Department of Cardiology, Hopital Cochin, Paris, France

Plethysmography of the lower limbs is probably the most suitable method for quantifying venous function, particularly for physiological and pharmacological studies.

Electrical impedance and mercury-strain gauge have been widely used for many years. Compared to them, the more recently developed air plethysmography (APG) has two advantages: the measurement involves the whole leg, from the knee to the ankle; and the results are expressed directly in volume units.

The objective of this work was to check: (1) the accuracy of volume measurements by APG, and (2) the reproducibility of venous compliance measurements, by this method.

1 - IN VITRO CALIBRATION OF APG

Methods: the leg was replaced by a Douglas bag, 33 cm long, modified into the shape of a truncated cone in order to resemble a leg. The bag was filled with water and suspended by its two ends, then surrounded by the air cuff of the plethysmograph (APG 1000, ACI medical, Sun valley, USA). The usual APG calibration was performed, as for in vivo measurements, by injecting 100 ml of air into the cuff. Then, the bag volume was changed by 20 ml square-wave steps, by means of a water-filled syringe.

Results:

When the bag volume is increased rapidly (20 - 30 sec between steps) until 380 ml, and then decreased in the same way:

- the linearity of APG response is rather good, with a maximum deviation from linearity of 1.5% and 5%, for increasing and decreasing volumes respectively

- the slope of the curve is 0.73 for increasing and 0.83 for decreasing volumes, instead of the expected value of 1; so that volume decreases are overestimated by about 10% when compared to volume increases (fig 1).

When the bag volume is increased and decreased slowly, that is allowing enough time between steps for the tracing to stabilize (60 to 90 sec), the APG response remains linear and the discrepancy between increasing and decreasing volumes has been reduced to 5%. The slope is 0.73 for increasing, and 0.79 for decreasing volumes (fig2). Similar results were obtained when the above experiment was repeated after the bag volume had been increased by 200ml. In other words, the gain of APG was not significantly modified when a bigger bag was used.

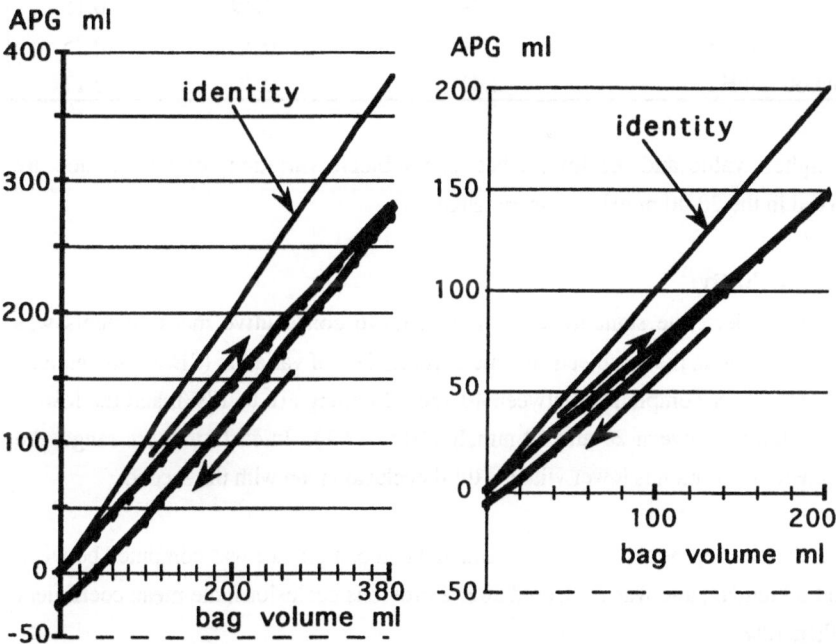

fig 1: fast volume changes   fig 2: slow volume changes

We conclude that APG, thank to its good linearity, is accurate for measurement of slow volume changes, as those induced by venous occlusion for the measurement of venous compliance.

2 - VENOUS COMPLIANCE OF THE LEG

Methods: We measured venous compliance of the right leg with APG, by the venous occlusion technique, in 8 normal male subjects, 20 to 44 years old, on 6 different days each, in similar timing and temperature ($22 \pm 0.5°C$) conditions.

Three occlusion manoeuvres were performed: the first one at 60mmHg; the second one at 60, then 40, and then 20 mmHg; and the third occlusion manoeuvre was identical to the second one. For each occlusion pressure, the plateau value of the volume record was used to compute compliance. According to the results of in vitro calibration, all APG readings have been multiplied by 1.3.

Results:

Table 1: Mean values and between subjects scattering of compliance.

Compliance between:	mean (ml/mmHg)	SD	SD/mean
20 and 40 mm Hg	3.34	0.49	0.15
40 and 60 mm Hg	1.48	0.37	0.25
20 and 60 mmHg	2.41	0.42	0.17

The highest value and the lowest between-subjects variations of compliance are observed in the 20-40 mmHg pressure range.

Reproducibility:

For each subject, the standard deviation of the 6 consecutive measurements was computed. Among the 8 subjects, the mean coefficient of variation (SD/M) so obtained is the highest for compliance between 40 and 60 mmHg (16 to 18%), and the lowest for compliance between 20 and 60 mmHg (10 to 13%). In each pressure range, the coefficient of variation is lower with the third occlusion than with the second.

So that the best reproducibility was obtained when compliance was computed between 20 and 60 mmHg, and was measured after 2 previous occlusions: the mean coefficient was then 10%.

These results are in good agreement with data from literature obtained by other methods.

Conclusion: the accuracy of APG is good enough for venous compliance measurements, the day-to-day variations of which are mainly due to physiological fluctuation.

Phlebology '95, D. Negus et al. (eds.). Phlebology (1995) Suppl. 1: 295

P053

Reliability Study of the Leg-O-Meter in Patients Suffering from Venous Insufficiency of the Lower Limbs

A. Bérard[1], X. Kurz[2], F. Zuccarelli[3], J.J. Ducros[3], L. Abenhaim[1] and the Veines Group

[1] Centre for Clinical Epidemiology and Community Studies, SMBD-Jewish General Hospital, McGill University, Montreal, Canada, [2] Laboratoire de Pharmacologie, Université de Liège, Liege, Belgium and [3] Hôpital St-Michel, Paris, France

Objective: To evaluate the reliability of the Leg-O-Meter, developped by F. Zuccarelli, which uses a tape measure to follow the clinical evolution of patients with venous insufficiency of the lower limbs.

Design: Inter-rater reliability study.

Patients: 39 patients (78 legs) consulting the phlebology clinic of the Hôpital St-Michel, Paris, France.

Intervention: Informed consent to participate in the study was obtained from patients consulting at the phlebology clinic. Participants were then asked to enter a closed room where 5 independent and blinded observers consecutively took measurements of both legs with the Leg-O-Meter. The order of the observers was randomized between patients.

Measurements: Measurements were performed using the Leg-O-Meter. The Leg-O-Meter is an instrument specifically designed to measure the calf circumference and consists of a tape measure fixed to a stand attached to a small board on which the patient is in standing position. For this study the tape measure of the Leg-O-Meter was fixed at 10 cm from the floor in order to standardize all measurements.

Results: Under the assumption of a two-way random effects model an Intraclass Correlation Coefficient (ICC) was used to determine the reliability or reproducibility of a measure with the Leg-O-Meter. The ICC for the right and left leg were estimated at 97.22% [95.55%;100%]$_{95\%}$ and 97.12% [95.25%;100%]$_{95\%}$, respectively.

Conclusions: The Leg-O-Meter gives a standardized and reliable measure of the circumference of the calf. Furthermore, it is not invasive or costly and is an alternative to the traditional tape measure which has been proven unreliable.

P054

How good are ankle pulses at detecting arterial disease?

CJ Moffatt MI Oldroyd RM Greenhalgh PJ Franks

Centre for Research & Implementation, 5-7 Parsons Green, London SW6, & Dept of Surgery, Charing Cross Hospital, London W6.

Objective : to investigate the ability of district nurses to detect lower limb arterial disease by palpation of ankle pulses

Design: Ankle pulse palpation of patients presenting with leg ulceration, compared with ankle brachial pressure index.

Patients: Sequential patients attending community leg ulcer clinics main outcome measures: Sensitivity and specificity of pulse palpation to detect arterial disease compared with the gold standard of ABPI.

Results: Of 533 limbs with ulceration in 462 patients (mean age 74 years, 67% female), 167 (31%) had no detectable pulses at the ankle. Of the 93 limbs with with ABPI <0.9, 34 (37%) had detectable pulses. Of those with ABPI>0.9, 108 of 440 (25%) had no detectable pulses. Sensitivity for lack of pulses as a predictor of arterial disease (ABPI<0.9) was 63% with a specificity of 75% and a positive predictive value of just 35%. Using only the lack of pulses as an indicator of arterial disease would lead to 37% of patients with arterial disease being treated inappropriately.

Conclusion: Palpation of pedal pulses by community nurses is a poor predictor of arterial leg disease, and must be used in combination with ABPI. Only when significant arterial disease is excluded should compression be applied.

P056

ALGORHYTHM FOR THE DIAGNOSIS OF SAPHENO-FEMORAL JUNCTION INCOMPETENCE.

M. Cappelli, R. Molino Lova, S. Ermini, A. Turchi and G. Bono.

Phebologic Surgery, 46 Datini Street, Florence, Italy.

Objective: Simple Doppler examination of sapheno-femoral junction (SFJ) may often lead to wrong conclusions due to the presence of pelvic shunts (PS). The aim of this work is to provide a simple method to correctly assess SFJ haemodynamics.

Design: Retrospective review of consecutive series.

Patients: 500 patients suffering from varicose veins whose SFJ was tested by Valsalva manoeuvre (VM) and muscle compression-relaxation test (CR test) in upright position.

Measurements: Presence of detectable Doppler speeds at Echo-Doppler examination.

Results: 1) VM and CR test negative: neither SFJ incompetence, nor PS. 2) VM and CR test positive: locate the Doppler sample on every SFJ collateral; if positive: PS with subterminal valve incompetence; if negative: SFJ incompetence with terminal valve incompetence.3) VM positive and CR test negative: proceed as point 2; either little PS or slight SFJ incompetence. 4) VM negative and CR test positive: Deflux, i.e. downflow without any escape point. (Calculation in course).

Conclusion: The study provides a simple algorhythm to assess SFJ haemodynamic. Such assessement can be carried out only by Echo-Doppler.

P057

DOES THE MEASUREMENT OF BRACHIAL VENOUS PRESSURE AT HIGH ALTITUDE ENABLES US TO DETECT THOSE SUBJECTS WHO RISK A.M.S.?

Todini A.R., Pillon S., Bartolo M.

Department of Angiology, s.Camillo Hospital, Rome, Italy

Objective: To evaluate whether there exists a simple method to detect patients risking pulmonary oedema at high altitude (due to Acute Mountain Sickness) and better understand its physiopathology.

Design: Monitoring of healthy subjects using a Doppler c.w. at an altitude of 5,050 mt. to measure (according to Bartolo's method) venous pressure and velocity on the brachial vein.

Patients: 26 subjects (21 M. - 5 F.), age 27-57.

Measurements: Measuring of base and under stress venous pressure and flow of the upper limbs. Comparing these measurements to those at sea-level.

Results: An increase in arm venous pressure (0-5 mmHg at sea-level) up to 20 mmHg and in flow velocity was observed in all the 26 members of the expedition. Venous pressure, which furtherly increased when these subjects moved at higher altitude (5,500 mt.), reached 40-45 mmHg in 6 subjects and was accompanied by signs of pulmonary embolism. Venous pressure increased up to 30 mmHg in 3 subjects with headache and tachycardia.

Conclusions: An increase in venous velocity and brachial venous hypertension at high altitude is a complementary reaction in all the subjects. When this increase is higher than 20 mmHg, the risk of pulmonary embolism increases. Thus, for subjects risking pulmonary embolism, measuring venous pressure could be essential.

P058

THE TEST OF BRACHIAL VENOUS PRESSURE DURING PULMONARY EMBOLISM

Todini A.R., Antignani P.L., Bartolo M.

Department of Angiology, s.Camillo Hospital, Rome, Italy

Objective: Clinical evaluation of pulmonary embolism can often be difficult: alongside the classic symptoms, the reability of non-invasive criteria needs to be researched, as they could back up clinical suppositions with quantitative data.

Design: To this end we used non-invasive testing of venous pressure with the Doppler c.w. method according to Bartolo and the measurement of venous velocity in the upper limbs.

Patients: 36 subjects with phlebothrombosis of the lower limbs with the suspicion of pulmonary embolism. 10 patients were asymptomatic.

Measurements: Using Doppler c.w. we measured venous pressure and the velocity of flow in the brachial vein in clinostatism comparing the data with the perfusional and ventilatory pulmonary scintigraphy.

Results: Venous pressure during a pulmonary embolism, as ascertained scintigraphically, varied from 15 to 40 mmHg (the physiological values in these zones vary between 5 and 10 mmHg).
Venous velocity was found to have increased, with evidence of fairly characteristic "torrential" flow.
Pulmonary hypertension has a variable duration in the postembolic period in relation to the degree of residual obstruction.

Conclusions: The method of measuring venous pressure in the case of pulmonary embolism was found to be of great importance both in hospitals and for home visits, by virtue of the inexpensiveness of the equipment and the reliability of the method.

OP/2.4
REPRODUCIBILITY OF TWO VOLUMETRIC INSTRUMENTS

van der Kley A M I, Veraart J C J M, Neumann H A M

Department of Dermatology, Academisch Ziekenhuis, Maastricht, The Netherlands

Objective: It was the aim of this study to compare the reproducibility and accuracy of two optoelectronic volume measuring instruments. Both the instruments can be used in the diagnosis and follow-up of patients with oedema of the leg and arm.

Patients: 15 volunteering individuals (12 females, 3 males, age range 22 -35 years; mean 26.5 years) entered the study. None of the persons had leg oedema or other signs of chronic venous insufficiency.

Measurements: After good fixation of a leg the total volume over a distance of 40 centimeters from the lateral malleolus was defined. The total volume of this distance was first measured with the "Volometer" (Bösl, Aachen, Germany) and then with the "Perosystem Messgerate GmbH, Wupperthal, Germany). Measurements were performed 3 times as instructed by the factory.

Results: The mean coefficient of variation of the results when measured with the Perometer was 01.5% (range 0.003 - 0.43%); and when measured with the Volometer 1.3% (range 0.072-6.5%).

Conclusions: The results show a clear difference in accuracy between the two instruments. The reproducibility of the Perometer seems to be greater than the Volometer. This is probably due to the easier frame drive of the Perometer.

P063
H-R TELETHERMOGRAPHY IN ASSESMENT OF HAEMODIALISYS VASCULAR ACCESS

Pillon S., Antignani P.L., Todini A.R., Carnabuci A°., Bartolo M.

Department of Angiology, s.Camillo Hospital, Rome, Italy
° Department of Nephrology, s.Camillo Hospital, Rome, Italy

Objective: Satisfactory function of vascular access is essential for adequate haemodialisys in patients with chronic renal failure. We propose a new non-invasive method to assess function of some vascular accesses. Thermal imaging is hightly developed with contemporary infrared imaging systems having thermal and spatial resolution far in excess of the earlier systems of 1960s. Real time imaging together with efficient processing has improved the ease of use and quality of information.
Design: We used a new High Resolution Teletermograph (ADIR,Eltcom, Italy) with a digital subtraction tecnique, prototype developed in Italian National Antarctic Research Program. This tecnique uses the difference in temperature between the superficialized arterial blood and surrounding tissue to obtain a thermal topographic map of blood flow.
Patients: 95 subject (74 m. and 21 w.).
Measurement:
1) Preoperative evaluations:
  a) Anatomical: we studied, the anathomy of peripheral arterial vessels to chose the best vessel to use, both to use a good vessel for satisfactory funcion fistula and to prevent ischemic complications of the hand, in patients at the first fistula or in patients with an history of multiple access.
  b) Functional: we studied the average and standard deviation of temperatures of both hands to compare with post operative results to assess the modifications of peripheral perfusion.
2) Postoperative evaluations:
  a) Anatomical: we observed the early morphological modifications in blood flow after the fistula, to evaluate the result of surgery.
  b) Functional: we observed the modifications due to fistula in peripheral perfusion of the hand comparing pre-operative temperature evaluations.
3) Follow-up:
  a) Time monitoring: we observed after different periods the evolution, to try to prevent or evaluate the modification with use, of vascular access.
  b) Study and evaluation of complications: we considered stenosis, escape pathways and obstacoled outflow.
Results: The results, after 12 months of study, were good.
Conclusions: The method is absolutely non invasive, easy to use, low cost. The general limits are the poor resolution on deep vessels, and needs experienced phisicians in the vascular aspect of hemodialysis.

OP/16.7
AUDIT OF INTRODUCTION OF HAND HELD DOPPLER INTO THE ASSESSMENT OF VARICOSE VEINS.

Pleass H, Holdsworth J.
Wansbeck Hospital, Woodhorn Lane, Ashington,
Northumberland, NE63 9JJ.

Objective: To assess how the routine use of hand held Doppler (HHD) in the initial clinical examination of a patient affects requests for secondary investigations, patients being placed directly on the waiting list (WL) without investigation and the type of operative procedure.

Design: Retrospective review comparing the year before HHD (1992) with the subsequent 2 years.

Patients: All new venous outpatients and operations for the years 1992 to 1994 inclusive.

Results:

		1992	1993	1994
New outpatients	n =	194	193	222
Discharged, no treatment		23%	19%	23%
Secondary investigation		16%	40%	41%
Placed directly on WL		47%	29%	29%
Others		12%	12%	7%
Operations	n =	123	134	121
Long saphenous		84%	83%	76%
Short saphenous		6%	10%	21%
Others		10%	7%	3%

Conclusion: HHD has led to a marked increase in secondary investigations and fewer patients being listed directly for surgery. Operations for short saphenous disease have increased from 6% to 21%.

P066
"CLAUDICANCE WALKING AND HAEMODYNAMIC CHANGES OF PLANTAR VENOUS FLOW"

Brizzio E O, Cumaldi A.

Vascular Peripheral Laboratory - 965 San Martin St, 1st Floor, Capital Federal, Bueno Aires, Argentina
Argentine School of Phlebology of AMA - 1171 Santa Fe, Av. Buenos Aires, Argentina

Objective: To compare haemodynamic changes in originates in the plantar pump during spontaneous walk and in a simulated claudicance walking in the same person.

Design: 1) to evaluate plantar pump with bidirectional continuous doppler connected with a plate explorer of continuous monitoring of tibial posterior vein and internal supramaleolar saphenous vein. 2) to examine a subject with 'ordinary' plantar gait and non-claudicance walk. 3) to compare different walking forms: non-claudicant and simulated claudicant walking, barefooted and wearing shoes with different heels.

Patients: The study was performed on 29 patients

Measurements: Calibration of the input doppler and registrator standardized to conduct the evaluation and comparison of the different curves obtained.

The plate explorator of the continuous monitoring allowed a better reception without any kind of shifting, rhythm, frequency and regularity of the obtained curves were evaluated.

Conclusions: 1) The claudicant walk are responsible for the haemodynamic alterations of the plantar pump. 2) Claudicants walks do modify the distributive potentials between the internal saphenous veins and the posterior tibial vein. 3) Footwear is variable which may effect the effectiveness of the plantar pump.

P067

ABSOLUTE VOLUME CHANGES. CORRELATION WITH MAXIMUM ISOMETRIC FORCE AND LEAN MASSES.
Ortega F, García Manso JM, Sarmiento L, Mompeó B, Centol A
Departamentos de Morfología y Educación Física. Universidad Las Palmas G.C. Spain

Objective: To evaluate the relationships between the venous Volume (VV), Ejection Fraction (EF) and Residual Volume (RV) and both the Maximum Isometric force (Fmax) and the Lean masses (L).
Subjects: 19 young subjects (10 males, nine females, 20+- 1.5 years) physically active without chronic venous insufficiency.
Measurement: 1) Volume changes in both legs were measured with an air plethismograph APG-1000; 2) The Lean mass of both lower extremities was measured with a Hologic QDR-1500X^R tomodensitometer; 3) Maximum isometric force developed during plantar flexing was obtained with a force transducer (Sensotec 3132^R). 4) T-Test for paired samples and Pearson Correlation matrixes were used in the statistical analysis of data.
Results: All groups show a statistically significative difference (p<0.05) between the mean values of Fmax, EF, RV and VV of both extremities. Males show a high correlation between Fmax and EF and also between L and both RV and VV. If we take in consideration only the dominant side (larger Fmax), correlations increased: 0.867, -0.833 and 0.806 respectively. Females show a positive correlation between both the Fmax and both EF and VV (0.8, 0.867) and between L and VV (0.947). However if we look for the dominant side, this correlation is only between L and VV (0.944).
Conclusions: The data of this study suggest the strong relationship between the lean mass and the maximum isometric force, mainly in the male group. However the different behaviour of the female is unknown but this could be, according to recent works, due to more lineal response of them during the determination of the maximum isometric force.

P076

PLANTAR PUMP:ANATOMY,PHYSIOLOGY AND NON-INVASIVE FUNCTIONAL EXPLORATION

BRIZZIO,E.O,CUMALDI ALBERTA
VASCULAR PERIPHERAL LABORATORY-965 San Martin St.1st floor.Capital Federal Buenos Aires .Argentina
ARGENTINE SCHOOL OF PHLEBOLOGY of AMA.-1171 Santa Fe Av.Buenos Aires. Argentina

OBJETIVE:1)Anatomic review of plantar pump.
2)Physiological review of plantar pump.
3)Evaluation of the effectivity of plantar pump
4)Evolution of the degree of alteration of plantar pump in venous insufficiency and determination of the result of treatments.
DESING:Anatomical structure of plantar pump.Physiology of the initial start of the plantar pump;effects of muscular contrations and of compression of anatomic structures on internal and external plantar veins flows.Functional exploration is made by means of bidirectional and continuoes doppler with plate continuoces monitoring explorator.
PATTENS:29 patients were evaluated
MEASUREMENTS:Ace the patients were evaluated with explorator placed on the posterior tibial vein and internal supramallcolar saphenous vein,together with test of total plantar load and extenso-flexion of the toes,with the aim of differenciating the flow of the external and internal plantar veins,walking both barefooted and wearing shoes.
CONCLUSIONS:The functional exploration of plantar pump provides the information that follow:1)Verification of its present condition;2)Verification of the haemodynamic effect of accurate or inaccurate footwear.3)Verification of the haemodynamic effect of the orthesic corrections.4)Follow-up of the results of plantar physiatry indicated.

The Investigation of Venous Disorders

Angiology

Phlebology '95, D. Negus et al. (eds.). Phlebology (1995) Suppl. 1: 302-304

PI/7.4

Evaluation of Spect Venography in the Management of Varicose Vein Disease

James Caplan, MD[1] and Edward A. Hanna, MD, FACS[2]

[1] Department of Radiology and [2] Department of Surgery, MacGregor Medical Association, Houston, Texas, USA

INTRODUCTION

Less invasive management of disease is preferred if it results in equally good outcome as conventional procedures. We investigated the utility of Three Dimensional SPECT Radio-nuclear Venography as an aid in assessing lower extremity varicosities and designing a less invasive plan of treatment.

DESIGN

Twenty patients with severe varicose vein disease of the lower extremities were initially evaluated clinically, underwent SPECT Venography and further studied by U.S. Duplex Scanning prior to surgical intervention.

METHODS

SPECT imaging of RBC's labelled with Tc-99 m pertechnetate in vitro with ultratag RBC's was performed on Trionix Biad gamma camera by the transpect technique from the waist to the ankles [1, 2]. 3-D reprojection images were generated on the screen as well as on hard copy. The images depict the anatomical outline of the venous pool and major superficial veins, their branches and communications are illustrated; Figure I.

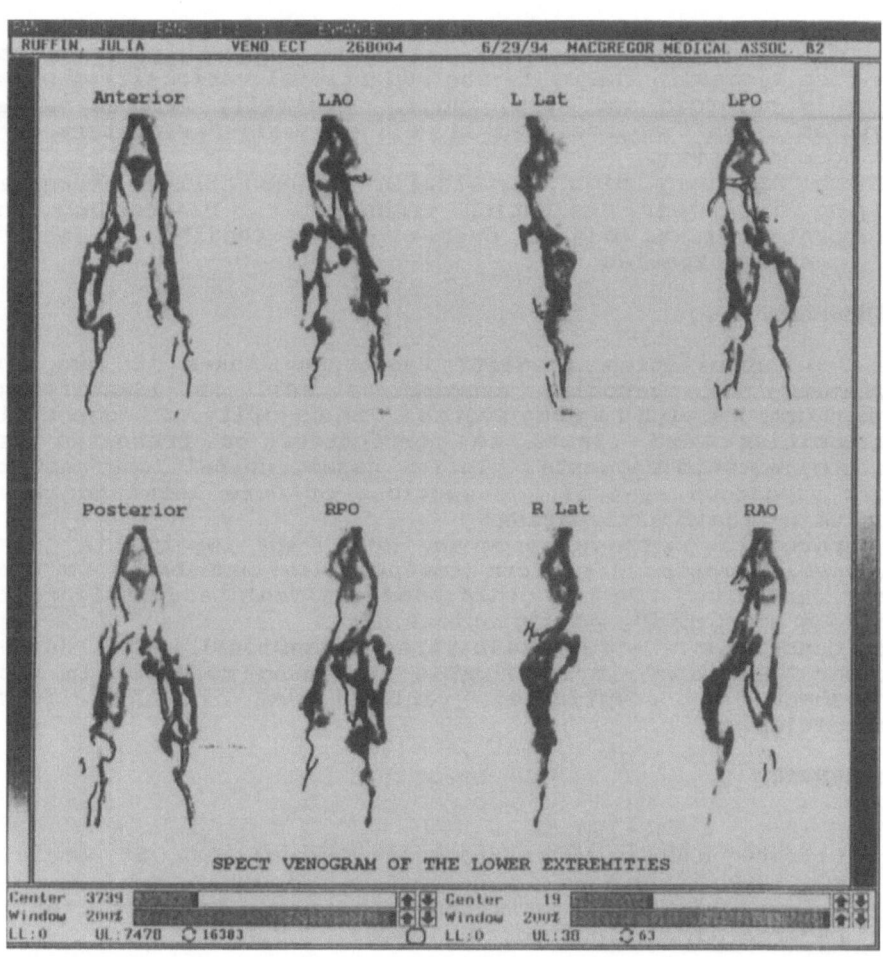

Figure 1.   Advanced Varicosities of Right Lower Extremity.

The image is compared to the physical findings and is used as a guide for Duplex Scanning. At surgery, the surgical findings were compared to the SPECT Venography views.

COMPLICATIONS

None, resulting from SPECT Venography.

RESULTS

The procedure confirmed the patency of the deep system in 18 out of 20 patients. Where it failed to adequately visualize the deep system in the calf, the superficial varicosities were markedly enlarged and serpiginous. Minimally dilated varicosities weren't appreciated as well as small perforators or A-V communications.

There was very high correlation between SPECT Venogram imaging and other evaluation techniques. Discrepancy in interpretation was due to over or under reading of images during early experience.

ASSESSMENT

The 3-D capabilities of SPECT Venography makes it easy to delineate the branching network as well as identifying communications with he deep system. The majority of sequential varicosities were traced to perforators or presented as isolated ectatic segments. In few cases, normal long and or short saphenous systems, or sections of were seen and thus precluding their extirpation.

Compared to contrast venogram, RBC SPECT imaging is less invasive, simpler to perform and provides comparison to the other extremity. On the other hand, it isn't a dynamic test and does not provide physiological data.

In conclusion, we feel that Three Dimensional SPECT Radionuclear Venography is a valuable screening modality in the management of complicated varicosities of the lower extremities.

REFERENCES

1. Ultratag RBC Kit by Mallinokrodt Medical Inc., St. Louis, MO, 63234, U.S.A.
2. Trionix Biad by Trionix Research Laboratory, Inc., 8037 Bavaria Rd., Twinsberg, OH, 44087, U.S.A.

Phlebology '95, D. Negus et al. (eds.). Phlebology (1995) Suppl. 1: 305-307

PI/7.3

Two Directional Phlebography and Ulcer Circumferential Phlebography

Hisashi Gotoh[1], and Hiroto Matsumura[2]

[1] Kanagawa Prefectual College of Nursing and Medical Technology, [2] Shakaihoken Sagamino Hospital, Yokohama, Japan

INTRODUCTION

Usually long saphenous vein insufficiency is the main cause of primary varicose vein. But in the cases of primary varicose vein with related symptoms such as ulceration, superficial thrombophlebitis, dermatitis and eczema, short saphenous vein insufficiency becomes one of the main causes of the varicosities.

The first purpose of this paper is to show you our methods of phlebography that demonstrate not only long saphenous vein insufficiency but also short saphenous vein insufficiency and that reveal the circumferential veins beneath the venous ulcer of the leg.

The second purpose of this paper is to show you our surgical management of patients with related symptoms. Operation was performed according to the findings by two directional phlebography and ulcer circumferential phlebography.

Materials and Methods

In these 18 years, Stripping operation was performed on 570 patients. In these cases, 126 patients (22%) had related symptoms [Table 1].

Table 1. Patients and related Syndroms

related syndroms	Male	Female	Total
Ulceration	42	24	66
Superficial Thrombophlebitis	12	28	40
Dermatitis, Eczema	17	3	20
Total	71	55	126

While in usual primary varicose vein, only 11% of the patients had short saphenous vein insufficiency, 37% of the patients with related symptoms had

306

short saphenous vein insufficiency(ulceration:41%,superficial thrombophlebitis :30%,dermatitis and eczema:35%) [1].
Because of high percentage of short saphenous vein insufficiency, special attention should be paid to the cases with related symptoms,especially to the cases of ulceration.

In 1974,at the 9th International Congress of Angiology held in Florence,Italy we already presented a paper of surgical management of venous ulcer of the leg using our two directional phlebography [2]. The outline of our methods are as follows:the patient is laid flat and rubber band tourniquet is tightened above the ankle to obtain sufficient obstruction of the superficial venous system. Forty ml of contrast medium is injected into the dorsal vein in 2 minutes,and the patient is forced to walk to a long casette which is set up in erect position. One minutes after the end of the injection, postero-anterior exposure is made and then lateral exposure is made with another casette. The postero- anterior exposure is useful in evaluating abnormalities of the long saphenous vein trunk ;meanwhile , lateral exposure is useful to evaluate abnormalities of the short saphenous vein trunk. This method consists of a combination of active movement of walking and standing on erect position. Our second method of phlebography is also very simple as follows:the patient is laid in supine position and rubber band tourniquet is tightened just below the knee. Before X-ray exposure, it is important to mark the ulcer with some marker like fuse. Thirty ml of contrast medium is injected into the dorsal vein in 30 minutes. After injection, postero-anterior and lateral exposure is made by using two X-ray cassettes in supine position. Comparing all these X-ray films, it becomes possible to evaluate the location of many abnormal veins beneath the ulcer with reasonable accuracy. We call this method ulcer circumferential phlebography.

Table 2. Surgical Procedure (126 Patients,146 Extremities)

Surgical Procedure	Ulceration	Superficial Thromboph-lebitis	Dermatitis Eczema	Total
St.of Long Saph. main Trunk	41 (59%)	40 (78%)	19 (73%)	100 (69%)
St.of Long Saph, short Saph.main Trunks	27 (39%)	8 (16%)	6 (23%)	41 (28%)
St. of Long Saph. main Trunk and Ligation of Short Saph. vein	0	3 (6%)	1 (4%)	4 (2%)
Ligation and Division of circumferential varicose veins	1 (2%)	0	0	1 (2%)
Total	69 (100%)	51 (100%)	26 (100%)	146 (100%)

The reason we tried this method is that the ulcerous area is always infected, indurated and edematous and it is difficult to observe varicose veins.This method of phlebography makes it easy to observe occult,hidden,pathological veins beneath the area of the ulcer. They became a good guide for us to perform a precise operation [3] .

Surgical procedures strictly indicated by our two directional phlebography are shown in Table 2.

Summary

Primary varicose vein is one of the most popular vascular diseases in Japan as well as many other countries in the world.
But management of the related symptoms such as ulceration,superficial thrombophlebitis,dermatitis and eczema is not so easy .We paid attention to the facts that in these cases ,varicose vein with related symptoms ,short saphenous vein insufficiency becomes one of the main causes of symptoms as well as long saphenous vein insufficiency .Our methods of phlebography are simple and useful to observe abnormality of those veins.We hope our methods would be widely tried and also hope many patients who suffered from those symptoms will be receive relief.

References

1. Gotoh H,Matsumura H,Soma T,Saitoh T,Matsumoto A,Wada T,Phlebography and surgical management of short saphenous varices.Geka 1977;39:479-82(in Japanese).
2. Gotoh H,Matsumura H,Yamamoto T,Yoshida S,Matsumoto A,Wada T,Surgical management of severe venous ulcer of the leg indicated by two-directional phlebography.Atti di Angiologica 1975;1:365-73.
3. Gotoh H,Matsumoto A,Matsumura H,Mou M,Surgical management of venous ulcer of the leg:follow-up studies of 100 patients. In:Raymond-Martimbeau P,editor. Phlebologie 92.John Libbey Eurotext,Paris 1992:1359-61.
4. Barker EC.Clinical and roentgenologic evaluatoin of phlebography based on an experience in 1,027 cases.Am J Roent Rad Theraphy 1947;58:603-16.
5. Scott HW,Roach JF.Phlebography of the leg in the erect position.Ann Surg 1951;134:104-9.
6. May R.Surgery of the Veins of the Leg and Pelvis.Stuttgart:George Thiema Publishers,1979.

Phlebology '95, D. Negus et al. (eds.). Phlebology (1995) Suppl. 1: 308-311

P152

Indirect Lymphography in Xeroradiography. Technique Evaluating Changes of the Lymphatics in Patients with Chronic Venous Insufficiency and Other Swellings of the Lower Extremities

S. Knorz and K.U. Tiedjen

Department of Radiology and Nuclearmedicine, Elisabeth-Hospital Bochum, Germany

INTRODUCTION

Chronic venous insufficiency causes both pathological-anatomical changes of the the limbs as well as functional. Isotope lymphography is suitable to investigate function of the lymphatic drainage [1]. Direct lymphography with oily contrast agent entails complications and aggravation of the lymphatic desease and should not be used in the examination of chronic edema [2].

Indirect lymphography with water-soluble contrast agent reveals morphological changes of the initial lymphatics and collectors. In combination with xeroradiography whitch is very suitable to analyse tissue demage of the cutis and subcutis chronicity of the desease can be evaluated [3].

PATIENTS AN METHODS

108 patients were investigated according to the clinical findings (8 normal legs, 84 legs with lymphedema, 18 legs with lipedema, 46 legs with lipo-lymphedema, 59 legs with chronic venous insufficiency). Indirect lymphography was performed by automatic injection of a water- soluble non ionic contrast agent (Iopamidol) in the interstitial space of the leg. Documentation with xeroradiography followed in definite intervalls.

Table 1. Morphological findings in indirect lymphography

	injection deposit compact	not compact	flamelike
normal n(=8)	87,5%	12,5%	-
lymphedema (n=84)	24,0%	76,0%	-
lipo-lymphedema (n=46)	17,5%	13,0%	69,5%
lipedema (n=18)	17,0%	5,0%	78,0%
cvi (n=59)	17,0%	76,0%	17,0%

(n = number of examined extremities)

Table 2. Morphological findings in indirect lymphography

	collectors middle diameter (in mm)	max.diameter (in mm)	detection of lymphatic capillaries
normal (n=8)	0,45	0,8	-
lymphedema (n=84)	0,6	1,05	89%
lipo-lymphedema (n=46)	0,5	0,85	85%
lipedema (n=18)	0,3	0,5	-
cvi (n=59)	0,6	1,1	49%

(n = number of examined extremities)

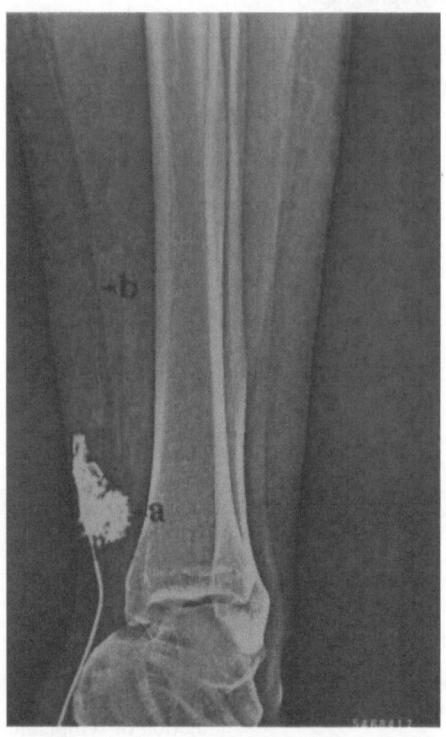

Fig.1 : lipedema
indirect lymphography with
Iopamidol (Solutrast 200®)
xeroradiography 3 minutes
after injection
a : flamelike compact
injection deposit
b : collector with normal width

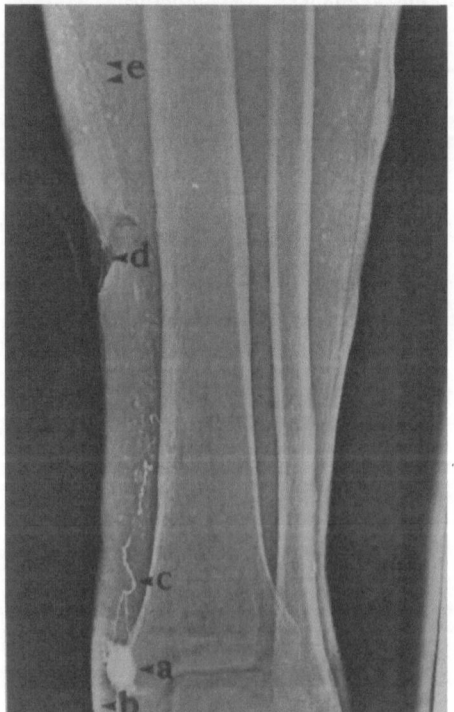

Fig. 2 : chronic venous
insufficiency
indirect lymphography with
Iopamidol (Solutrast 200®)
xeroradiography 6 minutes
after injection
a : the injection deposit is not
really compact because of
initial lymphatic vessels in the
surrounding
b : contrast agent in the cutis
indicating cutaneus
collateralisation
c : enlarged collectors
d : ulcus
e : phlebosclerosis

SUMMARY

Indirect lymphography evaluates the most dilated collectors in patients with chronic venous insufficiency (Table 2). In most of the Patients with CVI a not compact injection deposit appears (Table 1). Further on, the collectors often show a coiling and collateralisation (35%). In 49% of the patients pathologic initial lymhatic vessels are detected. 7% showed cutaneous backflow. Xeroradiography reveals chronicity of the desease in detecting increasing width of the cutis and subcutis, fibrosis and phlebosclerosis.

REFFERENCES

1. Weissleder M, Weissleder R. Lymphedema: Evaluation of Qualitative and Quantitative Lymphoscintigraphy in 238 Patients. Radiology 1988; 167:729-735.
2. Tiedjen KU. Nachweis von Lymphgefäßveränderungen bei Venenerkrankungen der unteren Extremitäten durch bildgebende Verfahren. Plebol.u.Proktol. 1989; 18:270-276.
3. Bruna J. Xeroradiography in Lymphedema. In: Weissleder, Bartos, Clodius, Malek (eds). Progress in Lymphology 1981;223-226.

Phlebology '95, D. Negus et al. (eds.). Phlebology (1995) Suppl. 1: 312-314

P064

Dynamic Radionuclide Venography in the Diagnosis of Chronic Venous Insufficiency

G. Menyhei[1] and M. Szabó[2]

[1] Department of Surgery I and [2] Central Radioisotope Laboratory, University Medical School of Pécs, Hungary

INTRODUCTION

Many diagnostic tests are currently used for assessment of patient with chronic venous insufficiency. Although dynamic imaging, in which changes in distribution of the radiopharmaceutical are monitored throughout the investigation, has been widely available in general hospitals for almost twenty years, radionuclide venography has been mainly applied for static imaging of the venous system [1,2,3]. Dynamic radionuclide studies for investigating venous diseases have been used only in a few centres [4,5].

A new technique of dynamic radionuclide venography has been developed in our institute, which provides functional as well as morphological information about the venous system of the lower extremity. The purpose of this study is to demonstrate this diagnostic method and to assess its value.

METHODS

The basic requirements for dynamic radionuclide venography are: a gamma camera supported by electronics, computing facilities and displays. Technetium-99m labelled diethyltriamine pentaacetic acid (DTPA) is used for the study, which is rapidly excreted by the kidney.

The detector under which the patient lies is placed over the thigh and pelvis. A cuff placed below the knee is inflated to a constant pressure of 80 mmHg. Maintaining the pressure for 3 minutes followed by releasing the cuff produces a test situation similar to venous occlusion plethysmography. Before releasing the cuff a bolus of 3 mCi of 99mTc labelled DTPA is injected into a dorsal foot vein and imaging is commenced. The course of imaging of femoral and pelvic veins can be seen on the screen simultaneously and a photographic device produces immediate polaroid pictures.

For investigating reflux the patient is asked to perform Valsalva manoeuvre starting as soon as the bolus of radiopharmaceutical enters the groin and to maintain it for 10 seconds.

A computer registers all the information detected. Registration is made in frame mode with 0.3 sec frame duration. The frames are visualised on a colour monitor and the calculations are made by computer software to analyse the dynamics of venous blood flow in the lower limb.

Continuous registration of venous imaging allows establishment of a time-activity curve over the desired regions of interest (ROI). The most commonly used ROIs for investigating the lower limb are located above the knee, at the ipsilateral groin, contralateral groin and on the common iliac vein. On the time-activity curve generated from the ROI the minimal transit time (MTT) of the bolus from the injection site in the foot to a region of interest is calculated. The velocity of the flow between two ROIs can also be calculated.

Application in venous insufficiency

All the information detailed below (morphological and functional) may be useful to make decision about treatment for patients with chronic venous insufficiency and helpful in postoperative monitoring after venous reconstruction.

Morphology

Polaroid pictures taken during imaging provide good quality information about the anatomy of femoral and pelvic veins. Obstruction of major veins can be clearly identified and compensating collaterals (e.g.suprapubic collaterals in unilateral iliac vein occlusion) can also be analysed. Recanalisation after deep venous thrombosis usually causes inhomogeneous segments of low radioactivity.

Functional assessment

Dynamic radionuclide venography provides the following functional information about the venous system of the lower extremity:
a) minimal transit time (MTT) at the groin, which indicates the capacity of the inflow and the capacity of the collaterals in chronic venous occlusion
b) crossover minimal transit time (CMTT) between the groins, which provides data about the functional capacity of major suprapubic collaterals or a femoral crossover (Palma) bypass in unilateral iliac vein occlusion
c) velocity of the venous flow in the test situation, which is standardized, therefore the numerical data obtained from the computer analysis are comparable
d) time-activity curve analysis of a region of interest, which shows the dynamic of the venous blood flow
e) density analysis, which gives information about the blood volume of two major venous pathways running parallel by comparing the amount of radioactivity within the two ROIs drawn over them.

Investigation of reflux

To analyse the recorded data during Valsalva manoeuvre ROIs are drawn over the image of the superficial femoral and the long saphenous veins at different levels and the time-activity curves generated from the ROIs are analysed. Drop in radioactivity on the curve at the beginning of Valsalva manoeuvre indicates valve incompetence at the level of the ROI. Placing several ROIs over the image of the femoral or saphenous veins the extent of reflux can be identified.

CONCLUSION

For investigating a patient with chronic venous insufficiency a morphological test and a functional test should be used[6]. Invasive tests such as phlebography are not necessary in most patients. Colour-Duplex Sonography has recently become a valuable tool, which combines the morphological information of sonography with the functional information of the Doppler ultrasound[7]. Dynamic radionuclide venography provides similar advantages. By this method the functional alterations and the morphological changes of the veins can also be studied. One of its disadvantages is that it can not provide suitable information about the veins below the knee.

This study appears to be suitable in the selection of patients for operations reconstructing veins and for postoperative monitoring. It is noninvasive, relatively quick, the radiation of the investigation is negligible and it can be performed in any hospital where a gamma camera is available.

REFERENCES

1. Dean RH. Radionuclide venography and simultaneous lung scanning: Evaluation of application. In: Bernstein EF, editor. Noninvasive diagnostic techniques in vascular disease. St Louis: CV Mosby, 1978.
2. Browse NL, Burnand KG, Lea Thomas M. Diseases of the veins. London: Edward Arnold, 1988.
3. Early PJ, Sodee BD. Principles and practice of nuclear medicine. St Louis: CV Mosby, 1985.
4. Zicot M, Guillaume M. The assessment of chronic venous obstruction by noninvasive haemodynamic studies and imaging with Kr-81m venography. Int Angiology 1986; 5:21-25.
5. Mark B, Szabo M, Nemessanyi Z. Testing minimal transit time in the venous system of the lower limbs using dynamic radionuclide venography. Int Angiology 1988; 7:231-233.
6. The Alexander House Group.Consensus Paper on Venous Leg Ulcers. Phlebology 1992; 7:48-58.
7. Rabe E, Fratila A, Stranzenbach W. Value of Color-Duplex-Sonography in diagnosis of chronic venous insufficiency. Phlebologie 1992; 21:130-133.

Phlebology '95, D. Negus et al. (eds.). Phlebology (1995) Suppl. 1: 315-317

P065

Ascending Dynamic Phlebography: Role in Recurrent Varicose Vein Diagnosis

J.E. Amorim, L.C.U. Nakano, M. Nunes, A. Lourenço, M. Reicher, R.P. Nunes Filho, M.C.J. Perez and E. Burihan

Division of Vascular Surgery, Department of Surgery - Escola Paulists de Medicina - Universidade Federal de São Paulo, São Paulo, Brazil

INTRODUCTION

Recurrence after initial surgical varicose vein treatment according to some reports, is variable around 4.6% to 25.8% (1,2,3,4). The sites of reflux in the superficial system, which are the cause of recurrence, also present great variation.

The objective of this work is to show the importance of ascending dynamic phlebography in the diagnosis and localisation of the reflux sites as a cause of recurrent varicose veins.

MATERIAL AND METHOD

From 1990 to 1994, 1063 dynamic ascending phlebograms were done in our department.

Recurrent varicose veins was the indication for phlebography in 260 extremities of 181 patients. The mean age of patients was 42.3 years (F: 149, M: 32) . Bilateral recurrent varices occurred in 79 patients (158 extremities) and was unilateral in 102.

Ascending dynamic phlebography was done by a Rabinov-Paulin modified technique, with the patient in foot down inclination. The reflux sites were evaluated by the Valsalva manoeuvre. The contrast medium used as the Uromiron^R.

All phlebograms were evaluated for the presence of deep venous system lesions and reflux in the superficial system, e.g.sapheno-femoral junction (SFJ), sapheno-popliteal junction (SPJ) and presence, number and site of incompetent perforating communicating veins (IPCV).

RESULTS

No lesions in the deep venous system was found in this group of patients.

In the bilateral recurrent varicose veins group, the right side (49.3%) and left side (50.3%) presented with similar frequency. The frequent cause of recurrence was the IPCV in 77.8%. Recurrence at the SFJ was 12.2% and 10.6% at the SPJ. (Table 1).

Table 1 - Bilateral recurrent varicose veins of the lower extremities. Distribution of reflux sites.

	SPJ	SPJ	IPCV lateral	IPCV medial	IPCV thigh	Total No.	%
Right	24	18	66	62	3	173	49.7
Left	19	19	64	65	8	175	50.3
Total	43	37	130	127	11	348	100

In unilateral recurrent varicose veins there was a predominance in the left lower limb (60.1%) however this was not statistically significant. The most frequent of reflux sites was due to the IPCV (78.1%), followed by SFJ (14.3%) and SPJ (7.6%). (Table 2)

Table 2 - Unilateral recurrent varicose veins in inferior extremities. Distribution of reflux sites.

	SFJ	SPJ	IPCV lateral	IPCV medial	IPCV thigh	Total No.	%
Right	14	07	35	36	03	95	9.9
Left	20	11	53	56	03	143	60.1
Total	34	18	88	92	06	238	100

In the phlebograms the frequency of SFJ reflux was 13.1% and SPJ reflux 9.4%. The frequency of IPCV reflux, was 37.3% on the medial aspect of calf, 37.2% on the lateral side and 2.9% in the thigh.

When evaluated unilateral and bilateral groups together we identified the perforating veins (77.8%) as the most important cause of recurrent varicose veins.

DISCUSSION

Recurrence after initial surgical treatment of primary varicose veins is a frequent problem facing the angiologist and vascular surgeon. The most recent reports present low rates of recurrent varicose veins perhaps due to greater knowledge of physiology, anatomic anomalies of the superficial venous system (5.6), associated with better diagnostic methods. In previous reports from our division (1), the results of determination of reflux sites by physical examination or phlebography were similar. The phlebography was important to detect IPCV on the lateral aspect of the calf.

The frequency of recurrent varicose veins according to the reflux sites reported in literature, was variable from 10.6% to 96% (2,7,8,9) at the SFJ and 4% to 96% at the SPJ (2,3,8,10,11). Concerning IPCV's the incidence of recurrence is high, with less variation of 81.1% to 92% (12, 13, 14).

Our results are according with the literature, confirming that the majority of cases with recurrence was due to IPCV (77.8%). Ascending dynamic phlebography is an accurate method in the diagnosis of the reflux sites.

REFERENCES

1. Silvestre, J M S. Contribution to study of recurrent varices diagnoses of inferior member Thesis presented to Escola Paulista de Medicina to obtain the Master's degree in the Pós-graduation Course of Cardiovascular Surgery 1986, São Paulo, Brazil.
2. Lofgren, E P & Lofgren K A. Recurrent of varicose veins after the stripping operation. Arch Surg 1971; 102: 111-5.
3. Perrin M, Becker F, Leclerq D, Laurent P. La chirurgie itérative dans les varices essentielles des membres inférieurs. Phlebology 1977; 30: 389.
4. Quiroli A, Stato attuale del problems delle recidive varicose postoperatorie degli arti inferiori. Revisione Bibliografica ed esperienza personale. Min Chr 1984; 39:35
5. Darke S G. The morphology of recurrent varicose veins. Eur J Vasc Surg 1992; 6(5):512-7
6. Urigo F, Pischedda A, Maiore M, Canalis G C. Role of phlebography in the study of recurrent leg varices. Radiol Med (Torino) 1993; 85(6):764-72.
7. Sheppard M. A procedure for the prevention of recurrent saphenofemoral incompetence. Aust N Z J 1944; 120:772-4
8. Dodd H. Persistent or recurrent varicose veins. Brit J Clin Prac 1963; 17: 501-3.
9. Rivlin S. Recurrent varicose veins. Med J Aust 1966; 1: 1907-9.
10. Lofgren K A, Myers T T, Webb Jr W D. Recurrent varicose veins. Surg Gynecol Obstet 1956; 102:729-30.
11. Frileux C, Gillot C, Le Bauer A, Pillot P, Cosson J Ph. La récidive variqueuse: opinion d'une équipe. Phlebologie 1982; 35:471-4.
12. Lea Thomas M, McAllister V, Rose D H, Tonge K, A simplified technique of phlebography for the localization of incompetence perforating veins of the legs. Clin Radiol 1972; 23:486-8.
13. Mathiesen F R. Clinical manifestations of primary varicose veins. An evaluation of incompetent communicating veins diagnosed by tilting phlebography. Arch Chir Scand 1959; 117:468-70.
14. Townsend J, Jones H, Willians J E. Detection of incompetent perforating veins by venography at operation. Brit Med J 1967; 3:583-6.

Phlebology '95, D. Negus et al. (eds.). Phlebology (1995) Suppl. 1: 318-320

OP/16.3

Dynamic Ascending Venography in the Management of Venous Disease

R.H. Fox[1] and P. Lewis[2]

[1] Department of Clinical Radiology and[2] Department of Vascular Surgery, Torbay Hospital
Lawes Bridge, Torquay, Devon, England

Most patients requiring superficial venous surgery have primary varicose veins. The site of valvular incompetence can be determined by bedside assessment, including the use of a Doppler-shift flowmeter. However when sapheno-popliteal ligation is planned, imaging of this area is necessary pre-operatively since there is great variability in the level of the sapheno-popliteal junction. Furthermore in recurrent varicose vein surgery, the point of incompetence can be more difficult to localise and dynamic imaging may be helpful. Finally, the results of superficial venous surgery in patients with venous ulceration or lipodermatosclerosis are closely related to the patency and competence of the deep venous system.

To provide a comprehensive venous service it is thus essential to be able to assess selected patients in more detail. Any investigation must provide haemodynamic information together with precise anatomical detail of the whole limb. Ideally the result should be recorded to enable discussion between the investigator and surgeon. The availability of a hard copy in the operating theatre is helpful to the surgeon.

Ascending venography is a gold standard investigation in venous disease. We have modified this investigation to study patients with more complex venous problems. This article describes the technique called " dynamic ascending venography " and our initial experience.

THE TECHNIQUE

Dynamic ascending venography (DAV) is performed in the Radiology Department. The patient is positioned supine on a screening table and a vein in the foot cannulated with a 21 guage 'butterfly' needle. A distal vein on the lateral aspect of the dorsum of the foot or a vein at the base of the hallux are the optimal sites for filling the deep plantar veins and ensuring proper mixing of contrast with the blood.

The table is then tilted until the patient is in the vertical position and a venous tournequet inflated at the ankle. With the screening carriage centered over the calf and the leg internally rotated some 20 degrees (to open up the interosseous space) non-ionic contrast medium is injected slowly by hand. The foot is squeezed to eject a stream of contrast into the calf and retrograde flow, via perforating calf veins, into the superficial veins indicates "perforator" incompetence. When compression is released, retrograde flow indicates incompetence of the deep calf veins. The screening carriage is then positioned over the thigh and the procedure repeated with further injection of contrast and squeezing and release of the calf. When contrast appears at the groin, the screening table is lowered to the horizontal position, the ankle tournequet released and the foot, calf and thigh squeezed to eject blood into the iliac veins. A Valsalva manoeuvre demonstrates sapheno-femoral incompetence.Non contrast filled areas are directly viewed to differentiate betwen thrombosis and turbulence.

Spot films are taken at relevant points of the examination and a video recording of the complete examination is taken with U-matic equipment.

THE PATIENTS

A consecutive series of 54 patients (63 limbs) were referred for DAV by one surgeon (PL). They comprised 31 females and 23 males with an average age of 40 years.The indications for DAV were either recurrent varicose veins (n=10) or lipodermatosclerosis / venous ulceration (n=36) or both (n=17). Information on these patients was derived retrospectively from case notes and information entered into a computerised database. No patient was diabetic or suffered with rheumatoid arthritis. Twenty patients had a past history of treatment for deep vein thrombosis. Previous venous surgery included sapheno-femoral ligation +/- strip of the long saphenous vein (n=22 limbs) and sapheno-popliteal ligation (n=5 limbs). No patient had undergone perforator ligation or deep venous surgery.

RESULTS

Only one patient could not be cannulated. No allergic reactions occured and no patient returned to hospital with a deep vein thrombosis. The procedure appeared to be well tolerated by all patients.
Popliteal vein incompetence was demonstrable in 40% of limbs. In most limbs this finding was associated with incompetence of the deep veins of the calf and thigh. When popliteal and deep thigh vein incompetence was associated with competence of the deep calf veins (n=6 limbs), the sapheno-popliteal junction was invariably incompetent. Most limbs with deep vein incompetence did not have post phlebitic changes.
Calf perforator incompetence was extremely common (84% of limbs), but was rarely an isolated finding .
Sapheno-femoral and sapheno-popliteal incompetence were also common findings. There was a clear difference between incompetence of an intact long saphenous vein and incompetence of one of its tributaries in the groin. The sapheno-popliteal junction was frequently 2-3 cms. above the knee joint and occaisionaly joined the side of the popliteal vein.

Sixty percent of patients with lipodermatosclerosis / ulcers had a superficial venous abnormality but competent deep veins.
In patients with recurrent varicose veins, sapheno-femoral incompetence remained a common problem (n=26 limbs) despite previous groin surgery in 6 limbs. Sapheno-popliteal incompetence was also common (n=17) but no patients with recurrent varicose veins had previosly undergone a sapheno-popliteal flush tie.

COMMENTS

This study was intended to assess the applicability of a dynamic form of a technique (ascending venography) that is widely regarded as a gold standard investigation. Dynamic ascending venography (DAV) was well tolerated and provided clear anatomical and haemodynamic resolution of the abnormalities which surgeons regard as being important in the treatment of chronic venous disease. Video recording enables a more timely review of the venous abnormalities in a limb; a complete view of the limb is obtained and the relative contributions of multiple abnormalities can be discerned. A surgeon can review the video recording if necessary to clarify the correct operative approach. Furthermore hard copy or spot films are invaluable to the surgeon preoperatively, when marking the patients leg, and in the operating theatre.
The results demonstrate that a high proportion of patients with the skin changes of chronic venous insufficiency (lipodermatosclerosis) have patent and competent deep venous systems indicating a favourable outcome after surgery. Simultaneous accurate definition of the superficial venous abnormality may thus be highly important in preventing the morbidity associated with recurrent venous ulceration.
The low number of patients with phlebitc changes despite a past history of deep vein thrombosis almost certainly indicates inaccuracy in the clinical diagnosis of this condition.
Isolated calf perforator incompetence was uncommon and is rare in the presence of intact and competent long and short saphenous systems. Conversely calf perforator incompetence was commonly found in the absence of an incompetent or phlebitic deep venous system. The significance of "concommitant" calf perforator incompetence must therefore remain conjectural.
Recurrent varicose veins due to sapheno-femoral incompetence despite previous groin surgery is a well recognised problem. Similarly recurrent varicose veins can be caused by sapheno-popliteal incompetence which is either missed or developed after the previous surgery.
Dynamic ascending venography has yielded no new information about chronic venous insuffiency. However we believe that it is an accurate and user friendly aid to the successful surgical management of this condition.

Biblioography: Gardner AMN, Fox RH. The return of blood to the heart. London: John Libbey, 1993.

OP/2.5

SPIRAL-CT-VENOGRAPHY OF THE LOWER EXTREMITY IN DEEP VENOUS THROMBOSIS: FIRST RESULTS

Baldt M, Zontsich T, A Stümpflen, Kontros M, Fleischmann D, Minar E, Mostbeck G H

University of Vienna, Währinger Gürtcl 18-20, A-1090 Vienna, Austria.

Purpose: To compare the diagnostic capability of spiral CT Venography (CTV) and conventional phlebography (CPH) in patients with deep venous thrombosis (DVT).

Material and Methods: In a prospective study 35 consecutive patients with clinical suspicion of uni- or bilateral DVT were studied by CTV and phlebography. CTV covering both extremities was performed on a spiral CT unit (Philips SR7000) by using 50 rotations (size thickness: 10mm, pitch = 2, reconstruction mode: 5mm) covering a 100cm section from the ankle to the inferior vena cava (IVC). 20ml contrast (Omnipaque 300, 300mgJ/ml) diluted with 80ml saline (=100ml) were injected (flow: 2ml/sec, delay 30 sec) in a dorsum vein of each foot. Images were reviewed using cine-mode and 3D capabilities of a remote console using standard criteria for the diagnosis of DVT. CTV was compared to conventional bilateral phlebography performed within 30 minutes.

Results: There was very good correlation compared to CPH in diagnosing DVT of the calf and thigh: sensitivity was 100%, specifity 94%, positive predictive value 92%, negative predictive value 100%. CTV better demonstrated extension of DVT into the pelvic veins and IVC and allowed better delineation of collateral veins.

Conclusion: These preliminary results suggest that CTV seems to be a valuable diagnostic tool in imaging of DVT. Using spiral CT might gain importance in detecting the cause of embolism without further patient mobilisation.

The Investigation of Venous Disorders

Angioscopy, Intravascular Ultrasound

Phlebology '95, D. Negus et al. (eds.). Phlebology (1995) Suppl. 1: 323-325

OP/2.1

Evaluation of Incompetent Valves by Angioscopy, Duplex Scan and Strain Gauge Plethysmography

T. Ogawa, S. Hoshino, A. Tsuda and H. Satokawa

Department of Cardiovascular Surgery, Fukushima Medical College, Fukushima, Japan

INTRODUCTION

The causes of varicose veins are chiefly valvular incompetence and incompetent perforating veins. In cases of valvular incompetence, we attempt to observe the morphology of valves and repair these valves under observation during the surgical procedure[1,2,3]. But,sometimes it is dificult to repair an incompetent valve becouse of bad morphology which is not predicted before an operation. Since it is important to predict the morphology of incompetent valves before operation,we compared data caluclated by Strain Gauge Plethysmography and Duplex scan before operation with the morphology of incompetent valves and evaluated incompetent valves.

METHODS

92 legs in 65 patients were diagnosed as valvular incompetence in Long saphenous vein (LSV) by angioscopy from September 1989 to September 1994 in our clinic and were studied. These patients consist of 22 males and 43 females,with an age range from 22 to 72 years old (mean age 51.7 y.o.).
Strain Gauge Plethysmography (Meda model SPG16) and Photo-plethysmography (Meda model PPG13) were used for assessment of venous function. Maximum venous outflow (MVO) and % venous capacitance (%VC) were measured by Strain Gauge Plethysmography. Venous refilling time (VRT) and % Expelled volume (%EV) were measured by Photo-plethysmography. Patients were first examined in the supine position,and then examined again after exercise of 5 times ankle dorsi-flexion in the sitting position.
Duplex scan (ALOKA model SSD-630) was used for assesment of diameter and reflux of LSV. The diameter of LSV was measured below the subterminal valve. Venous reflux flow pattern was measured using calf milking with cuff technique, which

was indeflated of 100 mmHg at standing position. The reflux time,and the peak
and average velocity of reflux flow were derived. The reflux mean and total
volume were then calculated as follows. Reflux mean volume = venous area $\times$
average velocity of reflux. Reflux total volume= reflux mean volume $\times$ reflux
time.

Angioscopy (OLYMPUS OES AF type 28C) was used for evaluation of subterminal
valve of LSV. We classify 3 types of morphology of incompetent valves as
observed by angioscopy [1,2,3,4]. Type I (35 valves) belong to incompetent
valves with elongated cusps. Type II (39 valves) belong to incompetent valves
with commisure expansion.Type III (18 valves) belong to incompetent valves had
cusps with other deformities.

The data was expressed as mean $\pm$ standard deviations. Analysis of variance was
used to assess significance of the 3 classifications. Intergroup comparisons
were made using Fisher's PLSD with significance taken at 5% level.

RESULT AND CONCLUSION

Compairison venous function as predicted by Plethysmography with morphological
classification of incompetent valves

Distinguished venous function by morphological classification is shown in Fig 1.
MVO tended to be high in the order of type I,II,III without significance. %VC of
type III classified valves tended to be highest of all types without significance. VRT of type I classified valves tended to be longest of 3 types without
significance. There is no significant diference of %EV classified 3 types.

From the data,it was determined that there is no significant diference in
venous function in the three types. However,we thought that venous function of
classified type I tended to be best of the 3 types because it was influenced by
not only the LSV,but also perforating veins,and other superficial and deep veins

Compairison data by Duplex scan with morphological classification of incompetent
valves

Distinguished data by morphological classification is shown in Fig 2. The
diameter of LSV classified type I was shortest of 3 types (P<0.01).The peak and
average velocity of reflux flow classified type III tended to be highest of 3
types without significance. The reflux mean and total volume classified type III
was largest of 3 types (P<0.01,P<0.05). Judging from data the morphology of
incompetent valves can be predicted with some certainty by examining the
diameter of LSV,the reflux mean,and total volume.

REFERENCE

1. Hoshino S, Satokawa H, Iwaya F, et al. Diagnosis and surgery of venous insufficiency : valvuloplasty using intraoperative angioscpy. Jpn. J. Phlebol. 199

3;4:1-8(in Japanese).

2. Hoshino S, Satokawa H, Ono t, et al. Surgical treatment for primary varicose veins of the legs using intraoprative angioscopy. In : Phlebologie 92. paris : John Libbey Eurotext, 1992:1083-1085.

3. Hoshino S, Satokawa H, Iwaya F, et al. valvuloplastie externe sous controle angioscopique Pre-operatoire. Phlebologie, Bulletin de la societe Francaise de phlebologie. 1993;46:521-530.

4. Satokawa H, Iwaya F, Igari, et al. Morphological studies of the venous valves by angioscopy. J. Jpn. Coll. Angiol. 1992;32:395-401(in Japanese).

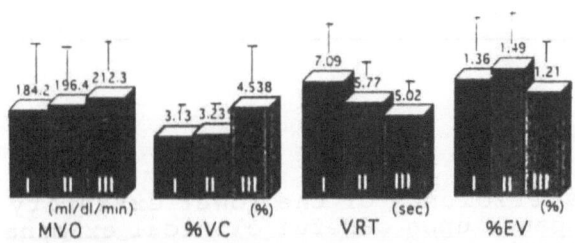

Fig 1 Comparison venous function by Plethysmograpy with morphological classification

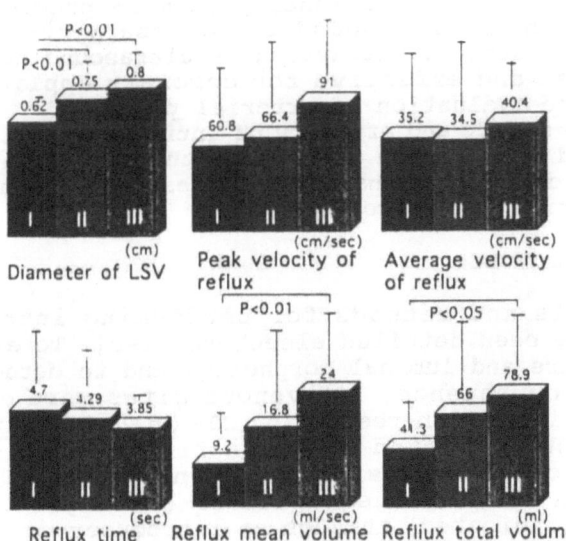

Fig 2 Comparison data by Duplex scan with morphological classication

Phlebology '95, D. Negus et al. (eds.). Phlebology (1995) Suppl. 1: 326-329

OP/2.2

Intravenous Ultrasound in the Management of Varicose Veins

Pauline Raymond-Martimbeau, MD

Dallas Noninvasive Vascular Laboratory, Dallas, Texas, USA

INTRODUCTION

In venous insufficiency of the lower extremity, proper management depends upon careful clinical examination and accurate diagnostic testing. A Doppler ultrasound instrument does not give precise information concerning the location of the insufficiency. When the anatomy is complex or uncertain, Doppler examination must be supplemented by duplex scanning, with or without color. Occasionally, a more precise method such as intravenous ultrasound is necessary for adequate diagnosis and treatment. Intravascular ultrasound has proved safe, accurate, and effective for coronary applications, particularly the evaluation of arterial plaque.

To assess the safety and efficacy of intravenous ultrasound, the author used this method to compare the wall architecture of healthy and diseased veins and to assess the venous wall's immediate reaction to sclerotherapy.

MATERIALS AND METHODS

Basic materials and methods for performing intravenous ultrasound have been detailed elsewhere [1-2]. To elucidate wall architecture and luminal morphology and to detect sites of valvular incompetence, 125 venous sites were examined endoluminally with a high-resolution 20- or 25-MHz ultrasound transducer, incorporated within a flexible 5.1-French catheter. The catheter was inserted into the target vein either through an 18-gauge needle or (if injection of a sclerosant was indicated) through an introducer-dilator set with a side arm. A standard (extravascular) ultrasound scanner was used to guide the intravenous transducer to the target site. Once correctly positioned, the transducer was oriented to yield 360° real-time, two-dimensional, cross-sectional images. The total venous diameter, tunical thickness, and luminal diameter were documented.

Twenty-five incompetent sites were subjected to sclerotherapy, which was performed according to a strict protocol: the most proximal site of reflux was injected first, followed by more distal sites; treatment was achieved by injecting a major sclerosant, iodine sodium iodide, intravenously at these sites, in concentrations of 1% to 3% and volumes of 0.25 to 1.0 mL. Selective dressings and graduated compression (20 to 30 mm Hg or 30 to 40 mm Hg) were applied after treatment, and patients were instructed to walk and engage in other forms of low-impact exercise. Intravenous ultrasound was performed before, during, and 1 to 60 minutes after sclerotherapy.

RESULTS

In all 125 cases, intravenous ultrasound proved efficacious and complication-free. It clearly delineated the wall structure and allowed the thickness of each tunica (intima, media, and adventitia) to be measured. The author was able to determine the luminal diameter, inspect the venous valves, and gauge the spacing of their leaflets. The venous lumen was echogenic in some cases and echofree in others. The intima and adventitia were echogenic, but the media was not. The valves were highly echogenic. Whereas normal blood flow was echofree, reflux flow produced a constant series of mobile echoes; the extent of echo-density correlated with the extent of reflux previously documented by means of bidirectional Doppler ultrasound.

At the 25 sites that underwent sclerotherapy, intravenous ultrasound revealed two different sclerosing reactions, which involved the following stages, in order of sequence:

Reaction 1: In 21 cases (84%), partial, echogenic intimal thickening was observed within 1 to 6 minutes (mean, 2.7 min). This response led to circumferential intimal thickening, which became evident within 3 to 26 minutes (mean, 6.2 min). At 5 to 21 minutes (mean, 11.9 min), the lumen was filled by heterogeneous, echogenic material that clung to the intima circumferentially; no echo-free space was left between this material and the intima. Ten to 29 minutes (mean, 18.4 min) after injection, a decrease in the luminal diameter of the vein was observed, while the intimal and endoluminal echogenic obstructive reaction increased in density. At 24 to 47 minutes (mean, 32.6 min) intimal disorganization was observed in seven cases (21%). An hour after injection, the perivenous tissue remained intact in all 21 cases. Within 1 week, the venous segment was transformed into a fibrous cord that showed no sign of inflammation, as observed clinically. Between the second and twelfth weeks after treatment, the vein disappeared completely.

Reaction 2: In four cases (16%), the vein exhibited weak or no intimal thickening within 20 minutes after injection; instead, partial adventitial thickening was observed at 2 to 5 minutes (mean, 3.6 min), and circumferential adventitial thickening was seen at 4 to 8 minutes (mean, 6.3 min). These stages were followed by the development of an echogenic, partially obstructive endoluminal reaction that did not adhere to the intima circumferentially. At 8 to 28 minutes (mean, 9.5

min), increased luminal diameter and increased echogenicity of perivenous tissue were noted. Twenty minutes to 1 hour after injection, the total venous diameter had increased, and the mobile endoluminal echoes were still present. Within 2 to 12 weeks, the vein underwent recanalization, as assessed by intravenous ultrasound. This sequence of events indicated the need for a second sclerotherapy session, which was performed with intravenous ultrasound monitoring and yielded successful results in all four cases.

DISCUSSION

Intravenous ultrasound can be performed in the presence of flowing blood. It provides highly accurate images of the morphology and dynamics of the vessel lumen and wall. It can also assess blood reflux, owing to the different sonic qualities of forward and reverse flow. Therefore, this technique can be helpful before, during, and after medical or surgical treatment. Because the equipment used in intravenous ultrasound is extremely fine, macroscopic incisions are eliminated, thereby enhancing patient comfort and minimizing the risk of complications. In the present study, intravenous ultrasound proved safe, effective, and entirely complication-free.

Intravenous ultrasound is particularly useful in sclerotherapy, allowing the physician to evaluate the composition of diseased veins, as well as the architectural changes that follow injection of a sclerosing agent. By instantaneously obtaining transverse sections of all the venous layers, one can immediately assess the sclerosing reaction and adjust the site of injection and/or dosage, if necessary. This capability reduces the number of sessions needed to complete the treatment and increases the likelihood of success. In most cases, only one session is necessary for permanent eradication of the diseased vein. Failure to obtain an ideal sclerosing reaction during the first session does not mean that the technique itself has failed; it simply means that further sessions will be needed in order to produce a permanent result. In the present series, sclerotherapy was successful in all 25 cases, yielding permanent eradication of the diseased vein after the first session in 21 cases (86%) and after the second session in 4 cases (16%).

In delineating the stages of a permanent, ideal sclerosing response (Reaction 1) and a failed or inadequate response (Reaction 2), the present study showed that the result of a specific therapeutic session depends upon the immediate intimal effect. The greater the intimal reaction, the better the chance of an ideal response. Therefore, the first few minutes after injection always indicate the final outcome: If the reaction is immediately satisfactory, it will lead to permanent obliteration of the vein. If the reaction is not immediately satisfactory, there is little chance that it will improve without further treatment; although flow through the vein may be temporarily disrupted, eventual recanalization is likely.

CONCLUSION

This study confirms that intravenous ultrasound not only is
unsurpassed in its ability to evaluate intraluminal components
and venous wall architecture but also is safe and
complication-free.

REFERENCES

1.   Raymond-Martimbeau P. Intravenous ultrasound: a new
      technology for venous assessment. Phlebologia Houston 91.
      Dallas: PRM Editions, 1991: 135-8.
2.   Raymond-Martimbeau P. Echographie endo-vasculaire. J Mal
      Vasc 1992; 17:123-6.

The Treatment of Varicose Veins and Telangiectasias

Lohmann Products for Phlebology and Lymphology

Bandages

Compression Bandages

Foam Rubber Bandages

Foam Rubber Pads

Elastic Adhesive Bandages

Thrombo-Prophylactic Stockings

Advanced Wound Management

Native Collagen

Hydroactive Wound Dressings

Antiseptic Wound Management

Ribbon Gauze

Paraffin Dressings

LOHMANN GmbH & Co. KG
P.O. Box 23 43
D-56513 Neuwied
Tel.: (0 26 31) 99-0
Fax: (0 26 31) 99-64 67

The Treatment of Varicose Veins and Telangiectasias

Surgery

Phlebology '95, D. Negus et al. (eds.). Phlebology (1995) Suppl. 1: 334-336

PI/9.2

A Modification to Improve the Aesthetical Result of Microincisions for Phlebectomy: Prospective Analysis of 615 Cases

A.P. Stehling[1,2], E. Vergara[1,3] and G.M. Morais[1]

[1] Department of Angiology and Vascular Surgery, Semper Hospital,
[2] Department of Angiology and Vascular Surgery, I.p.s.e.m.g. Hospital,
[3] Department of Emergency, João XXIII Hospital, Belo Horizonte, Brazil

INTRODUCTION

Regardless of whether saphenectomy or only phlebectomy is performed, surgery is the best method to eliminate varicose veins, in long term follow up[1,2,3]. Having this in mind, the routine at our institution is to operate every varicose vein, and sclerotherapy is performed only for minimal veins or telangiectasis.

OBJECTIVE

Not only the quality of the therapeutic treatment is important, but also a good final aesthetic result is mandatory nowadays. Therefore, the purpose of this study is to describe a modification of the traditional technique to improve the cosmetic result of the microincisions for phlebectomy [4,5,6].

TECHNIQUES

Because of the triangular shape of the blade, the incision usually turns out to be larger than necessary, leaving small, but visible scars. To reach deeper varicose veins, it is necessary to make a wider incision due to the shape of the blade. This does not happen when hypodermic needles are used, because they do not cut the skin and the underneath fibers, but divulse them in an uniform way (Fig. 1.).

The patient is precisely marked by the main surgeon in the upright position, and then submitted to the spinal bupivacaine anaesthesia or local anaesthesia with lidocaine 0.5% with adrenaline. The patient is then placed in Trendelemburg position to promote maximum emptiness of the vein. A perforation is performed on the skin and hypodermic tissues at the side of the varicose vein with a disposable hypodermic needle. Then the crochet needle is introduced into the hole to "hook" the vein off the tissue. The micro forceps is manipulated outside the skin, in delicate movements to take off the vein and to avoid injuries to the epidermis. After this, the vein is divided, trying to get the longest section of the vein as possible, in both sides.

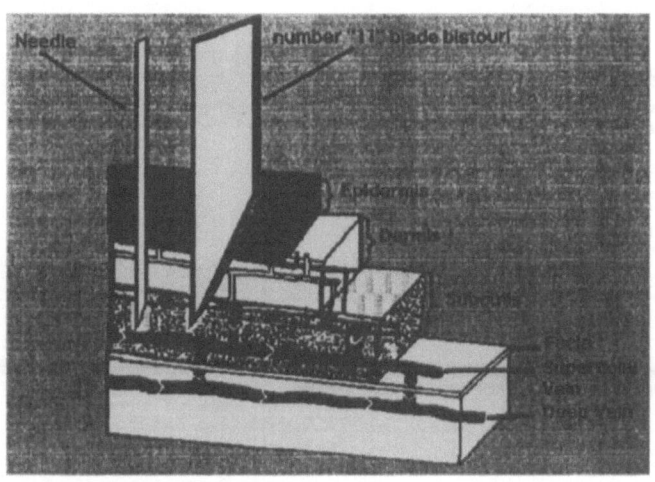

Fig. 1. Comparison between incisions

Different diameters of needles, according to the size of the varicose vein to be taken off, are used. We use the twenty one, nineteen, eighteen and sixteen "gauge" hypodermic needles, and 0.6 and 0.75 millimetres crochet needles. Some times, when varicose veins are considered too big to be taken out of the hole, an incision with number "11" blade bistouri is performed, to avoid injury to the borders of the skin.

After the procedure, the holes are covered with paper strips, and legs are compressed with crepe bandage for two days. When only phlebectomy is performed, after the initial two days, elastic stockings are worn up to fifteen days. Paper strips are removed in five days and patient can return to work after that period. When concomitant saphenectomy is performed, crepe bandage is used for seven days, and elastic stockings are worn for thirty days. Paper strips and stitches of the saphenectomy are removed after seven days, and patients can return to daily activities after fifteen days. If patients are submitted to local anaesthesia, they can be discharged immediately, and should remain in bed for twenty four hours only. When spinal anaesthesia is performed, patients are discharged the morning after. Patients of both groups can walk with moderation on the second day.

STUDY GROUP

From 1980 to 1991, we performed 1080 surgeries with number "11" blade bistouri only [4]. From 1991 to 1993, 615 patients have undergone this new procedure at our institution, for cosmetic reasons or because signs or symptoms of chronic venous insufficiency were present.

Group I consists of 300 patients (49 % of the total) who were submitted to spinal anaesthesia. Group II consists of 315 patients (51 % of the total) submitted to local anaesthesia only, without venous analgesia. In group I, 201 patients (33 %) were submitted to saphenectomy plus phlebectomy, and 99 patients (16 %) only to phlebectomy. In group II, all patients were submitted only to phlebectomy.

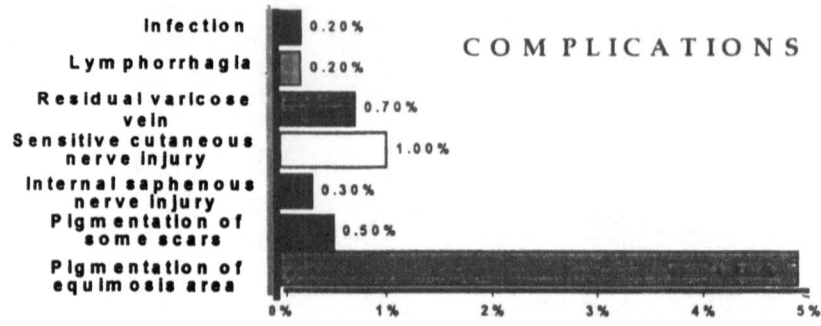

Figure 1. Complications observed on the study.

RESULTS

The cosmetic result was analysed subjectively. The purpose of the technique was to avoid visible scars, and this was achieved in all but five patients (0.5%) who showed permanent pigmentation on some scars of the surgery. Pigmentation of the scars and of the pre equimosed area (4.9%), can be a consequence not only of this technique, but also of all techniques of varicose veins surgeries and specially when large varicose veins are sclerosed. In almost all cases, stains disappeared after six to twelve months with local treatment, and thus, cosmetic results were excellent. Complications of this technique were extremely rare (Figure 1), and could have also occurred with other techniques [7].

CONCLUSIONS

This technique reduces manipulation and patients can be operated under local anaesthesia, without venous analgesia. Therapeutic treatment was achieved with very low morbidity. An excellent cosmetic result could be obtained, even for large tributaries. Cost is low, full recovery and return to daily activities can occur in a short time.

REFERENCES

1. Hobbs JT. Surgery and sclerotherapy in the treatment of varicose veins. Arch. Surg 1974;190:793-6.
2. Hamilton Jakobsen B. The value of different forms of treatment for varicose vein. Br J Surg 1979;66:72-6.
3. Einarsson E., Eklöf B., Neglén P. Sclerotherapy or surgery as treatment for varicose veins: A prospective randomized study. Phlebology 1993;8:22-26.
4. Muller R. Traitement des varices par phlébectomie ambulatoire. Phlébologie 1966;4:277-279.
5. Dortu J. La phlébectomie ambulatoire. Impact Médecin 1990;64:XII-XV.
6. Stehling A.P., Vergara E. Modificação da técnica de varicectomia por microincisões para seu melhoramento estético. Cirurgia Vascular e Angiologia 1992;8(3):4-6.
7. Keith ML, Jr; Smead WL. Complications of the saphenous vein stripping. The Surgical Clinics of North America 1983;63(6):1387-1396.

Phlebology '95, D. Negus et al. (eds.). Phlebology (1995) Suppl. 1: 337-338

P202

Crossectomy Elastic Compression and Sclerotherapy Treatment of Varicose Veins of the Lower Limbs in Outpatients

Dr L. Burgoa, Ms L. Burgoa and Dr L. Rotta

Private Practice, Ponte S. Pietro, Bergamo, Italy

From November 1990 to December 1994, 120 patients with primary varicose veins of the lower limbs were examined. They received outpatient treatment in the form of crossectomy, elastic compression and, subsequently, sclerotherapy.

Twelve percent of the patients were male with an average age of forty. Some patients (20%) had already undergone sclerotherapy but only on small veins and telangiectasis.

Besides the typical symptomatology for varices (oedematous limbs, a feeling of heaviness and heat in the legs, especially after prolonged standing and in summer) all patients had swollen varices, some more so than others, with dilation of the saphena and the collaterals concerned.

In all patients Doppler Fluximeter was used to investigate deep venous incompetence and the competence or not of the sapheno-femoral valve.

The indication for ligation of the great saphena and the collaterals at Scarpa Triangle is the incompetence of the above mentioned valve (reflux along the saphenous trunk 1o, 2o, 3o and 4o Hacks stage).

THERAPY

Besides the standard measures of hygiene and prophylaxis (keeping legs in a raised position, avoiding the heat and sedentary activities so that the muscular pump of the legs could be kept active and so on), the carefully screened patients underwent outpatients treatment. This involves local anaesthetic, approximately 15 cc. of carbocaina 1%, applied to the groin near the saphenous femoral inlet, followed by an incision of 2-3 cm, then a complete dissection of the crosse and ligation of all collaterals with rapidly absorbable thread. The saphenous vein is then tied both in the proximal direction at the precise point of the femoral attachment and distally with non-absorbable thread and then cut between the two ligatures.

Immediately afterwards elastic compression is applied to the upper thigh by means of an elastic stocking with 30-40 mm. Hg at the ankle.

In these cases the stocking was compound with polyamid 74%; Elastan 26% and was worn for approximately 2-3 months.

After about 4-8 weeks, a period during which the calibre of the varicose veins diminishes so that they become flabby and consequently less visible, the sclerotherapy session begins: the number of these sessions, the quantity injected, the concentration of the scleorsing solution and other less important variants will differ; followed by adhesive elastic compression , directed at times with sponges, especially on the most swollen veins.

12-13 weeks after crossectomy, but always based on the calibre, number and extent of the varicose veins, they completely disappeared. The results are excellent both from a physiopathological stand point (there's no longer a reflux from the femoral vein to the saphenous vein, there are no neurological complications or lymphatic complications) and a good aesthetic result since the limb bears no sign of scars.

Moreover, the patient is never disabled and can lead a normal life whether in the family, at work, socially, except for the first hours immediately after crossectomy.

Only rarely, and only when there is a marked dilation of the veins, does any residual thrombus form, and in this case it can be easily sucked out by means of a needle and syringe.

In cases with important incompetent perforating veins, it has been possible to sclerose the perforator itself, but this requires paying particular attention to the directed compression, thus the fibrosis of the incompetent perforating vein is achieved.

The only precaution needed to avoid possible damage to deep circulation (deep vein thrombosis) is to make the patient walk at length immediately after sclerotherapy.

In two cases we had to resort to surgical ligature of the perforating vein which had not been closed by sclerotherapy.

In conclusion we think that this is a good technique that avoids hospitalisation, operating theatre stress and relative morbidity.

REFERENCES

1. Apicella F. - Lombardi A. Crossectomy and sclerotherapy for treatment of varices of lower limbs out patient surgery. Abstracts V. European American Symposium on venous diseases 1990; 4, B3.
2. Patel J. - Leger L. Nouveau traitè de technique chirurgicale 1973; 5:513.
3. Holme J. B. - Holme K. - Sorensen L.S. The anatomic relationship between the long saphenous vein and the saphenous nerve. Phlebology digest 1989; 3:16.
4. Korstanje M.J. - Neumann H.A.M. Compression therapy with elastic stockings. Phlebology digest 1991; 2:21.
5. Kerner J. - Schultz - Ehrenburg U. Functional effects of different steps in the course of sclerotherapy. Phlebology Digest 1989; 2:28
6. Joseph G Sladen Compression sclerotherapy: preparation, technique, complications and results. Am. J. Surg 1983; 146:228.
7. Hubner K. (1993): The outpatient therapy of trunk varicosis of the great saphenous vein by means of ligation (crossectomy) and sclerotherapy. Phlebology digest 1, 21-25.

Phlebology '95, D. Negus et al. (eds.). Phlebology (1995) Suppl. 1: 339-340

OP/14.5

Comparison of the Results of Venocuff Implantation and High Ligation of the Sapheno-Femoral Junction: 1 Year Follow-Up

Gy. Acsády, Á. Laczkó, E. Turbók and A. Nemes

Department of Cardiovascular Surgery, Semmelweis Medical University, Budapest, Hungary

INTRODUCTION

40 patients with long saphenous vein reflux due to incompetence at the sapheno-femoral junction (SFJ) confirmed by duplex ultrasound scanning and PPG gave consent for inclusion in the study. Patients with post-phlebitic limbs, recurrent varicose veins or other venous pathology were excluded.

METHODS

The included 40 patients based on the type of surgical intervention were devided two groups (20-20):
I. group: Venocuff implantation,
II. group: high ligation of the SFJ.
The purpose of the study was to determine the recurrence-rate of the incompetence of SFJ after both surgical procedure.
All patients were re-examined by duplex scanning and PPG 1, 3, 6 and 12 months after surgery.

RESULTS

Table 1. Recurrence-rate of the incompetence of SFJ.

	1 month	3 months	6 months	12 months
I.group	-	-	1	1
II.group	-	-	2	1

The recurrence of the incompetence of SFJ were observed in 2 patients (10%) at the I. group, while 3 cases (15%) occurred in the II. group.

CONCLUSION

The external valvuloplasty with Venocuff device is a useful alternative for the treatment of varicosity of the lower limb's.

Phlebology '95, D. Negus et al. (eds.). Phlebology (1995) Suppl. 1: 341-343

PI/9.6

Endoscopic Perforator Interruption using Laparoscopic Equipment

Peter Conrad, FRCS, FRCS(Ed), FRACS, FACS(US)

Visiting Surgeon, Nepean Hospital, Sydney, Australia

INTRODUCTION

Perforating veins crossing the medial subfascial space of the calf when incompetent are believed to cause varicose veins in the leg. In particular, Boyd's perforator and those described by Cockett in the lower medial part of the calf can be mapped by Duplex scan and are readily visualised in the medial subfascial space of the leg by distending this space with CO_2 gas and introducing a camera by a technique similar to that developed for laparoscopic cholecystectomy. By means of a small five millimetre second port the perforating veins can be interrupted using diathermy.

In Australia laparoscopic cholecystectomy was popularised in 1991 and with increasing familiarising with the equipment used it was decided early in 1993 to apply this technique in exploring the medial subfascial space of the calf. The technique was first described in 1993 at the Pacific Vascular Symposium in Honolulu. Since that time the first seven legs treated by endoscopic exploration of the subfascial space using laparoscopic equipment and perforator division (EESSPD) have been written up and described in Phlebology in 1994 (Ref. 2).

Up till the time of this paper nineteen legs altogether have been treated by EESSPD and the technique appears to be safe and effective in dealing with varicose veins in the lower leg due to perforator incompetence. In particular, it is very effective in dealing with large perforators in the lower third of the leg associated with induration and ulceration.

SURGICAL TECHNIQUES

Pre-operatively the perforators are mapped by Duplex ultrasound. The equipment required is that used for laparoscopic cholecystectomy and a ten millimetre port is inserted below the knee in the region of Boyd's perforator. This entry point is also used to ligate a Boyd's perforator if present and also can be used in groin to knee stripping of an incompetent long saphenous vein if necessary.

Through this incision the deep fascia of the leg is entered and the filmy adhesions between the deep fascia and the muscles are broken down by a long pair of artery forceps. The ten millimetre port is then introduced and the subfascial space is distended with CO_2 at a pressure of thirty millimetres Hg with the insufflator set at high flow. The telescope with its attached camera is inserted through the port into the subfascial space and a good view of the subfascial space up and down the medial compartment of the leg is obtained. Further filmy adhesions are divided using the telescope under direct vision and the incompetent perforating veins are now clearly seen on the video monitor.

A five millimetre port is now inserted into the distended subfascial space level with the site of entry of the ten millimetre port and about three finger breadths posteriorly. A dissecting forceps is introduced through the five millimetre port and breaks down the last filmy adhesions. The incompetent perforating veins are now grasped with the dissecting forceps and a diathermy current coagulates the incompetent perforator. It is important to carefully dissect away any artery or nerve which may be accompanying the perforator. Occasionally as well as using the diathermy forceps endoscopic scissors or a diathermy hook can be used.

Several interesting observations have been made on direct visualisation of the subfascial space and the perforating veins. Firstly, it is interesting that most perforating veins are not short straight veins however often have a convoluted S shaped course through the subfascial space. Secondly, they often break up into two or three terminal branches before exiting the subfascial space.

Thirdly, they are often accompanied by a small nerve or artery which must be carefully dissected away and preserved.

RESULTS

All nineteen patients were discharged from hospital within twenty-four hours and there were no significant complications or problems with this method. The varicose veins in the lower calf have resolved although small venules and spider veins had to be injected by post operative compression sclerotherapy.

DISCUSSION

It now appears that EESSPD is an accurate and technically easy way of quickly exploring the medial subfascial space of the lower leg and interrupting incompetent perforators by the time honoured technique of "high ligation" close to the entry of the perforator into the deep veins.

In particular, it has a strong place in the armamentarium of the surgeon in cases where large perforating veins are associated with induration and ulcer formation in the lower third of the leg.

The cases of EESSPD presented in this paper are part of a long term project where all cases will be followed up by clinical and Duplex studies at six months and, three years post operatively as part of an extensive follow up.

References

1. Conrad, P. What's new in phlebology. Proc Straub Pac Health Found 1993;57(2):56.
2. Conrad, P. Endoscopic exploration of subfascial space of lower leg with perforator vein interruption using laparoscopic equipment. A preliminary report. Phlebology 1994;9:154-157.

Phlebology '95, D. Negus et al. (eds.). Phlebology (1995) Suppl. 1: 344-346

P186

Day Case Primary Vein Surgery. Results of a Team

Pereira Alves, J. Neves, A. Formiga, A. Fernandes and Alves Pereria

Hospital Stº Antonio dos Capuchos, Serviço 6, Unidade Vascular, Lisboa, Portugal

INTRODUCTION

Varicose veins surgery is a frequent surgical procedure. To do it as a day case surgery is cost effective.

The objective of the present study is to analise post operative complications, analgesic consumption, mobilization, and return to work in a group of 200 patients with primary and uncomplicated varicose veins operated by a team as a day case surgery.

METHODS

All patients are studied with the following computerized protocol :
(see protocol page)

RESULTS

(1) - Post operative complications :
- no cases of infection or saphenous neuritis .
- 8 cases of cutaneous ecchymosis of the thigh .
- 1 case of stripper track fibrosis .
- 1 case of groin haematoma .

(2) - Analgesics consumption :
- 2 or 3 demands of light analgesics (paracetamol)

(3) - Mobilization and return to work :
- Patients mobilizing the evening of the operative day and leave hospital next morning .
- Better mobilization after regional anesthesia .
- Return to work - a media of 10 days .

CIRURGIA VENOSA

IDENTIFICAÇÃO

SERVIÇO: [ ] ARQUIVO: [0] BOLETIM: [0] CAMA: [0]

NOME: [ ] IDADE: [0] SEXO: [ ]

ESTADO CIVIL: [ ] PROFISSÃO: [ ]

RESIDÊNCIA: [ ]

INTERNAMENTO: [ ] ALTA: [ ]

PRIMÁRIAS: [ ]
DIAGNÓSTICO | VARIZES   SECUNDÁRIAS: [ ]
RECORRENTES: [ ]

OPERADO : ( ) OPERAÇÃO: [ ]

| TERRITÓRIO | SAFENA INTERNA | DIREITA [ ] | ESQUERDA [■] |
| | SAFENA EXTERNA | DIREITA [ ] | ESQUERDA [ ] |

QUADRO CLÍNICO :

PERNA PESADA ?	[ ]
DORES VESPERTINAS ?	[ ]
DORES NOCTURNAS ?	[ ]
EDEMA MALEOLAR ?	[ ]
EDEMA PERNA ?	[ ]
EDEMA COXA ?	[ ]

ALTERAÇÕES / COMPLICAÇÕES :

1 - PIGMENTAÇÃO	[ ]
2 - LIPODERMATOSCLEROSE	[ ]
3 - ECZEMA	[ ]
4 - ÚLCERA VENOSA	[ ]
5 - VARICORRAGIA ?	[ ]
6 - VARICOFLEBITE	[ ]

DOPPLER

REFLUXO CROSSA SAFENA

INTERNA DTª: ( )   EXTERNA DTª: ( )
INTERNA ESQ: ( )   EXTERNA ESQ: ( )

PLETISMOGRAFIA ? [ ]
ECO-DOPPLER ? [ ]   ORTIGÃO [ ]

ANESTESIA : GERAL [ ] EPIDURAL [ ] RAQUIANESTESIA [ ]

SAFENA INTERNA DTª [ ]
SAFENA INTERNA ESQ [ ]
1. LAQUEAÇÃO DA CROSSA
SAFENA EXTERNA DTª [ ]
SAFENA EXTERNA ESQ [ ]

LAQUEAÇÃO DE PERFURANTES

INFERIOR [ ] MÉDIA [ ] SUPERIOR [ ] OUTRAS [ ]

2. STRIPPING CURTO [ ]
LONGO [ ]

COMPLICAÇÕES :

INFECÇÃO [ ] NEVRITE [ ] FIBROSE DO TRAJECTO [ ] EQUIMOSE [ ] OUTRAS [ ]

CIRURGIÃO: [ ]
1º AJUDANTE: [ ]   3º AJUDANTE: [ ]
2º AJUDANTE: [ ]   4º AJUDANTE: [ ]

COMENTÁRIOS: [ ]

CONCLUSIONS

Day case surgery of primary uncomplicated varicose veins is cost effective in our experience , and :
- with negligible post operative complications .
- more confortable for the patient .
- permits early mobilization and early return to work .

REFERENCES

(1) - Stefano Ricci and Mihael Georgiev with Mitchel P. Goldman : Ambulatory phlebectomy . A practical guide for treating varicose veins . Mosby , 1995 .

(2) - Goldman MP , Bergan JB : Ambulatory treatment of venous disease : an illustrated guide . St Louis , Mosby , 1995 .

(3) - Norman L Browse , Kevin G Burnand and Michael Lea Thomas : Diseases of the veins - Pathology, diagnosis and treatment- Edward Arnold , 1988 .

(4) - Tibbs DJ : Varicose veins and related disorders . London , Butterworth Heinemann , 1992 .

(5) - Jakobsen B : The value of different forms of treatment for varicose veins. Br. J. Surg. 66 : 182 - 184 , 1976 .

Phlebology '95, D. Negus et al. (eds.). Phlebology (1995) Suppl. 1: 347-350

P206

Exophlébectomie Varices sous Cutanées Abdominales

J.M. Trauchessec[1] and R. Vergereau[2]

1 Dermatologiste á Paris, France
2 Angéiologue á Chamaliers-Royal, France

Exopulebectomy of Subcutaneous Abdominal Wall Varices

SUMMARY

Abdominal varicose veins can be exo-evaginated by phlebectomy under local anaesthetic, as ambulatory cases (day cases). Provided that rigorous indications are adhered to, this precise surgical technique seems to us to provide an effective and elegant solution.

EXOPHLÉBECTOMIE DES VARICES SOUS CUTANÉES ABDOMINALES

J.M. TRAUCHESSEC[1], R.VERGEREAU[2]

[1]Dermatologiste à PARIS FRANCE
[2]Angiéologue à CHAMALIERES-ROYAT FRANCE

INTRODUCTION

La phlébectomie des varices abdominales par exo-éveinage est marquée par la rareté de descriptif technique dans la littérature phlébectomique. Autre particularité : du fait de l'origine pathologique de ces varices, il faut s'assurer de leur non-vicariance avant de les enlever. Enfin, ces vaisseaux se caractérisent aussi par leur relation anatomique avec les veines profondes extrapéritonales. Un exo-éveinage méticuleux, un travail en douceur respectant les conditions anatomiques et physiopathologiques apportent très certainement une solution au problème thérapeutique posé par ces varices.

Les varices abdominales sont d'origine pathologique. Elles apparaissent en tant que déviation face à un obstacle sur le retour veineux (bloc hépatique, thrombose veineuse profonde du petit bassin). L'exo-éveinage ne se conçoit que sous deux conditions :

-l'absence d'évolutivité de l'obstacle veineux profond,

-la disparition de la vicariance étiologique autorisant l'éveinage avec la sécurité de ne pas induire de troubles hémodynamiques secondaires.

RAPPEL ANATOMIQUE

Le plan veineux cutané superficiel

Caractérisé par son extrême variabilité, le plan veineux cutané superficiel n'est pas calqué sur celui des artères et est difficilement injectable à contre-courant(1). Les veines circulent dans le tissu cellulaire sous-cutané. Certaines veines sont aussi volumineuses que les artères. D'autres, au contraire, ne sont pas macroscopiquement repérables. Dans ce cas, le retour veineux est assuré par les veines profondes, via le réseau veineux sous-dermique situé le long de la face externe du pannicule adipeux, puis via le réseau veineux perpendiculaire, éventuellement satellite des artères perforantes.

Deux veines superficielles sont habituellement retrouvées :

La veine sous-cutanée abdominale

Elle naît dans la région sous-ombilicale, parfois à partir de plusieurs branches (2) et se jette deux fois sur trois dans la crosse de la saphène interne. Dans 18% des cas, elle s'abouche dans un tronc commun réalisé avec la veine circonflexe iliaque superficielle et/ou la saphène crurale antérieure.

Dans quelques rares cas, elle se draine :
- dans le tronc des veines honteuses externes (3),
- dans la veine fémorale,
- dans la veine saphène crurale postérieure,
- dans la veine de Cruveilhier ou dans celle de Giacomini.

La veine circonflexe iliaque superficielle
Elle naît dans la région située entre les épines iliaques antérieures supérieure et inférieure, parfois par plusieurs branches et se jette une fois sur deux dans la crosse de la veine saphène interne. Dans près de 40% des cas, elle rejoint la veine saphène crurale antérieure, la veine sous-cutanée abdominale ou encore un tronc commun à ces deux dernières veines.

Dans quelques rares cas, elle s'abouche:
- dans la veine fémorale,
- dans les veines crurales antérieures non saphéniennes.

La veine para-ombilicale, reliquat foetal, n'est pas toujours perméable.
Il peut exister des anastomoses entre veines superficielles et veines profondes du petit bassin par le plexus veineux rétropubien.

LA METHODE D'EXO-ÉVEINAGE

CONDITIONS HEMODYNAMIQUES
L'exploration préalable de la circulation veineuse profonde est incontournable. Elle précise la liberté ou l'obstruction des voies de retour veineux. Elle visualise la topographie des voies de suppléances qui se sont établies.

Les analyses morphologiques, telle la phlébographie, visualisent les éléments architecturaux.

Les analyses dynamiques, tels l'échographie et le Doppler, renseignent sur des éléments fonctionnels : mesures de débit, de flux et surtout constatation du reflux à la relaxe de la pression d'amont. Celui-ci signe l'absence de vicariance et autorise l'éveinage.

Le test clinique de la compression des varices à éveiner est un argument fiable.
Une compression permanente préalable pendant trois jours ne doit déclencher ni signe clinique, ni oedème, ni douleur. Elle simule le port du pansement post-opératoire.

L'EXO-ÉVEINAGE (4)
Pour opérer, la position déclive est importante car les varices abdominales se vident incomplètement et gardent une turgescence résiduelle avec laquelle il faut savoir composer. Il faut veiller à réaliser une anesthésie locale en périveineux strict (lidocaïne 0,5% adrénalinée bicarbontée). En effet, l'injection intravariqueuse amènerait rapidement les produits dans les cavités cardiaques à des pics de concentration élevés et toxiques. Le test d'aspiration préalable à chaque injection devient primordial. La quantité de liquide injecté n'est pas très élevée car les trajets variqueux sont relativement courts.

Capture variqueuse : l'incision cutanée est la solution de continuité à choisir. Elle sera courte (3 à 4 millimètres) et cachée dans la pilosité pubienne, l'ombilic, sur une ancienne cicatrice d'appendicectomie ou de laparotomie ou, au moins, orientée selon les lignes de tension cutanée minimale.

L'incision est effectuée à l'aplomb d'une ampoule variqueuse. Elle est superficielle sur 0,5 mm d'épaisseur pour ne pas sectionner l'ampoule variqueuse située habituellement à un millimètre au-dessous de l'épiderme. A ce stade, le fond de l'incision montre le derme superficiel blanc, fibreux, exsangue.

Une pince à mors fins écarte délicatement les berges de l'incision. Les fibres collagènes dermiques s'étirent, se rompent et laissent apparaître la paroi variqueuse bombée, bleutée, turgescente. Une deuxième pince à mors plus larges saisit l'ampoule variqueuse en douceur, en évitant sa rupture.

Exo-éveinage de proche en proche : une traction minime, verticale et perpendiculaire au plan cutané est imprimée par une pince clamplant l'ampoule variqueuse. Une deuxième pince, plus fine ou éventuellement une paire de ciseaux à bouts ronds, dissèquent l'espace périvariqueux sur sa circonférence par des mouvements doux d'écartement des mors et des branches.

Plusieurs centimètres de varice s'extériorisent par le même orifice. Une deuxième incision est pratiquée à distance. Les gestes identiques y sont effectués. Les collatérales et les extrémités variqueuses sont suturées au fil résorbable. Si une rupture variqueuse se produit en cours d'exo-éveinage, deux clamps sont immédiatement posées de part et d'autre de la rupture. Une suture cutanée est préférable pour réunir les berges de chaque incision.

CONCLUSION

Les veines et varices de l'abdomen peuvent être exo-éveinées par phlébectomie sous anesthésie locale, en ambulatoire. Sous réserve d'indications rigoureusement posées, cette chirurgie précise nous semble apporter une solution efficace et élégante.

BIBLIOGRAPHIE

1. Elbaz J.S. Flageul G. Liposuccion et chirurgie plastique de l'abdomen. Editions Masson - Paris 1988.
2. Boyd J.B., Taylor G.I., Carlett R. The vascular territories of the superior epigastric and deep inferior epigastric system. Plast. Reconstr. Surj. 1984 ; 73 : 1 - 14.
3. Henriet J.P. Le confluent veineux saphénofémoral et le réseau artériel honteux externe : données anatomiques et statistiques nouvelles. Phlébologie ; 1987 ; 40 ; (3) ; 711 - 35.
4. Trauchessec J.M., Vergereau R. : Exo-éveinage des veines de l'abdomen ; La lettre de l'angiologie ; N° 52 ; 1994 ; 2 - 4.

Phlebology '95, D. Negus et al. (eds.). Phlebology (1995) Suppl. 1: 351-354

V/1.2

La Gévaudanaise Bilatérale

J.M. Trauchessec[1] and R. Vergereau[2]

1 *Dermatologiste á Paris, France*
2 *Angéiologue á Chamaliers-Royal, France*

The Bilateral Gévaudanaise Operation

The Gévaudanaise (named after the ancient province of Gévaudan, where it originated in 1986), or aesthetic/exo-saphenectomy of the internal saphenous vein, is based on the aesthetic procedure developed by Muller for reticular varices. Nine years' experience allows us to state that the bilateral Gévaudanaise is a varicose vein operation which is a reliable, efficient and elegant solution to the problem of the varicose internal saphenous system.

LA GÉVAUDANAISE BILATERALE

J.M. TRAUCHESSEC[1], R.VERGEREAU[2]

[1]Dermatologiste à PARIS FRANCE
[2]Angéiologue à CHAMALIERES-ROYAT FRANCE

DÉFINITION

La Gévaudanaise se définit comme une exo-saphénectomie de la saphène interne variqueuse sans contre-incision le long du membre inférieur. La Gévaudanaise a été réalisée pour la première fois en 1986, dans l'ancienne province du Gévaudan, d'où l'origine de son nom (1). La Gévaudanaise bilatérale, pratiquée à droite et à gauche simultanément quand elle est indiquée, apporte aux patients une commodité souvent demandée.

PRINCIPE

La Gévaudanaise correspond à la solution ultime recherchée par R. Muller (2) dermatologue suisse, depuis près de 35 ans : adapter au trajet saphénien interne crural les principes de la phlébectomie ambulatoire. Ainsi, la Gévaudanaise satisfait à la demande d'exo-éveinage, au caractère ambulatoire grâce à l'anesthésie locale (3), au résultat esthétique par l'absence d'incision à la cuisse ou à la jambe.

Pour arriver à concilier ces trois directives, il fallait élaborer un instrument breveté particulier : Le dissecteur éveineur de Trauchessec (D.E.T.) (4). Il s'agit d'une tige métallique rigide de 60 cm de long terminée par un anneau dont la lumière admet le tronc de la saphène interne crurale.

L'EXO-SAPHÉNECTOMIE SANS CONTRE-INCISION (5)

La partie distale du moignon de la saphène interne est introduite à l'intérieur de l'anneau du dissecteur éveineur dans le champ opératoire inguinal de la crossectomie. L'anneau, mobilisé par des mouvements de pulsion rotation sur son axe, descend progressivement autour de la saphène interne crurale.

Son extrémité biseautée tranchante dissèque les tissus entourant la saphène interne tronculaire.

La section de la saphène tronculaire surale s'effectue soit par étranglement dû à l'effet lasso du filin d'acier, soit par une section pariétale de la saphène qui est immobilisée entre, à l'intérieur, l'anneau tranchant du D.E.T. et, à l'extérieur, le doigt de l'autre main de l'opérateur. Aucune puncture ni incision cutanée à la jambe n'est réalisée.(6)

Le pansement compressif assure la déambulation immédiate. Le patient sort du bloc opératoire en marchant normalement.

RÉSULTAT

Résultat anatomique

La Gévaudanaise bilatérale prouve l'efficacité de sa conception, car l'exoéveinage enlève obligatoirement la totalité des trois tuniques pariétales, y compris l'adventice (7).

Les dédoublements saphéniens parallèles partiels ou totaux sont également inclus dans l'anneau, donc exo-éveinés, limitant en conséquence la possibilité d'éventuelles récidives. En cas de phlébite saphénienne, l'exo-éveinage enlève la totalité du vaisseau thrombosé sans risque de migration du caillot intra-luminal.

Résultat esthétique

Aucune incision le long de la face interne de la cuisse, ni de la face interne du genou, ni de la face interne du mollet garantit le respect de l'intégrité cutané du membre inférieur (8).

INCIDENTS - COMPLICATIONS :

Neuf années de suivi nous indiquent qu'ils vont dans le sens de l'innocuité technique. Nous les avons classés en 3 tableaux selon leur étiologie cutané, vasculaire ou autre.

Incidents cutanés	Gévaudanaises Bilatérales	
	700 cas	(%)
Phlyctènes mécaniques	62	8,8
Pigmentations transitoires	20	2,8
Eczémas	4	0,6
Infections	2	0,3
Désunions inguinales	2	0,3
Induration persistante	0	0
Cicatrice hypertrophique	0	0
Nécrose	0	0

Incidents vasculaires	Gévaudanaises Bilatérales	
	700 cas	(%)
Varicosités-Télangiectasies	13	1,8
Lymphorrhées transitoires inguinales (quelques jours)	7	1,0
Rupture de jonction saphéno-fémorale per opératoire	3	0,4
Saignement au test du 1er pansement	3 inguinal	0,4
Oedème malléolaire régressif en un an	1	0,2
Phlébite superficielle	0	0
Hématome important	0	0
Lâchage de suture vasculaire	0	0
Récidive variqueuse ou saphénienne	0	0

354

Autres incidents	Gévaudanaises Bilatérales 700 cas	(%)
Sensitifs :		
Spontanément régressifs	30	4,2
Définitifs	0	0
Osseux : Douleur de compression	0	0
Musculaires : Σ d de Volkman	0	0
Généraux :		
Lipothymie	0	0
Malaise d'anesthésie locale	0	0

CONCLUSION

La Gévaudanaise, ou exo-saphénectomie esthétique de la saphène interne, associe à l'exo-éveinage saphénien conçu par Mayo en 1906, la notion esthétique développé par Muller vis-à-vis des varices réticulaires. Une expérience de neuf années nous permet d'affirmer que la Gévaudanaise bilatérale constitue une intervention de chirurgie phlébectomique sûre, efficace, élégante, solutionnant au mieux les problèmes des saphènes internes variqueuses.

BIBLIOGRAPHIE

1. Trauchessec J.M. ; "La Gévaudanaise : Association en chirurgie phlébologique ambulatoire d'une crossectomie et d'une saphénectomie sans contre-incision". J. Méd. Esth. et Chir. Derm., Septembre 1986, Vol. XIII, n° 51 : 227-231.
2. Muller R. : "L'esthétique en phlébectomie ambulatoire". J. Méd. Esth. et Chir. Derm., Juin 1989, Vol XVI, n° 62 : 155-158.
3. Vidal-Michel J.P., Arditti J.,Boudon J.H. : "L'anesthésie locale au cours de la phlébectomie ambulatoire : appréciation du risque par dosage de la lidocaïnémie". Congrès mondial de phlébologie, Strasbourg 1989, Editions John Libbey Eurotext Ltd, n° 403 : 1118.
4. Trauchessec J.M. : "Comment se servir du D.E.T. (dissecteur-éveineur de Trauchessec)". Les Nouvelles Dermatologiques, 1987, Vol. 6, n° 4 : 352-353.
5. Choukroun P.L., Trauchessec J.M. : Saphénectomie esthétique ambulatoire sans contre-incision", Congrès mondial de phlébologie, Strasbourg 1989. Editions John Libbey Eurotext Ltd, n° 351 ; 977-979.
6. Trauchessec J.M. : " Pour une saphénectomie esthétique" ; Phlébologie, 1987, 40 ;(2) ;409-422.
7. Trauchessec J.M., Vergereau R. : "La Gévaudanaise", Congrès Mondial de Phlébologie ; Montréal 1992 ; John Libbey, Eurotext ; 1015-1017.
8. Trauchessec J.M. : Technique chirurgicale : "La Gévaudanaise", La lettre de l'angiologie, Juin 1987, n° 23 : 5-7.

Phlebology '95, D. Negus et al. (eds.). Phlebology (1995) Suppl. 1: 355-359

OP/14.3

Clip Percutané Endosaphène

J.F. Van Cleef

45 rue de la Chaussée d'antin

Percutaneous Endosaphenous Prosthesis

The author describes the V-shaped prosthesis he uses to obtain obliteration of the sapheno-femoral junction.

This is introduced by catheterising the long saphenous vein and positioning the prosthesis under image intensifier control. Some indications are discussed and it is suggested that positioning under echo-Doppler control may be preferable.

CLIP PERCUTANE ENDOSAPHENE

JF. VAN CLEEF,
45 rue de la chaussée d'antin
F-75009   PARIS
Fax: 33 (1) 42813482

Dès la fin du 19ème siècle, on mit du fil imbibé de teinture d'iode dans des varices puis on imagina des objets cylindriques pour obtruer les reflux liquidiens dans les veines variqueuses (bollonnets, valves mécaniques, ressorts de toutes sortes, double cônes...). Ces objets cylindriques présentent l'inconvéniant d'être volumineux sous la peau et d'engendrer des thromboses très inflammatoires.

Le clip présenté ici est plan, il aplati la veine. Il n'est pas volumineux sous la peau et le volume de la thrombo-sclérose autour du clip est faible. Sa forme en V évite les migrations.

La pose de clip endoveineux est particulièrement destinée à la jonction saphéno-fémorale. Elle est actuellement réalisée dans une salle possédant un ampli de brillance.

C'est un traitement endovasculaire percutané qui pourra dans l'avenir être effectué sous contrôle échographique.

RAPPEL ANATOMO-PHYSIOLOGIQUE

La veine saphène a un axe préférentiel d'aplatissement parallèle à la surface de la peau et présente ainsi deux faces (wallsides) et deux bords (borders), une face interne ou aponévrotique, une face externe ou cutanée réunies par les deux bords de la paroi veineuse.

Les bords libres des bivalves sont parallèles à l'axe d'aplatissement de la lumière veineuse, les commissures valvulaires sont sur les bords de la veine. Dans les veines variqueuses la création d'un espace intercornéal sur les bords des veines favorise les reflux liquidiens. (Pour mémoire notons que les orifices des perforantes principales sont sur les bords de la veine).

LE MATÉRIEL

Il comprend:
-un cathéter court 16 G, un introducteur 8 F court
-un cathéter porteur ou lanceur dans lequel est chargé en position de prélargage à quelques millimètres de son extrémité distale le clip endoveineux. Ce clip métalique en forme de V est composé de deux branches soudées ensemble à l'une de leur extrémité et présentant à l'autre extrémité deux crochets de fixation. Il s'agit donc de deux tiges d'appui relié par un écarteur formant ressort (figure 1). Il est destiné à aplatir la lumière veineuse selon son axe physiologique .
-un cathéter poussoir
-un agent sclérosant IODE 3% (Meram Melun)

-des bandes collantes de contention
-en réserve: un doppler de poche, un guide de 0,020, une boite de
dénudation avec un crochet de dentellière en inox n° 3.

LE GESTE

-Le patient est en décubitus latéral sur la table de radiologie.
-Repérage par la palpation- percution de la saphène interne gonale,
rarement à l'aide du doppler de poche.
-Ponction de la saphène à l'aide du cathéter court 16 G, vérification à
l'aide d'un peu de produit de contraste si nécessaire.
-Introduition du J de l'introducteur dans la saphène et ablation du cathéter
-Anesthésie locale au point de ponction: xylocaine 1% adrénalinée 2 cc
-Petite incision verticale de 3 mm à la lame de scalpel
-Mise en place de l'introducteur
-Opacification de la saphène , repérage de la crosse
-Montée du lanceur à **13 cm** de la jonction saphéno-fémorale
-Injection d'une bulle d'air puis par exemple de 3 cc d'iode à 3% puis
d'une bulle d'air et de 1,5 cc de serum physiologique par le lanceur
-Orienter le clip pour une ouverture parallèle à la surface de la peau
-Largage du clip **à 10 cm** sous la jonction saphéno-fémorale: le poussoir
restant **fixe**, on tire **doucement** sur le lanceur, les branches du clip
s'appliquent sur la ligne des commissures, sur les bords de la saphène; le
clip aplati la lumière veineuse parallèlement à la surface de la peau .
-Scléroses étagées de la saphène tronculaire (Iode 1% par exemple) si
nécessaire en faisant bien attention de vidanger la lumière de lanceur
après chaque injection.
-Pose d'une contention fixe, par exemple bande collée de 8 cm de large
sur deux compresses pliées en quatre en regard du clip, bande
pratiquement circulaire autour du haut de la cuisse et au niveau du point
de ponction.
-Contention par bande collée ou amovible de l'ensemble du membre
(collant de contention).

QUELQUES ASTUCES

-La première difficultée est le cathétérisme de la saphène. La saphène
interne au genou a une topograhie constante à la face latérale du condyle
interne et est peu mobile devant l'aiguille. En cas de difficultées une
ponction sous échographie est possible. En cas d'échec même à un
autre niveau de la saphène, éventualité rare, une dénudation s'impose à
l'aide d'un crochet de dentellière en inox n°3, puis couper légèrement la
veine sur son bord et introduire directement le lanceur dans la lumière
veineuse.
-Si le lanceur bute sur une sinuosité variqueuse malgré les manoeuvres
habituelles, s'aider du guide 0,020 après avoir retiré le clip, plus tard
remettre le clip par l'extrémité proximale du lanceur en ayant dévissé la
valve d'étanchéïté.

-Après largage si le clip ne s'ouvre pas, il est sans doute bloqué dans une branche car le lanceur n'était pas parfaitement en place. Mettre alors un deuxième clip sous la crosse après vérification.

-Il est impératif que le clip s'ouvre parallèlement à la surface de la peau et parallèlement à l'axe préférentiel d'aplatissement de la veine. En effet une ouverture perpendiculaire à la peau expose à un sens d'aplatissement non physiologique de la veine. Une bonne orientation de la caméra et du clip est indispensable.

SOINS POST-OPERATOIRES

Ablation de la bande collée vers le quatrième jour.

Dès cette date, la jonction saphéno-fémorale étant obstruée, le traitement complémentaire du réseau sous jacent incontinant est impératif. Des scléroses ou des phlébectomies ambulatoires sont indispensables pour parfaire le résultat. Le bas ou le collant de contention est porté pendant un mois.

Des contrôles réguliers permettent de suivre cette maladie variqueuse évolutive.

INDICATIONS

Idéalement c'est le sujet âgé avec un ulcère de jambe que le chirurgien ne désire pas mettre sur une table d'opération pour un stripping et plus généralement ce sont les indications de la crossectomie.

Certains patients ne peuvent pas ou ne veulent pas interrompre leur activité professionelle; cette technique peut être comparée à la sclérose simple de crosse et permet d'obstruer une crosse même de 15 mm de diamètre en une seule séance, puis de traiter dans de bonnes conditions hémodynamiques le réseau sous jacent par phlébectomies ou par scléroses et ce, sans arrêt de travail.

Une femme ayant une maladie variqueuse sévère, dont la grand-mère et la mère ont eu un mauvais résultat après stripping pourrait être aussi une bonne indication pour ce geste minimal percutané.

A l'inverse une femme de 40 ans, ancienne sportive ayant une varicose tronculaire avec une saphène à paroi épaisse semble une très belle indication à un stripping par invagination.

Ces indications assez schématiques seront à modifier selon le pourcentage de récidives voire de reperméabilisations à 6 mois, à un an, à cinq ans de cette technique percutanée.

Actuellement la tolérence de ce dispositif intra-veineux est jugée très bonne: on note aucune infection, aucune déchirure de la paroi, aucune migration et aucune thrombose extensive. Avec son faible volume et sa forme plate il est très difficilement palpable sous la peau et au niveau de l'aine il ne gène pas les patients.

359

BIBLIOGRAPHIE

Van Cleef JF., Valves in varicose veins and external compression studied by angioscopy; PHLEBOLOGY (1993) 8: 116-119.

Van Cleef JF., A vein has a preferential axis of flattening
J. Derm Surg. Onc. (1993) 19, 5, 468-470.

Van Cleef JF., Clip percutané endosaphénien
PHLEBOLOGIE (1994), n°3, 251-254.

FIG:1
Clip endosaphène et son cathéter porteur

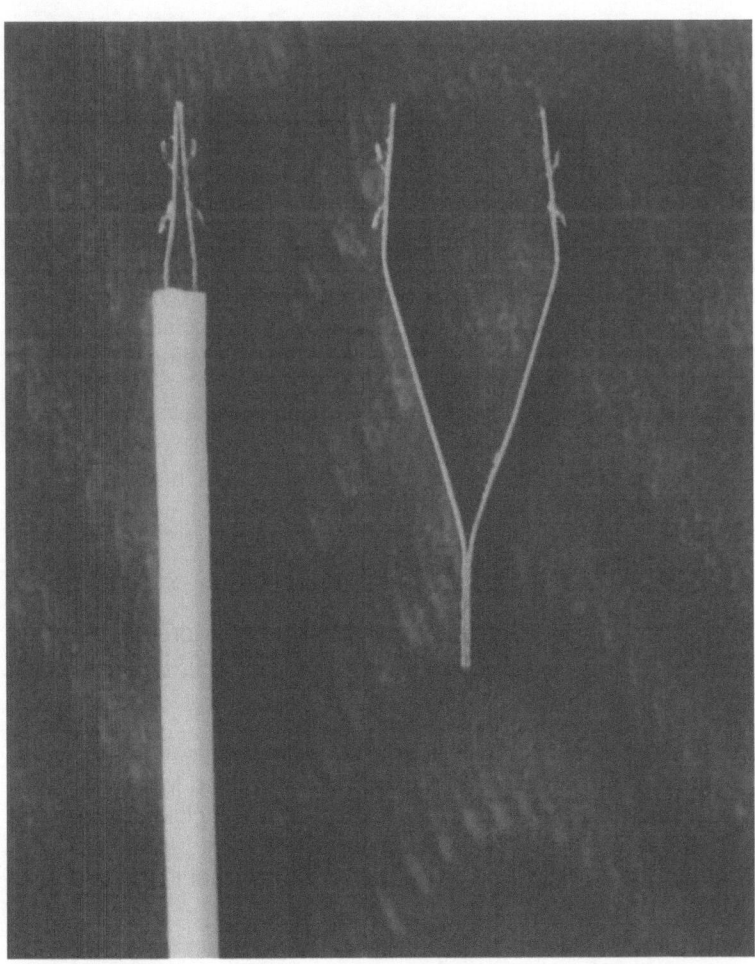

Phlebology '95, D. Negus et al. (eds.). Phlebology (1995) Suppl. 1: 360-361

V/1.3

Ecophlebectomy of Perforator Veins on Fat Patients (Video film)

J.F. Van Cleef

Phlebology Unit, N.D. de Bon Secours, Paris, France

In 1989 Knight R.M. et all. published "Echosclerotherapy" (1). Since Vin (2), Miserey (3), Schadeck, Cales, Ermini, Forestal, Grondin...and myself use this advisable practice.

For the cockett veins, Fischer, Hauer (4), Sattler or Conrad suggested the perivenous endoscopy performed in the cleft between the facia crutis and the flexor muscles.

With R. Muller (5) we have studied the phlebectomy technique, but sometimes it is difficult to find the vein with a hook trough a skin incision of 3 mm, especially if the vein is situated in a deep area like perforator veins of the thighs on fat patients.

With the help of the duplex it is easier to find one's way about.

Objective: to study a preliminary clinical experience of a new method of duplex guided phlebectomy.

Design: prospective investigation in single patient group.

Patients: 20 fat patients (20 thighs) with insufficiency of perforator veins and with a distance, between the surface of the skin and the vein, larger than 20 mm proved by duplex sonography.

Intervention: under local anaesthesia and after skin desinfection the probe (7,5-10 MHz) is packed with a plastic film, sterile gel and fixed just on the skin above the perforator, with a metal flexible. Through a skin incision of 3 mm a hook is passed. On the screen of the duplex we can see the perforator and the distal extremity of the hook. It is easier to hook the good vessel in the middle of the fat and to pull out the perforator vein.

Results: this technique is easy. The operator loses 5 minutes to put the probe, but afterwards the operator gains time and keeps his composure in finding easily the good vessel.

Conclusions: sometimes it is difficult to find a varicose vein at the level of the thigh on fat patients. With this technique, it is easy through a very small incision to pull out a perforator in a deep area.
With this technique we probably have less lymphatic and nerve lesions.
This technique is also interesting for the terminal portion of the short saphenous vein.

REFERENCES

1- Knight R.M., Zigmont J.A.- Echosclerotherapy- Proceeding of the Phlebology Society of America, 1989.
2- Knight R.M., Vin F., Zigmont J.A.- ultra sonor guidance of injections into the superficial venous system- Phlébologie 1989. A. Davy, R. Stemmer eds, 1989 J. Libbey Ldt pp. 339-341
3- Miserey G., Reinharez D., Ecalard P.- Scléroses sous échographie dans certaines zones à risques- Phlébologie, 1991, 44, n°1, 85-96.
4- Hauer G. - Operationstechnik der endoskopischen subfascialen discision der perforansvenen. 1987, Chirurg 58: 172-175.
5- Muller R. - Les varices de la cuisse, indépendantes de la crosse de la saphène interne- Phlébologie, 1993, 46, n°1, 117-122.

Phlebology '95, D. Negus et al. (eds.). Phlebology (1995) Suppl. 1: 362-363

OP/14.7

A Study of the Mechanisms by Which the Hemodynamic Function Improves Following Long Saphenous Vein-Saving Surgery

J. Hammarsten[1], P. Bernland[2], M. Campanello[1], M. Falkenberg[1], O. Henrikson[2] and J. Jensen[2]

[1] *Department of Surgery and* [2] *Department of Radiology, Varberg Hospital, Varberg, Sweden*

INTRODUCTION

Long saphenous vein-saving surgery is an interesting alternative to standard stripping for treatment of varicose veins[1,2,3,4,5]. The most apparent advantages of using long saphenous vein-saving surgery the way we perform it at our institution are: the long saphenous vein is preserved for future arterial reconstruction[4,5] and the operation creates less subjective postoperative discomfort[5]. The obvious drawback is the necessity of a thorough preoperative mapping of insufficient perforators.

However, the issue whether long saphenous vein-saving surgery can offer as good long-term results as standard stripping for long saphenous vein insufficiency is still controversial. Three reports have suggested that long saphenous vein-saving surgery can offer as good long-term results as standard stripping in patients suffering from varicose veins.[3,4,5] However, these results have not ben confirmed by other authors.[1,2]

Following conventional stripping procedures, the hemodynamic function is improved.[4,5] One reason is obvious - the leaking long saphenous vein has been removed. The present study deals with the question of why the hemodynamic function in our previous studies improves as much following long saphenous vein-saving surgery as following standard stripping.[4,5]

PATIENTS AND METHODS

20 patients, 14 women and 6 men, with a combination of perforator insufficiency and long saphenous vein insufficiency were investigated before and three months after long saphenous vein-saving surgery. The preoperative investigation included physical examination, strain gauge plethysmography, phlebography and measurements of the long saphenous vein diameter at four different locations using high resolution, real time ultrasound. Three months following vein-saving surgery, the patients were reassessed with a physical examination, strain gauge plethysmography and measurements of the long saphenous vein diameter.

RESULTS

All patients but one showed excellent or good results following surgery. The preoperative diameter of the long saphenous vein was reduced about 40 % at four different levels in the operated legs (p<0.01). The venous return time of the same legs increased 2.44 times (p<0.001). The decrease of the long saphenous vein diameter was positively correlated to the increase in venous return time (t-50), (r=0.50, p=0.04).

CONCLUSION

The results suggest that the development of insufficient perforators is an early major event in the formation of primary varicose veins. The results also support the hypothesis that the improvement of the hemodynamic function following long saphenous vein-saving surgery at least partly is due to a reduction of the long saphenous vein diameter, which in turn makes the previously insufficient valves of this vein sufficient.

REFERENCES

1.   Jacobsen BH. The value of different forms of treatment for varicose veins. Br J Surg 1979;66:182-4.
2.   Munn SR, Morton JB, McBeth WAAG, McLeish AR. To strip or not to strip the long saphenous vein? A varicose vein trial. Br J Surg 1981;68:426-8.
3.   Woodyer AB, Dormandy JA. Is it necessary to strip the long saphenous vein? Phlebology 1986;1:221-4.
4.   Hammarsten J, Pedersen P, Cederlund C-G, Campanello M. Long saphenous vein-saving surgery for varicose veins. A long-term follow-up. Eur J Vasc Surg 1990;4:361-4.
5.   Campanello M, Bernland P, Henrikson O, Jensen J, Hammarsten J. Standard stripping versus long saphenous vein-saving surgery for varicose veins. A prospective randomized study with the patients as his own control. Submitted for publication.

Phlebology '95, D. Negus et al. (eds.). Phlebology (1995) Suppl. 1: 364-366

OP/22.2

Standard Stripping versus Long Saphenous Vein-Saving Surgery for Primary Varicose Veins - A Prospective, Randomized Study with the Patients as Their Own Controls

M. Campanello[1], J. Hammarsten[1], C. Forsberg[1], P. Bernland[2], O. Henrikson[2] and J. Jensen[2]

[1] Department of Surgery and [2] Department of Radiology, Varberg Hospital, Varberg, Sweden

INTRODUCTION

The question whether long saphenous vein saving surgery can offer as good long-term results as standard stripping for long saphenous vein insufficiency is still unanswered. The matter is of clinical importance since the availability of an autogenous vein is of utmost importance in any patient who is going to be subjected to peripheral arterial reconstruction or coronary by pass surgery. It is therefore obvious that unnecessary surgical removal of the long saphenous vein should be avoided.

This study was undertaken in order to once more evaluate the long-term results of long saphenous vein-saving surgery compared with standard stripping using the patients as their own controls. This also made it possible to compare the subjective postoperative discomfort caused by the two surgical procedures.

METHODS

Eighteen patients with primary bilateral varicose veins were investigated. The material comprised 5 men and 13 women aged 32 - 72 years (mean $\pm$ SD = 51 $\pm$ 12). In the group of legs subjected to vein-saving surgery, three legs had superficial insufficiency, the rest a combination of superficial and perforator insufficiency. All legs subjected to stripping had a combination of superficial and perforator insufficiency.

The patients were prospectively randomized for stripping or long saphenous vein-saving surgery. The patients served as their own controls. The leg causing most discomfort was operated first. The other leg was operated using the alternative method. The preoperative investigation included a physical examination, strain gauge plethysmography and phlebography. After a mean follow-up period of four years, the patients were evaluated with clinical assessment and strain gauge plethysmography. The saved long saphenous vein was examined with ultrasound.

RESULTS

Table I shows the clinical assessment of the results of the operation at a mean follow-up period of four years. Excellent or good results were found at follow-up in all stripped legs. One of the legs subjected to vein-saving surgery had evidence of recurrence (n.s).

After the above follow-up period, VRT (t-50) was prolonged in both the vein-saving surgery group and in the stripped group compared with the preoperative values, while the preoperative and postoperative VRT (t-50) values did not differ between the groups (Table II).

Ultrasound examination of the preserved long saphenous veins in the high ligation group showed that the vein in all limbs was open, compressible, non-sclerotic and free of intraluminal echoes.

To sum up, the present study confirms our previos study that long saphenous-vein saving surgery can be performed with as good results as standard stripping, provided that insufficient perforators are thoroughly mapped preoperatively and ligated at surgery. The subjective postoperative discomfort inflicted to the patient by long saphenous vein-saving surgery was less than that caused by standard stripping. The preserved long saphenous vein can be used for future arterial reconstruction.

Table I.   Assessment of outcome

	Vein-saving surgery		Stripping	
	n	%	n	%
Excellent	14	77	17	94
Good	3	17	1	6
Poor	1	6	0	0
Worse	0	0	0	0
Total	18	100	18	100

Table II. Venous return time (t-50) before and at a mean follow-up period of four years after operation for varicose veins (Mean $\pm$ SEM)

	Preoperative values s	Postoperative values s	p<
Vein-saving surgery n=18	4.7 ± 0.6	9.9 ± 1.0	0.003
Stripping n=18	3.8 ± 0.3	10.3 ± 1.1	0.001
p<	n.s	n.s	

n.s=not significant

REFERENCES

1. Jacobsen B.H. The value of different forms of treatment for varicose veins. Br.J.Surg. 1979; 66: 182 - 184.
2. Munn S.R, Morton J.B, Mc Beth W.A.G and Mc Leish A.R. To strip or not to strip the long saphenous vein? A varicose vein's trial. Br.J.Surg. 1981; 68: 426 - 428.
3. Hammarsten J, Pedersen P, Cederlund C-G and Campanello M. Long saphenous vein-saving surgery for varicose veins. A long term follow-up. Eur.J. Vasc.Surg. 1990; 4: 361 - 364.
4. Woodyer A.B and Dormandy J.A. Is it necessary to strip the long saphenous vein? Phlebology. 1986; 1: 221- 224.
5. Holm J, Nilsson N.J, Scherstén T, Sivertsson R. Elective surgery for varicose veins. A simple method for evaluation of the patients. J.Cardiovasc.Surg. 1974; 15: 565 - 572.
6. Nylander G. Venography of the lower extremities. In: H.C Abrams, ed. Angiography 2nd Edition. Boston: Little Brown & Co, 1971; 1251 - 1271.
7. Hammarsten J, Bernland P, Campanello M, Falkenberg M, Henrikson O, Jensen J. A study of the mechanisms by which the hemodynamic function improves following long saphenous vein-saving surgery. Submitted for publication.
8. Blebea J, Schomaker WR, Hod G, Fow, RJ, Kempczinski RF. Preoperative duplex venous mapping: A comparison of positional techniques in patients with and without atherosclerosis. J. Vasc. Surg. 1994; 20: 226-234.

Phlebology '95, D. Negus et al. (eds.). Phlebology (1995) Suppl. 1: 367-369

OP/22.3

Comparison of Externally Banded Valvuloplasty to Interruption of the Saphenofemoral Junction by Clipping-in-Continuity for Isolated GSV Incompetence

L.L. Tretbar

Department of Surgery, University of Missouri-Kansas City School of Medicine, USA

INTRODUCTION

Few studies have compared different methods of correcting reflux at the saphenofemoral junction (SFJ) of the incompetent greater saphenous vein (GSV). When comparisons are made they usually report only the presence or absence of reflux as tested by augmentation methods. This study attempted to evaluate not only this function but the hemodynamics of the GSV in response to exercise, a more physiologic approach.

METHODS AND MATERIALS

For inclusion into this study the subjects had to demonstrate reflux limited to the GSV as determined by duplex ultrasonography and directional Doppler. There global venous functions were also evaluated with air-plethysmography (APG) and light reflection rheography (LRR). In addition the vein walls and junctional valves must have appeared normal with dilatation of the GSV to no more than 10 mm.

Twenty four limbs of 23 patients met these criteria and were randomly assigned for correction of reflux at the SFJ by clipping - in - continuity or by externally banded valvuloplasty. There were 21 women and 2 men whose ages ranged between 19 and 43 years included in this study. Symptomatic truncal varicosities were present from 3 - 21 years.

Blood flow was tested in the GSV 3 cm below the SFJ and in the low thigh. It was quantified by multiplying the mean velocity, as obtained by directional CW Doppler, by cross sectional area, as obtained by ultrasonography. A flat Doppler probe was secured over the appropriated vein site for exercise testing.

With the patient standing blood flow was augmented in the mid calf with a rapid inflation /deflation cuff at 80 mm Hg pressure. Exercise - induced blood flow was obtained by tiptoeing at 1 second intervals.

After randomization there were 14 SFJs clipped with tantalum hemostatic clips without cutting the vein. Ten patients had external banding of the incompetent SFJ valves using a commercially - produced device (Venocuff, Vaso Products, Inc. Somerville, NJ USA) which calibrates and secures a reinforced silicone rubber cuff around the vein.

Surgery was performed under local anesthesia as an outpatient. Residual varicosities were treated after 3 weeks by sclerotherapy or phlebectomy.

Patients were tested before and after surgery and followed from 18 - 24 months.

RESULTS

The figures show blood flow from a single pre - operative test and from multiple post - operative tests carried out every 3 - 6 months. The test results from all of the subjects are composited to provide a general view. It is obvious from the post - operative test results that there are marked variations of blood flow during the testing period. This is true for one individual from test to test and between the various subjects.

Figure 1 shows the pre and post - op flow at the SFJ in response to augmentation. Both groups have marked pre - op reflux. The clipped group of course has neither prograde flow nor reflux after treatment. On the other hand the valvuloplasty group showed enormous prograde flow. There was some residual reflux during an occasional examination however.

Figure 2 demonstrates blood flow in the low thigh also induced by augmentation. Except for 1 limb both groups have an equal amount of pre - op reflux. Both groups show a marked decrease in post - op reflux with many exams showing prograde flow. This suggests that correction of reflux by either method helps restore competence to the lower vein. Concomitant shrinkage of the vein wall was noted with the improvement of flow.

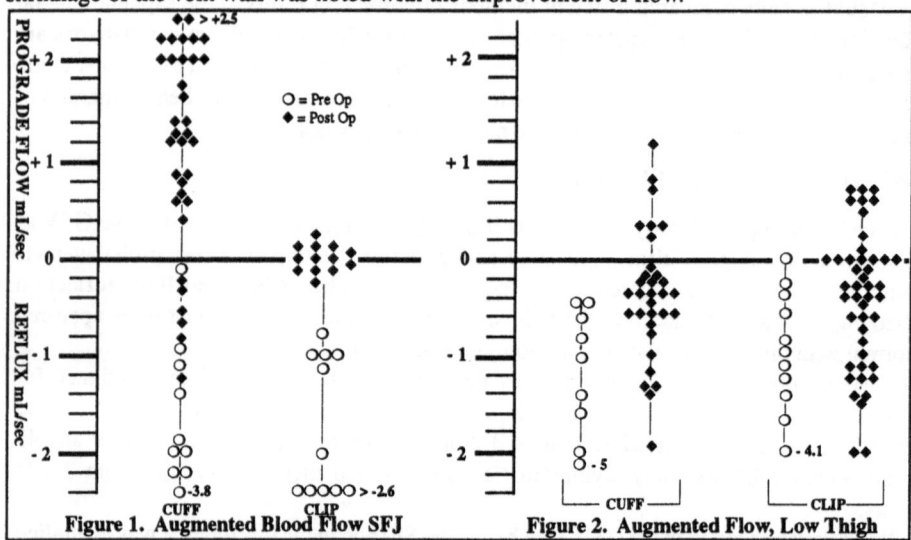

Figure 1. Augmented Blood Flow SFJ   Figure 2. Augmented Flow, Low Thigh

Figure 3 records the changes in blood flow in the low thigh in response to exercise. Pre- op flow ranged from zero to marked in both groups. Following surgery both groups also showed an equal improvement with some upward flow from time to time. The majority of tests show no or minimal reflux with exercise. Compared to the flow shown in Figure 2 there is greater prograde flow created with exercise than with augmentation.

Figure 4 records the most interesting data of the study. It shows flow induced by exercise at the SFJ. Obviously it includes only the valvuloplasty group inasmuch as there is no flow through the clipped SFJ. Although only 2 pre -op tests were performed the post - op results are astounding. Except for the immediate post - op period there was either no reflux or active upward flow. All limbs at one time or another showed prograde flow through the SFJ. This too showed variation in flow from one examination to another. Another amazing finding was that there might be low grade reflux in the low thigh during the exam while there was prograde flow through the SFJ.

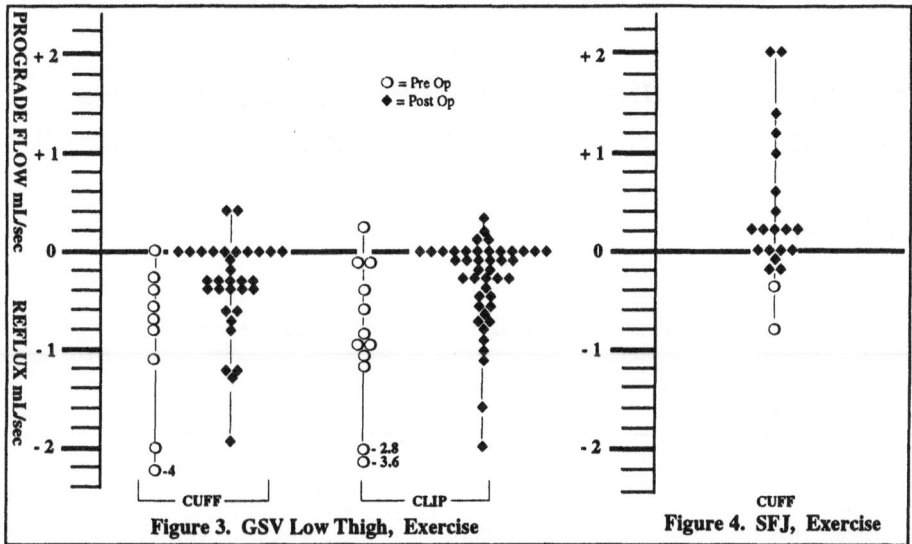

Figure 3. GSV Low Thigh, Exercise

Figure 4. SFJ, Exercise

Post - operative duplex scanning showed no evidence of thrombosis in any of the vessels even at the top of the veins which were clipped. All vessels decreased in diameter both at the SFJ and in the low thigh. No new sources of reflux were identified at the SFJ nor any new varicose veins developed during this brief study.

Global hemodynamic functions as determined with APG and LRR showed normalization in both groups with little if any difference between the two.

SUMMARY / CONCLUSIONS

These data show for the first time that the valvuloplasty is not just an anti - reflux mechanism but a method which does restore competence to the valve and lower vein. Not all of the valvuloplasty group were completely normalized but all showed flow through the SFJ with exercise. We consider this to be the supreme physiologic test. Clipping the SFJ was also an effective method of improving and restoring function to the GSV in the lower thigh.

It may be necessary to reevaluate whether augmentation can be considered the standard by which competence or incompetence of a vein is judged. For example we found reflux with augmentation but no reflux with exercise. This is particularly important as the discussion continues about the efficacy of preserving the GSV.

These data also raise a question about the method of SFJ ablation. Many other reports of high ligation and division show an unacceptable degree of thrombosis in the preserved GSV. It may be that clipping - in - continuity provides a less traumatic method than the traditional technic wherein the vein is divided.

Obviously greater followup is necessary to evaluate the durability of the valvuloplasty. At this time it is a useful method of restoring form and function to the incompetent GSV in this highly selected group of patients.

Phlebology '95, D. Negus et al. (eds.). Phlebology (1995) Suppl. 1: 370-372

OP/14.4

Hemodynamic Alterations in the Greater Saphenous Vein Before and After Externally Banded Valvuloplasty

L.L. Tretbar

Department of Surgery, University of Missouri-Kansas City School of Medicine, USA

INTRODUCTION

When treating varicose veins, preservation of structure and restorration of function has always been an objective for the surgeon. In recent years there has been an increasing interest in not only preserving the incompetent greater saphenous vein (GSV) but in restoring function to the valves at the saphenofemoral junction (SFJ). In most cases restoration of valvular competence has been obtained by narrowing the dilated valves with a plastic band, a procedure termed external banded valvuloplasty.

Most studies reporting the results of valvuloplasty record only the presence or absence of reflux as determined by augmentation technics. This study evaluates the hemodynamic changes observed in the incompetent GSV before and after reflux had been trreated at the SFJ by externally banded valvuloplasty.

METHODS AND MATERIALS

Ten limbs of 10 patients met the criteria for inclusion in the study; a duplex ultrasonographic examination which demonsrated a normal saphenous vein wall with a diameter no greater than 10mm, normal-appearing ostial or pre-ostial valves and reflux limited to the GSV alone.

Included in the study were 9 women and 1 man whose ages ranged between 19 and 43 years and whose varicose veins were present from 3 - 19 years.

Blood flow was tested in the GSV 3cm below the SFJ and in the low thigh. Flow was induced by augmentation and by exercise. With the patient standing flow was augmented in mid calf with a rapid inflation/deflation cuff set at a pressure of 80mm Hg. Tip toeing at one second intervals was the form of exercise use to create a dynamic type of flow.

Blood flow was quantified by multiplying mean velocity by the vein's cross sectional area. Velocity and direction of flow were determined by directional Doppler. A flat Doppler probe was secured over the appropriate vein for the exercise testing.

Each incompetent valve was banded with a commercially -produced device (Venocuff, Vaso Products Inc. Somerville, NJ USA) which secures a polyester-reinforced silicone rubber cuff around the vein. Surgery was performed as an outpatient under local anesthesia. The residual varicose veins were treated 4 - 6 weeks later. All patients have been followed from 18 -24 months.

RESULTS

Figure 1 shows the quantified blood flow at the SFJ as produced by augmentation of the calf. The pre-operative reflux marked in all but one. patient (5). This degree of rreflux is not unusual especially when large varices are included in the augmentation cuff. Their evacuation during compression provides a large, low resistance reservoir into which the augmentated blood can rapidly reflux. It also helps explain the large post-operative prograde flow before the varices have been treated. Once they have been abolished the enormous prograde flow decreases to a level of about 1 mL/sec. It is noted that patient (2) demonstrates some residual reflux although it is markedly diminished.

Figure 2 shows blood flow at the SFJ induced by exercise. An interesting finding in patient (2), who had persistent reflux with augmentation, is the vigorous prograde flow through the valve with exercise. Although only two patients had pre-operative exercise testing at the SFJ all but two have demonstrated upward prograde flow through the banded valve. The diversity in the flow pattern is similar to that seen in the SFJ of normal subjects. It is also noted that the flow varies rather markedly form one examination to another a response also seen in normal limbs.

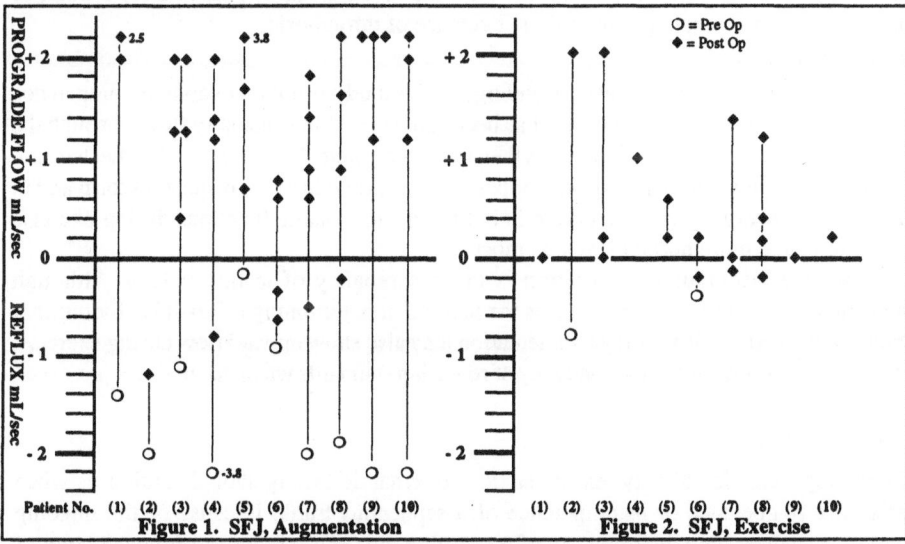

Figure 1. SFJ, Augmentation   Figure 2. SFJ, Exercise

Figure 3 demonstrates blood flow through the GSV in the low thigh. The large reflux patterns are typical for veins with a larger diameter and large varicosities. All but patient (1) had a diminution of reflux after treatment. Although valvular competence was restored at the SFJ the lower GSV did not compensate as well as the others and allowed reflux with augmentation. Half of the limbs did however show mean upward flow with compression of the calf.

Figure 4 records the changes in flow induced by exercise in the low thigh. Reflux actually increased in a few patients immediately after surgery but returned to pre-treatment levels. Although patients (2) and (10) continued to reflux in the low thigh they can be seen to have priograde flow through the SFJ.

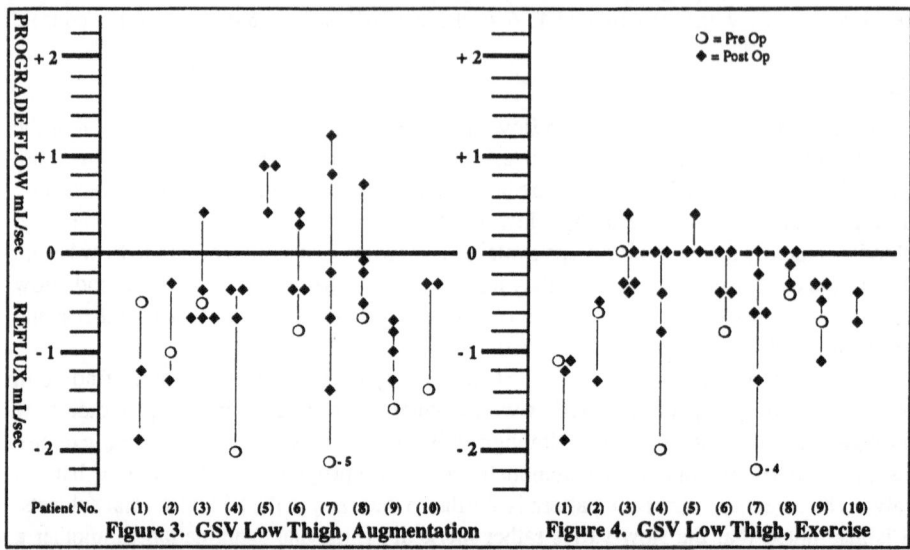

Figure 3. GSV Low Thigh, Augmentation   **Figure 4. GSV Low Thigh, Exercise**

Post-operative sonographic evaluation show all of the treated veins to have shrunk in diameter. They all remain patent with no evidence of thrombosis.

DISCUSSION

Probably one of the most interesting findings of this study is not that valvular competence can be restored, but for the first time it has been shown that blood actually flows through the SFJ during exercise. The valvuloplasty is then something more than an anti-reflux mechanism. In many of the limbs the restored GSV appears to be normal in every respect; in form and in function. Not every vein works perfectly but the hemodynamic functions during exercise have returned to more normal levels in most..

The second fascinating observation is the descrepancy of testing values. Although augmentation has always beend used as the method of determining reflux it is obvious that a vein which might reflux during augmentation may also show upward flow during exercise. This raises a question about the adequacy of reporting standards when discussing a preserved saphenous vein.

CONCLUSIONS

The discrepancies found between the static and dynamic testing should caution us when describing competence or incompetence of a saphenous trunk. Because of the ongoing controversy concerning the propriety of preserving the GSV a more careful evaluation of the vein is mandatory. Simple augmentation testing does not appear to be adequate in every case to fully evaluate the potential funtion of the vein.

Continuing evaluation of this small and rather select group of patients is essential to test the durability of function.

Phlebology '95, D. Negus et al. (eds.). Phlebology (1995) Suppl. 1: 373-375

PI/9.3

Correction of Reflux at the Saphenofemoral Junction by Clipping-in-Continuity: A 6 Year Follow Up in 476 Patients

L.L. Tretbar

Department of Surgery, University of Missouri-Kansas City School of Medicine, USA

INTRODUCTION

The controversy of preserving the incompetent greater saphenous vein (GSV) is far from being resolved. The newer, less-traumatic stripping methods have renewed interest in ablative approaches for the correction of reflux at the saphenofemoral junction (SFJ). However phlebologists agree that there are many GSVs which do not need to be sacrificed for proper control of truncal varicosities. As is always the case in this kind of situation the problem is one of selection and individualization of treatment.

This study has attempted to identify which GSVs can be preserved and which GSVs should receive more than simple clipping of the SFJ for adequate control of reflux. It also compared the results of SFJ obliteration by the use of hemostatic clips, without transection, to the traditional method of ligation and division.

METHODS AND MATERIALS

All patients seen during the 6 year period 1989 - 1995 who presented with truncal varicosities due to GSV incompetence at the SFJ were included in the study. Although patients with incompetence due to previous DVT were treated in a similar fashion their results are not included in this report.

All were evaluated preoperatively with directional CW Doppler and some with duplex scanning. Significant reflux at the SFJ was defined as that persisting for 1 second or longer. Reflux lasting from 1/2 - 1 second was included if the velocity was 10 cm/sec or greater.

Followup evaluations included duplex scanning to help evaluate recurrent reflux at the SFJ, the presence of incompetent perforating veins and the character and patency of the GSV. The development of new varicosities was particularly noted.

Surgery was performed as an outpatient under local anesthesia. The SFJ was interrupted with 3 - 4 large hemostatic clips and the tributaries with smaller clips. The GSV was not divided but left intact. The exception was for large SFJ varices which were overclipped or removed. Associated truncal varicosities were treated at a later date with sclerotherapy or phlebectomy.

Included in the study were 476 limbs of 408 patients. There were 302 women and 106 men whose ages ranged between 19 and 78 years. Varicose veins had been present from 3 - 53 years and measured up to 2.5 cm in diameter. The incompetent SFJ measured from 6 mm - 2 cm in diameter.

The data presented here report the findings from the evaluation of all the 136 limbs treated from 5 - 6 years and a random assay of 156 limbs treated between 1 and 5 years. Reoperative data are for the entire patient population of 476 limbs.

RESULTS

General. There were no major complications from treatment, e.g., death, infection, DVT or prolonged disability.

Reoperation. A second operation was required in 4 patients for refluxing tributaries not identified at the original procedure. Reoperation was also required in 3 patients in whom only 1 clip was applied. After this experience in the first 10 patients 3 or more clips were applied to the vein and a more thorough search made for tributaries at the femoral junction.

Reflux at SFJ. Low grade reflux, i.e., less than 5 cm/sec velocity or less than 1/2 second duration, has been observed in 20 % of all patients. These limbs have not been associated with a higher frequency of new varicose vein development when compared to those without reflux. The exact source of this reflux has not been identified. Major reflux causing varicosis was found in the 4 patients mentioned above who required reoperation. Major reflux was also found in the clipped tributaries but was not associated with an increase of varicosis. There was very little other reflux at the SFJ causing major varicosis.

Reflux in the low thigh. Of the group of 6-year limbs 70/136 showed no reflux in the GSV in the low thigh when tested by augmentation. This group consisted primarily of pregnancy-related varicosis and those with a SFJ measuring 8 mm or less in diameter. Complete competence varied from examination to examination. This group also developed the fewest new varices, about 4 % over the course of 6 years.

Reflux in the low thigh due to incompetent perforating veins presented in only 5 limbs and was corrected by excision of the perforator.

New varicosis / diameter of SFJ. Saphenous veins measuring more than 1 cm at the time of surgery demonstrated significantly greater new varicosis than smaller ones. 8 % of these limbs required repeat sclerotherapy or phlebectomy. Almost all of the patients have returned for treatment of smaller non-truncal varices. This represents the basic nature of the venous disease.

New varicosis / duration of disease. The diameter of the vein and the duration of disease are directly related. Almost all of the patients mentioned above had their disease for 15 years or more; most were men. Four men have had a limited invagination stripping of the GSV to limit the large volume reflux in the low thigh causing extensive, recurring varicosis. The reflux here did not come from incompetent perforators or the SFJ but from large lateral and medial saphenous branches.

Fate of GSV. Ultrasonographic evaluation of the preserved saphenous vein showed complete patency of all veins. There was no thrombosis found, even in the highest segment between the clips and the first perforator. 80 % of the veins shrank in diameter, especially those with a smaller diameter to start.

CONCLUSIONS

Although these data are not presented for statistical analysis the following observations seem pertinent:

♦ Clipping - in - continuity is a safe and effective method of correcting reflux at the SFJ.

♦ The preserved GSV remains patent without thrombosis using this method of venous interruption.

♦ Judging from the absence of thrombosis in the GSV and the minimal degree of new reflux at the SFJ there must be a basic difference between the technics of clipping - in - continuity and the traditional ligation and division.

♦ Between 4 - 8 % of all patients will develop new varicose veins.

♦ The shorter the duration of varicose disease and the smaller the GSV the lesser the development of new varicosis and vice versa.

♦ Men with a vein measuring more than 1 - 1.5 cm in diameter with a duration of disease of 15 years have the greatest degree of varicosis development. This group might be considered for a limited stripping of the GSV, or at least interruption of the GSV at the knee along with SFJ interruption.

♦ The development of incompetent perforating veins along the course of the preserved GSV was a minor cause of new varicosis. This argument for stripping the GSV does not appear valid.

♦ The development of new reflux at the SFJ was also a minor cause of new varicosis. This argument for stripping also appears invalid.

Phlebology '95, D. Negus et al. (eds.). Phlebology (1995) Suppl. 1:376-378

P105

Haemodynamic Correction of Proximal, Non-Ostial, Refluxes of the Greater Saphenous Vein. Preliminary Report of 5 Cases

A. Vannuzzi[1], A. Pieri[1], A. Duranti[2] and S. Innocenti[3]

[1] *III Cardiology Unit - Diagnostic Angiology Module,* [2] *General and Vascular Surgery Unit - Az. Osped. CAREGGI,* [3] *II Surgery UnitUSL 10/H FIRENZE (I)*

Sapheno-femoral junction (SFJ) incontinence is generally held to be the only cause in the genesis of the varices of the greater saphenous vein (GSV). Recent observations (1,2) have denied such an hypothesis, showing that sapheno-femoral junction continence is still present in a large percentage of varicose patients (3,4,5).

Primitive incontinence of the second valve (pre-ostial valve) of the GSV (fig 1) occurs as the main cause of the varices in up to 38% of the patients (6).

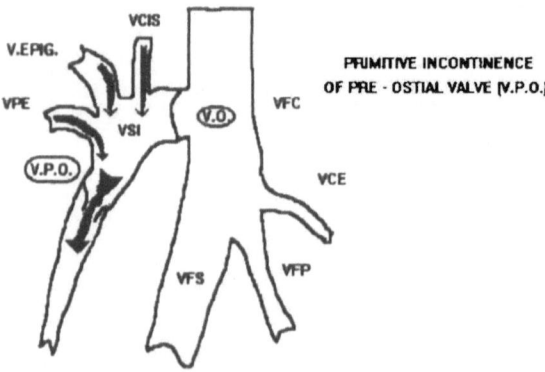

Fig. 1 - Genesis of varices of the greater saphenous vein (VSI) in the cases of sapheno-femoral junction (V.O.) continence and primitive incontinence of pre-ostial valve (V.P.O.).

Further ultrasounds (U.S.) investigations of the GSV allowed to understand that most of the varices at thigh level are not related to saphenous trunk but to its' afferents, because the middle saphenous valve (a constant valve of the saphenous trunk, located at the crossing of GSV with medium abductor muscle) is generally continent, and that the GSV at calf level is often not involved in varicose disease.

Such observations allow to understand that a demolitive intervention of the GSV (stripping) can be avoided in many patients, after an accurate haemodynamic study.

Recent interest in conservation of the GSV for by-pass surgery supports the authors' aim.

A sure and correct diagnosis of primitive incontinence (or agenesia) of the pre-ostial valve is now possible by Color Doppler examination of SFJ that shows the normal continence of the first saphenous valve and the incontinence of the second one (fig. 2).

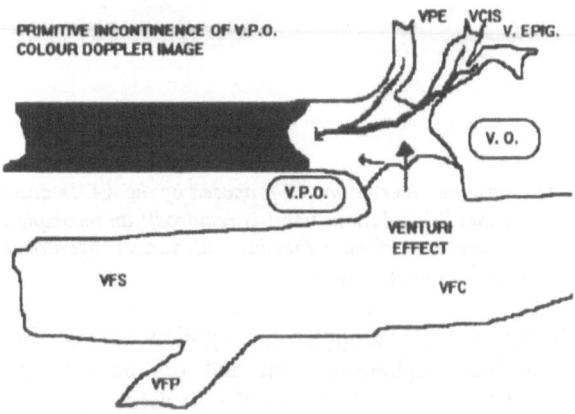

Fig. 2 - Color-Doppler imaging of primary incontinence of the pre-ostial valve (V.P.O.) of the greater saphenous vein.

Color-coded reflux appears only beneath pre-ostial valve and sometimes only a tiny reflux, of very short duration, passes through the ostial valve owing to Venturi's suction effect on valvular edges. Color-Doppler imaging of SFJ incontinence is quite different: a great and prolonged amount of color-coded reflux, coming from common femoral vein, passes through ostial valve.

These observations lead to consider the opportunity of an aethiological intervention aimed to preserve GSV's function.

The AA. propose therefore a new haemodynamic interventional technique in the correction of this kind of pathogenesis (fig 3) in accurately selected cases.

The AA report the preliminary results of the first 5 cases (follow up >6 months).

Operations were performed under local anesthesia. Mean duration was less than two hours. All the 5 patients immediately walked up after surgery with a second class compression elastic stocking.

In these cases the simple ligature of the crosses' afferent veins (preserving the SFJ that was continent) and of the varices lead to the disappearance both of

378

varices and symptoms, according to the hypothesis, and it was followed by a full recovery of the saphenous function.

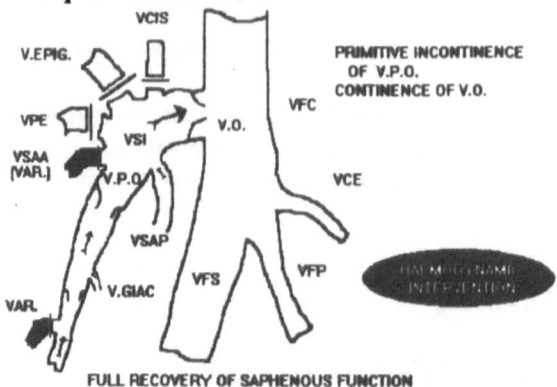

Fig. 3 - The new haemodynamic intervention as proposed by the AA. in accurately selected cases (the continence of the middle saphenous valve is required): an incomplete crossectomy, saving sapheno-femoral junction, is performed together with varices' ligature. A full recovery of the greater saphenous vein function is achieved.

After surgery Duplex or Color Doppler examination showed the suppression of the reflux in proximal saphenous vein and the absence of thrombotic complications. Pain and symptoms relief were always achieved.

Complementary sclerotherapy or Müller phlebectomy of collateral varicose veins, were sometimes necessary to get a fully aesthetical result.

Six months follow-up was quite satisfactory both for functional and aesthetical results.

REFERENCES

1 - Pieri A., Vannuzzi A., Duranti A., Vin F., Benelli L., Michelagnoli S., Caillard Ph., De Saint Pierre G.: Rôle central de la valvule pré-ostiale de la veine saphène interne dans la gènese des varices tronculaires des membres inférieurs. Démonstration diagnostique par echo-doppler couleur. Phlébologie 1995; (48), 2: in press.

2 - Pieri A., Vannuzzi A., Duranti A., Caillard Ph., Marcelli F., Michelagnoli S., Vin F.: Physiopathologie de la valvule pré-ostiale de la veine saphène interne. Angiologie 1995; in press.

3 - Franco G.: Echo-Doppler couleur et exploration veineuse superficielle. - Phlébologie 1994; 47 (1): 63-75

4 - Sarin S.: Sapheno-femoral junction recurrence after surgical ligation.:Abstracts Book of the 7th annual congress of the North American Society of Phlebology Maui (HAWAI) 21-23 Feb. 1994

5 - Campbell W.: Sapheno-femoral reconnection in recurrent varicose veins.: Abstracts Book of the 7th annual congress of the North American Society of Phlebology Maui (HAWAI) 21-23 Feb. 1994

6 - Pieri A., Vannuzzi A., Duranti A.: L' incontinence primitive de la valvule pré-ostiale dans la gènese des varices: demonstration par echo-doppler couleur. J Mal Vasc 1994; 19 (suppl. B): 246

Phlebology '95, D. Negus et al. (eds.). Phlebology (1995) Suppl. 1: 379-381

P208

Treatment of Chronic Venous Insufficiency in Patients with Severe Cardiac Problems

A. Faenza, S. Faenza, S. Selleri and F. Catena

Department of Surgery and Transplantation - University of Bologna, Italy

INTRODUCTION

Patients with Chronic Venous Insufficiency (CVI) on anticoagulant therapy for cardiac problems (artificial valves or arrhythmia for example) often are treated only with elastic stockings and the indication to avoid the standing position as much as possible because the correction of the CVI with a stripping needs the restoration of a normal clotting time which is unsafe and appears unadvisable for a small problem .
Since people with artificial heart valves can lead a normal life if the CVI is severe this becomes the main limit to a normal life style .
In case of chronic non treatable cardiac failure , swollen legs are part of the disease but if the patient has also a CVI , large ulcers can develop sometimes causing more pain than cardiac failure itself . In those cases many doctors think that even a compression bandage should be avoided for the risk of pulmunary edema [1].
The aim of this paper if to demonstrate that an active treatment in both types of patients is possible , advisable and safe and to present the techniques used and their limits .

MATERIALS AND METHODS

Group I : 8 patients (11 legs) No or mild cardiac failure but chronic anticoagulant therapy (7 for artificial heart valves and 1 for multiple pulmonary embolism) .
8 legs had normal deep veins but in all there were a troublesome edema with trophic changes in 4 . Three had postphlebitic changes with two painful ulcers
All these cases had a flush ligation of the incompetent saphenous veins under local anesthesia without modifying medical treatment .Sclerotherapy of the remaining varicose veins , including the saphenous veins below the ligation , was than performed and repeated when necessary to complete and mantain the eradication of the varices .
Group II : 4 patients with severe cardiac failure all over 70 years old .
Case 1 Congenital untreated interatrial septal defect with pulmonary hypertension and tricuspid regurgitation .She had a pulsating internal left saphenous vein with a 20 years old ulcer which recently became so painful to oblige the patient to keep the leg in elevation and bedridden..She had a short stripping of the internal saphenous vein under light general anesthesia and compression bandages .

Case 2 Aortic stenosis with a secondary tricuspid insufficiency. He had bilateral leg ulcers larger and more painful at the left which opened 4 years ago .
Recently he wasn't able to walk for the pain . He had a saphenous vein ligation under local anesthesia followed by compression sclerotherapy
Case 3 and 4 had advanced cardiomiopathy and where on chronic anticoagulant therapy .. Case 3 had bilateral postphlebitic legs with painful ulcers and he too underwent a saphenous vein ligation uder local anesthesia and compression sclerotherapy .
Case 4 had only compression sclerotherapy
In all the patients of this group, the swelling of the legs progressively decreased with frequent repeated applications of pneumatic devices mantainig them seated because they did not tolerate the recumbent position .Between two subsequent sittings the temporary result was mantained with normal low elasticity bandages .
The ligations under local anesthesia were always performed in the semi seated position and therefore they were made as high as possible but not flush. : the technique was considered safe because they all were on oral anticoagulants.

RESULTS

Group I One patient bled at the site of the saphenous vein ligation and it was necessary to stop anticoagulants and he was reoperated ..
None had problems with sclerotherapy
The two ulcers healed and did not recur . All the patients are symptom free after 2 to 5 years but in all the legs a short period of repeated sclerotherapy was necessary
Group II Case 1 The pain disappeared after the treatment but the patient died one month later of pneumonia and cardiac failure. Case 2. The right ulcer healed but not the left which became painless in two months and decresed in size but did not heal within the two and a half years of survival during which he could lead a nearly normal life .Case 3 Is the best result . Both ulcers healed and the patient resumed his work as an architect
.
Case 4 Survived one year .Two of three ulcers healed and the patient was and remained pain free after three months of treatment .

DISCUSSION

Group I We have already reported the possibilty of an active treatment of CVI in patients with artificial heart valves mantaining anticoagulant therapy [2] .
The increased experience confirms the data : the single case of bleeding obliged to stop the anticoagulants for three days which seems to be a limited risk .
Chronic anticoagulation does interfere with sclerotherapy : the quantity and concentration of the sclerosing agent must be increased and varicose veins recurrence appears more likely but the final result is nevertheless good and in all of our cases it has been mantained with a few repeated injections once a year .
Group II In patients with chronic untreatable heart failure leg swelling reduction with compression bandages or mechanical devices can induce a pulmunary edema .
A slow progression of this treatment can abolish the risk and the advantage for the patient of a "dry leg" is often dramatic . Sclerotherapy and the ligation of the large veins are possible in these patients preferably under local anesthesia : this obliged us to tie the saphenous vein below the sapheno - femoral junction without complications .
In all the 4 patients the healing or the simple improvement of the ulcers which followed the treatment was accompanied by an early disappearance of the pain with the possibilty of resuming a better life style .

REFERENCES

1   Wiehert V Vansheidt W Rabe E et al   Leg ulcers due to venous insufficiency
     In Leg Ulcers Westerhof W Editor   Elsevier   Amsterdam 1993 : 108-118
2   Faenza A Faenza S Selleri S et al   Place of surgery and sclerotherapy in the
     therapy of chronic venous insufficiency In Davy A Stemmer R   Editors
     Phlebologie 89 John Libbey Eurotext   London Paris 1989 : 830-832

Phlebology '95, D. Negus et al. (eds.). Phlebology (1995) Suppl. 1: 382-384

P209

External Valvuloplasty of the Preterminal Valve of the Long Saphenous Vein. Result After 5 Years

G. Botta[1], S. Massi[1], C. Ruggiere[1], R. Intermite[1], G. Vedovini[2], M. Bicchi[2] and S. Mancini[1]

[1] III Department of General Surgery and Pabisch Center of Phlebology, [2] Institute of Medical Semeiotics, Faculty of Medicine and Surgery, University of Siena, Italy

INTRODUCTION

The anatomical studies carried out by Edward (1937-1940) and Cotton (1961) on 200 veins removed, mainly from the lower limbs, of patients suffering with varicose veins revealed the following:

1) structural alterations in the valves were never found, whereas alteration in the nearly parietal tissue were costantly found.

2) the only valve which is costantly present is the preterminal valve (98.1% compared with 93.4% for the ostial valve).

Psaila and Melhuish (1989) compared normal and varicose veins, finding in the latter an higher elastic fibre content and a smaller amount of collagen, wich explains the lower resistance of the vein walls to stretching forces. In the normal veins both the perivalvular and non valvular walls were analysed. It was established that the perivalvular wall had a lower resistance than the non-valvular wall, calculated to be 50% lower.

Recently Curri (1992) examined the microcirculation and confirmed that the varicose desease depends not so much on an initial alteration of the valves but more so on alteration of the wall, particularly perivalvular wall.

If this is true, for those patients in the initial stages of the desease it may be concluded that we can stop the pathological reflux by correction of the ectasia of the parietal-valve complex.

MATERIALS AND METHODS

The external valvuloplasty of the preterminal valve of the long saphenous vein is a surgical technique aimed to stop the reflux, that is caused by an insufficient sapheno-femoral junction, leaving in situ the long saphena.

The operation, performed with local anaesthesia, consists of bandagin with autologous (ring or peduncle) or prosthetical material around the preterminal valvular complex. This is than tightened around the vein until the edges of the valve come closer together, regaining the valvular continence at the doppler control. Finally it is firmly attached to the perivein by stitches.

The operation is proposing in all cases which the long reflux does not exceed the lower third of the thigh.

Our technique was undoubledly developed in such a way that the anatomical composition of the junction remains unaltered, but in the first operated cases we had prefer stop the collaterals for two reasons:

1) in the eventuality that a new reflux was revealed after a period of time, we had to be sure that this must proceeded to an alteration in the continence of the valve, or to a perforant lower down which did not allow the collateral circles to be fed from above.

2) in case conditions were created which did not allow the valvuloplasty to remain in situ (eg. thrombosis), the stripping of the long saphenous vein would be much quicker and simpler.

RESULTS

At the Center of Phlebology "W. Pabisch" of the University of Siena, directed by Prof. Sergio Mancini, from march 1989 to march 1994 31 varicose patients has been subjected to surgical intervention of external valvuloplasty of the preterminal valve of the long saphenous vein.

In the first 3 cases cribrose ring bandages were used and the reflux reappared after two months. An echograpic aspect of the last section of the saphenous vein revealed that the diameter of the vein had returned to it's previous dimention. This signified that the external valvuloplasty had subsided, probably due to a vascular sufference, which brought necrosis with time.

This first partial success urged us to experiment with a vascular peduncle taken from a cribrose strip, which was used as the plastic. This new idea was tried on four patients, however it did not resolve the problem, but did make a significant improvement. In fact, these four patients after approximately 2 months of having received the operation showed a reflux which slowly reduced until it was eventually present only at the start of the test of Valsalva. This ultrasonographic aspect, present in all cases, was explained by a partial suffering of the more distal part of the peduncle (this intending the part fixed to the perivein). In effect total coliquification of the plastic was not produced, as in the previous operations, only a relaxing of the binding around the valvular complex. Occurred in actuality, the part of the cribrose bandage which was not affected by suffering, remained attached to the walls creating, with time, a perivenous sclerosis capable of returning partial control to the valvolar apparatus.

Afterwards we have rested the use heterologous material obviously without problems about vascularity. We have choosed first of all a segment of PTFE and after of Dacron 10-15 mm lenght and with a diameter of 4-5 mm, applied using the same method. The 24 patients, till now operated all with local anesthesia, were discharged on the day of the operation and they don't present postoperative complications. The clinic and instrumental control, variable since 1 to 5 years, with a Doppler examen shows a completely valvular continence in the 62.5% of cases. So the strain-gauge plethysmoghraphy has proved a condition of normality in 70% of controled patients, with a moderate superficial venous insufficiency in the remaining 30%. Howewer the varicose sympthomatology and recurrence is absent in everybody.

CONCLUSIONS

The results obtained emphasizes the necessity for a rigid selection of the patients, with precise indications to the intervention and a perfect technical execution of the same. Morevor a rigid follow-up program is necessary for early detection of failure of the operation and for rapid treatment of eventual recurrent varices.

REFERENCES

1. Basmajan JV. "The distribution of valves on the femoral external iliac and common iliac vein and their relationship to varicose veins". Surg Gyn Obst 1952, 95:537-541.
2. Bihari I. "Saving the greater saphenous vein for prothesis, with incomplete ligation, in varicose cases". In: A. Davy, R. Stemmer (Eds), Phlebologie 1989, John Libbet Eurotext, Paris 1989; 1018-1020.
3. Camilli S. "La venoplastica esterna nella terapia dell'insufficienza valvolare venosa profonda primitiva. Risultati a distanza in 21 casi operati." Cardiovasc.Surg. 1988, 2:83-88.
4. Curri SB. Significato delle alterazioni anatomo-funzionali dei "vasa venarum" nella patogenesi del danno parietale venoso e della malattia varicosa. Flebolinfologia, 1991; 2/2:23-42.
5. Curri SB. The role of vasa venarum in the genesis of varicose veins. In P.Raymond-Martinbeau et al. (Eds), Phlebologie, 92, John Libbet Eurotext, Paris 1992; 38-40.
6. Mancini S, Mariani F. "La valvola esterna della valvola pre-terminale della vena grande safena. Follow-up dei pazienti operati". Atti 1° Congresso Nazionale Società Italiana Flebolinfologia, Cortona sett.1990.

Phlebology '95, D. Negus et al. (eds.). Phlebology (1995) Suppl. 1: 385-387

P106

The Short Stripping Saphenectomy in the Varices of the Lower Limbs

G. Botta, S. Massi, C. Ruggieri, R. Intermite, St. Mancini and S. Mancini

III Department of General Surgery and Pabisch Center of Phlebology, Faculty of Medicine and Surgery, University of Siena, Italy

INTRODUCTION

New techniques have been proposals in the last years in the surgery of the varicose veins of the lower limbs. Some techniques of conservative type, as the external valvuloplasty of the sapheno-femoral junction, are set the objective of take care of the varicose veins, eliminating the long reflux, so that prevent the progressive parietal expansion of the long saphenous vein, that comes left in situ. Other techniques, as the haemodynamic surgery, are maintaining the long saphenous vein in situ, eliminating the reflux equally and transforming the points of escape in points of reenter of the blood in the deep circulation. Other techniques, as the laser or the cryotherapy, propose to reduce the own complications of the intervention of saphenectomy for stripping, as haematomas and nervous injuries.

To this Holme intention [1] it bring again areas of anesthesia or of postoperative hyperaesthesia in the anteromedial region of leg in the 39% of the patients subjected to long stripping and in the 7% to short stripping immediately under of the knee. In the experience of Agrifoglio [2] with the long stripping the persistence of paraesthesia or perimalleolar anesthesia is reported in the 16,3% of the cases at distance of three years from the intervention.

But is the long stripping always necessary to apply ? The response according to our experience is negative for the physiopathological and clinics considerations [3]. In fact the long saphenous vein in more of the 70% of the patients that arrived at to our observation has interested from the long reflux only to the third superior of leg, to the junction of the Boyd's perforans with the collateral anterior, posterior and arciform veins.

Moreover the sapheno proximal trunk to the alleged crossroad is isolated as haemodynamics and not interested from the medial perforans veins, that instead connects the arciform vein of Leonardo with the venous deep system.

In third place the long saphenous vein in his proximal trunk is situated in a splitting of the band at narrow contact with the saphenous nerve, for which the possibility of peripheral neurophaties after long stripping is increased.

MATERIALS AND METHODS

From 1985 to 1994 at the Center of Phlebology "W. Pabisch" of the University of Siena, directed by Prof. Sergio Mancini, 1010 varicose patients were subjected to saphenectomy for stripping. Of them 731 is female (72,4%) and 279 males (27,6%), with middle age of 47 years. Of these 1010 patients 710 (70,3%) has been operated of inside saphenectomy for short stripping, 231 (22,9) of inside saphenectomy for long stripping, and 69 of external saphenectomy All has been subjected to anamnestic-clinical examination and to instrumental investigations, as Doppler c.w, Echography and more recently Echocolordoppler. In some selected cases also an ascending phlebography has been effected.

The intervention has been performed usually in hospital day regimen and more recently in ambulatorial regimen, using the block in the best part of the cases. The surgical times of the saphenectomy for short stripping are those ancient. Performed the crossectomy and the ligation of the sapheno-femoral junction, the long saphenous vein is cannulated from the top toward the lower part in the same direction in which take place the stripping. The distal contraincision, 10-15 mm. small, is practiced to level of Boyd's perforans. For the extraction of the long saphenous vein a stripper with the smallest ogive is used. It is retired after the phleboextraction with the technique of the towing and it come out the inguinal region by the same incision. The intervention is completed with phlebectomy of the varicose collateral trunks and with ligation and section of the incontinent perforans veins. Always a compressive bandage of the limb take place after the intervention. Foreseeable it is the mobilization of the patient, once exhausted the anesthesia.

RESULTS

For as concern the immediate results, complication as haematoma has been appeared in the 0,5% of the patients. Paraesthesia and transitory anesthesia of the anteromedial region of leg has been appeared in the 6% of the patients after short stripping and in the 15% after long stripping. In the 4% of the patients operated of long stripping we had observed peripheral irreversible neuropathies for injuries of the saphenous nerve, instead they don't appear in the patients subjected to intervention of short stripping.

At the follow-up from 1 to 9 years the patency and the absence of reflux in the residual saphenous vein has highlighted with clinical examination and ultrasonographic investigations in all the patients subjected to saphenectomy for short stripping. The percentage of varicose recurrence to the leg (19%) for appearance of new incontinent perforans veins is similar to that of the long stripping (21%).

DISCUSSION

Is very important a precocious diagnosis of the varicose disease, because the presence of a vein saphenous reflux with normal valves consents to effect surgical conservative interventions, as the external valvuloplasty of the preterminal valve, adopted by the our School now from years and with acceptable results.

For the more advanced stages, with more extensive vein saphenous reflux, the validity of the saphenectomy for stripping remains intact. In this case we prefer perform a short stripping, because it involves a less surgical trauma and allows a better check of the complications, above all neurological, while the intervention preserve a character of radicality To that must be added the necessity of preserve for as possible the venous patrimony of the subject in prevision of a possible use of the long saphenous vein for interventions of peripheral or myocardiac revascularization.

All that has brought to a progressive reduction of the technique of saphenectomy for long stripping until to arrive to our more recent experience to an annual percentage of patients subjected to short stripping equal to 79 % of all cases.

REFERENCES

1. Holme JB, Skajaa K, Holme K. Incidence of lesions of the saphenous nerve after partial or complete stripping of the long saphenous vein. Acta Chir Scand 1990; 156: 145-9.
2. Gabrielli L, Martelli E, Corsi G, Falautano M, Perduca R, Agrifoglio G. Chirurgia delle varici degli arti inferiori mediante metodica CHIVA: risultati e prospettive. Flebologia, 1993, 4:81-85.
3. Ruggieri C, Massi S, Peccatori A, Mariani F, Mancini S. Lo stripping corto della vena grande safena: Follow-up dei pazienti operati. Atti III Congr Intern Soc It Flebolinfologia, Ferrara settembre 1991.

Phlebology '95, D. Negus et al. (eds.). Phlebology (1995) Suppl. 1: 388-390

The Recurrent Varicose Veins of the Lower Limb

G. Botta, S. Massi, C. Ruggieri, St. Mancini, M. Poggialini and S. Mancini

III Department of General Surgery and Center of Phlebology "W. Pabisch", Faculty of Medicine and Surgery, University of Siena, Italy

INTRODUCTION

The incidence of the varicose recurrence after surgical therapy of the essential varicose veins of the lower limbs is quantifiable with a certain difficulty, being a great deal varying the data brought again in literature. It is gone from the 1 to the 15-20% with an average that is attested currently in the principal worlds case-records around to the 10% [4,7,8].

This inhomogeneity of the data depends overall from the different interpretations that the Authors bugger to the concept of varicose recurrence [1,2], for which the digital varied datum to second that the recurrence of the varicose veins is advised a "real recurrence" or a "pseudorelaps."

According to the other Authors we consider like "real recurrences" the varicose veins tie to the evolutivity of the desease, while "pseudorelaps" are the varicose veins ties to a preceding inadequate treatment, surgical or sclerosant, don't complete or not correct.

In the second place the patients with varicose recurrence return difficultly from the surgeon that the first time has operated them, unless it doesn't establisched a rigid protocol of follow-up, that foresees the control of the own operated at fixed expirations [3].

MATERIALS AND METHODS

From 1 July 1985 to 31 December 1994 at the III Department of General Surgery and the Center of Phlebology "W. Pabisch" of the University of Siena, directed by Prof. Sergio Mancini, had been effected 1291 surgical interventions for venous superficial insufficiency of the lower limbs, of which 1219 (94,4%) for essential varicose veins, and 72 (5,6%) for varicose veins recurrences. The first intervention had been effected in other places in 67 cases and at the our Center in 5 cases. Of these 72 patients, 65 (90,3%) belonged to the female sex and 7 (9,7%) to the male sex. The age varying from a minimum of 30 to a maximum of 74 years, with a middle age of 51 years. The break of time passed among the preceding surgical intervention and the appearance of the varicose recurrence changed from a minimum of 9 months to a maximum of 16 years. The varicose recurrence interested the inferior right limb in 37 cases (51,4%) and that of left in 35 (48,6%).

All the patients have been subjected to an accurate clinical and instrumental preoperative study. Particularly the ultrasonography doppler and the echography has been essential for a complete and precise exploration of the sapheno-femoral junction with finding of the saphenouses reflux, for the search of insufficient perforans veins responsible of brief reflux, for the search of residual saphenouses trunks, for the discovery of anomalies or anatomical variations. The phlebography has been required and effected only in select cases, when the patients reported a evocative symptomatology for previous deep venous thrombosis.

Of the 72 patients observed in 30 (41,6%) the crossectomy had been performed already with stripping of the long saphenous vein, but the presence of a tall reflux to level of the sapheno-femoral junction was tied to the absent or incomplete section of one or more periostial collateral veins. It joined the incontinence of one or more perforans. In 11 of these the echography of the inguinal region documented the presence of a cavernoma to level of the triangle of Scarpa. In 32 (44,4%) patients, in which the simple ligation of the long saphenous vein had been performed 2-3 cm distant from the cross associated complemental scleroterapy of the saphenous trunk, the present reflux to the cross was continued along the recanalized saphenous vein or a her collateral insufficient, that passing along the inside region of the thigh achieved the leg. In 7 (9,7%) cases for been missing recognition of the doubleness of the long saphenous vein a part of the saphenous axis had been left in seat. In 2 (2,7%) patients the binding of the cross associated heamodinamic correction of the venous reflux according to the technique of Franceschi had been performed. Finally in 1 (2%) patient, already subjected to stripping of the external saphenous vein, the recurrence concerned the external territory of the leg, with presence of a big cavernoma to level of the popliteal hallow.

All the 72 patients have been subjected to surgical reintervention, that has been practiced in hospital day regime for 25 (34,7%) operated in block and 32 (44,4%) in epidural or spinal anesthesia, and with hospitalization for the 15 (20,9%) operated in general anesthesia because of attendant cardiovascular pathology. About the type of operation 71 (98,6%) revisions of the sapheno-femoral junction has been performed according to the technique of Li [6], with stripping of the residual saphenous vein in 41 (56,9%) cases, because it has been subjected to simple ligation or any had not been recognized the doubleness. In 1 (1,3%) case the revision of the sapheno-popliteal junction is been practiced . In almost all the patients are been associate the ligation and the section of one or more insufficient perforans of leg, responsible of the recurrence of new varicose veins to this level.

The operative mortality has been absent. The postoperative complications has represented from 3 cases of abiding lymphorrhoea for more of 30 days. The hospitalization has been of few hours in the patients operated in hospital day regime and of 2-3 days in the other patients.

CONCLUSIONS

The recurrent varicose veins recognize their origin in errors of technical nature frequently: ligation of the saphenous trunk not very near to the femoral vein, not correct or don't complete crossectomy, not preoperative identification of a dual long saphenous vein, missing recognition of insufficient perforans veins.

It's necessary to pay attention to the clinical and instrumental preoparative evaluation of the varicose patient and then to the indication and the performance of the surgical intervention.

It's essential that the operated patients are subjected to periodic controls, so that to recognize pseudorelaps varix as soon as possible and set remedy to them with a following and adequate treatment.

On the other hand the evolutivity of the varicose disease imposes a continual vigilance, being in these patients the risk of a recurrence always possible, also if the operation has been technically perfect. Of this the patients must be informed in order to have their consent to a following treatment of the varicose disease of which carriers are.

REFERENCES

1. Agus GB, Castelli P, Sarcina A. Varici recidive degli arti inferiori. Patogenesi e indicazioni al trattamento chirurgico. Min Cardioang, 1982, 30,25:27
2. Bassi G. Le varici safeniche recidive. Compendio di terapia flebologica. Ed Minerva Medica 1985, pag 93-98.
3. Davy A. Ce qu'il faut connaitre du problème de la récidive variqueuse. Phlébologie 1986, 39,315:318.
4. Frileux Cl, Gillot CL, Le Baleur A. Les limites de l'indication chirurgicale dans les récidives post-opératoires des varices. Phlébologie, 1982, 35:865.
5. Gadeddu A, Marongiu GM, Bacciu PP, Chessa B, Lamberti MA, Cossu ML, Nemati Fard M. Le varici recidive. Eterno problema. Atti Giornate Flebolog Algheresi, 21-24 giugno 1989 CIC Ed Inter, Roma 1990, pag 435-439.
6. Li A. A technique for re-exploration of the saphenofemoral junction for recurrent varicose veins. Br J Surg 1975, 62, 745:746.
7. Lofgren EP. Treatment of long saphenous varicosities and their recurrence: a long-therm follow-up. In Surgery of the veins, Bergan & Yao, Ed Grune & Stratton, 1985, pag 285.
8. Wallois P. A propos d'un essai statistique: réflexions sur le devenir du variqeuux. Phlébologie, 1982, 35:467.
9. Vigoni M. Une conception personelle des facteurs qui conditionnent la récidive des varices essentielles après leur traitement. Phlébologie 1982, 35, 523:528.

Phlebology '95, D. Negus et al. (eds.). Phlebology (1995) Suppl. 1: 391-393

OP/5.7

Selection of Treatment Options for Patients with Surgical Type Superficial Venous Insufficiency

S.J. Simonian

Department of Surgery, Georgetown University Medical Center, Washington, DC, USA

INTRODUCTION

The purpose of this study was to quantitate reflux by noninvasive hemodynamic testing using air plethysmography (APG) [1-4] in postoperative residual incompetent varicose veins (VV), which were untreated by surgery and to determine the extent of their hemodynamic improvement after their treatment with compressive sclerotherapy [5].

METHODS

Noninvasive hemodynamic testing was performed with continuous-wave (CW) Doppler [2,6], air plethysmography (APG) (ACI Medical, Inc, Sun Valley, California) and duplex scan [7,8].

The study design was a retrospective review of a consecutive series of 77 patients with VV in 117 lower extremities (limbs). Fifty-nine limbs (50%) had primary VV, without prior surgery and without skin changes of hyperpigmentation or lipodermatosclerosis and these 59 limbs were the subject of this analysis.

The remaining 58 limbs were excluded from these measurements because 28 limbs (24%) had recurrent VV at the saphenofemoral junction (SFJ) and 30 limbs (26%) had both primary and secondary VV with skin changes of hyperpigmentation, lipodermatosclerosis and/or healed ulcer and with saphenofemoral incompetence (SFI) and/or saphenopopliteal incompetence (SPI).

Of the 59 limbs with primary VV included in this analysis 51 limbs had SFI and 8 limbs had SPI.

Of the 51 limbs with SFI, 30 limbs (60%) had greater (or long) saphenous vein (GSV) trunk incompetence from groin to ankle with additional tributary and perforator incompetent VV (peripheral VV). These 30 limbs were surgically treated with ligation, division and excision of the GSV trunk from groin to ankle and stab avulsion of the peripheral VV. Sixteen limbs (30%) had GSV incompetence from groin to knee with additional peripheral VV. These 16 limbs were surgically treated with ligation, division and excision of the GSV trunk from groin to just below the knee and stab avulsion of the peripheral VV. Five limbs (10%) had no GSV trunk incompetence but had only peripheral VV. The 5 limbs were surgically treated with ligation, division and excision of the SFJ locally and stab avulsion of the peripheral VV. Of the 8 limbs with SPI, 4 had incompetence from knee to ankle. These were treated with standard removal of the whole lesser saphenous vein. Four limbs had incompetence at the popliteal area only and the lesser saphenous vein was locally excised in the popliteal area.

Four weeks following surgical treatment the 59 limbs were evaluated clinically, with CW Doppler and APG for any postoperative residual, incompetent VV which were not treated by surgery. The postoperative residual, incompetent VV found were additionally treated with compressive sclerotherapy using sodium tetradecyl sulfate.

Student's t test was used for statistical analysis.

RESULTS

Of the 59 limbs evaluated clinically, with CW Doppler and APG testing, four weeks after surgery, 50 (85%) of the limbs had no residual incompetent VV. The remaining 9 limbs (15%) were found to have residual, incompetent, peripheral VV with reflux.

Before surgery the 59 limbs tested with APG had a venous filling index (VFI) of 5.3 ± 0.5 ml/sec, (the normal being < 2.0 ml/sec). After surgery, the 50 limbs without reflux had a reduction in the VFI to 1.6 ± 0.3 ml/sec ($p < 0.01$). After surgery the 9 limbs with peripheral reflux had a reduction in the VFI to 2.2 ± 0.2 ml/sec ($p < 0.01$). Following additional treatment with compression sclerotherapy, the reflux was abolished and the VFI was further decreased to 1.5 ± 0.3 ml/sec ($p < 0.06$) (compared to VFI of 2.2 ± 0.2 ml/sec).

DISCUSSION

In 50 of 59 limbs, the postoperative decrease to normal in VFI observed (from 5.3 ± 0.5 ml/sec to 1.6 ± 0.3 ml/sec) indicated the venous reflux has been eliminated, as confirmed by CW Doppler.

In 9 of the 59 limbs, the VFI decreased after operation but did not normalize is likely to be due to reflux, as confirmed by CW Doppler. Additional treatment with compression sclerotherapy in the 9 limbs with postoperative, residual reflux eliminated the reflux as confirmed by CW Doppler and reduced the VFI to normal (from 2.2 ± 0.2 ml/sec to 1.5 ± 0.3 ml/sec). The apparent hemodynamic improvement achieved by compression sclerotherapy was not significant in this small series.

393

CONCLUSION

In 50 of 59 limbs reflux was eliminated and significant hemodynamic improvement of the VFI resulted following standard surgical treatment. In the remaining 9 limbs persistent peripheral VV reflux was found. Additional compression sclerotherapy abolished the reflux. There appeared to be hemodynamic improvement which was not statistically significant.

REFERENCES

1. Christopoulos D, Nicolaides AN, Szendro G, Irvine AT, Bull ML and Eascott HHG. Air plethysmography and the effect of elastic compression on venous hemodynamics of the leg. J Vasc Surg 1987;5:148-59.

2. Schull KC, Nicolaides AN, Fernandes e Fernandes J, et al. Significance of popliteal reflux in relation to ambulatory venous pressure and ulceration. Arch Surg 1979;114:1304-6.

3. Katz ML, Comerota AJ, Kerr R. Air Plethysmography (APG): A new technique to evaluate patients with chronic venous insufficiency. J Vasc Technol 1991;15:23-7.

4. Christopoulos D, Nicholaides AN, Galloway JMD, Wilkinson A. Objective noninvasive evaluation of venous surgical results. J Vasc Surg 1988;8:683-87.

5. Goldman MP. Sclerotherapy: treatment of varicose and telangiectatic leg veins. St. Louis: Mosby Year Book, 1995.

6. Barnes RW, Ross EA, Strandness DE Jr. Differentiation of of primary from secondary varicose veins by Doppler ultrasound and strain-gauge plethysmography. Surg Gynecol Obstet 1975;141:207-11.

7. van Bemmelen PS, Bedford G, Beach K, Strandness DE. Quantitative segmental evaluation of venous vulvular reflux with duplex ultrasound scanning. J Vasc Surg 1989;10:425-31.

8. Vasdekis SN, Clarke GH, Nicolaides AN. Quantification of venous reflux by means of duplex scanning. J Vasc Surg 1989;10:670-7.

Phlebology '95, D. Negus et al. (eds.). Phlebology (1995) Suppl. 1: 394-396

P282

Vulval Varicose Veins by Phlebectomy of Muller

P. Sainte-Luce[1] and M. Samake[2]

[1] Angeiology, [2] Gynecology, Basse Terre, Guadeloupe, F.W.I

INTRODUCTION :

Vulval varicose veins, are found, rather during pregnancy, in 5 to 10 % of cases (1).

Pregnancy is often a period when varicose veins appear, in 4 to 25 % of cases during the first pregnancy, twice more in the second. 20 % of women will keep a definitive venous insufficiency.

In my own experience of 4000 patients with varicose veins only three of them have been treated by surgery for vulval varicose veins.

This disease is common after the sixth month of pregnancy which induce problems during delivery.

CLINIC :

The physician may examine vulval varicose veins in several cases:

- during pregnancy
- when patients consult for premenstrual syndrome
- patients with saphenous insuficiency

Women can show :

- pain before or during the menstrual period.
- vaginal itching or burning
- feeling of heaveness in the pelvic area
- dyspareunia

ANATOMY :

The anatomy of the vulval- vaginals veins is very different between women because of many anastomotics of the perineal area.

However, there are two principal venous systems : one anterior, one posterior.

- the veins of the posterior part of minus and major labia go to the "internal private vein"
- the veins of the clitoris go to the internal saphenous vein.

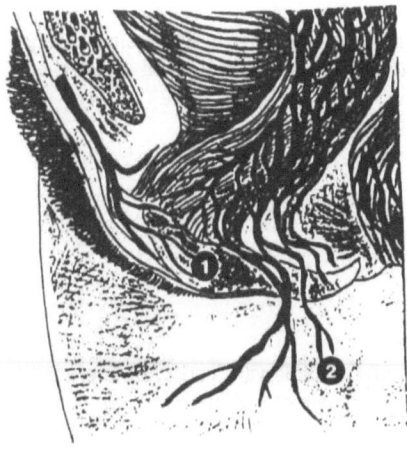

1.   Vulval veins
2.   Perineal veins

PHYSIOPATHOLOGY :
Pregnancy has many consequences on the venous hemodynamics :
   - the venous pressure in the inferior legs increase lineary during pregnancy. With normal subject on the back, the mean venous blood pressure is about 11,43 cm of water in the femoral vein.(2, 3)
During a normal pregnancy, the venous blood pressure can be very high and reach 24,37 cm of water at the end of the pregnancy.
   - the decrease of the flow has been detected by marked natrium(91 92 93 ng)
This decrease of the speed flow has been also seen by duplex scan at the end of the pregnancy. (4)
   - the sexual steroids hormone have a direct effect on the wall of vessels by specific receptors.

Other factors : case history, way of life, hormonal statute

SURGERY :
The patient is in gynecologic position. The vulval varicose veins are marked on the skin. Local anesthesia is done with Xylocaïn.
Incisions are very small, about 2 or 3 mm, spaced according to the size of the varicose veins.
The veins are stripped off by the hook of Muller as completely as possible.
Incisions are closed by cicagraf* or by one stitch.
The sound of Folley is removed after 12 hours.
It is very important to put a lateral compression to prevent the bleeding.
The dressing is done every day until the sixth day when it is definitively removed

396

DISCUSSION :
This desease is not very common and the treatment described here is interessing because one-day-surgery, made in a single time, without risk of thrombosis and pain versus sclerotherapy.
The cosmetic result is perfect.

BIBLIOGRAPHY :

1 - Mullane , D.J. :
    Varicose veins of pregnancy .
    Am. J. Obstet. Gynecol. , 1952 , 63 : 620

2 - Bouvier, C.A., Engels, E. , Grobety, J. , Niebes, P. : La maladie variqueuse. Différences structurelles, histochimiques et chimiques entre veine normale et variqueuse.
Ed. Zyma Nyon , 1977.

3 - Mac Lennan , C.E. :
    Anticubital and femoral venous pressure in normal and toxemic pregnancy.
    Am. J. Obstet. , 1943 , 45 : 568 .

4 - Wright, H. Payling, osborn, S.B. , and Edmonds D.G. :
    Rate of flow of venous blood in the legs.
    Lancet, Lond. , 1948, 255 : 767.

Phlebology '95, D. Negus et al. (eds.). Phlebology (1995) Suppl. 1: 397-399

OP/17.5

Stripping From the Groin to the Knee: A Plethysmographic Assessment

R. Zinnari, G. Lucertini and P. Belardi

Cattedra di Chirurgia Vascolare, Università Degli Studi, Genova, Italy

INTRODUCTION

Treatment of primary varicose veins with valvular incompetence of the sapheno-femoral junction has received a lot of attention in recent years. There are many options for the treatment of this venous disease. Stripping of the greater saphenous vein is the standard procedure and the effectiveness of this has been widely stressed. After the success of total stripping (TS), from the groin to the ankle, there has been an increasing interest in partial stripping (PS), from the groin to the knee, in recent years. However this procedure has been assessed mainly on clinical criteria [1], while very few investigations have been carried out on the hemodynamic aspects [2,3].

The purpose of the present study is the hemodynamic evaluation of partial stripping.

MATERIAL AND METHODS

From January 1993 to December 1994 a prospective study was carried out to compare PS with TS. A series of consecutive patients showing primary varicose veins with valvular incompetence of the sapheno-femoral junction was randomised to either PS or the TS group.

The PS group consisted of 41 patients (10 males and 31 females; ages ranging from 30 to 70 years, mean 46.6), of which 50 legs required PS. They underwent downward stripping from the groin to the knee. Avulsion of collateral varicose veins was associated in all cases. Interruption of one or more incompetent communicating veins at the calf or/and the thigh was required in 14 cases. Stripping of the lesser saphenous vein was carried out in 1 leg.

There were 40 patients in the TS group (10 males and 30 females; ages ranging from 28 to 69 years, mean 49.8), of which 48 lower limbs required TS. They underwent downward stripping from the groin to the ankle. All cases also required avulsion of collateral varicose veins. Interruption of incompetent communicating vein(s) at the calf or/and thigh was performed in 19 cases. Stripping of the lesser saphenous vein was performed in 1 case.

Each lower leg was assessed before surgery by strain-gauge plethysmography examination, which was repeated at intervals of 1, 3, 6, 12 and 18 months after surgery. Strain-gauge plethysmography examination was carried out using Periquant 3800 plethysmograph (Gutmann, Germany). The patient lies in a supine position with his/her elevated feet, with a strain-gauge placed around the calf at its greatest circumference. A pneumatic cuff is applied to the mid-thigh and inflated to 40 mmHg and held for 1 min, increased to 60 mmHg and held for another 1 min, increased to 80 mmHg and held for a further 1 min (a total of 3 min) and then deflated quickly. Venous capacity is calculated at 1 min, 2 min and 3 min intervals during the inflation of the pneumatic cuff. Maximum venous outflow is determined after rapid deflation of the pneumatic cuff.

The two groups were compared taking into account male:female ratio, mean age, preoperative and postoperative hemodynamic data.

Comparison between ratios was made using χ^2 method and Mann-Whitney U test were applied for comparison between means.

RESULTS

The two groups were matched for sex, age and preoperative hemodynamic data. Statistical differences were not found (p > .05) between the two groups for venous capacity (4.2 +/- 1.3 versus 4.3 +/- 1.5 ml/100 ml after 3 min) and maximum venous outflow (47 +/- 19 versus 55 +/- 23 ml/100 ml/min) before surgery.

After surgery at 1 month interval 50 and 48 cases were assessed, at 3 months 43 and 41, at 6 months 35 and 34, at 12 months 34 and 31 and at 18 months 14 and 12 in the PS and TS groups, respectively.

For statistical purposes, we compared the two groups taking into account venous capacity 3 min after inflating the pneumatic cuff and maximum venous outflow. There were not statistical differences (p > .05) between the two groups regarding both venous capacity and maximum venous outflow up to 18 months after surgery. The hemodynamic results of the both groups did not vary significantly during follow-up assessment.

DISCUSSION

The present study shows that partial stripping gives results equivalent to those of total stripping.

Although total stripping has been the main technique for many years, this procedure is limited and leads to possible complications. It requires avulsion of the entire saphenous trunk, which can be utilised as autologous graft. The procedure can also cause lesion of the saphenous nerve, resulting in pain and dysaesthesia.

By partial stripping it is possible to prevent some of these problems. The distal segment of the saphenous trunk (from the ankle to the knee) is spared. This procedure is justified by the observation that the below-knee segment of the greater saphenous vein becomes neither dilated nor varicose. Varicose veins at the calf are usually tributaries of the greater saphenous vein. After partial stripping the residual segment of the saphenous trunk remains patent in most cases [4].

Moreover this procedure avoids lesion of the saphenous nerve, because the two structures are related only from the knee or the upper calf to the ankle [5].

Helsted and coworkers [2] and Sarin and coworkers [3] have investigated the hemodynamic aspects of stripping from the groin to the knee, studying venous refilling time using strain-gauge plethysmography. Hemodynamic results 3 months after surgery did not differ significantly between the two procedures. The present study leads to the same outcome with a longer follow-up (18 months). Strain-gauge plethysmography was utilised for objective assessment of venous surgery for its reliability and the quantitative investigation of the venous system. Hemodynamic parameters investigated were venous capacity and maximum venous outflow, which evidence the state of the whole venous system of the calf.

At the present time several procedures are suitable for the treatment of valvular incompetence of the greater saphenous vein. The clinical evaluation and hemodynamic assessment enables us to consider partial stripping as the most effective option of excisional surgery for the valvular incompetence of the greater saphenous vein.

REFERENCES

1. Negus, D. Should the incompetent saphenous vein be stripped to the ankle? Phlebology 1986;1:33-6.
2. Helsted M, Hesselfeldt-Nielsen J, Mathiesen FR. Partial versus total stripping of the great saphenous vein evaluated by strain-gauge plethysmography. Phlebology 1991;6:227-32.
3. Sarin S, Scurr JH, Coleridge Smith PD. Assessment of stripping the long saphenous vein in the treatment of primary varicose veins. Br J Surg 1992;79:889-93.
4. MacFarlane R, Godwin RJ, Barabas AP. Are varicose veins and coronary artery bypass surgery compatible? Lancet 1985;2:859.
5. Ramasastry SS, Vick GO, Futrell JW. Anatomy of the saphenous nerve: relevance to saphenous vein stripping. Am Surg 1987;53:274-7.

Phlebology '95, D. Negus et al. (eds.). Phlebology (1995) Suppl. 1: 400–403

PI/4.5

Get it Right First Time: The Results of Primary Operations for Varicose Veins are Better than for Recurrence

H. Hafez[1], S. Jarvis[2], A. Harvey[3] and R. Corbett[1]

[1] Department of Surgery, The Princess Royal Hospital, Haywards Heath, West Sussex, UK
[2] Epsom General Hospital, Epsom, Surrey, UK
[3] Cuckfield Medical Practice, Cuckfield, West Sussex, UK

INTRODUCTION

Any experienced surgeon who is honest and who carries out some follow-up of patients operated on for recurrent varicose veins will have been disappointed on occasion by the result. In some unfortunate patients, further recurrences are apparent within months. This is despite full clearance of all visible varices. This observation led us to a systematic study of 44 patients who, overall, had a moderate result after a mean follow-up of 33 months [1]. Suspecting that the results of operations for recurrence might be worse than after a primary operation, we have carried out a retrospective review of results, both in patients having operation for the first time and those operated on for recurrence. While there is now a wealth of information available on the pre-operative investigation of recurrent varicose veins [2,3,4], there is very little written about the results of operation for recurrence. On primary operations, reported results have in the past been largely subjective, but more reliable information is emerging [5].

METHODS

Patients

We set out to examine about 30 patients in each group, choosing a minimum follow-up of 2.5 years since operation. Patients were selected at random from a list by one author who had never seen the patients (MH). In all, 95 patients were invited to attend for review, of whom 67 (71%) accepted. Details are shown below.

Table 1. Patients studied

	Patients	M	F	Mean age at review	Mean follow-up in months
Primary	36	7	29	57.8	58 (31-109)
Recurrent	31	9	22	59.7	63 (30-106)
Combined	67	16	51	58.6	60 (30-109)

The 36 patients in the primary group were drawn from a pool of 497 patients in our database who had been operated on more than 2.5 years before the review. The 31 patients from the recurrent group were drawn from a total of 97 who were eligible. As expected, patients with recurrence are older. There were more patients in the primary group, as we decided to include in this group patients whom we recalled because of operation for recurrence, but who had a primary operation on the contralateral leg at the same time. Indeed, this would have been an alternative way to conduct the study. Later analysis of this sub-group of 5 patients confirmed that they behaved the same way as the remainder.

Investigation

In all patients, initial assessment had consisted of careful clinical examination, supplemented by use of hand-held Doppler ultrasound. Imaging was by varicography, as duplex ultrasound was not available to us when these patients had their operations. Varicography was used in all patients with recurrence and additionally in primary cases where hand-held Doppler suggested short saphenous incompetence.

Operative

Standard operative methods were used, with emphasis on careful flush saphenofemoral ligation in primary cases. Most long saphenous veins were stripped. Complete clearance of varices by phlebectomy was always the aim. Formal perforator ligations were not done. Short saphenous ligation was not accompanied by short saphenous stripping. In recurrent cases, flush saphenofemoral ligation was done whenever it was deficient in an earlier operation. Residual long saphenous veins were stripped, thereby avulsing any incompetent thigh perforators, but formal perforator ligations were not done. Short saphenous ligation was performed where indicated, without stripping.

Clinical review

Each patient was examined by two independent observers (HH and RC) and an estimate of the degree of recurrence made, (minimum score 0, maximum score 8). Venous refill times were measured by photoplethysmography (PPG). Patients completed self-assessment scores for cosmetic appearance and degree of discomfort, using linear analogue scales (0-10). Height and weight were recorded to calculate body mass index (BMI).

Statistical methods

Correlation was established between the scores for recurrence given by each observer, (r=0.77, p <0.001). Accordingly, the two scores for each limb were averaged to give the mean recurrence score and this was used for further analyses. Mean scores between primary and recurrent groups were compared by Student's unpaired t test. Student's t test was also used to compare PPG refill times and patients' self-assessment scores between the two groups. We looked for correlations between recurrence scores on the one hand and age, PPG refill times and BMI on the other.

RESULTS

The mean recurrence score for patients after a first operation was 1.8, (95% confidence interval 1.47 - 2.13) and 2.68 after operation for recurrence (2.16 - 3.21). The means were significantly different (unpaired Student's t test, p <0.01). PPG refill times were shorter in the recurrent group at 19.9 sec compared with 21.8 sec in the primary group, but this was not significant. In terms of self-assessment, patients scored their limbs more favourably after first operation than after one for recurrence, but the differences were not significant. Combining both groups, we found that recurrence scores correlated positively with age (r=0.31, p <0.01) and recurrence scores correlated negatively with PPG refill times, (r= -0.54, p <0.001). Recurrence scores did not correlate with body mass index.

DISCUSSION

The results confirmed our expectation that limbs operated on for recurrence fared worse than those having an operation for the first time. The results were comparable to those reported earlier for recurrent cases in whom the mean recurrence score was 2.85 [1]. As in many areas of surgery, the best time to achieve a good result is at the first operation. We consider the key elements are careful initial assessment, with imaging where necessary, accurate flush saphenofemoral ligation and long saphenous stripping. This study was not designed to clarify whether or not the long saphenous vein should be stripped but we are fully in agreement with Sarin et al [6] that long saphenous stripping lowers the risk of recurrence.

REFERENCES

1.   Maw A, Handa A, Harvey A, Harvey M and Corbett CRR. The results of operating for recurrent varicose veins. Phlebology 1993;8;41.

2.   Loveday EJ, Lea Thomas M. The distribution of recurrent varicose veins: a phlebographic study. Clinical Radiology 1991;43:47-51.

3.   Bradbury AW, Stonebridge PA, Ruckley CV and Beggs I. Recurrent varicose veins: correlation between pre-operative clinical and hand-held Doppler ultrasonographic examination, and anatomical findings at surgery. Br J Surg 1993;80:849-851.

4. Redwood NFW and Lambert D. Patterns of reflux in recurrent varicose veins assessed by duplex scanning. Br J Surg 1994;81:1450-1451.

5. Laurikka J, Sisto T, Salenius J-P, Tarkka M and Auvinen O. Long saphenous vein stripping in the treatment of varicose veins: self- and surgeon-assessed results after 10 years. Phlebology 1994;9:13-16.

6. Sarin S, Scurr JH and Coleridge Smith PD. Stripping the long saphenous vein in the treatment of primary varicose veins. Br J Surg 1994;81:1455-1458.

Phlebology '95, D. Negus et al. (eds.). Phlebology (1995) Suppl. 1: 404-406

P189

A New Technique for Ambulatory Treatment of Varicose Veins Associating Surgical Section and Sclerotherapy of Large Saphenous Veins (3S Technique)

F. Vin[1], F. Chleir[1] and F.A. Allaert[2]

[1] *Service de Phlébologie, Hôpital Notre Dame de Bon Secours, Paris*
[2] *Département de Biostatistique, CHRU du Bocage, Dijon*

The principle governing treatment of varicose veins in the lower limbs is the removal of leaking points and their tributaries (1). Sclerotherapy is an ambulatory method which has been found most satisfactory in the great majority of cases, its efficacy having been proven over several decades.Several studies show that sclerotherapy is less efficacious for greater saphenous veins of over 8-9 mm in diameter (2). A considerable number of patients refuse surgical treatment by ligation-division and stripping, even as ambulatory treatment, for medical, social or professional reasons. For several decades, patients with considerable trophic troubles and highly developed varicose veins have been seen in daily practice, and surgical treatment has been offered which does not, in fact, offer a cure.In order to ensure swift, efficient, and economical treatment, and to avoid failures in sclerotherapy or ligation-division, we have conceived of a new technique for treating large, incontinent greater saphenous veins. It is a combination of two well-known techniques : section and ligation, together with sclerotherapy carried out at the same time.This technique, called 3S (saphenous - section - sclerotherapy) takes its inspiration from the SAVAS method described by BELGARO (3) .

DESCRIPTION OF THE TECHNIQUE
In a patient presenting trophic troubles and a large varicose vein, in which a swift, efficient result is desired , a clinical examination plus Duplex exploration and mapping is essential (4). All patients for whom the 3S technique is suitable must be subjected to complete exploration of the deep and superficial venous systems.
One criterion is greater saphenous veins with diameter of more than 8 - 9 mm at 3 cm from the ostium, in which it is known that sclerotherapy is less successful. The great majority of these patients have also refused traditional surgical treatment for several years. Most of them also present trophic disorders.
The 3S technique is an ambulatory treatment with no pre-medication, administered to a fasting patient. The basic principle is suppression of reflux by section and ligation, together with a sclerosing injection up- and down-stream of the surgery, the whole at some 10 cm distance from the saphenofemoral junction. After skin disinfection of the teguments, local anaesthetic - an injection of 5 to 8 cc of 2% lidocaine with adrenaline - is injected in the subcutaneous tissu.

Then a 2 - 3 mm incision is made with a n° 11 lancet at some 10 cm distal to the saphenofemoral junction. Using a special hook with a rounded end, the greater saphenous vein is hooked out and exposed. Its position outside the fascia thus avoids any risk of traumatising deep vessels. It is then placed between two clips and two lakes, and the adventitia is dissected at the level of the proximal and distal segments. A horizontal incision is made to enable a catheter to be slipped in. Initially, the proximal segment is catheterized and 2cc of a 3% iodine solution is injected with a syringe . Then the distal segment is catheterized and injected with the same amount of 1% iodine solution . A ligation transfixing the vein using non-resorbable thread is then performed, and the venous segment of about 2 cm between the proximal and distal ligations is resected. The two stumps are then replaced in the hypodermis and a single stitch closes the skin. The entire lower leg will be compressed by an elasto-adhesive bandage, from the root of the toes to the inguinal region, for a period of ten days.

Criteria for Evaluation of Effectiveness.
The main criterion is collapse of the varicose veins in the operated leg.Clinical examination and pulsed Doppler scanning enable the efficacy of the technique to be assessed.The duplex-scan examination is performed with an ESAOTE AU 530 with 7,5 -10 Mhz probe.The first examination and controls are made by the one physician and the intervention was made by another. Effective sclerosis occured when flow stopped or reflux was no detectable in a vein which was shown to be non-collapsible in ultrasound scan, and presented endo-parietal echogenous matter at the junction level segment , and in the median and lower thirds of the thigh.No deep venous thrombosis, nor neurologic complications were observed .The scar is invisible in the great majority of cases, comparing favorably to phlebotomies.

RESULTS
The initial descriptions of the clinical, Doppler, Duplex scanning characteristics of the patients in this study, and the results of these examinations, were compared at one month, 6 months and each year after the operation. To date, 100 patients have been operated on by the 3S technique, of which 93 at least one year ago. The number of patients lost to follow up is 7.One patient died in a car accident after 6 months' follow up, 3 had good results and din't see the interest to come back for a control.One patient had the same varicose vein on the foot and 2 were unhappy because the 3S technique had not resolve the whole part of their sikness.
We describe the characteristics of patients in the study successively, and the results obtained at one month, and one year of follow-up.

Description of Patients
Most were female (79,5%), aged 64 +/- 13 years, and weighing 70 kg +/- 14 kg, for a height of 1 metre 67 +/- 7 cm. Clinical examination showed marked varicosity in 80,4% of cases. The classic functional symptoms of venous insufficiency (pain and heaviness) and oedema were of moderate intensity in half of the patients, and of considerable intensity in a little more than a quarter of them, 16,1% presented varicose ulcers.
Pulsed Doppler examination revealed flow with reflux in all patients and ultrasound scanning revealed collapsible veins with no endoluminal or parietal modifications with a mean diameter of 9.9 , mim : 8 mm - max : 20 mm.

Results of Follow-up at 1 Month, and One year
Varicose veins which were present at Day 0 has totally disappeared in 59,8% after 1 month and in 76,2% after one year and only collateral veins remained to be treated in 40,2% of cases at 1 month, and 23,8% of cases at one year .The heaviness and pain in the lower limbs followed similar curves with rates of total permanent disappearance in 76,2 and 91,7% of patients at one year of follow up. Under 2,4% and 1,2% suffered in any way at one year.

Oedema followed similar evolution curves . The number of ulcers fell by half in the first month, and healed entirely at six months with no reappearance at one year .The duplex-scan revealed veins which had remained non-collapsible without any flow with sclerosis at junction level in 97.7% of cases at one month, and 96,4% at one year . In the lower third of the thigh, sclerosis persisted with non-collapsed veins without any flow in 93.1% of cases at 1 month, and 91,6% after one year . We have never seen development of tributaries of the sapheno-femoral junction with reflux at the groin.Undesirable effects were monitored after 10 days, 1 month, 6 months and one year.

The scar was invisible, with no underlying inflammatory reaction in 77,5% of cases after the tenth day, the rate rising to 94,7% at the end of the sixth month and to 96,7% after one year, the other aspects being represented by a small scar with inflammatory reaction of the dermis or hypodermis.

DISCUSSION

This technique is primarily aimed at aged patients presenting venous insufficiency with leg ulcers, or exposed to the risk of trophic disorders. Our first observation was that recurrences at the level of the saphenofemoral junction were few, where the Duplex examination plus mapping had eliminated other leaking points such as vulvo-pudendal varicose veins or incontinent anterior or posterior collaterals flowing into the saphenous crook. Only 3,6% of failures with reflux at the level of the junction were found after one year, despite the fact that the collaterals left in place could have led to recurrences. In fact, it would seem that these collaterals drain the blood from the superficial system into the common femoral vein, given the considerable pressure gradient VENTURI effect. This haemodynamic role has already been demonstrated in the CHIVA technique.

CONCLUSION

The 3S technique, which associates ligation and section with injection of the proximal and distal segments of the greater saphenous vein, gives satisfactory results. Reflux with collapsed veins persist in 3,6% of cases after one year at the sapheno-femoral junction. It represents an alternative to surgery in aged patients who have always refused surgery on large varicose saphenous veins with reflux, for which sclerotherapy is inadequate. The presence of reflux in the lower third of the thigh and in the lower leg is easily treated by additional sclerotherapy or ultrasound-guided sclerotherapy.

REFERENCES

1- VIN F PRINCIPES DE LA SCLEROTHERAPIE DES AXES SAPHENIENS DES MEMBRES INFERIEURS ET DE LEURS COLLATERALES A L'EXCEPTION DES VEINULES ET TELANGIECTASIES. PHLEBOLOGIE 1994, 47N° 4, 399-406

2 - SCHADECK M INDICATION ET RESULTAT DE LA SCLEROTHERAPIE A LA FRANCAISE PHLEBOLOGIE 1992,45 N°4 385-87

3 - BELGARO.G, CHRISTOPOLLOS.D, VASDEKIS.S TREATMENT OF SUPERFICIAL VENOUS INCOMPETENCE WITH THE SAVAS TEHNIQUE. JOURNAL DES MALADIES VASCULAIRES 1991, 16 23-27

4 - SCHADECK M DUPLEX PHLEBOLOGY ED GNOCCHI 1994

5 - WEINDORF N, SCHULTZ-EHRENBURG V CONTROLE DU TRAITEMENT PAR SCLEROSE DES VARICES PHLEBOLOGIE 1990, 43, N°4 681-688

Phlebology '95, D. Negus et al. (eds.). Phlebology (1995) Suppl. 1: 407–409

PI/4.6

Duplex Ultrasound Audit of Operative Treatment of Primary Varicose Veins

W.A. Campbell and A. West

Carrington Clinic, Brisbane, Australia

INTRODUCTION

The main causes of recurrent varicose veins are inadequate technical performance of the primary operation, and/or an incomplete or incorrect diagnosis of the venous pathology (1). Duplex ultrasound examination (DUE) allows the accurate identification of incompetent connections between the deep and superficial venous systems.

The objective of the present study was to assess the adequacy of a plan of operative treatment (OT) of primary varicose veins (VV), which had produced satisfactory early clinical results, by DUE during the early postoperative period.

METHODS

A prospective study was performed on a consecutive series of 144 patients in which 185 legs with VV were all treated by one of the authors (W.A.C.) by high saphenous ligation +/- saphenous truncal stripping, plus ablation of varices by a combination of intra-operative sclerotherapy and multiple stab avulsions. At operation, no intervention was directed specifically to incompetent perforating veins (IPV).

All patients had DUE on the day before operation, and two to six weeks after operation, by the same sonographer (A.W.). DUE was performed with the patient erect, standing or sitting as necessary, using manual compression of the thigh and calf to test the competence of valves. Reversed venous flow for a duration greater than 1 second was diagnosed as reflux. The duplex scanner was an ATL Ultramark 4 with a multi-frequency sector scanhead. (Advanced Technology Laboratories, Bothell, WA). The results of the postoperative clinical assessments were recorded by the operating surgeon independently of the DUE results, and later compared.

RESULTS

59 (31%) of 185 legs showed at least one area of persistent venous reflux (PR) on the postoperative DUE. The results were classified according to the outcome of the various OT and their effects on the reflux which had been demonstrated in the saphenous systems, gastrocnemial veins (GV), and IPV at preoperative DUE. In addition, the fate of IPV was analysed according to their site and size.

Long Saphenous Vein (LSV)

Table 1 shows the results of OT on 169 LSV with reflux.

Table 1. DUE result after OT for LSV reflux

Operation	no.	reflux	competent	no flow
High ligation LSV (HSL)	13	6	4	3
HSL & strip to knee	122	19	13	90
HSL & strip to ankle	24	1	0	23

Short Saphenous Vein (SSV)

Table 2 shows the results of OT on 54 SSV with reflux.

Table 2. DUE result after OT for SSV reflux

Operation	no.	reflux	competent	no flow
High ligation SSV (SSL)	5	0	2	3
SSL & strip to midcalf	12	4	0	8
SSL & strip to ankle	37	0	0	37

Gastrocnemial Veins (GV)

In 185 legs, reflux was found in only 6 GV at preoperative DUE. 1 became competent after HSL alone, and 3 after HSL and stripping of the LSV; only 2 showed PR.

Incompetent Perforating Veins (IPV)

226 IPV were diagnosed in 164 out of 185 legs. At the postoperative DUE the majority of IPV 182/226 (80.5%) had become competent or showed no flow. None of only 4 IPV in the thigh showed PR.

In the proximal half of the leg, PR was found in 9/47 (19%) IPV on the medial, 1/11 (9%) on the posterior, and 3/25 (12%) on the lateral aspect. In the distal half of the leg PR was found in IPV in 22/89 (25%) on the medial, 1/5 (20%) on the posterior and 8/45 (18%) on the lateral aspect.

IPV diameters ranged from 2mm to 6mm. In the leg, PR was present in 31/133 (23%) IPV of 2mm, in 10/59 (16.9%) IPV of 3mm, and in 3/28 (11.5%) IPV of 4mm diameter, but not in either the 2 of 5mm or the 2 of 6mm diameter.

Of the 59 legs with PR, 33 had residual saphenous reflux +/-IPV; 29 of these were clinically normal. 26 legs had persistent IPV only; 23 were clinically normal. Thus, 52/59 (88%) of legs with PR were clinically normal.

In summary, OT of VV on 185 legs was found to be inadequate in 59 legs (31%), with PR in at least one area; 88% were clinically normal.

CONCLUSIONS

Others (2) have recommended the ligation of every site of deep to superficial reflux for the prevention of recurrent varicose veins. Although this appears to be logical, findings in this study challenge that concept.

The fate of IPV is unpredictable from their size or site. Also, reflux in IPV is not necessarily pathological (3). Therefore, ligation of IPV at OT of VV is not indicated.

Incompetent GV were found in only 3% of 185 legs and most (4/6) (66%) became competent after operation on the LSV alone. It seems inadvisable to ligate these at OT for VV.

7% of SSV showed PR, and that was in the distal part following external stripping from knee to mid calf. Further improvement would require complete ablation of the SSV, and this has to be balanced against possible sural nerve injury.

15% of LSV had PR, 96% after either HSL, or HSL + stripping the LSV to the knee. Reduction in the incidence of PR would require complete ablation of the LSV, thus removing a potential arterial conduit.

Conventional OT of VV is followed by a substantial incidence of residual PR which often is not apparent on CE and is revealed only by DUE. Unrecognized PR may be important in the genesis of recurrent varicose veins. Routine postoperative DUE is recommended.

REFERENCES

1. Dodd, H., Cockett, F.B: The Pathology and Surgery of the Veins of the Lower Limb, 2nd edition, Edinburgh, London, New York, Churchill Livingstone, 1976, pp. 121-146

2. Juhan C, Haupert S, Miltgen G, et al. Recurrent varicose veins. Phlebology 1990;5:201-11.

3. Burnand KG, O'Donnel TF, Thomas ML, et al. The relative importance of incompetent communicating veins in the production of varicose veins and venous ulcers. Surgery 1977;82:9-14.

Phlebology '95, D. Negus et al. (eds.). Phlebology (1995) Suppl. 1: 410–412

P214

Varisectomy by Micro Insitions

V. Spano[1], and D. Duverges

[1] *Private Practice, Buenos Aires, Argentina*

Introduction

In this work we have placed emphasis on areas such as: foot, foot instep, backside of knee, sub-gluteal, to perform surgical operations using micro-incitions.

The nervous or lymphatic lesion adds morbidity and discredits our specialty.(8,29)

The ambulatory phlebectomy is therapeutic and esthetical. The cicatrices must not be worse than the esthetical problems created by the varices.(17).

Anatomy

The venous system of the foot is divided in dorsal and inferior. In the foot instep the superficial vessels are: in the dorsal side, the internal saphena, in the exterior side, the external saphena and in the interior side, the internal retromalleolars. In the thigh: the hypogastric, superior and inferior gluteous.(3,6,7,21,23,28,29).

Physiology

The venous circulation is centripetal, with valves and collateral circulation.

Intervene: vis a tergo, vis a fronte, vis a latere, capability, venous tone, tissue or outerwall pressure, Lejars sole, vein-muscle-articulate pump, valvular system. Physical alterations: pressure, flow, resistance, blood viscosity.(8,11,12,18,28)

Material and Methods

Between December 1989 and December 1994 the authors operated 220 patients with varices.

We found varicose pathology in:

foot	70 cases
foot instep	40 cases
backside of knee	80 cases
thigh	30 cases
Total	220 cases

We must point out that this type of surgery using micro-incitions has been performed since 1976. Due to the fact that the satistical data is incomplete, only the last 5 years have been calculated.

The patients were treated in hospital environments and private practices.

The patients' ages oscillated between 19 and 76 years, with an average age of 41; 172 were women (78.18%) and 48 were men (21.81%).

Evolution and Complications

Of the 220 patients operated, there were:

Four patients (1.81%) with serious, important hematomas (external face of the foot, backside of knee, sub-gluteous).

Six patients (2.72%) with small and circumscribed hematomas (sub-malleolar, external side of the thigh).

The hematomas added up to 4.33% and our opinion is that the majority were caused by the incorrect use of elastic compression.

Twelve pacients (5.45%) presented some light regional parestesis (irritation, traction or rupture of tiny nerve fillets in the foot, foot instep and back and lateral side of leg).

Eighteen pacients (8.18%) manifested pain of short duration.

Twenty four pacients (10.90%) continued with edema, attributed fundamentally to painless decubitus and poor application of the elastic bandage.

To resume:

154 Pacients .... 70% .... Very Good evolution
42 Pacients .... 19.09% .... Good evolution
24 Pacients .... 10.90% .... Regular evolution
Total 220 .... 99.99%

Discussion

Some authors use alternative or complementary techniques as: transcutaneous ligatures; the "take" is realized in blocks (skin, subcutaneous cells, vein, etc.) (closed operation). Microsurgery is an open operation and the lesions can be "controlled". Transitory ligatures, utilized during the surgery are excluded at its end. We think that is to add another "surgical step". Previous esclerotherapy: one must choose between one or the other treatment.

In the end, with the adecuate knowledge of anatomy and the apropriate technique, the complications as hematomas, neuralgias, linforragias, etc. will decrease considerably.(9,24)

Criteria of inclusion

Varicose tracts larger than 2 to 3 mm. in diameter, anarchaics, sistemics, affluent in both safenas (internal and external), residual and/or relapsed varices and perforant veins.

Criteria of exclusion

Advanced senility, malignant tumors, septic processes, cardiac insuffenciency and renal discompensation, dermatitis, vascular ulcers, fistulas.

Anaesthesia

We use a local anaesthesia. Lidocaine (ester) in concentrations that vary from 0.25 to 2%.(2,5).

Technique

A thorough bearing, antiseptic with iodine-povidona. Local anaesthesia.

A point-form incision with a surgical knife blade No. 11. Externalization of the vein with a Crochet needle. Next we take Halsted or Crile forceps. We divide at one proximal and

412

another distal end. We twist the tract over the forceps approximately one complete turn, we retake the vein at the edge of the skin with another forcep and make a counter turn. By this manner, we separate the vein on its upper and lower face or viceverse.

These movements are continued until the vein is cut. Control of the hemostasia. Gauze, cushioned with cotton and elastic compression.(4,16,22).

The post-operative controls are done after 24 hs., in seven days and in fifteen days.

In case of post-surgical pain, it is immediately indicated: Diclofemal in general manner and Ruscogenina in topic form. The post-operative ambulation is almost immediate.

References

1- Alegrotti, L. Anestesia en flebología, Escuela Argentina de Flebología, A.M.A., 1987.
2- Anafilaxia y riesgo vital durante la anestesia, Revista Panameriacana de Flebología y Linfología 4 - N° 3:47, 1994.
3- Andoniades, O. y col. Várices genitales femeninas. Flebología 4 - N° 2:9, 1992.
4- Bonvini, G. Nuevo instrumental y nuevas técnicas en flebología. Flebología 6 - N° 3:283, 1993.
5- Bacci, P. y col. La anestesia local en la cirugía flebológica ambulatoria. Flebología 5
6- Caldevilla H. Cirugía del linfedema. Tesis de Doctorado en Medicina. Universidad de Buenos Aires, 1982.
7- Delevaux, J. y col. Historia de la Flebología. Culturas precolombinas. Flebología 4 - N° 2:82, 1992.
8- Enrici, E.A.; Caldevilla H.S. Insuficiencia venosa crónica. Editorial Celcius: 16, 34, 35, 1992.
9- Fernández, J. y col. Ligaduras transitorias en cirugía fleboestética. Flebología 4 - N° 2:101, 1992.
10- García Méndez, A.G.; Iusem, M. Flebopatías: estudio y tratamiento. Editorial Adegraf:64, 1986.
11- García Méndez, A.G. y col. Circulación venosa y su patología. Flebología 5 - N° 2:59, 1992.
12- Giffoniello, A. y col. Relaciones entre alteraciones de la estática del pie e insuficiencia venosa. Flebología 3 - N° 1 a 3:231, 1990.
13- Haid Fischer, F.; Haid H. Enfermedades de las venas. Editorial Salvat:194-197, 1984.
14- Iusem, M. Historia de la Sociedad Argentina de Flebología y Linfología. Flebología 4 - N° 1:64, 1991.
15- Iusem, M. La flebología del futuro. Flebología. 6 - N° 3:249-259, 1993.
16- Iusem, M. y colaboradores. Nuestra experiencia de trabajo flebológico en el Hospital Penna y Sanatorio Municipal Julio Méndez. Flebología. 6- N° 2:120, 1993.
17- Iusem, M. Historia de la flebectomía ambulatoria. Flebología. 7 - N° 2:193, 1994.
18- Kaplan, G. Kinesioterapia en flebología. Flebología. 4 - N° 1:45, 1991.
19- Kira, F. Fisiatría flebológica. Flebología. 3 - N° 1 a 3:227, 1990.
20- Laurence, A. Várices del miembro inferior. El Ateneo:149, 1949.
21- Levin, M.; O'Neal, L. El pie diabético. Editorial Elicien. España:111, 1977.
22- Medina, C.; Beltramino, R. Posibilidades tácticas y técnicas en el tratamiento de la insuficiencia venosa superficial. Flebología. 6 - N° 1:45, 1993.
23- Odisio, A. Las várices de los miembros inferiores. Editorial Akadia:11, 1979.
24- Pace, F. y colaboradores. Ligadura transcutánea ambulatoria de várices gigantes.
25- Pietrovallo, A. Estética en flebología. 6 - N° 1:27, 1993.
26- Sevilla, M. Las várices. Editorial Lumen:118, 1992.
27- Simkin, R. Enfermedades Venosas. Editorial López:98, 1979.
28- Spano, V. Microcirugía de las várices del pie. Flebología. 6 - N° 3:266, 1993.
29- Testut, L.; Jacob, O. Anatomía topográfica. Editorial Salvat. Tomo 2:971, 1034, 1075, 1964.

Phlebology '95, D. Negus et al. (eds.). Phlebology (1995) Suppl. 1: 413-415

V/1.9

O.C.R.A.M: Operacion Correctiva de Reflujo Ambulatoria por una Mini-incision (Corrective Operation of Ambulatory Reflux by a Micro-incision)

Prof Dr Alberto Hugo Saúl Nasi

Adjunct Professor of Phlebology and Lymphology at the Argentine University Jon F Kennedy. Scientific Secretary of the Panamerican Society of Phlebology and Lymphology. Phlebology and Lymphology Specialist. (A.M.A.) Medical and Surgical Licenciate. (Madrid, Spain). Director of the Phlebology Department at "Sanatorio Privado del Sudeste (Marcos Juárez City, Còrdoba Province, Argentine.)"

INTRODUCTION

This intervention consists of correcting reflux at the sapheno-femoral junction and in the internal saphenous vein below Hunter's perforating vein. This operation is done in order to segment the long saphenous vein in an attempt to normalise the pressure in the superficial veins. This in turn favours a return of blood to the circulation by means of the deep venous system when this is competent and patent. This operation preserves the long saphenous vein for possible future use in arterial or coronary revascularisation or as an arterio-venous shunt in patients undergoing renal dialysis.

This type of surgery is inadvisable in the presence of deep venous system disease. This operation would overload the deep veins, worsening the symptoms of venous disease. It is also ill-advised to perform this procedure in the presence of an unrecanalised femoro-popliteal deep vein thrombosis since the vein collateral route of venous drainage (a long saphenous vein) would be divided.

Patients with varicose veins should be investigated by clinical examination and in the vascular laboratory before surgical intervention. It is helpful to use a bidirectional Doppler ultrasound probe with graphical display in order to assess the extent and location of incompetent valves. Photoplethsymography is helpful in assessing the venous function. Pre-operative results compared with those obtained during post operative follow up. Where the diagnosis can not be reliably established using Doppler ultrasound and photoplethysmography, colour Duplex ultrasonography may be employed to clarify the situation.

A number of surgical techniques may be used to correct venous reflux. These include external valvuloplasty.

a) - By means of localised post-operative fibrosis (Nasi)

b) - With the use of the fascia cribosa (Mancini)

c) - Application of P.T.F.E. synthetic graft to support the vein (De Anna, Donini, Gasbarro, Corcos)

FRAGMENTATION OF HYDROSTATIC PRESSURE IN SAPHENOUS VEIN

a) - C.H.I.V.A. (Franceschi)[3] Haemodynamic Preservative Cure of Venous Insufficiency by Ambulatory Surgery

b) - 'Operation of Rosario'. Ligature and Division of the Long Saphenous Vein at the Sapheno-femoral Junction combined with Ligation of its Tributaries.

c) - O.C.R.A.M. (Nasi). Corrective Operation of Ambulatory Reflux by Mini-Incisions.

The strategy of O.C.R.A.M. is mainly based on disconnection of the long saphenous below Hunter's perforating vein where the calf muscle pump is inefficient. The thigh perforating veins act as a re-entrant conduit for the blood in the superficial venous system. Blood from the superficial veins is passed to the deep venous system without overloading the deep veins. In most instances, this procedure brings about disappearance of superficial varicosities because

1) - A haemodynamic corrective operation has been achieved: and

2) - Complementary micro-surgical removal of varicose veins assists the results of haematoma haemodynamic correction.

The residual proximal section of the long saphenous vein from the sapheno-femoral junction to Hunter's perforating vein undergoes neither fibrosis nor thrombosis and remains patent in more than 90% of cases. The diameter of this vessel remains between 5 and 8 mm following immediate post operative follow up investigations. The average length of vein available as a vascular graft is 30cm.

The remaining long saphenous vein is suitable for use as an arterio-venous shunt in patients with chronic renal failure needing haemodialysis, as a vascular graft in patients requiring peripheral vascular reconstructive surgery or coronary artery surgery, and also as a venous bypass graft. It can be used in a Palma's procedure in the case of contra-lateral iliac vein thrombosis or a Dale's operation if a venous thrombosis occurs in the ipsi-lateral iliac vein. None of these operations would be possible if a CHIVA or 'Rosario's operation' had been done.

The results of the procedure have been found to vary according to the stage at which venous valvular insufficiency was detected by Doppler ultrasound. The operation was more successful in the earlier stages of venous valvular incompetence.

Many phlebologists do not consider examination using Doppler ultrasound to be important. However many patients presenting with minor varices or telangiectaces may prove to have incompetence in the long saphenous vein. These patients may be most suitable for O.C.R.A.M. Such patients have veins which have not lost their normal histological structure and may revert to normal size following suitable surgical intervention. Later stages of the development of varices result in degeneration and fragmentation of the muscular fibres of the tunica media of the veins and it is less likely that these veins will return to normal, even with haemodynamic correction of the venous abnormality.

REFERENCES

1. Nasi A H S, Altmann Canestri E, Sànchez C, Barceló R, y Ulla J.: Les throboses Variqueses, leur traitement ambulatoire. Phlebologie, 1993, 46, N°4, 711-717.

2. Nasi A H S: Mètodos no invasivos de diagnòstico. Tratado de Flebologìa y Linfologìa. Ed. Fundaciòn Flebològica Argentina-Enero 1995.

3. Francheschi C: Theorie et pratique de la cura SHIVA. Editions de L'Armançon, 1988

Phlebology '95, D. Negus et al. (eds.). Phlebology (1995) Suppl. 1: 416–418

P216

Conservative Surgery of the Abdominal Subcutaneous Vein (A.S.V.)

O. Andoniades[1], R. Mirabella[1], E. Brizzio[2] and M. Conzi[1]

[1] *Department of Surgery, Phlebology and Limphology Section, Bernardino Rivadavia Hospital associated with the University of Buenos Aires, Argentina*
[2] *Headmaster of the Argentine School of Phlebology, Buenos Aires, Argentina*

INTRODUCTION

Internal iliac vein thrombosis is a relatively frequent pathology that will eventually result in a post-thrombotic syndrome and presents the risk of thromboembolism, wich is not the subject of this paper.

In the presence of deep venous obstruction there is an urgent need to detour the venous return by alternative passages. These passages may be natural or may be a creation of surgery, beginning with by-pass techniques, such as the Palma operation, among others. [1, 2]

Within the natural derivation passages there is the collateral venous net with uncountable communications, which due to the high venous blood pressure produced by the stumbling block, will be useful to drain away blood flow and also rely on the help of extra pelvic branches of the hypogastric vein: gluteal, internal pudenda, ischiatic and obturator veins.[1, 2, 3, 4, 5, 6]

Based on this knowledge, our sugical team made the decision to preserve the A.S.V. through surgery on the long saphenous vein, defending in this way a natural by-pass between the upper cava and the lower cava systems. This anatomic disposition has been proved by embriologic investigation.

Likewise in case of obstruction of the superior cava system or in case of portal high blood pressure, this same circuit helps to drain the blood flow away to the inferior cava system, passing through external and internal mammary veins, abdominal thoracic veins, umbilical veins and epigastric veins to the A.S.V. and , finally, to the inferior cava system.[1]

These conditions have to be considered when an operation over the saphenofemoral junction is to be performed.

METHODS

Our findings are based on experience with fourty patients who were chosen 3 years ago. All of them had total insufficience in the saphenofemoral junction valve. The final decision of preserving the A.S.V. was adopted into the operation following two pre-established standards:

a) Outlet of the A.S.V. near the saphenofemoral angle, which was present in 92.5 % of patients (fig. 1).

b) Normal anatomic shape of the vein, which means without varicosities. Only one patient was eliminated for this circumstance. In fact, this patient represented only 2.5 per cent of the total (fig.2).

All patients were thoroughly studied with complete semeiology, Doppler ultrasound and Duplex scanning.

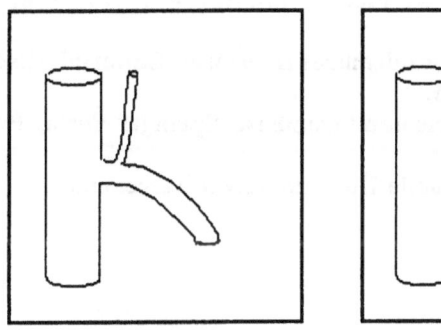
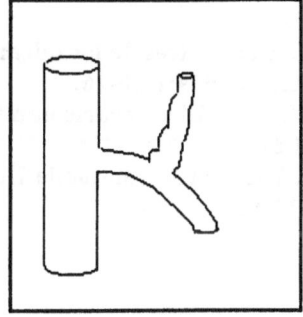

Figure 1   Figure 2

RESULTS

During these three years we lost contact with 11 patients. The other 29 have been examined every four months. None of them showed either relapses or pathology of the residual stump due to saphenoctomy preserving the A.S.V.

DISCUSSION

Our strategy was based on preserving an important natural by-pass [1, 2]. We made sure that this modification to the original saphenofemoral junction surgery technique did not produce any problems with the residual stump nor relapses. In addition, comparing patients operated on with or without the ecthomy of the A.S.V. there were no differences in further results.

CONCLUSIONS

We consider that modern surgery might avoid the ligation of the A.S.V. in long saphenous stripping, maintaining a natural by-pass that will behave as a two-way buffer circuit in case of obstruction, not only in the inferior cava system, but also in the superior cava and portal systems, and that the interruption of the A.S.V. is not necessary in the majority of cases.

REFERENCES

1. Albanese A. Flebopatias.Interamericana. Buenos Aires,1986; 2,3:22-27.
2. Albanese A. Anatomia venosa. International Congress of Angiology. London . 1965.
3. Garcia Mendez A., Iussem M., Flebopatias. Buenos Aires. 1986; 5,9:47-63.
4. Odisio A. Las varices de los miembros inferiores. Editorial Akadia. Buenos Aires, 1979;1:20-26.
5. Sigg K. Varizen-Ulcus cruris und thrombose. Springer Verlag.Berlin. 1968; 1:2-5.
6. Testut L. Latarjet A. Anatomia Humana.Salvat.Barcelona. 1954:3:512-518.

Phlebology '95, D. Negus et al. (eds.). Phlebology (1995) Suppl. 1: 419-421

OP/17.4

Surgery for Primary Troncular Varicose Veins Without Stripping the Saphenous Vein - Pre- and Post-Operative Evaluation by Duplex Scan and Photoplethysmography

F.P. Fonseca[1,3], A.L. Sarquis[2] and S.S.M. Evangelista[3]

[1] Department of Surgery, Federal University of Minas Gerais, Brazil, [2] Member and [3] Senior Member of the Brazilian Society of Angiology and Vascular Surgery

INTRODUCTION

We assessed the anatomy and function of the venous system before and after surgical treatment for primary troncular varicose veins, with preservation of the long saphenous veins (LSVs), which are of utmost importance to arterial reconstructions and coronary surgeries.

PATIENTS AND METHODS

This is a prospective study, in progress, of a consecutive series.
We operated on 28 limbs of 26 patients, 22 women and 4 men, age 47 $\pm$ 13, for treatment of primary troncular varicose veins without stripping the LSVs. We have excluded patients with deep venous thrombi, deep venous reflux or previous venous surgery. We assessed pre and post-operatively :

A) By photoplethysmography (PPG): the venous returning time (VRT) at the ankle level.

B) By duplex scan: the superficial and deep venous system studying the patency, direction of flow, site and extent of reflux, perforating veins and diameters of the LSVs at 7 levels. To mark these 7 levels we draw a line, with ink, on the cutaneous projection of the LSV, with the aid of a continuous wave Doppler. Then we ask the patient to flex the leg, and in the sulcus formed, at the knee, we draw a line crossing the LSV mark line. We call the intersection of these 2 lines the K point (K after knee). With a ruler we mark 3 points, on the LSV line, 10, 20 and 30 centimeters distal to the K point and another 3 points, 10, 20, and 30 centimeters proximal to K (Fig. 1). We draw these lines, which are precisely reproducible, before each duplex scan examination, done pre and post-operatively and before operation. They are used to make a cartography of the perforator veins, to compare the LSV's diameters pre and post-operatively at the 7 levels, to describe precisely the site and extent of the relfux and the reverse flow in the LSV pre and post-operatively, to relate precisely the site and extent of post-operative thrombi in the LSVs and to mark sites of transference of reflux, being all these parameters assessed by the duplex scan, and to place the photoplethysmography probe always in the same position.

In 23 limbs there was reflux at the saphenofemoral junctions (SFJ), and the LSVs were ligated and divided adjacent to the common femoral vein, with a flush ligation, after all the terminal tributaries (TT) of the LSVs had been ligated and divided [1]. In 5 limbs there was no reflux at the SFJs, but the LSVs were dilated and with reflux, distally to insufficient thigh perforating veins or insufficient external pudendal veins that joined the LSVs more distally. In these cases the SFJ and the TT were left in place. The long saphenous veins were never removed, insufficient perforators were ligated, and varicosities were avulsed through 1-2 mm incisions with crochet needle. The surgeries were done under local anesthesia on an outpatient basis.

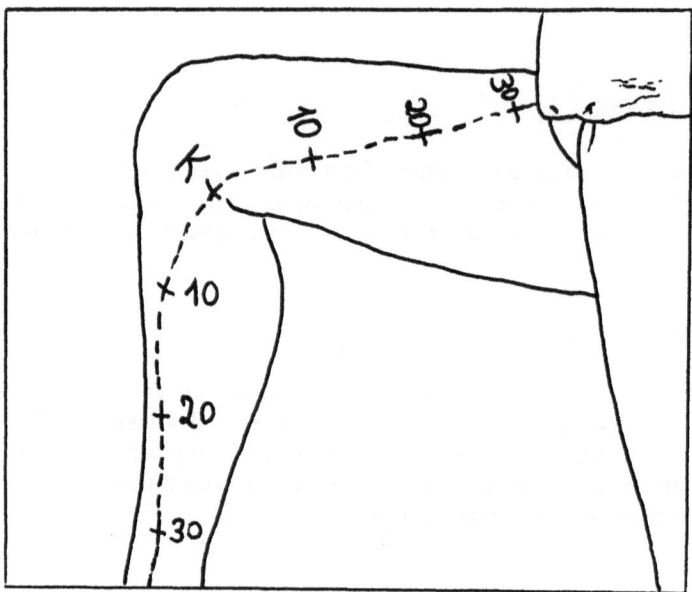

Fig. 1. LSV cutaneous projection drawn on the skin with the aid of a continuous Doppler. The point **K** is the intersection of the line formed by the sulcus at the flexed knee joint with the LSV's line. The 3 points on the thigh and the 3 on the leg are 10, 20 and 30 centimeters distant to **K**.

RESULTS

We assessed post-operatively all the 28 limbs, after 1 month. In 19 limbs (68%) the LSVs were preserved in almost their full extensions with small 1-2.5 cm thrombi at the saphenous high ligation level. In 9 limbs (32%) the thrombi were longer, 6-30 cm and the LSVs were 90 to 60% preserved respectively. The larger diameters of the LSVs, found pre-operatively, mean 6.0 ± 2.2 mm, were significantly different from the diameters, measured post-operatively at the same level, mean 3.6 ± 1.5 mm. (t= 7.72 p< 0.001). Before surgery the VRTs were normal in 7 limbs (24 ± 4.5 seconds), impossible to be measured in 1, and abnormal in 20 limbs. In the group of

20 limbs with abnormal VRTs, the mean VRT at the ankle level before surgery was 12.32 ± 3.9 seconds, and after surgery 19.77 ± 6.9 seconds (t=6.198, p<0.001).

We assessed post-operatively 11 limbs after 1 year, clinically, by duplex scan and by PPG. According to the clinical classification of Chronic Venous Insufficiency (CVI) [2] of these 11 patients:

Six patients changed from Class 1 (mild CVI), before operation, to Class 0 (asymptomatic) 1 year after operation, 1 patient change from Class 3 (severe CVI with ulcerative skin changes) to Class 0, 3 patients remained in Class 0 (esthetic improvement), and 1 patient remained in Class 1.

In 8 patients there was no reflux in the LSVs, but there was a reverse flow characterized by low and constant velocity. In 4 limbs (36.5%) the velocity of the reverse flow did not increase with Valsalva's maneuver, and in 4 limbs (36.5%) the velocity increased slightly. The mean VRT at ankle level was 23 ± 3.8 seconds in this group.

In 3 limbs (27%), there was reflux in the LSVs, characterized by rapid and large reflux velocity increase with Valsalsa's maneuver. In one limb, 2 TT of the LSV were left in place and in 2 limbs, there were thigh perforators with reflux, one per limb. The mean VRT at ankle level was 13.7 ± 1.7 seconds in this group (t= 3.6, p<0.01).

The diameters of the LSVs were found the same 1 month and 1 year post-operatively (t = 1.06, p>0.1) and significantly smaller than the diameters found pre-operatively (t = 3.09, p<0.005).

CONCLUSIONS

We found that this saphenous saving surgery has the following advantages:

1) Preservation of the LSVs in almost its full length in 62% of the cases, or in a significant extension, 60- 90%, in 32% of limbs.

2) Abnormal VRTs increased.

3) Surgery is less traumatic, can be done under local anesthesia on an outpatient basis.

4) Approval of patients for surgery is easier.

REFERENCES

1. Hammarsten J, Pedersen P, Cederlund CG, Campanello M, Long saphenous vein saving surgery for varicose veins. A long-term follow-up. Eur J Vasc Surg 1990;4:361-4.
2. Porter JM, Rutherford RB, Clagett GP, Cranley JJ, O'Donnel TF, Raju S, et al, Reporting standards in venous disease. J Vasc Surg 1998;8:172-81.

Phlebology '95, D. Negus et al. (eds.). Phlebology (1995) Suppl. 1: 422-424

P084

The Method of Anaesthetic Application in Surgical Treatment of Varicose Veins in Day Surgery

Z. Fiutek, K. Twardowska-Saucha and D. Czaczka

Varicose Vein Clinic "Medservice", Zabrze, Poland

INTRODUCTION

Surgery of varicose veins may be successfully performed in general, intraspinal, block and local anaesthesia. Each method has certain advantages and disadvantages.

Adequately administered regional anaesthesia provides excellent intraoperative pain control, with normal ventilatory or cardiovascular function, causing fewer bowel problems, fewer postoperative lung complications and less mental confusion than conventional analgesia [1]. In addition to improved pain control, the endocrine metabolic responses to postoperative pain may be significantly reduced by neural blockade [2]. The patients receiving local anaesthesia may have shorter recovery time than those requiring general anaesthesia [3].

The essential for the regional blockade is an adequate knowledge concerning femoral nerve topography.

The femoral nerve is the component of the lumbar plexus. It emerges from the lateral border of the psoas major muscle in the iliac fossa, running down in the groove between psoas and iliacus deep to the iliac fossa, and entery the thigh passing beneath the inguinal ligament to the lateral side of the femoral arthery. It is the nerve of supply to the muscles in anterior compartment of the thigh and supplies the skin over the anterior aspect of the thigh. The terminal branch of the femoral nerve is the saphenous nerve, which passes through the canal under the sartorius muscle to reach the medial aspect of the knee. Then it is usually divides into two nerves, which accompany the saphenous vein on either side and pass anterior to the medial malleolus, sometimes extending the proximal phalanx of the great toe. Thus, the saphenous nerve supplies a strip of skin down the medial side of the leg from knee to great toe [4].

MATERIAL AND METHODS

In Varicose Vein Clinic „Medservice", 320 operations were carried out between October 1993 and December 1994. The age of patients ranged from 13 to 78 (mean 43). The interventions concerned the complete and limited stripping of LSV, ligation and stripping of SSV. These interventions were often accompanied by the local phlebectomy. All these patients received the type of anaesthesia presented below. In addition to standard preoperative evaluation, an examination of the landmarks needed for the particular technique is necessary. The preoperative sequence includes: premedication, transport to the operating room, positing for the nerve block, intravenous needle insertion and infusion of 0.9% NaCl. The injection of local anaesthetics is often associated with severe discomfort. Therefore, the use of adjunctive intravenous sedative and analgesic drugs (so called conscious sedation technique) during the local anaesthesia is highly recommended. In our Clinic intravenous Diazepam (5-10 mg) and Pentazocine (15-30 mg) are usually administered just after the preparation of the operation site.

Next, using 22 G needle, 1-2 cm. laterally from the femoral artery, close to the inguinal ligament 10 - 15 cc. of 1% Xylocaine is injected. The skin at the region of planned incision site is infiltrated with the use of 0.5% Xylocaine. If there is a need of some extra incisions in the region out of the femoral nerve supply this site is also infiltrated with the same solution. During the intervention we have a constant verbal communication with the patient and continuous monitoring of ECG parameters. It makes it possible to react immediately if any discomfort occur.

RESULTS

By means of this type of anaesthesia we managed to carry out all the surgical interventions with positive effects. This type of anaesthesia was highly accepted by the patients. 5% described the analgesic effect as acceptable, 23% as sufficient, 72% as good or very good.

Patients remained in the recovery room until stable cardiovascular and respiratory variables were monitored for at least one hour. Patients were mobilised 1-2 hours after operation what is crucial in deep venous thrombosis prophylaxis. We observed the nausea and vomits in 8% of subjects. These troubles retreat spontaneously or after Metoclopramide administration. The pain in the operated leg was alleviated with the help of Tramadol in 37% of patients. It was administered from 1 to 4 days after surgery 2-3 times daily.

DISCUSSION

Regional anaesthesia in general and nerve block in particular provide ideal operative conditions. However, there are disadvantages to the use of nerve block. The most significant is failure of the nerve block to provide analgesia. The incidence of failure depends on the skill and experience of anaesthesiologist. The success with regional nerve block depends on insertion of the needle near the target nerve and keeping it there during injection [4].

The knowledge about the pharmacology and farmacokinetics of the drugs used in the whole process is mandatory.

The intravascular injections may sometimes produce some systemic adverse reactions. Such a crises are best prevented by frequent aspiration tests to ascertain that the needle has not entered the lumen of a blood vessel and by strict observation of safe limits of injection dose.

The sedative hypnotics can also be administered through intravenous catheter placed in a site that will not interfere with the planned surgery. Diazepam is very useful in the situations in which local anaesthetic agents are used. Diazepam 0.15 mg/kg raises the convulsion threshold to local anaesthetics by about 50% and therefore is a useful premedicant [3]. The combination of Diazepam with Pentazocine often produces a very tranquil and co-operative patient who can often give rational feedback with regard to paresthesias or pain but not experience to emotional reaction reflexes.

Regional anaesthesia has then the significant benefits in both the intervention and the postoperative period but success strongly depends on having a good rapport with the patient and the ability to have the psychological contact with the patient in the immediate preoperative period.

CONCLUSIONS

In our opinion femoral nerve block accompanied by local anaesthetic infiltration and intravenous premedication with Diazepam and infusion Pentazocine is a method of choice in varicose vein surgery. It is safe, sufficient and convenient. It is also well tolerated by the patients. Such an anaesthesia is also the best prophylaxis against deep vein thrombosis.

REFERENCES

1. Widlund L Regional blockade vs analgesic therapy. Acta Anaesth Scand 1980; 74:169.
2. Kehlet H. The endocrine-metabolic response to postoperative pain. Acta Anaesth Scand 1982;74:173.
3. De Jong RH. Local anaesthetics. Springfield, JLL, Charles C. Thomas, 1977
4. Terence N, Murphy MB. Nerve block. In: Anaesthesia. Second Edition. Churchill Livingston 1986, 1035.

Phlebology '95, D. Negus et al. (eds.). Phlebology (1995) Suppl. 1: 425-426

P217

Cosmetic Phlebectomy Hooks

Z. Fiutek, and K. Twardowska-Saucha

Varicose Vein Clinic "Medservice", Zabrze, Poland

INTRODUCTION

The purpose of every operation of varicose veins of lower extremities is the achievement of the best possible effect both functional and cosmetic. This is possible due to prior careful diagnosing and precise marking of varicosities before the operations. The use of specific instruments such as Varady hooks [1] makes it possible to remove them through very small incisions. Since we had no access to the hooks used for "microphlebectomy" we constructed the hook which is a bit different from the ones used so far but still we might find it useful.

In the course of preparing the present paper we found the information about the offer of International Medi Surgical, Dallas Tx, Dortu-Raymond Martinbeau blunt hook [2], which is very similar to ours.

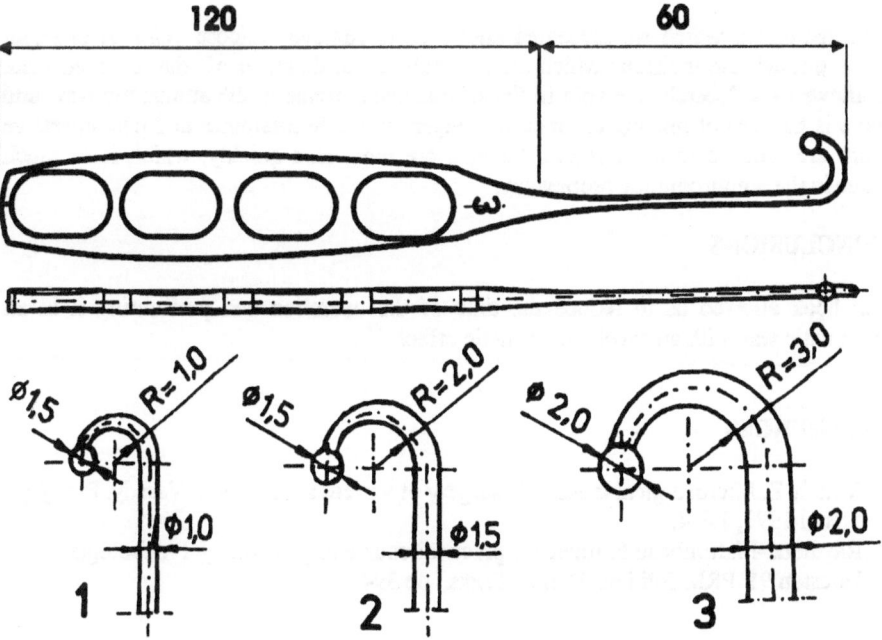

MATERIAL AND METHODS

Our instrument is made of non-flexible stainless steel. It is made in three different sizes (Fig 1).

Its characteristic feature is its blunt tip which takes the form of a ball. Due to this tip we can easily penetrate the subcutaneous tissue, the size of the cut depends on the diameter of the vein which has to be removed, but not on the size of the hook. The hook has a universal character; it can be used for the removal of very big as well as very small veins. It is of particular importance that the distal incision while removing long saphenous vein (LSV) may not be longer than 0,5 cm (invagination method).

From November 1993 to December 1994 we performed 300 operations of varicose veins of lower extremities, including full stripping or limited stripping of LSV, and/or stripping or ligation of short saphenous vein (SSV), with local phlebectomy.

The invagination method was always used for stripping of LSV or SSV. Only the wounds in the groin and in popliteal fossa were sutured. The remaining incisions were dressed with adhesive tapes "steri-strip". The average time of operation was 1 h.

Follow up observations was performed one week and 1-3 month after the operation.

The patient was asked whether, in their opinion, the cosmetic effect had been very good, good or bad. The patient was also asked whether he was taking analgesic drugs.

RESULTS

80% of patients did not take any analgesic drugs after the operation. 60% of patients evaluated the cosmetic effect as very good, 40% as good, none of the patients estimated the cosmetic results as bad.

DISCUSSION

The instrument which we presented allows us to remove varicose veins of different sizes, through the incisions which are as small as the diameter of the removed vein. It allows us to "search" the vein in the subcutaneous tissue in the atraumatic way, and there is no need of making the incision longer. so the haematomas and post operative pains are reduced to a minimum. Patients are mobilized quickly, which is of much value in the antithrombotic prophylaxy.

CONCLUSIONS

Our hook allowed us to reduce the time of the operation and performing it in an atraumatic way with an excellent cosmetic effect.

REFERENCES

1. Varady Z. Microsurgical method in surgery of varices according to Varady. Przegląd Flebol 1993; 1:3-4.
2. Raymond-Martimbeau P. Intensive practical course in phlebology. Phlebologia Houston 91 PRM Editions Dallas, Texas, pp 394

Phlebology '95, D. Negus et al. (eds.). Phlebology (1995) Suppl. 1: 427–428

P218

Limited Stripping and Phlebectomy in Varicose Vein of Lower Extremities in Day Surgery

Z. Fiutek, K. Twardowska-Saucha J. Kuleszynski and M. Filipowski

Varicose Vein Clinic "Medservice", Zabrze, Poland

INTRODUCTION

Surgery was one of the early treatments of varices. Hippocrates described puncturing of varices and pressing out the contents from below. Celsus coined the term cirsotomy, which is still in use today, and means incision of varix, and Galen described the extirpation of varices as well as their cauterisation. Ligation of vessels was first practised by Antillus and was then reintroduced by the Pare. Oribasius made short incisions over varices and removed the dilated vessels bit by bit. Operations for varicose veins are associated with famous names including Schede, Trendelenburg, Perthes, Rindfleish, Mayo and Babcock. The work of Babcock marks the beginning of modern surgical treatment of varices. A number of surgical methods are available; the choice is normally governed by the type of varix [1]. The innovations introduced by Van der Strichta [2], Mullera and Varady'ego [3] require particular attention. We operate our patients in the way shown below, according to both, the achievements of past, and those of contemporary surgeons.

MATERIAL AND METHODS

All patients (230 persons) admitted to Varicose Vein Clinic "Medservice" were carefully diagnosed by means of physical and Doppler examinations. Patients with long saphenous vein (LSV) and/or short saphenous vein (SSV) insufficiency were directed to surgery. In the operating day all varices which were decided to be removed were marked with indelible pencil. After vein catheterization Diazepam in dose 5-10 mg was administered. Patients were placed on the operating table in Trendelenburg position. When the operating field was prepared, femoral nerve block and local infiltration anaesthesia had been performed. Pentazocin in dose 15-30 mg was administered intravenously. A 5 cm transverse incision was made in the groin fold. The sapheno-femoral junction was carefully exposed. All the tributaries were ligated and cut off. Stripping of LSV was conduced by an invagination technique to the level below the knee. The distal part of LSV was fragmentarily removed only in the case when it was varicously changed. Femoral and crural varices were removed using special hook from a few minute incisions. Only incision in the groin were closed with sutures. Another ones were spontaneously supplied with adhesive tapes. SSV operation was conduced in abdominal position. A 4 cm incision above the place of sapheno-popliteal junction was made. SSV was removed by invagination technique or only ligated. The operation was finished by the application of the sterile dressing and elastic compression bandage. After 3 hours patients were able to walk and were released home from the clinic. Control investigations were performed in the 1st week and 6th month after surgery.

RESULTS

Purulation of post-operative wound was present only in one case. There were no post-operative paraesthesia. No patient got worse after the operation. Additional post-operative sclerotherapy was performed in 20% of cases.

Cosmetic effect was estimated as very good by 60% patients, and good by 40% after 6 month. Only 4 subjects (about 2%) assessed cosmetic results as bad. Only 20% patients took analgesics in after operation. The others did not need them.

DISCUSSION

Careful clinical and Doppler examinations allowed us to give up X-ray picture in most cases. Local anaesthesia together with intravenous injection of analgesic and sedative drugs allowed us to perform every operation on varices of lower limbs. Negative side of this method is that right blockade of n.femoralis demands much experience. Atraumatic stripping LSV by invagination causes that haematomas and following post-operational pain are minimal. Patients are willing to walk a lot. Anaesthesia without myorelaxation, quick mobilization, atraumatic surgical proceeding; all of this make excellent antithrombotic prevention. Surgery and sclerotherapy form the combination of complementary treatment for incompetent main superficial valves, that give the varicose vein patient better functional and cosmetic results than could be obtained using either method alone [4].

Incidence of injury to the saphenous nerve may be reduced if the LSV is stripped from groin to upper calf only [5].

CONCLUSIONS

The procedure that we presented here is radical, safe and gives very good cosmetic results in the treatment of varices in the conditions of one day surgery.

REFERENCES

1. Large J. The treatment of varicose veins; a personal review. Phlebology 1990;5:141-146.
2. Van Der Stricht H.R. Saphenectomiae par invagination sur fil. La presse Medicule Paris 1963;1081-82.
3. Varady Z. Microsurgical method in surgery of varices according to Varady. Przegl.Flebol. 1993;1:15-18.
4. Walter P., de Groot M. D. Treatment of varicose veins; modern concepts and methods. J Dermatol Surg Oncol 1989;15:2.
5. Browse NL, Burnand KG, Lea Thomas M. Diseases of the veins. Pathology, diagnosis and treatment. London; Edward Arnold 1988:214.

Phlebology '95, D. Negus et al. (eds.). Phlebology (1995) Suppl. 1: 429-431

V/1.7

New Technique of Varicose Vein Operation by Means of the "Varix-Set" Instruments

M. Pardela[1], P. Matousek[2] R. Zbronski[1], K. Twardowska-Saucha[1] and A. Urbaniak[1]

[1] Varicose Vein Clinic "Medservice", Zabrze, Poland
[2] Municipal Hospital Kromeriz, Czech Republic

INTRODUCTION

Regular care of the patients with varices of lower limbs and efforts to perform radical operation with good and long lasting functional and cosmetic effects were the reason to invent and design new original technics and instruments for varicose vein operation [1,2,3]. One of them is new technique of varicose vein operation by means of new original instruments "VARIX-set" constructed in Czech Republic, produced by RYVA.

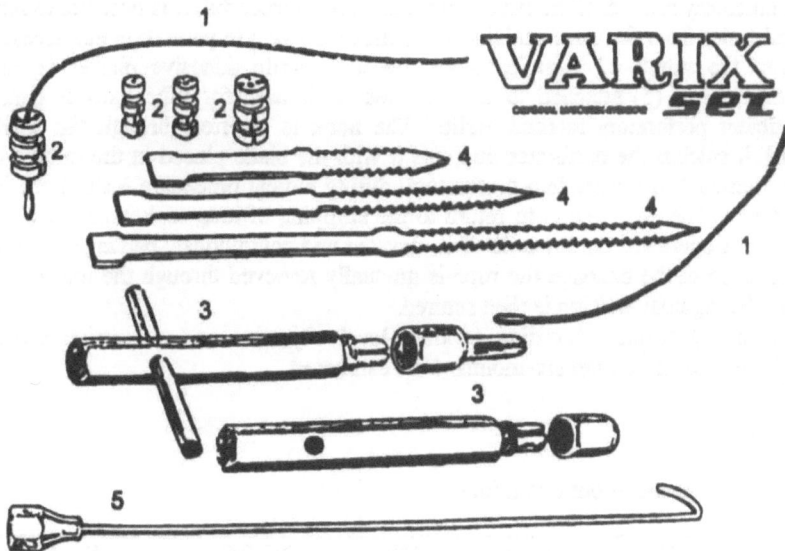

Fig 1. Components of „VARIX-set"
1. stripper rope, 2. cylindrical stripper, 3. universal handle, 4. extirpator,
5. communicotom

430

MATERIAL AND METHODS

428 patients (487 limbs) with primary varicose veins were operated on in two centres: Varicose Vein Clinic in Zabrze, Poland and Municipal Hospital in Kroměříž, Czech Republic. All of them had complete physical and Doppler examination of veins and arteries of lower limbs. The range of surgical procedure was determined strictly individually and surgeons had to mark all varices and insufficient perforators before the operation [4,5]. We used general or local anaesthesia of femoral nerve.

The description of the surgical procedure:

Short incision in inguinal fold is made and then sapheno-femoral junction and saphenous vein branches are dissected, ligated and finally transsected. Distal part of saphenous vein is exposed from short incision nearby medial ankle. After transsection of this part, the distal one is ligated and stripper rope (1) is inserted into the vein lumen and pushed through it proximally to the groin and its end is extracted out of the vein. Universal handle (3) is attached to the bullet at the proximal end of the rope. The cylindrical stripper (2) is attached to the distal end of the rope. Then, universal handle and rope are gently pulled and cylindrical stripper is located subcutaneously in the lumen of long saphenous vein. The incision is sutured. Now this step is finished and it is ready for stripping of saphenous trunk.

The next part of the surgery is to remove varicosities and phlebectasies by means of extirpator (4) of the necessary length through minimal (3-4mm) skin incision. Extirpator is inserted percutaneously in the direction of varicose area marked on the skin before operation. It is not necessary to insert extirpator strictly intraluminally. Meanders are punctured directly. The instrument is then rotated several times around longitudinal axis until only slight resistance is felt. Then extirpator is extracted and simultaneously rotated. If the first extirpation is not successful, it is possible to repeat it several times from the same incision and remove the vein in parts. It is not necessary to perform the suture of these incisions, because sterile adhesive plaster is enough. Communicotom (5) attached to universal handle is used for subcutaneous cutting of insufficient perforators marked earlier. The hook is inserted through the small cut (2mm), it catches the perforator and cuts it with the blade placed in the middle of the hook. Removal of insufficient perforators during radical procedure lasts short, is safe and simple. The next step is to return to the stripping of long saphenous vein. During this step of operation the lower limb is elevated and continuously bandaged. Under the compression of the bandage the rope is gradually removed through the incision in the groin. The inguinal incision is then sutured.

Effects of operations according to three-level objective and subjective score and Doppler examination after six-months were estimated.

RESULTS

Table 1. The results of our operations

RESULTS	VERY GOOD	GOOD	BAD
OBJECTIVE	364 (75%)	107 (22%)	16 (3%)
SUBJECTIVE	352 (72,3%)	123 (25,3%)	12 (2,4%)

CONCLUSIONS

New operation technique by means of „VARIX-set" instruments has given very good objective and subjective as well as cosmetic effects. These new instruments are easy in used and safe in the hands of an experienced surgeon and can be use in outpatient procedure. They allow to reduce markedly operation time.

The most common indication for radical surgical treatment using „VARIX-set" instruments are varices of long saphenous vein and collateral branches, true and false recurrences and residual varices.

REFERENCES

1. Babcock WW. A new operation for the extirpation of varicose veins of the leg. N Y Med J 1907; 86: 153
2. Muller R. La phlebectomie ambulatoire. Phlebologie 1978; 31: 273.
3. Pardela M, Lityński W. Surgical treatment of lower limb varices with preservation of femoral segment of the long saphenous vein. Ann Acad Med Siles. 1981; 5: 1.
4. Bishop C.C.R., Jarrett S. Outpatient varicose vein surgery under local anaesthesia. Br J Surg 1986; 86: 153.
5. Clinton O, Negus D. Suitability for day-care varicose vein surgery. Phlebology 1990; 7: 20.

Phlebology '95, D. Negus et al. (eds.). Phlebology (1995) Suppl. 1: 432-434

P219

Operations of Varicose Veins of Lower Extremities Using Babcock Procedure vs "Varix-Set"

R. Zbronski[1,2], P. Matousek[3], M. Pardela[1,2], K. Twardowska-Saucha[1] and A. Urbaniak[1]

[1] Varicose Vein Clinic "Medservice", Zabrze, Poland
[2] Clinical Department of General and Vascular Surgery Silesian Medical School, Zabreze, Poland
[3] Municipal Hospital Kromeriz, Czech Republic

INTRODUCTION

Varicose vein of the lower extremities and their complications are the most popular disease in developed countries. We observe it in 5-30 % of European population. Many authors emphasize more frequent occurrence of this disease in women [1].

Among operative manners Babcock operation is dominant. A functional result of this method is very good but cosmetical one is worse because collateral varicose veins are removed by numerous incisions (multiple, visible scars) [2, 3].

MATERIAL AND METHODS

We compared two groups of surgically treated patients: I - 300 patients (386 legs) operated on Babcock procedure and II - 428 patients (487 legs) using „VARIX-set" instruments (new original instruments for varix operation produced by RYVA in Czech Republic).
In patients from the first group only basic clinical examination was completed and patients from the second group had also Doppler examination. Both groups had veins and arteries assessed.
In the first group we used general or subspinal anaesthesia and in the second group general anaesthesia or femoral nerve blocking with 1% xylocaine was performed.

Clinical outcome using simple scoring system to assess functional and cosmetic results (three grade score: very good, good and bad) was documented three and six months after the operation.

RESULTS

Table 1. Results of our operation.

Type of operation	Very good	Good	Bad
Babcock procedure	222 (57,4%0	120 (31%)	44 (11,6%)
VARIX-set operation	364 (75%)	107 (22%)	16 (3%)

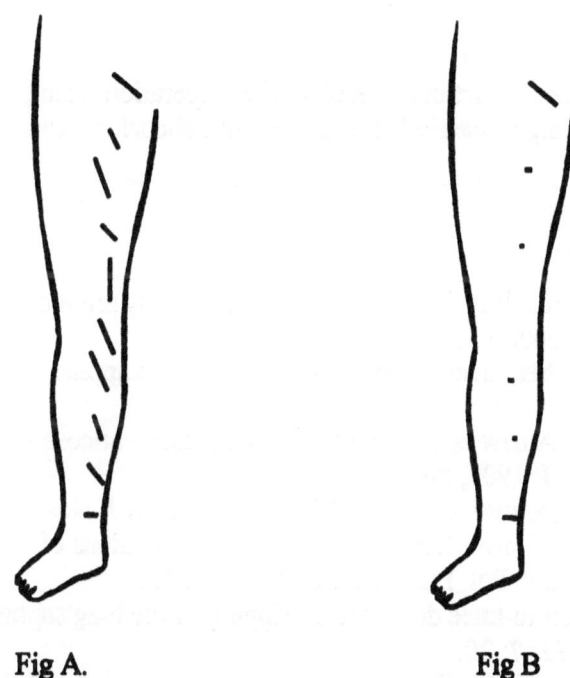

Fig A.   Fig B

Fig. A. Cosmetic effect after Babcock procedure.
Fig. B. Cosmetic effect after operation by means of VARIX-set operation.

DISCUSSION

The most popular surgery in treating varicose veins of lower limbs is Babcock surgery. Since females happen to suffer from varicose veins and undergo surgery veins more frequently often for cosmetic reasons as well, the most efficient surgery should then provide good functional and cosmetic results [1, 4].

Surgery including partial removal of long saphenous vein or its ligation have been performed and surgical treatment has been combined with obliteration.

Common features of these methods are the attempts to limit the number and size of skin incisions in order to reduce the number of scars [5.6].

However, in operation by means of VARIX-set instruments in about 30% of cases relatively large haematomas are being observed (though they are absorbed without major complications). Still this method seems to be a better solution then increasing the number of incisions in order to reduce bleeding like in Babcock's method.

CONCLUSIONS

Both cosmetic and functional results after operation using „Varix-set" instruments were significantly better than after Babcock procedure.

REFERENCES

1. Beaupre L, Perreault A. Varicose veins: a frequently missed diagnosis. The Canadian Journal of Diagnosis 1990; 10: 153
2. Hobbs JT. The treatment of venous disorders. J.B. Lippincott Cie. City. 1977
3. Babcock WW. A new operation for the extirpation varicose veins of the leg. N.Y. Med. J. 1907; 86: 153.
4. Andziak P, Witkowski M, Sowiński A, Ruszkowski J, Noszczyk W. Varicose veins of lower limbs in the selected populations of Bródno district in Warsaw. Pol. Przegl. Chir. 1987; 59: 465.
5. Conrad P. Grion-to-knee downward stripping of the long saphenous vein. Phlebology 1992; 7: 20.
6. Zbroński R, Pardela M. Early results of limited stripping of LSV and additional sclerotherapy in the treatment of primary varicose veins. Phlebological Review 1993; 1: 40.

Phlebology '95, D. Negus et al. (eds.). Phlebology (1995) Suppl. 1: 435–438

OP/5.4

Intérêt de L'Incision à L'Aiguille Fine au Cours des Phlebéctomies Ambulatoires Selon Muller

N. Fays-Bouchon and S. Fays

40 Avenue Foch, 54000 Nancy, France

Fine Needle Incisions in Out-Patient Micro Phlebectomies

SUMMARY

The authors report on the result of a telephone enquiry involving 81 practitioners on the instrument used for the incision in micro phlebectomies. Two out of these used a fine scalpel, one out of six a needle and the other one out of six used either.

INTERET DE L'INCISION A L'AIGUILLE FINE AU COURS DES PHLEBECTOMIES AMBULATOIRES SELON MULLER

FAYS-BOUCHON N. [1], FAYS S [1].

[1] 40, Avenue FOCH, 54000 NANCY, FRANCE

INTRODUCTION

Dans le but constant d'améliorer la qualité esthétique des cicatrices cutanées des phlébectomies selon la technique de Muller [1], on a été amené à utiliser des bistouris de plus en plus fins, et finalement à recourir à une simple aiguille. L'incision à l'aiguille, publiée la première fois [2] au X° Congrès Mondial de l'Union Internationale de Phlébologie (Strasbourg 1989), reprend en fait une idée de Ricci [3], [4]. Ce procédé est loin d'avoir séduit tous les opérateurs. Pour évaluer l'intérêt de l'aiguille par rapport au bistouri classique, nous avons entrepris une enquête auprès de Phlébologues français pratiquant la phlébectomie ambulatoire.

MATERIEL ET METHODE

Nous avons sélectionné 81 praticiens représentatifs des différentes régions françaises, lesquels ont été interrogés non pas par questionnaire comme c'est traditionnel mais par entretien téléphonique, ce qui nous a permis d'obtenir un chiffre inespéré de réponses (80 sur 81). Les questions posées ont été les suivantes : instrument habituel d'incision (aiguille ou bistouri)? - type et nature des crochets utilisés? - moyen de fermeture cutanée? - qualité de la cicatrisation (complications et réactions locales)?

RESULTATS DE CETTE ENQUETE

Le bistouri est l'outil de la majorité des opérateurs. Pour 66% de ceux-ci, c'est le seul instrument d'ouverture. Certains (17%) utilisent tantôt l'aiguille, tantôt le bistouri auquel ils ont recours pour extraire les varices de gros calibre. Il y a seulement 15% des opérateurs qui emploient exclusivement l'aiguille. Rares sont les praticiens (2%) qui font appel à des instruments particuliers (crochet acéré, poinçon).

Les aiguilles, à quelques exceptions près, sont de type microlance BD ®. L'aiguille rose (40mm de long, 12/10 mm de diamètre) est employée par les 3/4 des utilisateurs. Les autres calibres sont d'usage plus restreint : blanche (11/10) : 8% - jaune (9/10) : 4% - verte (8/10) : 8% - noire (7/10) (4%).

Le plus souvent les crochets sont de type Muller et fabriqués par les établissements Padulli (Pont-Astier 63190 Lezoux - France). Après avoir interrogé les 12 premiers confrères, il nous est apparu que le modèle de crochet semblait dépendre de l'outil d'incision. En conséquence, aux 69 collègues interrogés par la suite nous avons posé cette question :"Pensez-vous que le mode d'incision cutané influence le choix de votre crochet?". Les résultats sont répertoriés dans le tableau 1. Les praticiens sont répartis en trois groupes, groupe I : bistouri uniquement - groupe II : aiguille uniquement - groupe III : les deux instruments utilisés en complémentarité. Les tailles de crochet dans chaque catégorie sont réparties selon les 4 modèles classiques de Muller, du plus fin au plus gros, en 1, 2, 3 et 4. On constate que les opérateurs à l'aiguille n'em-

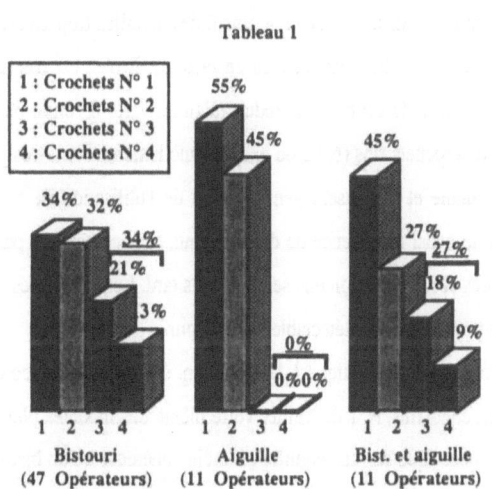

Tableau 1

1 : Crochets N° 1
2 : Crochets N° 2
3 : Crochets N° 3
4 : Crochets N° 4

Bistouri (47 Opérateurs)
Aiguille (11 Opérateurs)
Bist. et aiguille (11 Opérateurs)

ploient jamais de crochet de calibre supérieur au N° 2, alors que ces derniers sont utilisés par 34% des opérateurs du groupe "bistouri" et 27% de ceux du groupe "bistouri et aiguille". On peut en conclure que ce sont les opérateurs qui recherchent les incisions les plus réduites qui emploient le plus fréquemment les crochets les plus fins.

Pour la fermeture cutanée, 80% des utilisateurs de bistouri ont recours à des Steristrips ou exceptionnellement à des fils, tandis que ce pourcentage tombe à 58% chez les opérateurs à l'aiguille. Or l'abstention de toute fermeture cutanée aux Steristrips est un point important qui sera développé par la suite.

La cicatrisation est toujours jugée satisfaisante par les opérateurs, quel que soit leur mode d'incision. Toutefois, si la fréquence des pigmentations résiduelles ou des taches rosées ne dépend pas de l'instrument de puncture, ces traces sont nettement moins apparentes après une incision à l'aiguille.

Les varicosités sont considérées comme fréquentes chez 11% des utilisateurs de bistouri seul et 8% des utilisateurs de "bistouri-aiguille", alors qu'elles sont constatées exceptionnellement par les opérateurs travaillant à l'aiguille.

Les phlyctènes, qui laissent parfois des séquelles disgracieuses, sont essentiellement liées à la mise en place de Steristrips. Cet incident est constaté chez 68% de utilisateurs de ce ruban adhésif contre 25% seulement chez les opérateurs qui écartent ce procédé d'occlusion. Il découle donc que l'incision à l'aiguille fine, qui rend inutile la fermeture aux Steristrips, peut être considérée comme un intéressant moyen de prévention des phlyctènes.

Quant aux infections et aux séromes, ils ne sont signalés qu'exceptionnellement et leur apparition n'est pas liée au mode d'incision cutanée.

COMMENTAIRE

De cette étude il se dégage que tous les opérateurs sont satisfaits des résultats esthétiques de leur méthode, quel que soit le mode d'incision pratiqué et en dépit des rares incidents cutanés toujours bénins. Personnellement cependant nous déplorons que la technique des micro-incisions attire peu d'adeptes (15% seulement des opérateurs interrogés). Beaucoup de collègues nous ont confié qu'ils n'avaient pas été enthousiasmés par leurs premiers essais d'incision à l'aiguille. Utilisant malheureusement des aiguilles trop grosses (roses 12/10mm), ils n'ont pas obtenu de cicatrices plus discrètes qu'avec un bistouri. Il n'en est pas de même pour les ouvertures effectuées avec une aiguille de calibre plus réduit (aiguille verte 8/10mm ou aiguille noire de 7/10 mm) qui implique l'usage de crochets fins (N° 1 ou exceptionnellement N° 2). Ces petits intruments limitent ainsi le traumatisme cutané et dispensent généralement de l'utilisation de Steristrips. Compte tenu de ces mini-incisions et de la quasi inexistence de délabrements sous-cutanés, le pansement compressif peut être enlevé après 48 heures et la contention une semaine plus tard. Les ouvertures cutanées sont oblitérées au 2° jour et les cicatrices pratiquement indétectables au 10° jour.

L'intérêt de cette méthode est avant tout d'ordre esthétique à la condition, soulignons-le encore, d'employer des aiguilles fines et surtout des crochets fins, N°1 de Muller voire même des modèles plus réduits. En second lieu, son intérêt est d'ordre économique car une aiguille est moins onéreuse qu'un bistouri et le pansement est simplifié, d'éxécution rapide et bien évidemment plus confortable pour le patient.

BIBLIOGRAPHIE

1.   Muller R. La Phlébectomie Ambulatoire. Helv chir Acta 1987;54:555-558.

2.   Trauchessec J.M., Choukroun P.L.. Phlébectomie Ambulatoire.   in :10ème Congrès Mondial de l'Union Internationale de Phlébologie. Strasbourg. Phlébologie 1989 A Davy, R. Stemmer eds, 1989 John Libbey Eurotext Ltd.1081-1082.

3.   Ricci S. - Communication orale 1987.

4.   Ricci S., Georgiev M. - Office varicose surgery under local anesthesia. J Dermatol Surg Oncol 1992;18:55-58.

Phlebology '95, D. Negus et al. (eds.). Phlebology (1995) Suppl. 1: 439-442

OP/5.5

Le Bloc Crural est-il Supérieur a L'Anesthésie Locale lors du Stripping de la Saphène Interne?

H. Campaert

Waregem, Belgium

Is Inguinal Block Better than Local Anaesthesia for Stripping of the Long Saphenous Vein?

SUMMARY

The author discusses the advantages of regional anaesthesia over local anaesthesia in the out-patient treatment of varices.

LE BLOC CRURAL EST-IL SUPERIEUR A L'ANESTHESIE LOCALE LORS DU STRIPPING DE LA SAPHENE INTERNE ?

H. Campaert

Waregem, Belgium

INTRODUCTION

L'importance d'une mobilisation postopératoire précoce dans la prévention de la thrombose veineuse profonde ayant été largement démontrée, nous réalisons depuis 1976 systématiquement les opérations de chirurgie veineuse sous anesthésie locale, nous travaillons depuis une dizaine d'années exclusivement sous bloc crural. L'objectif de cette étude était d'évaluer les avantages du bloc crural par rapport à l'anesthésie locale.

METHODES

Un total de 1.995 patients ont subi une crossectomie de la jonction saphéno-fémorale, suivie d'un stripping de la saphène interne, avec ligature des collatérales et suppression des éventuelles perforantes insuffisantes. Le traitement a dû être bilatéral chez 733 de ces patients, ce qui porte à 2.850 le total des interventions pouvant être incorporées dans cette étude. Les cures bilatérales n'ont jamais été réalisées des deux côtés en une seule intervention.

L'intervention a été réalisée sous anesthésie locale pour 1.061 patients, et sous bloc crural pour 1.789 patients.

Lorsque l'intervention était réalisée sous anesthésie locale, on procédait dans un premier temps à une infiltration de la région de l'aine au moyen de lidocaïne à 1 %, suivie d'une crossectomie réalisée dans les règles de l'art. Le trajet de la veine saphène interne au niveau de la cuisse était ensuite infiltré au moyen de lidocaïne à 1 %, à l'aide d'une aiguille longue (16 cm) de 1,2 mm de diamètre (18 G). Cette infiltration était réalisée distalement par l'incision inguinale et en direction proximale via une petite incision au niveau du canal de Hunter ou à un niveau distal par rapport au genou. Le stripping était généralement de type court, les varicectomies supplémentaires étant réalisées au moyen d'infiltrations locales de petites doses de lidocaïne 1 %.

En cas de bloc crural, une quantité de 8 à 12 ml de lidocaïne à 1 % était infiltrée au niveau du pli de l'aine, latéralement par rapport à l'artère fémorale. Le nerf fémoral comprenant des fibres motrices et sensitives, on peut observer une parésie temporaire du quadriceps, de résolution spontanée. Aucune autre injection d'anesthésique n'est nécessaire le long du trajet de la veine saphène interne au niveau de la cuisse. Des injections locales de lidocaïne sont uniquement réalisées lorsqu'une incision cutanée est nécessaire. La suite de l'intervention ne diffère pas par rapport à la technique utilisée sous anesthésie locale. Des quantités moindres d'anesthésique sont toutefois nécessaires, en raison d'une nette diminution de la sensibilité au niveau de la face interne du mollet. En aucun cas, nous n'avons utilisé de l'adrénaline en association avec de la lidocaïne.

Afin de pouvoir comparer les deux techniques d'anesthésie, nous avons d'une part noté la dose totale de lidocaïne nécessaire, ainsi que la dose / kg de poids corporel dans les deux groupes, et d'autre part demandé aux patients d'indiquer sur une échelle visuelle allant de 1 à 100 un score correspondant à l'intensité de la sensation douloureuse et au vécu subjectif de l'intervention.

RESULTATS

Dans le groupe sous anesthésie locale, la dose totale de lidocaïne nécessaire atteignait en moyenne 515 mg, avec des valeurs extrêmes de 310 et 720 mg. Par rapport au poids corporel, la dose moyenne utilisée était de 9,75 mg/kg, avec des valeurs extrêmes de 6,5 et 13 mg/kg.
Le score moyen de satisfaction était de 43 sur l'échelle de 1 à 100.
Dans le groupe sous bloc crural, la dose totale de lidocaïne nécessaire atteignait en moyenne 325 mg, avec des valeurs extrêmes de 420 et 230 mg. Par rapport au poids corporel, la dose moyenne utilisée était de 6,33 mg/kg, avec des variations entre 8,07 et 4,6 mg/kg. Les patients indiquèrent un score de 23. Mentionnons encore que plusieurs patients ont signalé que l'intervention pratiquée suivant cette technique était totalement indolore.

DISCUSSION

Les résultats de la présente étude montrent clairement que le bloc crural permet, de l'avis même des patients, d'obtenir une meilleure analgésie, et ce au moyen de doses d'anesthésique inférieures. La supériorité de cette technique porte principalement sur le trajet de la saphène interne au niveau de la cuisse, où le stripping est nettement moins douloureux, en tout cas par la technique d'invagination. Le patient perçoit uniquement une douleur lors de la traction sur une branche latérale.

Bien que la dose maximale de lidocaïne généralement admise soit de 400 mg, on peut souligner qu'aucun patient de notre série n'a présenté de signes d'intoxication, même légers. La durée de la période d'administration de l'anesthésique joue sans doute un rôle à cet égard. Un facteur à prendre en considération dans le cas du bloc crural est toutefois la parésie du quadriceps, empêchant le patient de se lever pendant un certain temps après l'intervention. Ce signe, qui ne peut être considéré comme une complication, mais bien comme une conséquence normale d'une éventuelle infiltration des branches motrices du nerf fémoral, a été observé chez 356 patients au cours de notre étude, soit dans environ 12,5 % des cas. La durée de cette parésie, calculée à partir de l'injection de l'anesthésique, variait entre 1 et 3 heures, de sorte que le phénomène peut parfois être perceptible pendant quelque temps après l'intervention. Cette parésie disparaît spontanément durant la période de surveillance postopératoire. La perception par le patient d'un lancement douloureux au niveau du genou lors de l'infiltration indique généralement un contact avec les branches motrices. Il est alors préférable de déplacer l'aiguille avant de poursuivre l'infiltration.

CONCLUSION

Le bloc crural présente nettement plusieurs avantages par rapport à l'anesthésie locale dans le traitement chirurgical ambulatoire des varices, tout particulièrement dans le territoire de la veine saphène interne. Le seuil de toxicité est moins rapidement atteint, les doses utilisées étant plus faibles. L'analgésie obtenue, évaluée par les patients eux-mêmes, est en outre supérieure.

Phlebology '95, D. Negus et al. (eds.). Phlebology (1995) Suppl. 1: 443-445

OP/20.4

Recurrent Varicose Veins and Perforating Veins Insufficiency

C. Setacci and G. Sozio

Chair of Vascular Surgery, University of Siena, Italy

Primary varicose veins affect between 10 and 12 per cent of the adult population [1,2,3]. Recurrence is defined as persistence or reemergence of varicosities after previous operative treatment. Varicose veins that recur after treatment remain one of the greatest challenges facing modern phlebology practice. The 5-year recurrence rate after surgery is around 50% and for compression sclerotherapy around 90% [4]. Although inadequate evaluation and surgical techniques are common causes of recurrent varicose veins (early recurrence within 6 months of surgery is generally owing to incomplete surgery or incorrect diagnosis), new varices may develop after technically correct surgery.

There are four common sources of reflux associated with late recurrence of varicose veins after 6 months: 1) recurrence of significant reflux at the saphenofemoral or saphenopopliteal junction because of neovascularization or inadequate ligation; 2) incompetent thigh or calf perforating veins; 3) incompetent gastrocnemius veins; 4) persistent varicose tributaries or duplication of the long saphenous vein in the thigh.

In our experience all the recurrences are associated with incompetent perforating veins.

The ideal method of investigation of recurrent varicose veins should be capable of diagnosing all of the above recognized entities in order to obtain the better surgical treatment.

MATERIALS AND METHODS

1653 primitive varicose veins and 203 recurrent were observed (162 surgical procedures in other institutions and 41 in our department) from January 1982 to December 1993. There were 1126 woman and 527 men with a mean age of 47 years (range 29-78). In relation to recurrences there were 141 women and 62 men with a mean age of 49 years (range 31-78).

A history was obtained in order to ascertain whether the long saphenous vein or short saphenous vein had been stripped. If the patient had a groin incision it was presumed that the saphenofemoral junction had been ligated.

Only 39 patients had, preoperatively, a correct method of investigation (continuous-wave doppler ultrasound, ecodoppler, strain-gauge plethysmography, ascending or popliteal phlebography and eventually varicography) combined with clinical examination. 166 patients had a previous stripping operation of the long saphenous vein, 32 had a previous surgical ligation of the saphenofemoral junction, 2 a stripping operation of the short saphenous vein and 3 had an avulsion of varicose tributaries with the Müller technique.

RESULTS

All the patients with recurrent varicose veins had a new surgical procedure. 102 patients with an inappropriate initial treatment in the groin were treated with re-exploration of the sapheno-femoral junction with removal of the persistent communication with the common femoral vein. 6 patients with an unsuitable initial treatment in the popliteal fossa with a consequent stripping of the short saphenous vein. All the 203 patients with the presence of incompetent thigh or calf perforating veins had a subfascial nonadsorbable ligature in order to avoid the pathological reflux and multiple phlebectomies.

DISCUSSION

The results of this study indicate that incompetent perforating veins are the most common source of reflux from the deep system in patients presenting with recurrent varicose veins following the stripping operation. However, there was also a significant incidence of reflux from recurrent communication with the common femoral vein, from the saphenopopliteal junction, and from incompetent pelvic veins, indicating inadequate incomplete surgery and/or inadequate primary evaluation.

Incompetent gastrocnemius perforating veins were the most frequent incompetent perforating veins encountered. It is reasonable to suggest that varicose veins recur because the primary etiologic factor has not been addressed by initial treatment. High recurrence rates after the stripping operation may be owing to incorrect assumptions on the etiology of varicose veins or inadequate surgical technique.

Burnand et al. [5] have shown that flush ligation of the long saphenous veins at the saphenofemoral junction appears to be more important than perforating vein ligation in reducing foot vein pressure on exercise. Cotton [6] was unable to attribute the development of varicose veins to a descending incompetence originating at the saphenofemoral junction. Thibault et al. [7] indicates that an ascending pattern of incompetence was more common with the saphenofemoral junction becoming the ultimate secondary source of reflux. Lofgren et al. [8] observed that with recurrent veins following surgical stripping, recanalization began distally and moved progressively in an upward direction when the ligation at the saphenofemoral junction had been performed satisfactorily.

The other major assumption of the stripping operation is that it will avulse incompetent perforating veins. This has proved to be an incorrect assumption [9,10], as only one or two perforating veins (Hunterian and Boyd) communicate directly with the long saphenous vein. In particular, the posterior tibial perforating veins usually communicate directly with the posterior arch complex, which is not removed by the stripping operation [11]. In addition incompetent gastrocnemius perforating veins will not be removed by the stripping operation. Advocates of the stripping operation often postulate that the operation will result in a lower incidence of recurrence compared with saphenofemoral ligation alone because incompetent thigh perforating veins will be removed by the procedure of stripping [12,13].

The relatively high incidence of incompetent thigh perforating veins found in many studies does not support this view. Instead, it is possible that the stripping operation may predispose to the development of incompetent thigh perforating veins by causing trauma to the hunterian perforator during the procedure.

CONCLUSIONS

Varicose veins pose many problems for the clinician, not the least of these being the high incidence of recurrent varicosities following therapy. In the past, new method of treatment have been advocated and altough initial results have often indicated a low recurrence rate, with the passage of time the recurrence rate is still found to be significant.

It is presumptuous to conclude that because control of the high pressure leak from the incompetent perforating vein has immediate beneficial effects on patient symptoms that the incompetent perforating vein is the primary cause of the varicose disorder.

There is no doubt that in order to avoid subjecting the patient to multiple secondary treatment procedure with their associated costs, including absence from the workforce a correct approach with a precise and extensive mapping of incompetent superficial veins by through clinical examination combined with duplex evaluation prior to initiation of surgery is mandatory.

REFERENCES

1. Darke SG. Recurrent varicose vein and short saphenous insufficiency: evaluation and treatment. In: Bergan JJ, Yao JST, eds. Venous disorders. WB Saunders Company 1991; 217-232.
2. Thompson H. The surgical anatomy of the superficial and perforating veins of the lower limb. Ann R Coll Surg Engl 1979; 61:198-205.
3. Goren G, Yaellin AE. Primary varicose veins: topographic and hemodynamic correlations. J of Cardiovascular Surgery 1990; 31:672-677.
4. Thibault PK, Lewis WA. Recurrent varicose veins. Part 1: Evaluation utilizing Duplex venous imaging. J Dermatol Surg Oncol 1992; 18:618-624.
5. Burnand KG, O'Donnel TF, Thomas ML et al. The relative importance of incompetent communicating veins in the production of varicose veins and venous ulcers. Surgery 1977; 82:9-14.
6. Cotton LT. Varicose veins: gross anatomy and development. Br J Surg 1961; 48:589-598.
7. Thibault P, Bray A, Wlodarczyk J et al. Cosmetic leg veins: evaluation using duplex venous imaging. J Dermatol Surg Oncol 1990; 16:612-618.
8. Lofgren KA, Myers TT, Webb WD. Recurrent varicose veins. Surg Gynecol Obst 1956; 102:729-736.
9. Lofgren EP, Lofgren KA. Recurrence of varicose veins after the stripping operation. Arch Surg 1971; 102:111-114.
10. Thompson H. The surgical anatomy of the superficial and perforating veins of the lower limb. Ann R Coll Surg Engl 1979; 61:198-205.
11. Hobbs JT. Surgery and schlerotherapy in the treatment of varicose veins. Arch Surg 1974; 109:793-796.
12. Juhan C, Haupert S, Miltgen G et al. Recurrent varicose veins. Phlebology 1990; 5:201-211.
13. Rivlin S. The surgical cure of primary varicose veins. Br J Surg 1975; 62:913-917.

Phlebology '95, D. Negus et al. (eds.). Phlebology (1995) Suppl. 1: 446-450

P222

Approche Thérapeutique de la Maladie Variqueuse à Propos de 4 Cas

L. Tessari

Studio Medico 'Dott. Glauco Bassi' Trieste, Italy. Casa di Cura 'Dott. Pederzoli' Peschiera del Garda, Itlay

Therapeutic Approach to Varicosis: A Report on Four Clinical Situations

SUMMARY

The author discusses 'functional therapeutic phlebological methods' and illustrates this approach to treatment in four different situations.

APPROCHE THERAPEUTIQUE DE LA MALADIE VARIQUEUSE A PROPOS DE 4 CAS

Lorenzo Tessari

Studio Medico "Dott. Glauco Bassi" Trieste ITALY
Casa di Cura "Dott. Pederzoli" Peschiera del Garda ITALY

INTRODUCTION

La stratégie thérapeutique que chaque phlébologue utilise découle de nombreuses "petites découvertes", fruit d'autant de petites expériences et observations faites dans le cabinet de consultation.

Pour les phlébologues de l'école fonctionnelle toute la pathologie veineuse des membres inférieurs est dominée par la stase, c'est-à-dire par des phénomènes liés au ralentissement de remontée du sang et de la lymphe.

La stase est en même temps une cause primordiale de phlébopathie et son principal facteur aggravant.

Celle-ci est à l'origine d'une cascade d'évènements en chaîne qui confinent progressivement à la cellulite, aux capillaires et aux varices aboutissant aux troubles trophiques.

Pour vaincre la stase la nature a pourvu l'homme depuis qu'il est denvenu "homo erectus" d'un système de pompe aspirante et de pulsion du sang, dont les plus importante, sont celui du pied de la cheville du mollet et du diaphragme et d'un système de faisceaux rigides qui ici et là compriment les veines en y accélérant le flux.

La phlébologie "fonctionnelle" a comme rôle principal de réactiver ces mécanismes. Comme ces systèmes naturels entrent en action seulement lors de mouvements du membre, ses armes principales sont la marche et la élastocompression de la jambe.

Pour cette raison, je m'attacherai plus à décrire la stratégie thérapeutique fonctionnelle relative aux soins du 1er cas clinique qui est la base, comme les fondations d'une maison, sur laquelle on construit ensuite la thérapeutique des autres cas cliniques.

CAS CLINIQUES

La description de 4 cas cliniques volontairement schématiques, correspondant à des situations phlébologiques de pratique quotidienne, permet de faire le point sur la stratégie thérapeutique dans plusieurs pays du monde.

1er cas: jeune femme de 30 ans, 1 enfant, secrétaire, sans antécédents notebles; consulte pour une apparition progressive de varicosités et lourdeurs de jambes en fin de journée. Il n'existe aucun point de reflux; *insuffisance veineuse fonctionnelle*.
Ce cas est typique d'une insuffisance veineuse fonctionnelle. J'entends par là, non pas la

présence d'une maladie valvulaire, mais la non utilisation des mécanismes normaux hémato-propulsifs habituels, dont la membre inférieur est doté, qui aboutit à la symptomatologie typique de pensateur et de lourdeur des membres inférieurs en fin de journée.

Dans ce cas, le rapport de collaboration et de confiance médecin-malade est déterminant pour convaincre le patient de modifier ses propres habitudes de vie sédentaire et se conformer à l'exercice quotidien à l'extérieur.

Comme il n'est pas facile de modifier ses habitudes de vie, il convient d'aller dans le détail. Par exemple, en conseillant de laisser la voiture à 1 km de son lieu de travail ou de descendre 3 arrêts avant sa destination finale, ou de faire 3 fois par jour des semi-flexion sur les genoux, ou des exercices sur la pointe des pieds.

Il ne faut jamais cesser de répéter que le fait de marcher dans la maison ou le jardin fait plus de mal que de bien, puisque la réduction de la pression veineuse commence des les premiers pas et acquiert son efficacité seulement si la marche est rythmique, continue et prolongée.

Il faut également recommander le port de chaussures adaptées, éventuellement corrigées, par des semelles faites sur mesure par les podologues "fonctionnalistes". Eviter l'immobilité en station debout, ne pas rester longuement assis, couper les journées avec une demi-heure de repos, en position allongée, avec les pieds surélevés de 8 cm, faire de longues promenades bi-quotidiennes. Ce sont les règles indispensables qui pourraient être complétées ou dans certains cas particuliers substitués par la gymnastique à la chambre, par des flexions sur les genoux, et dorsi-flexions de pieds, exercices de pédalage, gymnastique respiratoire et avec du yoga ou avec l'usage d'instruments particuliers, comme le tapis hémo-phlébo dynamique anti-stase.

Ces soins ne sont très efficaces que si les patients ont compris que "sans leur collaboration" rien de durable ne peut être obtenu.

L'utilisation de phlébotoniques, qui trouvent leur pleine indication dans ce cas, porte souvent le malade à ne pas écouter les conseils de marche, donc le rôle du phlébologue est d'être extrêment clair et attentif. Dans le cas en question qui présente des capillaires avec un intérêt esthétique, il est utile d'établir une stratégie thérapeutique qui peut se servir, de la diathermocoagulation pulsée, la micro-injection chromatique, différents types de laser, la mésothérapie, la pressothérapie, l'électrothérapie et la créno-thérapie.

2ème cas: femme de 45 ans, 3 enfants, ouvrière d'usine, avec une hérédité maternelle variqueuse franche; consulte pour des phlébectasies disgracieuses; *reflux saphénien interne bilatéral marqué, avec du côté droit, une écharpe antérieure bien développée.*

La stratégie thérapeutique que j'utilise dans ce cas, qui est celle de mon Maître Glauco BASSI est la suivante:

1. crossectomie bilatérale de la saphéne interne en anesthésie strictement locale et stripping court invaginé sur fil suivant la technique de J. Van der Stricht.

2. période d'attente de 1 à 2 mois pour donner la possibilité aux varices les moins altérées de récupérer leur élasticité, leur calibre et donc leur fonction. Pendant cette période, le malade portera un bas elastique.

3. micro-intervention aux crochets de Bassi sur les éventuelles perforantes incontinents résiduelles, mises en évidence, une fois éliminés les reflux longs verticaux.

4. après une période supplémentaire d'attente d'un mois environ, on pratiquera une sclérothérapie des troncs variqueux résiduels.

LE MAINTIEN DES RESULTATS OBTENUS DOIT ETRE ASSURE AVEC UNE HYGIENE VEINEUSE CONSTANTE COMME DECRITE DANS LE CAS N. 1. ET JE REMARQUE TOUJOURS CE POINT, PARCE QUE C'EST AVEC UNE BONNE HYGIENE DE VIE ET UN FORT RAPPORT MEDECIN-MALADE QUE ONT

AURA LE BENETRE ET LA SANTE DES JAMBES.

3ème cas: homme de 45 ans, restaurateur, sans antécédents; consulte pour des carices de la face postérieure d'un mollet; *insuffisance saphénienne externe unilatérale marquée, avec un reflux ostial et tronculaire isolé.*

Le problème de la saphène externe est beaucoup plus complexe. Quand le reflux vien d'une crosse incontinente j'utilise la ligne de conduite suivante:

1. varicographie dynamique fonctionnelle suivant J. Van der Stricht per-operatoire;
2. crossectomie et eventuel stripping court jusqu'au ras de la perforante du soléaire suivant la technique de Bassi en anesthésie strictement locale;
3. période d'attente de 1 à 2 mois (cf. cas n. 2);
4. micro-intervention avec les crochets de Bassi (cf. cas n. 2);
5. éventuellement sclérothérapie des troncs variqueux résiduels.

DANS CE CAS AUSSI LE MAINTIEN DES RESULTATS OBTENUS DOIT ETRE ASSURE AVEC UNE HYGIENE VEINEUSE CONSTANTE COMME DECRITE DANS LE CAS N. 1.

4ème cas: Femme de 65 ans, 2 enfants, en retraite, sans antécédents; consulte pour un ulcère de type veineux, sur la face interne de la cheville, centré sur la veine tibiale postérieure, avec une perforante de Cockett à deux travers de doigt au dessus. *Il existe une insuffisance saphénienne interne homolatérale majeure.*

La présence d'une lésion trophique (ulcère, hypodermite, atrophie blanche, etc...) pose au phlébologue 2 problèmes:

1. soigner le trouble trophique cutané;
2. soigner la cause:
• s'il s'agit d'un syndrome post-phlébitique;
• s'il s'agit de pathologie variqueuse stade 3 de la saphène interne ou de la saphène exsterne.

1. Pour le trouble trophique cutané, j'adopte le schéma thérapeutique suivant:
• contention élasto-adhésive inamovible rigide suivant la technique de Bassi-Stemmer, appliquée par le phlébologue lui-même. Dans le cas d'une ulcère il faut preparer la lesion avec une correcte toilette medical;
• hygiéne veineuse scrupuleuse et persistante;
• interruption des perforantes péri-ulcéreuses avec le crochet de Bassi;
• sclérothérapie des veines qui nourissent la zone ulcéreuse.
2. Dès guérison du trouble trophique:
• s'il s'agit d'un syndrome post-phlébitique, je conseille l'utilisation constante du bas élastique, sur mesure type Varitex Néodurelna;
• s'il s'agit de pathologie variqueuse stade 3 de la saphène (interne ou externe), soin de cette pathologie comme cas n. 2 ou 3.
3. Thérapeutique associée des éventuelles pathologies concomitantes, qu'elles soient de type artériel ou de type ostéo-articulaire (vasoactifs, kinésithérapie).

DANS CE CAS AUSSI, LE MAINTIEN DES RESULTATS OBTENUS DOIT ETRE ASSURE AVEC UNE HYGIENE VEINEUSE CONSTANTE.

CONCLUSION

Il est toujours difficile d'unifier une thérapie phlébologique en rapport à des cas cliniques bien souvent divers entre eux; il y a toutefois un commun dénominateur qui est le fondament de la thérapie phlébologique "fonctionnelle". Ce denominateur commun repose sur la collaboration du "phlébologue fonctionnaliste" avec les systèmes naturels de défense érigés par l'organisme du "phlébopatique" qui devra de plus se rendre compte que sans sa participation rien de durable ne peut être obtenu.

REFERENCE

Bassi Gl., Les varices des membres inférieurs. Editions Doin 1967.
Bassi Gl., Compendio de terapia flebologica. Edizioni Minerva Medica 1985.
Bassi Gl., Stemmer R., Traitements mecaniques fonctionnels en phlébologie. Editions Piccin 1983.

Phlebology '95, D. Negus et al. (eds.). Phlebology (1995) Suppl. 1: 451-453

PI/9.5

A Randomised Trial to Compare Standard and Invagination Stripping of the Long Saphenous Vein i the Thigh

M.R. Tyrrell, M. Rocker, N. Maisey, R. Marshall, P.R. Taylor and D. Negus

Department of Surgery, Lewisham Hospital, London, UK

Surgical excision of the long saphenous vein in the treatment of varicose veins was first practised by Madelung in 1884[1]. Stripping was introduced by Babcock in 1907[2]. Many surgeons now practise stripping limited to the thigh with reduction in the incidence of long saphenous nerve injury[3].

"Standard" stripping is associated with a significant incidence of thigh bruising and discomfort. This can be avoided by sequential avulsion of the long saphenous vein[4] and might also be reduced by invagination stripping. The latter was described by van der Stricht in 1963[5]. This paper reports the results of a study to compare post-operative blood loss and thigh bruising after "standard" stripping (using a stripper head) and invagination stripping.

PATIENTS AND METHODS

Sixty consecutive legs in 55 patients with primary long saphenous varicose veins were randomised (after informed consent and ethical committee approval) to receive limited stripping of the long saphenous vein in the thigh by either the standard or invagination techniques. Randomisation was by computer generated random numbers. Flush sapheno-femoral disconnection and multiple avulsions of varicosities were conducted in the conventional manner. Closed suction drainage was applied to the long saphenous vein track after stripping. The drained volumes were recorded at four and twenty four hours post-operatively. All legs were bandaged with crêpe bandages, which were replaced with class I compression stockings prior to discharge from hospital. The stockings were worn day and night until the first clinic review on the fourteenth post-operative day. Both the patient and the independent observer were "blind" to the stripping technique that had been employed. Both assessed thigh bruising and the presence of residual varicose veins on a simple scale: none; mild; moderate; severe; very severe. All complications were noted.

Statistical comparisons were made using the Chi^2 and Mann-Whitney-U tests as appropriate.

Surgical techniques:

1. Standard stripping: following flush disconnection of the sapheno-femoral junction, a stripper wire was passed down the long saphenous vein to the knee. A second incision was made at this point, through which the wire tip was extracted. The vein was firmly tied (1 vicryl) to the wire at the groin and a 9mm stripper head screwed onto the wire. One end of the tie was kept long and used to tow the suction drain into the long saphenous vein track after stripping. All veins were stripped from the groin to the knee.

2. Invagination stripping: the sapheno-femoral disconnection and wire passage were carried out as above. A long tie was fixed to the proximal end of the wire and towed through the long saphenous vein until it appeared at the knee incision. The tie was then fixed to the proximal long saphenous vein by two half hitches. Invagination was initiated by traction of the tie at the knee and triangulation of the vein 1-2cm distal to the securing half hitches. Once invagination was established, the haemostats used for triangulation were removed and stripping effected by continued traction on the tie at the knee. The long proximal end of the tie was used to tow a suction drain into the long saphenous vein track.

RESULTS

Thirty two patients were randomised to receive standard and 28 invagination stripping. The groups were well matched for sex (19 male: 41 female legs), age (median 50, range 26 to 82 years) and height/weight ratio (median 2.29, range 1.79 to 3.04 cm/kg).

It proved impossible to pass the wire in 5 legs in the standard group. Stripping failure occurred in 8 of the invagination group - 4 because of inability to pass the wire and 4 because of vein breakage. The incidence of technical failure is not statistically significantly different between the groups.

The drained volume at 24 hours is shown in figure 1. Both groups achieved low drainage volumes, with no significant difference between the groups.

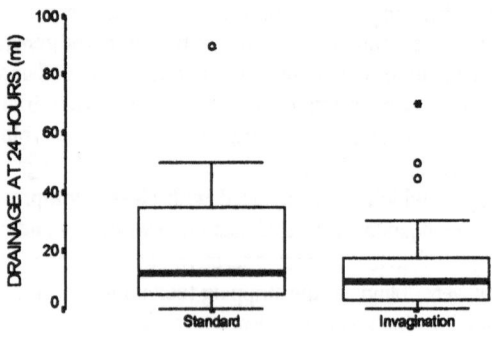

Figure 1. Drained volume at 24 hours. Median (solid bar), interquartile (box) and full ranges are shown.

There was no difference between the groups in terms of thigh bruising and residual varicose veins. Overall, 67% and 98% of cases were assessed as "none" or "mild" by the observers.

DISCUSSION

Stripping of the long saphenous vein has been controversial , particularly with regard to (i) the requirement for stripping to reduce varicose vein recurrence rate, (ii) long saphenous nerve injury and (iii) the possibility of arterial bypass utilisation of non-stripped veins. The current consensus is that stripping of the long saphenous vein in the thigh reduces varicose vein recurrence rates. Calf stripping adds little to this reduction but increases the risk of long saphenous nerve injury. Limited stripping is therefore common practice. Little attention has been paid to the influence of various stripping techniques on post-operative morbidity.

Intuitively, the van der Stricht method has the advantage of limited tissue trauma along the stripped long saphenous vein track and also avoids the multiple thigh incisions associated with sequential stripping. It is therefore surprising that we have been unable to demonstrate a significant advantage for invagination stripping. A possible explanation is that both groups drained considerably less than expected. Previous studies led us to expect an average drained volume of 50ml in the "standard" group[6]. However, our "standard" patients only drained between 0 and 90ml with a median value of 12.5ml. This small volume may be a function of the protocol which specified a 9mm stripper head. It is possible that larger heads result in more bleeding.

The technical failure rate is slightly (but not statistically significantly) greater in the invagination stripped group. This reflects the "double opportunity" for technical failure inherent to the technique - namely wire passage and vein breakage. The latter is usually recoverable providing a long tie is used. This allows for wire retrieval and conversion to conventional or sequential stripping. Even where this proves impossible a largely invaginated but incompletely removed vein is unlikely to recanalise and the procedure should still fulfill its primary purpose.

Overall we feel that invagination stripping is an elegant surgical technique with little disadvantage, but no identifiable advantage when compared with standard stripping using a small stripper head. The apparent advantage of sequential stripping may be related to the use of a large stripper head in Khan's study[4]. Nevertheless, sequential stripping is probably a useful salvage technique for those cases where the passage of a stripper wire proves impossible.

REFERENCES

1. Kohler H. History of venous diseases of the legs and of methods for their treatment. *Orthopädie Technik* 1/87
2. Babcock WW. A new operation for extirpation of varicose veins of the leg. *NY Med J* 1907;**86**:153-156.
3. Negus D, Nichols RWT. Is it necessary to strip the incompetent saphenous vein to the ankle? In *Phlebology '85* (eds Negus D, Jantet G) London, Libbey. pp 148-150, 1986.
4. Khan RBN, Khan S, Baxter MG, Greaney S, Blair D. Prospective randomised trial of stripping versus sequential avulsion of the long saphenous vein. *Br J Surg* 1995;**82**:549.
5. van der Stricht J. Saphenectomie par invagination sur fil. *Presse médicale Paris* 1963/21 pages 1081-1082.
6. Negus D. Suction drainage after stripping of the long sapehnous vein. *Lancet* 1979;**2**:334

Phlebology '95, D. Negus et al. (eds.). Phlebology (1995) Suppl. 1: 454–457

P226

Radical Varicectomy in Ascending LSV Varicophlebitis

A. Hetenyi

Department of Vascular Surgery, Magyar Imre Hospital, Ajka, Hungary

INTRODUCTION

We suggest to distinguish the "acute superficial thrombophlebits" (phlebitis of non-varicose veins) from the "acute varicophlebitis" (localised or extended phlebitis of varicose veins) because of the different causes and the strategies of treatment. The superficial thrombophlebitis is the conservative therapy, in the varicophlebitis the surgical treatment (depends on the status) is the choice of therapy.

The ascending varicophlebitis (AV) is a frequent and potentially serious complication of varicosity. Most of the authors agree with the high ligation of the long saphenous vein (LSV), if the large veins in the thigh are filled with thrombus, but the saphenofemoral junction is not yet affected. Others perform an acute varicectomy: flush ligation of the LSV and excision of all thrombosed veins.

DIAGNOSIS

The extension of the thrombus is always further then the visible signs of the inflammatory process, therefore the exact clinical diagnosis is dubious. The CW-Doppler examination can be helpful: if we can produce at the saphenofemoral junction or under it a retrograde flow in the saphenous trunk - only until the thrombus hasn't propagated to the groin. The negative finding is acceptable. If we can't induce a retrograde flow, we perform a Duplex-scan examination. It represents not only the blood flow alterations, but the really anatomical situation, too. If there is a suspition of DVT, we perform an ascending phlebography.

SURGERY

We report the most essential details of radical varicectomy of ascending LSV varicophlebitis. The method of crossectomy and stripping isn't the same, as at the conventional varicectomy.

The first step is the approach of the saphenofemoral junction from a groin incision. The tributary veins are divided and the saphenous vein is isolated, lifted up carefully with a dissector and two rubber-loops. If the thrombus doesnot extend all the way to the junction, we ligate the saphenous vein flush with the common femoral vein. If we are informed before the operation, that the thrombus propagated near to the junction, or we palpate the vein, we incise longitudinal the anterior wall of the LSV and remove the thrombus with suction, or with a Fogarty balloon-catether - and only after this clamp, ligate and cut through the saphenous vein (fig.1.).

Fig. 1

We introduce the Nabatoff-stripper from above into the mass of the thrombus, filling in the saphenous trunk, as far as it is possible, without difficulty (usually to the level of the knee). Here we make a long incision on the LSV and than stripp out the trunk downwards, so, that the thrombus can bubble out without mishap from the vein.
  This is very important, otherwise we can introduce the thrombus into the deep veins across the perforating veins (fig. 2.). If it is necessary, we draw a suction-drain into the bed of the LSV.
All varicose tributary veins (thrombosed or not) are dissected out directly through multiple small incisions.
Compression bandage, subcutaneous Heparin, early mobilisation, but no antibiotics are necessary.

RESULTS

From 1985-1994 we operated AV of LSV on 148 patients. 134 patients were controlled one year after surgery. We report only the controlled patients.
We have made only crossectomy (C) on 10, crossectomy + stripping above the knee (CS) on 24, and a radical varicectomy as writen above (RV) on 100 patients. There was no death, we observed one postoperative bleeding (transfused 2 units of blood) and one infection in the groin in the RV group.

456

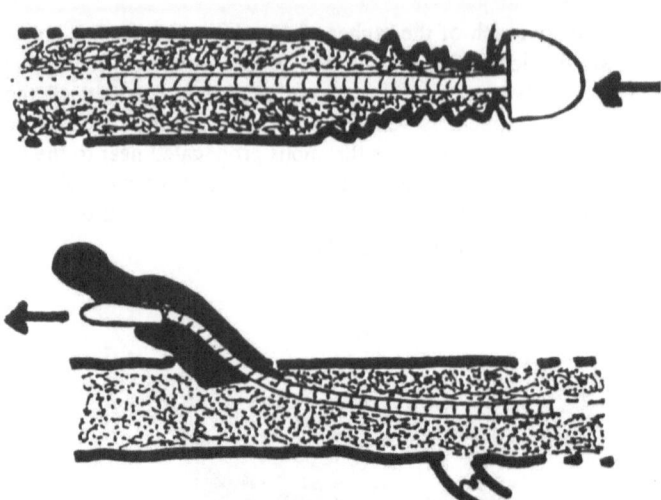

Fig. 2

During the controlling of patients we observed the residual varices, edema, skin alterations and subjective complaints. The Fig. 3. represents the incidence of the residual varices and painful conglomerates in the different groups. The patients with residual thrombosed veins complainted of permanent pain and were unable to work. We have seen the best results after the RV. These patients return to their own daily activities and were able to return to work far more earlier then others. The mean hospitalisation time in RV group was 5.8 days (including investigation time before operation).
In the C and CS groups we have made 12 secondary varicectomies, more than one year after the first operation.

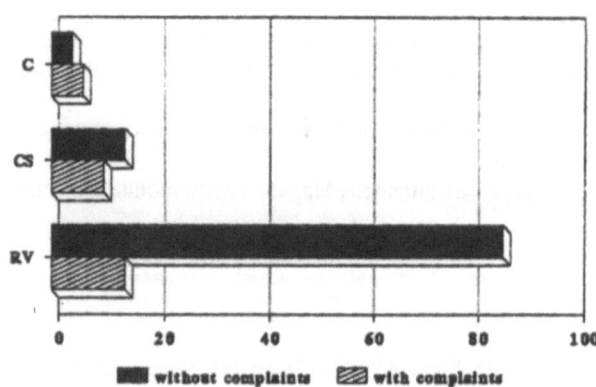

Fig. 3

CONCLUSIONS

The conservative treatment of the AV, there is a danger of direct pulmonary embolism or DVT. If the patient is lucky, the patient is only left with complaints and the varicosity remains unchanged. The inability to work is long-lasting, and the medical treatment is expensive. The simple flush ligation of the LSV protects the patients from the embolism and DVT, but not from other disadvantages. Frequently a second "a froid" varicectomy is necessary, with additional inconveniences and costs. The suggested radical surgical treatment is a definitive, simple and not dangerous, but much more economical then other methods. The acute radical surgery of AV of LSV is equally important, as the acute appendectomy: both diseases have critical consequences and both can be cured with a simple, secure active method.

Phlebology '95, D. Negus et al. (eds.). Phlebology (1995) Suppl. 1: 458-460

OP/14.1

Pin-Stripping: A Two Year Review

A. Oesch

Kramgasse 16, 3011 Bern, Switzerland

INTRODUCTION

The technique of ambulatory phlebectomy has demonstrated that varices can be removed in a simple and atraumatic way. This method is not very appropriate in eliminating the long (LSV) or the short saphenous vein (SSV). Here, stripping is the better choice. However, classic text-book stripping is still a crude procedure and now accounts for most postoperative complications and complaints. A desideratum of veinous surgery is to render stripping as simple and inoffensive as ambulatory phlebectomy.

METHOD

Perforation-INvagination-Stripping ('PIN-Stripping') is a minimal invasive method enabling selective removal of incompetent segments of the LSV or the SSV without preparation of the vein at the distal end point. The vein is stripped out by invagination.
After saphenofemoral junction or saphenopopliteal junction ligation, the vein is secured with two clamps and a semi-rigid stripper is introduced to the distal point of incompetence. Here, external pressure is applied and the tapered tip of the instrument is forced through the vein wall. The stripper is pushed down to distend the skin and passed through it by making a stab incision of 3-4 mm over its tip. After fixing a long ligature to the upper end, the stripper is completely pulled into the vein. The protruding ends of the ligature are passed beneath the clamps around the vein and tied firmly.The vessel is not fixed directly to the stripper as in conventional stripping. By pulling on the distal end of the stripper, the vein is invaginated.

Only little force is needed in stripping. Increasing resistance indicates the presence of a tributary and may be the harbinger of vein rupture. Small tributaries may be avulsed by kneading the skin, stronger ones should be disconnected using the hook phlebectomy technique. The invaginated LSV or SSV is delivered through the stab incision.The following distal segment is torn off without ligature or, if necessary, stripped by repeating the same procedure.

PIN-strippings of the SSV are easily done under local anesthesia. This has become the routine even for bilateral SSV-strippings, whereas LSV-strippings (or combinations of LSV and SSV treatments) are normally performed under regional anesthesia.

INSTRUMENTS

The instrument was developped in 1992 and was first presented in 1993 (2).Basically,it is a specially shaped stainless steel wire of less than 2 mm in diameter, manufactured now in two lengths for the LSV and the SSV*. Since the first presentation, the design has been slightly changed: the upper end is provided with an eye instead of a knob, ensuring a faster and more reliable fixation of the ligature.

RESULTS

From January 1992 to December 1993 the PIN-Stripper has been used 506 times in 399 operations of mono- or bilateral varices, including 376 partial or complete strippings of the LSV and 155 partial or complete strippings of the SSV. 16 SSVs and 76 LSVs were removed under local anesthesia. Patients were controlled 2 months postoperatively with special regard to the typical side-effects of stripping: sensory losses, lymph cysts or fistulas, delayed or unsightful healing.

Complications: Two - painless- sensory losses were seen in the territory of the sural nerve, none in the area of the saphenous nerve. There was one partial thrombosis of the popliteal vein with a benign evolution. All these complications occurred in the smaller group of SSV-strippings. There were no visible signs of lymphatic disruptions. Wound healing was uneventful,the cosmetic result good.

Technical problems: Incomplete stripping due to rupture of the vein or slipping of the vein out of the ligature occured in 66 of the 506 stripping manoeuvres. The risk of slipping is lower when using the new stripper model; it can be prevented by a transfixation instead of a ligature(1).

*Manufacturer: T&R Tüscher AG, Bern, Switzerland

CONCLUSIONS

The low invasiveness of PIN-stripping is demonstrated by the few complications and the feasibility of SSV-strippings under minimal anesthesia. The low incidence of nerve lesions must be attributed in large part to the selective stripping. However, in my own patients treated with selective stripping, lesions of the saphenous nerve dropped from 2.8% to zero and lesions of the sural nerve from 9.1% to 1.3% after the introduction of PIN-stripping. In several other points (blood loss, operation time, postoperative pain) PIN-stripping seems to perform better than conventional stripping, but these parameters have not yet been evaluated exactly.
Initially, PIN-stripping was recommended mainly for short veinous segments (2).In the meantime, we have completely abandoned conventional stripping or cryostripping and it has become our routine method.

REFERENCES

1. Goren G, Yellin A.Invaginated axial saphenectomy by a semirigid stripper: Perforate-invaginate stripping. J of Vasc Surg 1994; 20: 970-977.
2. Oesch A.'Pin-Stripping':A Novel Method of Atraumatic Stripping.Phlebology 1993; 8: 171-173.

Phlebology '95, D. Negus et al. (eds.). Phlebology (1995) Suppl. 1: 461–464

P231

Our Experiences with Different Surgical Methods in the Treatment of Acute and Chronic Illnesses of the Leg Veins

S. Baricza, J. Pfeiffer, F. Tollas and L. Pataky

Department of General Surgery, Town-Hospital, Várpalota, Hungary

INTRODUCTION

In general surgery departments it is frequently necessary to find surgical solutions to the problem of varicosity of the lower limb Elective operations are predominantly used, but "semiacute" operations are also in practice in the case of the most simple and most frequent complication of varicosity, ascending superficial thrombophlebitis.

In this paper we will give a summary of our seven years' experience in surgery.

METHODS

In our department we have introduced Vàrady's microsurgical method instead of the Madelung-operation in 1988. The essence of the method is that we draw out the veins through a prick of 2 mm with the phlebextractor produced by Aesculap. The phlebextractor can be used for the extraction as well as for the preparation. The instruments are shown in Figure 1.

462

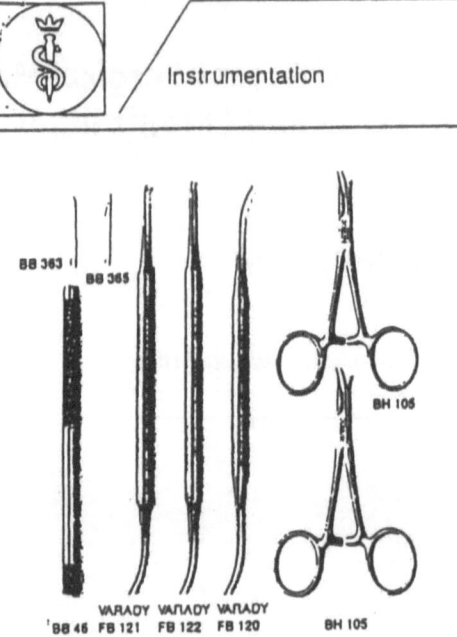

We use this method for the extraction of branches in itself or combined with crossectomy or stripping. In the case of inflamed thrombosed varices, however, the minichirurgical method cannot be used, so in such cases we make transverse incisions over the thrombosed vein to remove it. In this case the operation should be completed as soon as possible. In such situations we always do crossectomy, but we use stripping only when circumstances make it possible.

We do not recommend either the elective nor the semiacute operations done ambulantly, since the anaesthesia as well as the post-operative observation make hospitalisation necessary.

RESULTS

In the years 1988-1994 there were 990 elective varix operations completed in our hospital while we did 92 for ascending thrombophlebitis.

The following table (Table 1) shows the number of operations and the changes in our attitude towards surgical methods.

Table 1

Summary of Statistical Data

Operation	Várady's method	Madelung's method	Thrombophlebitis
1988	20	37	8
1989	45	18	11
1990	87	5	3
1991	213	9	15
1992	232	6	12
1993	191	12	24
1994	114	1	19
	902	88	92
	990	(8,8%)	

As the table shows, after two years the minichirurgical method started to attract patients, who did not feel awkward about it any longer. During this period the average inpatient time decreased by 3-4 days compared to that of the Madelung-operation. The patients left the hospital sooner, did not complain of pain and were able to return to work earlier.

DISCUSSION

Our seven years experience convinced us that the cosmetic results of the Várady-operation are excellent. The 2 mm pricks do not need suture so in the majority of cases they leave no trace after healing. If the vein is broad, the prick opens wider and a stitch might be necessary. In the healing of the wounds no complication arose. After the operation a local erythema sometimes developed; this later disappears. Haematoma occurred infrequently, only in the case of 5-10 mm veins - that means 5-6 occasions a year. This can be cured with a prick of the 2 mm lancet and with the extrusion of the haematoma post-operatively bleeding only occurred in 2 cases.

We undertake this operation with epidural anaesthesia, using crossectomy, stripping and the Várady-method together, as well as in cases of thrombophlebitis. If we remove only the branches, we use local anaesthesia plus Dormicum.

After the operation the patient gets up. In case of thrombophlebitis we used to use Calcium-Heparin for prevention, but now we use Fraxiparin. In the post-operative period, thromboses of the deep veins subject to operation occurred only once. Non-fatal pulmonary embolisation occurred also in one case, with a patient who had no observable thromboses of the deep veins.

Lymphorrhoea that ceased without treatment occurred in four cases.

The acute operation of thrombophlebitis lessens the patient's pain, restores the ability to walk and prevents the process spreading on to deep veins.

CONCLUSIONS

Our results, the infrequent occurrence of complications, and the good cosmetic results all helped us to convince the GPs as well as the patients that the operation is the best solution for varicose problems. Since at one time we operate only one of the patients' legs, our patients always return to have their varicose veins on the other leg operated as well.

REFERENCES

1.Baricza S Fünfjahrige Erfahrungen mit der Minichirurgie.
7. Internationaler Frankfurter Workshop für Plebologie 53.p. 1993

2. Fratila A. Rabe E Blitz H n Kreysel A W. The significance of Modern Varicectomy in the Surgical Treatment of Varicose Veins.
Zitschrift für Hautkrankheiten 65. 487-491, 1990.

3. Fratila A. Outpatient microsurgical varicectomy -Phlebology Digest 3. 1-3. 1990

4. Goren G.: Ambulatory star evulsion phlebectomy for fruncal varicose veins. American Journal of Surgery 3. 162.1991

5. Szabó T.. Heténvi A. és Nagy L.: A ascendálo varicophlebitis mütéti kezeiésével szerzett tapasztaiataink. Orv. Hetil. 133.3019-3021, 1990.

6. Várady Z.: Möglichkeiten der Varizenoperation Angio 8. 385-389, 1986.

7. Várady Z.: Technics of Esthetic Varicectomy. Érbetegségek 1.229-33. 1994.

Phlebology '95, D. Negus et al. (eds.). Phlebology (1995) Suppl. 1: 465–467

OP/14.6

Varicose Vein Surgery Under Local Anaesthesia. Complications of n= 47.057 Operations

N. Frings[1], B. Wagner[2] and M. Brus[3]

1 MOSEL-EIFEL-KLINIK, D-56864 Bad Bertrich Kurfürstenstr.,
2 KLINIK IM PARK, D-40721 Hilden, Hagelkreuzstr. 37
3 D-45130 Essen, Bertholdstr. 1-3

INTRODUCTION

Nihil nocere - above all else do not harm. This has always been a watchword in medicine and should especially be considered in the treatment of varicose veins, because usally these do not represent a life-threatening disease.
Unfortunately varicose-vein-surgery carries with is a number of complications regarding the operation itself as well as the anaesthesia. Therefore we have tried to develop a modified procedure enabling us to reduce the complication rate.

METHODS

From 1986 to February, 15th, 1995 there have been done 47.057 operations (ligation of the saphenofemoral or - popliteal junction and/ or stripping) under local anaesthesia in 3 centres. The management of therapy was performed in a successsive way:

1st day:	Ligation of the junction, stripping of the saphena and ligation/ dissection of perforating veins; only one saphena was operated per day.
2nd (and 3rd) day:	Varicectomy in the non invasive technique by Muller.

Only experienced surgeons did the operations.

Afterwards sclerotherapy of the remaining varicosities. As local anaesthesia we used Mepivacain 0,5-0,25% in a dosage of 80 ml (400mg) - 120 ml= 600 mg or sometimes even more.

RESULTS

I. Anaesthesia

We found no one major anaesthesia complication, though our dosages of local anaesthesia were quite high. They exceed the recommended maximum dose (60ml=300 mg Mepivacain 0,5%) by far.
However an own study of 9 patients, where we investigatet the plasma-level of Mepivacain, revealed that in no case the toxic level of 5 µg/ ml was reached.
All plasma-levels were below 3,5 mg/ ml.

In literature the complication rate of general or epidural anaesthesia comes to:

General anaesthesia

	n/ 10.000	%
Letal outcoms	1,9	0,019
	- 0,05	0,0005

Epidural anaesthesia

Severe neurological complications

1 : 65.304

1 : 10.000

II. Operation:

I would like to compare our complication rate with the two other huge statistics I have found: Balzer 1991, n= 25.457 and Waibel 1974, n= 87.665; see figure.

CONCLUSION:

We think, that operating under local anaesthesia is a very fine technique consisting an extremly low complication rate; especially when it is done by experienced surgeons in the described step-by-step-way.

REFERENCES:

1. Balzer, K., Die Tageschirurgie in der Venenchirurgie. Chirurg 1991; 62: 598-603

2. Marnitz, U., Narkosefähigkeit; in: Der ältere Patient. Urban und Schwarzenberg, 1995

3. Waibel, p., Fehler, Gefahren und Komplikationen bei der Varizenchirurgie. Phlebol. u. Proktol. 1974; 3: 134-139

Fig.	Frings %	Balzer %	Waibel %
Mortality	0	0,0039 (n=1)	0,02 (n=18)
Pulm. emb.	0,013 (n=6)	0,019 (n=5)	0,114-0,6 (lit.)
Deep vein thromb. (in clinic)	0,014 (n=7)	0,027 (n=7)	0,3-4,0 (lit.)
Deep vein thromb. (total)	0,049 (n=23)	?	?
Maj. artery inj.	0,002 (n=1)	0,011 (n=3)	0,057 (n=5)
Deep vein inj.	0,008 (n=4) (2 without sequelae)	0,0078 (n=2)	0,0022 (n=2)
Deep wound inf./Septicemia	0	0,051 (n=13)	?
Superf. wound inf.	0,15 (n=72)	?	0,1-8,0 (lit.)

Phlebology '95, D. Negus et al. (eds.). Phlebology (1995) Suppl. 1: 468–471

P248

Dans Quels cas la Chirurgie Itérative est-elle Utile Après Mauvais Résultats Suivant le Traitement Chirurgical des Varices?

M. Perrin[1] and J.L. Calvignac[2]

Unité de Chirurgie Vasculaire [1] Clinique du Grand Large, DECINES, [2] Clinique Ste Marie-Thérèse, BRON, France

When is Further Surgery Useful in the Presence of Bad Post-Surgical Results in Varicose Veins?

SUMMARY

Further surgery in the presence of failure of previous surgery is indicated: <u>absolutely</u> when there is major reflux from the deep system, <u>sometimes</u> in the presence of very large superficial varices even if no major reflux is present, <u>rarely</u> when there is deep venous insufficiency, <u>never</u> on superficial varicosities if there is an associated deep venous insufficiency. In all cases follow-up sclerotherapy will be necessary.

DANS QUELS CAS LA CHIRURGIE ITERATIVE EST-ELLE UTILE APRES MAUVAIS RESULTATS SUIVANT LE TRAITEMENT CHIRURGICAL DES VARICES ?

M. Perrin[1] , J.L. Calvignac[2]

Unité de Chirurgie Vasculaire [1] Clinique du Grand Large, DECINES,
[2] Clinique Ste Marie-Thérèse, BRON. FRANCE.

INTRODUCTION

Après traitement chirurgical des varices, des signes d'insuffisance veineuse persistent ou réapparaissent environ une fois sur 4 (1).

Les causes d'échec sont multiples et n'imposent pas toutes une réintervention. Les données cliniques, essentielles pour éliminer les plaintes d'origine non veineuse, sont souvent insuffisantes pour préciser la nécessité ou non d'une nouvelle opération.

Dans quelques cas, la décision nécessite mesure des pressions veineuses, plethysmographies et/ou phlébographies ; mais en règle le Duplex-Scan suffit à déterminer le mécanisme en cause donc l'indication thérapeutique, en précisant notamment :
- L'existence ou non d'une insuffisance veineuse profonde (IVP) ;
- La présence éventuelle, et l'importance d'un reflux au niveau d'une crosse ou d'une perforante.

INSUFFISANCE VEINEUSE SUPERFICIELLE (IVS) SANS IVP

1) Présence d'un reflux majeur
On peut distinguer 3 situations :
- Reflux persistant : Le reflux existait au moment de l'intervention précédente mais n'a pas été supprimé, ou pas complètement.
. Faute technique = Reflux mal supprimé : C'est la situation la plus fréquente représentant 73 % de nos cas dans une étude antérieure (2). La ligature du réseau superficiel n'a pas été effectuée au ras de la voie profonde.

L'illustration typique est représentée par les crossectomies incomplètes où reste en place un moignon résiduel (Fig. 1).

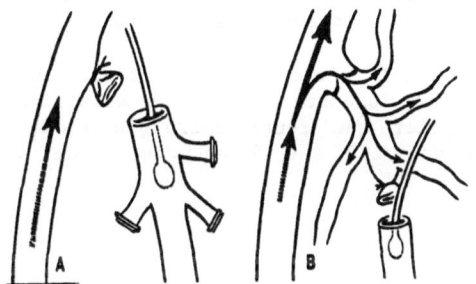

Fig.1 Chirurgie de la jonction saphéno-fémorale.

A Crossectomie SI correctement exécutée supprimant toutes les connexions entre la VFC et la VSI et ses collatérales.
B Crossectomie SI incorrectement exécutée. Le moignon de crosse SI laissé en place laisse persister à travers la valvule ostiale incompétente un reflux pathologique. Celui-ci va progressivement dilater les collatérales de la VSI (3).

Il faut reprendre chirurgicalement la jonction entre réseaux profond et superficiel pour placer correctement la ligature destinée à la supprimer.

. Faute tactique = Reflux non supprimé : Un bilan incorrect n'a pas révélé un point de fuite avant l'intervention, souvent au niveau du système saphène externe ; la persistance ou la réapparition précoce des signes peut être un élément d'orientation.

Le geste "oublié" lors du premier acte chirurgical doit donc être réalisé au cours d'une nouvelle intervention.

- Reflux "de novo" : Le reflux n'existait pas au moment de l'intervention précédente, mais la compétence valvulaire s'est dégradée au niveau d'une jonction entre réseaux superficiel et profond précédemment expertisée comme normale.

Ici encore, le chirurgien doit réintervenir pour supprimer le reflux.

- Reflux récidivé : Le reflux, supprimé par la précédente intervention, s'est progressivement reconstitué.

Cette question des néovasculogénèses reste posée au plan de leur nature exacte (4) et en conséquence de l'indication chirurgicale à retenir ; en général, ce n'est qu'après échec de la sclérothérapie, qu'on discute la réalisation d'un enveloppement prothétique de la voie profonde après dissection itérative élargie.

2) Absence de reflux majeur :

Un reflux de débit modéré qu'on ne parvient pas à supprimer par sclérothérapie peut conduire à une reprise chirurgicale comme dans les cas précédents.

En l'absence de reflux, le choix entre phlébectomie et scléroses est surtout fonction du calibre des varices présentes ; s'il est important, la phlébectomie nous parait à privilégier. Cependant, d'autres éléments sont à prendre en compte : Aspects topographiques et morphologiques des varices, contexte socio-professionnel du malade, compétence du chirurgien et du phlébologue.

Un tronc de saphène laissé en place après chirurgie d'une saphène dédoublée est un cas particulier non exceptionnel ; il n'est cependant pas formellement chirurgical s'il n'y a pas sur son trajet une perforante (DODD) avec reflux majeur.

INSUFFISANCE VEINEUSE PROFONDE (IVP) AVEC OU SANS IVS

La fréquence de découverte d'une IVP dans le bilan des mauvais résultats après chirurgie des varices est diversement appréciée ; ALMGREN (5) l'a retrouvée chez 29 % des malades ayant été réopérés une fois, et jusqu'à 50 % après 2 réinterventions ou plus.

Le traitement est essentiellement médical mais la chirurgie restauratrice du réseau veineux profond est indiquée chez des patients jeunes, invalidés par des troubles trophiques malgré une compression élastique correcte ; une anomalie de la coagulation ou une pompe musculaire insuffisante contre-indiquent une telle intervention.

Au traitement chirurgical de l'IVP s'ajoutera si besoin une réintervention sur une éventuelle IVS associée, selon les modalités déjà décrites; mais il faut rappeler que toute chirurgie isolée d'une IVS récidivée associée à une IVP est à nouveau vouée à l'échec.

CONCLUSION

Les mauvais résultats après chirurgie des varices constituent un problème préoccupant par leur fréquence et leur compréhension (6,7). Quoique non absolue, une certaine prévention est possible, aux conditions d'un acte chirurgical validé par l'expérience, rigoureusement réalisé, et encadré par un bilan pré-opératoire précis et un suivi médical correct.

Une nouvelle intervention de chirurgie veineuse peut être nécessaire:
- Toujours en cas de reflux majeur à partir du réseau profond ;
- Parfois en cas d'IVS isolée sans reflux majeur mais avec volumineuses varices ;
- Rarement face à une IVP de traitement généralement médical ;
- Jamais isolément sur une IVS si une IVP y est associée.

S'il est nécessaire, ce nouvel acte chirurgical n'est jamais suffisant et doit être complété par une sclérothérapie.

Il faut cependant souligner que, malgré une prise en charge post-opératoire correcte par un phlébologue compétent, la chirurgie itérative ne permet pas d'obtenir des résultats aussi bons que la chirurgie de première intention correctement exécutée (8).

REFERENCES

1   Darke SG. The morphology of recurrent varicose veins. Eur J Vasc Surg 1992 ; 6 : 512-517.
2   Perrin M. Les échecs. La chirurgie itérative. In :
    L'insuffisanc veineuse chronique des membres inférieurs.
    M. PERRIN. ARNETTE. 1994. PARIS : 81-86.
3   Perrin M, Gobin JP, Calvignac JL, Grossetête C, Lepretre M.
    Comprendre les mauvais résultats après chirurgie de
    l'insuffisance veineuse superficielle. J. Mal. Vasc 1994 ;
    19 : 265-271.
4   Couffinhal JC, Franco G. Récidive variqueuse. La
    néovasculogénèse existe-t-elle ? Act. Vascul 1992 ; 1 : 37-43.
5   Almgren B. Non thrombotic deep venous incompetence with special
    reference to anatomic, hemodynamic and therapeutic aspects.
    Phlebology 1990 ; 5 : 225-270.
6   Davy A, Ouvry P. Tentative d'explication des récidives
    variqueuses. In : Phlebology 1985 (Negus D and Jantet G Eds).
    J. Libbey. LONDON. 1986 ; 14-17
7   Negus D. Recurrent varicose veins : a national problem.
    Br J Surg 1993 ; 80 : 823-824.
8   Perrin M, Gobin JP, Grossetête C, Henri F, Lepretre M. Valeur
    de l'association chirurgie itérative - sclérothérapie après
    échec du traitement chirurgical des varices. J Mal Vasc 1993;
    18 : 314-319.

Phlebology '95, D. Negus et al. (eds.). Phlebology (1995) Suppl. 1: 472-473

OP/22.4

Comparison of Valve Reconstruction and Routine Saphenoctomy (Clinical and Functional Results)

F. Lurie and N.P. Makarova

Sverdlovsk Vascular Centre, Yekaterinburg, Russia

OBJECTIVE

To compare long term clinical and functional results in patients undergoing valve reconstruction (VR) with those treated with vein stripping and perforator ligation(SPL).

DESIGN

Randomised parallel group open clinical trial.

INTERVENTION

Patients randomised to valve reconstruction or to routine saphenectomy with perforator ligation. Operations performed by two surgeons using standartised surgical techniques and anaesthetic methods.

MEASUREMENTS

Standart clinical examination, ascending and descending venographies, duplex ultrasound scanning and venous pressure measurement performed either before operation and 1, 12 and 60 month after operation.

RESULTS

The two groups (67 VR and 62 SPL)were matched for age, sex, severity of disease and valvular reflux degree (measured by doppler reflux index RI) .
VR patients reported both increase and decrease dynamics of reflux degree. No spontaneous RI decrease was found in the SPL group. No significant differences were found in results between the two groups with RI<10%. The SPL patients with RI>10% had significantly higher levels of ulceration and recurrences of varicose veins (p<0.001) when compared to the patients of the VR group with the same value of RI.

CONCLUSIONS

Results of treatment depend on preoperative valvular reflux degree.
SPL and VR show the same clinical and functional results in patients with low level of reflux. VR shows better results in patients with high level of reflux.

Phlebology '95, D. Negus et al. (eds.). Phlebology (1995) Suppl. 1: 474-476

P249

Recurrence Following Combined Surgery and Postoperative Sclerotherapy of Varicose Veins

S.J. Simonian

Department of surgery, Georgetown University Medical Center. Washington, DC, USA

INTRODUCTION

The purpose of this study was to evaluate the incidence and etiology of recurrence of varicose veins, 3 years following treatment with surgery and adjunctive post-operative compression sclerotherapy for residual incompetent peripheral small varicosities which were left untreated at surgery.

METHODS

This is a retrospective review of a consecutive series of 117 lower extremities of 77 patients. Ages were 16 to 86 years, mean 50 years. Eighty-two were female limbs and 35 were male limbs with a female to male ratio of 7:3. Ninety-four limbs (80%) had saphenofemoral incompetence (SFI), 15 limbs (13%) had saphenopopliteal incompetence (SPI), eight limbs (7%) had both SFI and SPI, 105 limbs (90%) had saphenous vein trunk and tributary incompetence, 30 limbs (26%) had perforator or communicating vein incompetence.

Patients were evaluated clinically and with noninvasive vascular diagnostic laboratory testing including continuous wave (CW) Doppler, volume plethysmography, photoplethysmography, air plethysmography, duplex scan, and in selective cases color flow imaging.

All 117 limbs had surgery. SFI or SPI was surgically ligated and excised in all cases. The incompetent saphenous trunk was excised in all cases. The long saphenous vein trunk was excised from groin to below knee in 57 limbs (61%), from groin to ankle or foot in 19 limbs (20%) and when the trunk was competent it was locally excised in the thigh in 18 limbs (19%). The short saphenous vein trunk was excised from knee to ankle in 7 limbs (47%) and when the trunk was competent, was excised locally in the calf in 8 limbs (53%). Microsurgical stab avulsion (ambulatory phlebectomy) was performed in all tributary, communicating and peripheral varicosities in 105 limbs (90%).

Residual incompetent peripheral small varicose veins which were not excised by surgery were treated with compression sclerotherapy using sodium tetradecyl sulfate, 0.25% to 0.3% and 30-40 mm Hg gradient compression stockings.

MEASUREMENTS

Follow-up period was from 33 to 51 months, mean 3 years, at annual intervals. Patients were evaluated clinically, with CW Doppler and in selected cases with color flow imaging and phlebography.

RESULTS

The initial treatment with surgery and adjunctive compression sclerotherapy completely eliminated all incompetent varicose veins in all 117 limbs.

One hundred and one limbs were followed up.

There were six limbs (6%) with new recurrent peripheral varicose veins. The etiology of these recurrences appeared to be due to the ongoing chronic nature of the disease. Treatment with compression sclerotherapy eliminated the varicosities.

There were two limbs (2%) with recurrent SFI. One of these two limbs had a new recurrent small "neovascular type" SFI, plus a new SPI with an iliofemoral, deep vein incompetence, on color flow imaging. The etiology of the superficial recurrent SFI and the new SPI were probably secondary to the chronic deep vein incompetence.

The second limb had a second anomalous incompetent long saphenous vein trunk which joined the common femoral vein at a separate more cephalad position, as shown by colour flow imaging and phlebography. The etiology of the recurrence in this second limb was due to a technical error of having overlooked the second saphenous vein trunk at the first surgery.

The last two limbs were treated by surgery again which eliminated the SFI and SPI.

DISCUSSION

In a three year follow-up period of 101 limbs with varicose veins adequately treated by standard methods, there were 6 limbs (6%) with peripheral recurrent varicose veins, probably secondary to the chronic nature of the disease. The varicosities were abolished by compression sclerotherapy. There was one limb with recurrent SFI and new SPI probably secondary to deep iliofemoral vein incompetence. A second limb had recurrent SFI probably due to a technical error of having overlooked a second saphenous trunk entering the common femoral vein separately, at the first surgery. The last two limbs (2%) were treated by surgery again which eliminated the SFI and SPI [1].

CONCLUSION

One hundred and seventeen limbs with SFI and SPI were treated by standard surgical and sclerotherapy techniques, which abolished all incompetence satisfactorily. One hundred and one limbs were followed-up for a mean of 3 years. There were 6 (6%) peripheral recurrences, probably due to the recurrent nature of the disease. These were eliminated by compression sclerotherapy. There was one limb with recurrent SFI and new SPI due to deep iliofemoral vein incompetence. Another limb had SFI due to a second saphenous trunk which entered the common femoral vein separately and which was overlooked at the first surgery. These last two limbs (2%) were treated by surgery again which eliminated the SFI and SPI.

REFERENCES

Browse NL, Barnard KG, Lea Thomas M. Diseases of the Veins, London: Edward Arnold, 1988: 233-48.

Phlebology '95, D. Negus et al. (eds.). Phlebology (1995) Suppl. 1: 477-478

P250

Effectiveness of Therapeutic Association Redosurgery-Sclerotherapy in the Treatment of Varicose Veins Recurrence After Surgery

M. Perrin[1], J.P.Gobin[2], C. Grossetete[2], M. Lepretre[2] and F. Henri[3]

[1] Chirurgie Vasculaire, LYON, France
[2] Angéiologie, LYON, France
[3] Laboratoire BEAUFOUR, PARIS, France

OBJECTIVE :
To control effectiveness of redo surgery + sclerotherapy after failure of initial surgery $\pm$ sclerotherapy by a retrospective study with a follow-up at 5 to 6 years.
MATERIAL and METHODS :
From January 1985 to December 1986, one hundred and five patients (145 lower limbs) have been treated by redosurgery followed by sclerotherapy.
Inclusions criteria were :
- Persistence or recurrence of varicose veins after initial surgery followed or not by sclerotherapy.
- Persistence of major reflux between deep and superficial systems assessed by Doppler (100 %) and Phlebography (80 %).
- Decision to perform redo surgery made conjointly by the surgeon and the phlebologists.
Redo surgery was performed by the same surgeon (MP).
The sex ratio was F/M = 3.
Mean age at the first surgical procedure were 39.4 years and for redo surgery 52 years.
86 % of patients had primary varicose veins and 14 % secondary varicose veins (Post Thrombotic Syndrome or Primary Deep Vein Insufficiency).
Anatomical topography of surgical procedures was :

	LSV	SSV	LSV + SSV
First surgical procedure	92%	4%	4%
Redo surgery	59%	24%	17%

Failure mechanisms of initial treatment had been classified in three groups :

. Group 1	Incorrect Surgery	73 %
. Group 2	Incomplete Surgery	28 %
. Group 3	Evolution of the disease	12 %

Analysis of patients symptoms who leaded to redosurgery were :

Functional problems (Pain, Heaviness, etc)	80 %
Aesthetic problems	52 %
Edema	52 %
Ulcer	10 %

After the first surgical procedure, 36 % of patients were not improved and 64 % had noted a progressive recurrence of their symptoms or signs.

Mean interval between first and second surgical procedures was 12 years (6 months - 22 years).

After the initial surgery 51 % of patients and after redo surgery 83 % of patients have had complementary sclerotherapy.

Results were assessed by a questionnaire sent to the patient and by an independant audit.

Answers to the questionnaires were obtained for 105 extremities out of 145 : 71 %.

Only fifty patients (66 extremities) were examined by the audit.

Results after redo surgery were estimated as follows in percentage :

Patient estimation :

Symptomatic improvement	65 %
Cosmetic improvement	68 %

Audit estimation :

Symptoms :	Excellent or good results	78 %
	Fair results	12 %
	Poor results	10 %
Cosmetics:	Excellent or good results	30 %
	Fair results	53 %
	Poor results	17 %
	Ulcer recurrence	1 %

82 % of patients were satisfied by redo surgery completed by post op sclerotherapy.

CONCLUSIONS :

Redo surgery followed by post operative sclerotherapy appears to be a reliable therapy after failure of surgical treatment in varicose veins in this selected group. The main feature of this group was presence or persistence of a major reflux from deep to superficial venous system.

Phlebology '95, D. Negus et al. (eds.). Phlebology (1995) Suppl. 1: 479–481

P196

Local Anaesthetic and Outpatient Phlebectomy

P. Ilieff

Phlebological Praxis, Steinbühlweg 15, 4123 ALLSCHWIL, Switzerland

The ambulant phlebectomy is limited by the dosage of local anaesthetica (LA). The partially irrational determination of the limiting doses varies in different publications. The addition of Epinephrin lowers down the toxication of known LA, however there are to expected various complications due to type of application. Our experience in ambulatory phlebectomy is with Mepivacain-Adrenalin 0,25% and the lower dosage of LA allowed:

- to cover larger areas
- to increase the amount induced up to 120 cc
- inspite of small dosage a satisfactory anaesthetica activity is achieved
- the low dosage did not eliminate the sensitivity of the subcutaneous nerves

THE GOAL

- to investigate the possible factors that might increase the chance of complication during surgery
- to investigate the possible difference in change in systolic, diastolic blood pressure (BP) and heart rate (HR) from baseline due to age, gender and amount of LA.

METHOD

For the evaluation of complication a logistics regression was carried out using amount of LA, age, gender and leg as explanatory variables. For the evaluation of BP and HR analysis of covariance were carried out on the change in BP and HR (before and after). The model included pre-operation value, age, gender, leg and amount of LA as explanatory variables compared with the post operation values of BP and HR.

GROUPS OF PATIENTS

There were 766 patients treated having mean age of 50,5. The oldest patient was 84 years old, the youngest was 22 years old. There was a total number of 355 right legs and 411 legs treated.

480

RESULTS

The logistic regression on complication shows that age, amount of LA or the leg have significant effect (p = 0.97, 0.17 and 0.89).

However, there is a gender effect. The female patients is expected to have more complication (36.9% vs 17.5% for males). The observed average change in systolic blood pressure (SBP) in female was -2.46 mm Hg (n=586) and -1.97 (n=79) for male. The average pre-operation values were 130.16 and 136.67 (for female and male, respectively). There was a significant gender effect (p= 0.05). In addition there was a significant age effect (p=0.0001). Every 10 years increase in age added 3.22 mm Hg to the change. 10 cc of LA is expected to increase the change by 1.1 mm Hg (p=0.039). However, the effect of leg was not evident (p=0.95).

Basically the same results were observed in the of change of diastolic BP. The observed average change in pulse rate were -8.7 and -8.87 beats/minute for female and male, respectively. The influence of age, leg and amount of LA were not evident. However, due to the difference in pre-operation value between the gender, there was a significant difference between gender (p=0.03). For a male and a female patient at same age, the female's post operation HR is expected to be about 2 beats higher than that of the male.

CONCLUSIONS

- Female patients is expected to have more complication after phlebectomy. However, age, leg and amount of local anaesthesia do not have significant impact on the chance of having complications
- There was a significant gender effect in the change of systolic BP, diastolic BP and heart rate.
- There was significant effects due to age and amount of local anaesthesia in the change in systolic BP and also significant effect due to age in the change in diastolic BP.
- The amount of LA has no significant effect of on the pre and post operative changes of BP and HR.
-
-

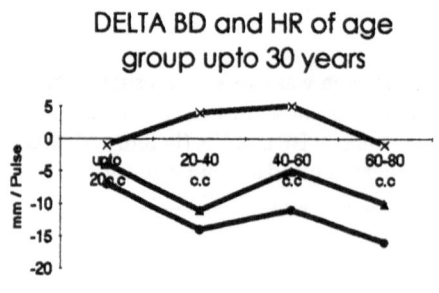

DELTA BD and HR of age group upto 30 years

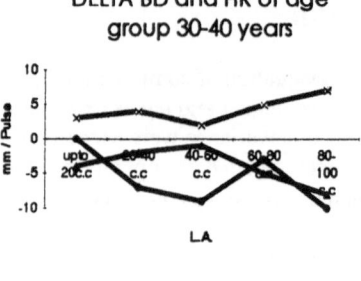

DELTA BD and HR of age group 30-40 years

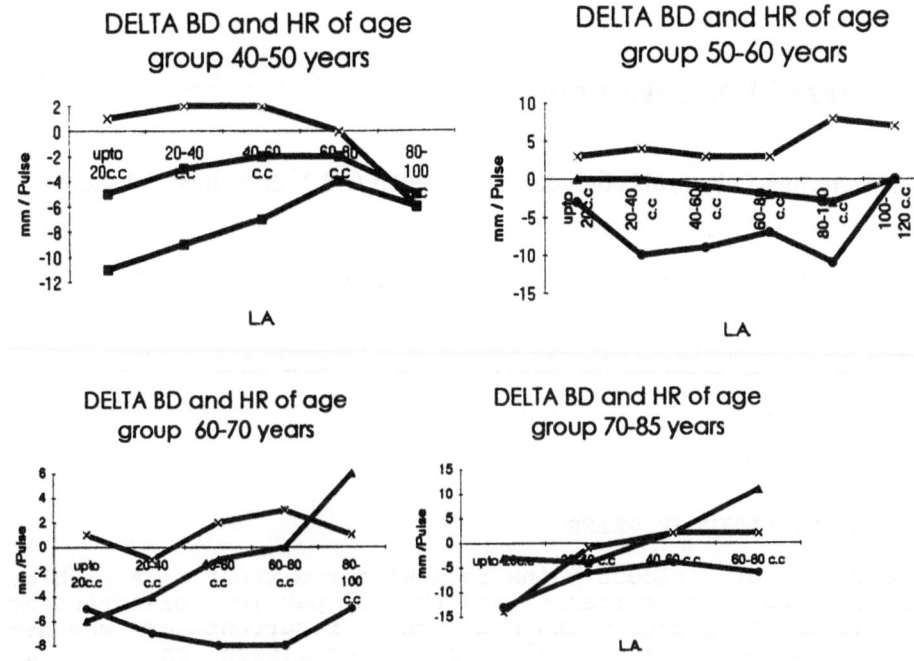

Δ - **Delta systolic blood pressure**
X - **Delta diastolic blood pressure**
O - **Delta heart rate**

REFERENCES

1. H.C. Niesel und H. Kaiser: Grenzdosis für Lokalanästhetika. Anästhesieabteilung, St. Marien und St. Annastiftskrankaenhaus, Ludwigshaven/Rh. Reg Anaesth (1991) 14: 79-82.
2. J. P. Vidal-Michel et all: L'anesthésie locale au cours de la Phlébektomie ambulatoire selon la Methode de R. Muller. Phlebologie, 1990, 43, no 2, 305-315.
3. M.Tryba: Klinik für Anaesthesiologie, Intensive medizin und Schmerztherapie, 1989, Gustav Fischer Verlag - Stuttgart-New York.

Phlebology '95, D. Negus et al. (eds.). Phlebology (1995) Suppl. 1:482–483

PI/4.1

Recurrent Varicose Veins

D.J. Moore, M.P. Colgan, M.C. Grouden, S.J. Sheehan, G.D. Shanik and G. Donald

Department of Vascular Surgery, St James' Hospital, Dublin 8, Ireland

RECURRENT VARICOSE VEINS

Recurrence of varicose veins following treatment is a problem familiar to both surgeons and the lay public. Disagreement exists about whether veins are truly recurrent or whether they have been inadequately treated or missed on previous treatments. Lack of clear pre-operative definition of the disease and inadequate or ineffective treatment are widely implicated as the commonest cause of so called recurrence.

Our experience with recurrent venous disease following surgery relates to 600 varicose veins operations which were performed between August 1993 and June 1994. Recurrent varicose veins accounted for 90 cases (15%) of the work load. All patients in this group had venous duplex scanning pre-operatively and the site or sites of valvular incompetence were recorded for each patient.

The distribution of sites of valvular incompetence compared to previous surgical treatment is shown in the table below.

Previous Ligation	n	Sites of Incompetence			
		SFJ	SPJ	LSV	Unidentified
SFJ	62	37	28	20	1
SPJ	2	0	2	0	0
Avulsion	26	21	1	6	2
TOTAL	90	58	31	26	3

A number of patients presenting with recurrent disease had incompetence at more than one site. The extra time involved in these patients compared to those undergoing surgical treatment for primary varicose veins was almost 50% longer (85 minutes versus 57 minutes:p>0.001).

A 12 month prospective follow up on all patients who have undergone surgery for recurrent varicose veins has revealed that the long saphenous vein and short saphenous vein junctions remain occluded in all patients and with the exception of occasional missed scattered varicose veins in the thigh or calf, all the patients remain symptom free from their varicose veins. A number of post-operative complications were recorded; these included lymphoedematous swelling (1), peripheral nerve neuropraxia (3), DVT (1).

We conclude that recurrent varicose veins contribute significantly to both the case load and operating time in a vascular surgical practice. With few exceptions, in our experience the need for further surgery was due to apparent inadequacy of the technique or incompleteness of the primary procedure carried out before.

We feel that duplex scanning is mandatory to plan treatment for recurrent varicosities. Furthermore, if it had been available pre-operatively on the first occasion the need for further surgery might have been avoided. We believe complete stripping of the long saphenous vein is essential to avoid recurrences, and close attention must be paid to the short saphenous system. Attention to pre-operative diagnosis, careful advice to patients about aims of treatment and meticulous surgical technique should establish quality control for treatment of this common disorder and reduce the incidence of recurrent disease.

OP/14.8

ENDOSCOPIC LINTON PROCEDURE IN PATIENTS WITH SEVERE VENOUS ULCERATION

Bogachev V, Reviakin V, Vasilenko Yu.

Department of vascular surgery, State Russian medical University, Moscow, Russia

Objective: To develop low traumatic method of surgical treatment for severe venous ulceration resistant to conservative therapy.

Design: Retrospective review of a consecutive series.

Patients: 11 patients with severe venous ulceration.

Intervention: Communicating and perforating veins were interrupted by endoscopic coagulation and clipping. Compression sclerotherapy was used in seven cases after intervention.

Measurements: Duplex-scanning and emission computed tomography were applied for detection of the insufficient communicans and perforants after intervention.

Results: All our patients demonstrated complete liquidation of pathological venous reflux on calf. We achieved complete healing of the venous ulcer in 100 per cent of our patients in a mean 1.5 months.

Conclusion: Endoscopic Linton procedure is an effective method for treatment of patients with severe venous ulceration.

P231

OUR EXPERIENCES WITH DIFFERENT SURGICAL METHODS IN THE TREATMENT THE ACUTE AND CHRONIC ILLNESSES OF THE LEG VEINS

Baricza S., Pfeiffer J., Tollas F., Pataky L.
Dept. of General Surgry, Town-Hospital Várpalota, Hungary

Objective: To evaluate the chainges and the results in the surgical treatment the venous diseases of the leg in the last seven years.

Design: Retrospective clinical analysis

Patients: 990 primary chorinc varicous and 92 acute varicophlebitic patients who had different surgical interventione.

Interventions: Since 1988 at chronic varicous vein interventions we could introduce the Várady's microsurgical methods and instruments and since the same year we began the active surgical treatment of acute ascendent thrombophlebitic patients with crossectomy, stripping and radical removal of the inflammed veins.

Measurement: The data of distribution the different surgical methods, the hospital stay, the complications by years.

Results: The new methods after two years became the first choice of surgical resolution of the problems. The hospital stay, the complication rate are very low. The patients left the hospital earlier, and they have possibility to begin their work earlier too, the cosmetic results are more better than was before.

Conclusion: Our results can convince the patients and their general practitioners to sent or come for surgical resolution at both forms (acute and chronic) venous diseases of their legs.

P205

"CHIVA" CARE vs STRIPPING IN THE TREATMENT OF VARICOSE VEINS: THREE-YEARS FOLLOW-UP.

M. Cappelli, S. Ermini, R. Molino Lova, A. Turchi, and G. Bono.
Phlebologic Surgery, 46 Datini Street, Florence, Italy.

Objective: To compare three-years results following classical "CHIVA" Care to those, from Reading, following Stripping Operation, with the same follow-up.

Design: Retrospective review of consecutive series.

Patients:105 procedures in 84 consecutive patients

Intervention:The classical "CHIVA" Care described by Franceschi.

Measurements: The presence of visible varicose veins, graded in three categories ("Cured", "Improved" and "Failed"), according to Hobbs and Einarsson, as shown by clinical examination.

Results: In the "CHIVA" group very good results were obtained in 60 cases ("Cured"), who presented without any finding of varicose veins throughout the follow-up. Of the remaining 45, in 1 case we found a new varicose network completely developed ("Failed"), whereas in 44 cases only some visible vessels were present ("Improved"). Statistical analysis showed significantly better results in the "C.H.I.V.A." group compared to Stripping groups (Pearson "Chi Square" Test: P=0,006).

Conclusions: According to this study, patients suffering from varicose veins could be better treated by "CHIVA" Care than by Stripping Operation.

P207

THE MICRO-PHLEBO-ASPIRATION WITH THE PHLEBOCURETTES

Umansky S. Umansky M.

Dept. of Plebology.Clinicas Mepres.

España.

Objective:To review the different chirurgical technics and show the instruments we used to realized the Micro-Phlebo-Aspiration to treat dilated venules and collaterals V.V.

Design:Compare of Phlebocuretage and Aspiration with other technics.

Patients:20 patients with other technics and 53 with the Micro-Phlebo-Aspiration that we choose in dilated venules and collateral V.V.

Intervention:with our instrument,the PHLEBOCURETTES,we can perform the operation:1-by Extraluminal Disection,2- by Phlebodestruction,3-by Phlebocuretage,and 4-by Aspiration or Phlebosuction. (Description of the apparatus).

Measraments:The tecnique that we propone with the use of the PHLEBO CURETTES and PHLEBOSUCTION,differ from others: in post operative reactions,time of operation,minor number of punctures,four differents techniques, bring to perfection the same purpose,put into practice a better esthetic point of view for the Phlebologue and the Patient.

Results:The primitive technics gave good results in training hands but in our opinion we can have a better one for this reasons:Les punctures,les hematomas,les time for recuperation,les operative time,able to operate venules and collaterals at same time.All that with local anesthsia.

Conclusion:The MICRO-PHLEBO-ASPIRATION,gives better results,and impruves many other possibilities than the primitive ones for the reasons above mentioned.

V/1.1

THE SHORT STRIPPING OF SAPHENA AND THE CORCOS' EXTERNAL PHLEBOEXTRACTOR IN OUR EXPERIENCE.
F. CAMPITIELLO, F. PACIFICO, D. SEPE, V. LAULETTA, F. LUMINELLO, S. CANONICO

VI DIVISION OF GENERAL AND GERIATRIC SURGERY
SECOND UNIVERSITY OF NAPLES - ITALY

OBJECTIVE: To evaluate our experience with Corcos' external phleboextractor in the day case surgery of varicose veins.
DESIGN: Retrospective review of a consecutive series.
PATIENTS: 478 patients underwent operations for varicose disease of lower limbs; 283 subjects (59.2%) had a local anaesthesia as day cases. 88 (31.1%) were operated on by means of the Corcos' external phleboextractor. These patients were suffering from sapheno-femoral ostium (crosse) insufficiency and thigh saphena incompetence without reflux along the leg segment; a varicectomy was performed according to Muller technique in 37 patients (42%).
INTERVENTION: After classic crossectomy a short saphenectomy by means of the Corcos' external phleboextractor was carried out. With this phleboextractor the cutaneous distal cut can be avoided since the leg saphena is dissected by careful finger pressures, while the extracted segment can be removed from the inguinal incision. When it was indicated, a varicectomy according to the Muller technique completed the intervention. A 40 mmHg compression by idealbinde elastic bands was applied on the leg.
MEASUREMENT: An immediate and long term follow-up was programmed.
RESULTS: No problem occurred during the intervention and no patient required to be readmitted to hospital for complications; only 5 patients (5.7%) required paracetamol as postoperative analgesia for 48 h. After 12 months 7 patients (7.95%) had new small varicouse veins on the postero-lateral aspect of leg: none required fourther surgical treatment and all were treated by injection sclerotherapy.
CONCLUSION: This experience support the evidence that in selected patients the ambulatorial short stripping of saphena by Corco's external phleboextractor is a better procedure than the Mayer's internal instrument.

V/1.4

PREVENTION OF SAPHENOUS NERVE INJURY CAUSED BY STRIPPING OF THE LONG SAPHENOUS VEINS: PERSONAL APPROACH

Sorrentino P, Baccaglini U, Castoro C, Spreafico G, Giraldi E, Pavei P, Penzo S, Ancona E

University of Padua, Italy, 2nd Dept. of Surgery

Objective: To prevent sensory impairment in the cutaneous distribution of the saphenous nerve (S.N.) caused by stripping of the long saphenous vein (L.S.V.).

Design: Prospective study of a consecutive series of patients.

Patients: 75 patients with primary L.S.V. varices and reflux from sapheno femoral junction (S.F.J.) to the ankle, documented by duplex. 7 patients were excluded: 5 for the presence of trophic cutaneous lesions due to chronic venous insufficiency and 2 for previous sclerotherapy treatments below the knee.

Intervention: 68 patients underwent high ligation and stripping of the L.S.V. from groin to ankle, under monolateral subanacnoidal anaesthesia on day on day surgery. Surgical Technique: After high ligation of the S.F.J., an internal modified metallic stripper was inserted from groin to ankle where a short incision anteriorly and below the internal malleolus was performed. An external modified Mayo stripper used to dissect the vein and to section collateral branches from groin to upper calf, and then an upwards invagination of the vein from ankle to upper calf completed the stripping. Adhesive short stretching bandage was applied immediately after the operation and patients were allowed to walk.

Measurements: Intraoperative and post-operative complications were recorded. All patients were evaluated for subjective and objective sensory impairment in the cutaneous distribution of the S.N., one, three and six months after the operation.

Results: No intraoperative complication was recorded. Stripping of the L.S.V. was complete in all cases without any rupture of the veins and no post-operative haematoma. 1 out of 68 (1.5%) evaluable patients had an area of 16cm^2 of hypoaesthesia persisting at 6 month follow-up.

Conclusions: Incidence of paraesthesia varies wildly in published series. Best results have been reported with the invagination technique by Van Der Strichts, but this technique is associated with a high incidence of vein rupture and incomplete stripping. Our approach combining external and invaginated stripping resulted in a very low complication rate without any rupture of the vein.

OP/14.2

ALTERNATIVE SAPHENOUS VEIN SPARING SURGERY

Zamboni P, Marcellino M G, C Feo, L Pisano, A P Murgia

Department of Surgery, University of Ferrara, Ferrara, Italy

Objective: To evaluate long saphenous vein sparing surgical procedures alternative to high ligation and distal stab avulsion.
Design: Prospective evaluation of 135 operations for primary varicose veins.
Patients: 58 patients (62 limbs) with duplex scanning evidence of mobile valve leaflets underwent external valve-plasty of the saphenofemoral junction (EV-SFJ) (42 performed using the hand sewing technique and 20 using the Veno-cuff device). 61 patients (73 limbs) were operated on using the hemodynamic correction technique (CHIVA).
Interventions: EV-SFJ restores valve function correcting vein wall dilatation by applying an external prosthesis. CHIVA consists of selected ligatures that allow the superficial blood aspiration in the deep veins through the re-entry perforators.
Measurements: The outcome was evaluated with clinical and ultrasonographic examinations, ambulatory venous pressure (AVP) and light reflex rheography (LRR-RT) measurements.
Results: Long saphenous vein patency registered after EV-SFJ (mean follow-up 40 months) and CHIVA (mean follow-up 36 months) was 93.5% and 89.1%, respectively. Varicose veins recurrence rate was 11.3% and 12.3%, respectively. Pre and post-operatory reduction of AVP and prolongation of LRR-RT were significantly different (P 0.001, Student and Wilcoxon tests).
Conclusions: These two alternative procedures seem to be more effective in varices treatment than high ligation and have the advantage of preserving a higher rate of saphenous vein patency.

P210

LIMITED STRIPPING OF THE GREATER SAPHENOUS VEIN; SAVING THE SAPHENOUS VEIN WHEN POSSIBLE

Noppeney T., Noppeney J., Kurth I.
Department of vascular surgery. Clinic Hallerwiese, Nuremberg, Germany

Objective: One of the principles of modern varicose vein surgery is to save the saphenous vein when ever possible.

Patients: In 1993 366 patients with 585 legs were operated. In these data the ambulatory cases are not included. 357.legs presented with varicose vein degeneration and insufficient valves of the greater saphenous vein (GSV) down to the ankle.

Intervention: The operative strategy is to ligate and divide all tributaries at the saphenofemoral junction (SVJ), to remove the GSV totally.

Results: In more than 100 legs we could save either parts of the GSV or could leave the GSV totally in site. Our complication rate was very low. 3.6 % of the legs presented with a saphenous or suralis nerve damage, 1.9 % showed a pigmentation, 2.7 % an edema of the leg. A deep veinous thrombosis did not occur.

Conclusions: The advantages of our procedure is to save the GSV totally or in parts, you also have a lower incidence of complication an a low recurrence rate. Other operation procedures to save the GSV, e.g. ligation of the GSV, can not be recommended generally, because of a much higher recurrence rate of varicosities. During the follow-up period we could show by colour-duplex-imaging, that the remaining GSV is suitable for other vascular procedures like bypass-operations.

V/1.5

VIDEO ASSISTED VENOUS SURGERY

Welch H J, Iafrati M D, O'Donnell T F

Division of Vascular Surgery, New England Medical Centre, Boston MA, USA

Objective: To illustrate video assisted techniques of vein valvuloplasty and subfascial vein ligation.

Design: Videotape of intraoperative techniques

Patients: Selected intraoperative procedures

Intervention: Improved video technology has revolutionized all types of surgery, including vascular. Angioscopy is extremely helpful in the intraoperative diagnosis of primary valvular insufficiency. Repair of a floppy, incompetent valve is technically demanding, and angioscopic guidance of the repair allows for a closed valvuloplasty (ie, without venotomy) with proven competence and hemodynamic improvement. Subfascial ligation of incompetent perforating veins often requires an incision through skin and tissue ravaged by the changes of chronic venous insufficiency, which may result in significant post-operative morbidity. Entering the subfascial space away from damaged tissue and using video-assisted techniques to locate and divide the incompetent perforators can avoid the wound morbidity often seen. This videotape will illustrate these techniques, and describe the benefits and pitfalls of this new technology.

Conclusions: Video assisted technology significantly aids the surgical procedures of vein valvuloplasty and subfascial ligation.

P081

ELASTIC COMPRESSION TIGHTS: THE BEST DRESSING AFTER VARICOSE SUGERY?

Lefebvre-Vilardebo M., Uhi J.F., Lemasle Ph.

Centre de Chirurgie Esthétique des Varices, 75116 Paris, France.

Objective: To clinically evaluate the interest of elastic compression tights as unique dressing after varicose surgery.

Design: Retrospective review of a consecutive series.

Patients: 1000 patients with varicose veins operated under local anaesthesia.

Intervention: Whatever the type of varicose sugery was performed, all patients had elastic tights pulled on without other dressing on the operating table. 2 superposed tights (P≥40mmHg) during 4 days, then 1 tights (P=20mmHg) during 25 days. The superposition improves the rest pressure and the work pressure with a higher efficiency against bleeding and clinical discomfort.

Results: Stay of 6-24 hours after surgery. No hematoma in avulsion paths, but only minor to moderate skin ecchymosis. No deep vein thrombosis. 83,78% patients were hostile to compression before operation, but 86,48% were favourable after 1 month. 76,7% patients came back to work before D12.

Conclusions: Elastic compression tights is a very simple post-operative dressing, with an accurate efficiency for a quick recovery and with a major part to teach the interest of compression therapy in day-life.

P211

SAPHENO-FEMORAL JUNCTION: ANATOMY AND SURGICAL CONSEQUENCES

Lefebvre-Vilardebo M

Centre de Chirurgie Esthétique des Varices - 75116 Paris (France)

Objective: To study anatomy of the sapheno-femoral junction and evaluate the risk of recurrence on account of forgotten tributaries (because ending subfascially in the junction).

Designs: Anatomical study during long saphenous vein surgery.

Patients: 200 patients with primary varicose veins proposed to a first hand surgery.

Intervention: After dissection of the common femoral vein (CFV) for a correct flush ligation, all the tributaries were noted, with their relation to the deep fascia.

Results: 63 patients (31.5%) had one or more tributaries ending under the Allan Burns fascial hole, but above the ostial valves: 57 collaterals on the medial side which could reintroduce varicose network on medial and posterior side of the thigh - 47 collaterals on the external side. Among them, the anterior saphenous vein (13 = 6.5%) is the most difficult to discover. And in 94 cases (47%) this vein ended with abdominal and inguinal veins through a common trunk. If not separately ligated, they could be the way of secundary indirect reflux.

Conclusions: Careful dissection of the CVF 1 or 2 cme around the junction must be done to be sure to perform a correct flush ligation, especially in bad surgical conditions (e.g. obesity).

P082

VARIX SET-INSTRUMENTARIUM FOR VARICOSE VEINS OPERATION.

Matousek P.

Department of Surgery, Kromeriz, Czech Republic.

Objective: Author presents a new modern instrumentarium designed and constructed by himself for the purpose of varicose veins operation.

Design: During the development of the instrumentarium I wanted to reach:
1. Perfect cosmetic effect without disfigurements with scars.
2. Maximal radicality of the operation.
3. Unexacting and safe intervention for a patient.
4. Simplicity, speed and comfort of the intervention for a surgeon.

Intervention: The basic part of the instrumentarium is the universal holder to which surgical equipments such as knives, strippers etc can be separately fixed. The set of instruments provides economical and cosmetically advantageous stripping of the vena saphena magna and vena saphena parva and the extirpation of collateral branches and epifascial section of the perforators.

Results: The operation with this newly-modelled VARIX SET instrumetarium is easy, safe, does not damage the surrounding tissue and also bleeding is minimal. There were no peroperative and postoperative complications.

Conclusions: The VARIX SET gives the possibility to be assembled variably according to the surgeon's needs during each separate phase of the operation. Manipulation with the instruments and technical performance of the operation is simple, quick and with very good cosmetic effect.

V/1.6
A NEW VARIANT OF SAPHENOUS STRIPPING
E. DALLA VALLE^ and F. CORTESE*

Casa di Cura Villa Aprica^ (CO) Italy
Ospedale di Saromno* (VA) Italy

Objective: A new simple variant of the traditional saphenous vein stripping technique is evaluated.
Design: Retrospective review of a consecutive series.
Patients: 223 patients with varicose veins due to valve incompetence at the saphenofemoral junction by Doppler examination underwent g. sapenous stripping, limited in 177 cases and total in 46, between May 92 and October 94.
Intervention: After ligature and division of the g. saphenous vein and its tributaries at the inguinal crease a stripper is passed through the vein retrogradely to knee or ankle level where its tip is pulled through a 2 mm incision. The vein is hooked, ligated and divided distally and, at the inguinal site, a stripper head with a 1,2 mm hole at its tip, is screwed, the vein is ligated around the stripper and a 0 silk suture of appropriate length threaded and ligated through the stripper head hole. The stripper then is pulled distally until the head is subcutaneous, the vein is gradually extracted around the stripper through the distal mini-incision and the head retrieved through the inguinal crease by pulling the silk suture upwards.
Results: Two patients had a transient postoperative paresthesia at the ankle level and in one case a cheloid was observed at the knee level.
Conclusions: Among the various techniques devised to minimize trauma at the site of extraction of the stripper head ours has the advantage to be effective in every case, even when the vein is sclerosed and postinflammatory. Less local trauma is produced and a better cosmetic result is achieved in any case.

P212
AESTHETIC TREATMENT OF LOWER EXTREMITIES VARICOSE VEINS
Rück F., Agostini M., Giansiracusa G.

Surgical Department, General Hospital, Lovere, Italy

Objective: To evaluate the results of the "cosmetic" treatment of varicose veins of the legs.

Design: A retrospective review of a consecutive series.

Patients: 300 patients, 80% women, 20% men, underwent the "cosmetic" surgical treatment of varicosities.

Intervention: The treatment consisted in the "crossectomie" of the large saphena; the short stripping of the large saphena; the ligature of the communicating veins; the phlebectomies according to Müller; the sclerotherapy.

Measurements: The patients who underwent a clinical control after 1 - 3 - 6 - 12 - 24 - 36 months.

Results: The patients presented good results in 98% of cases; no paresthesia observed.

Conclusions: This method is to be considered excellent from the functional and the aesthetic point of view.

OP/5.2

ONE DAY SURGERY FOR CHRONIC VENOUS DISEASE OF THE LOWER LIMBS

Constantinova G, Gradusov E, Alekperova T, Dudkin B

United Centre of Out-Patient Surgery, Moscow, Russia

Objective: To evaluate quality of one-day surgery method for treatment of chronic venous diseases of the lower limbs

Design: Retrospective analysis of a four year experience.

Patients: 614 patients with primary varicosity's, 9 patients with post thrombotic disease. Traditional veinectomy in 214 cases and 163 corrections of valve insufficiency were made, sclerosurgery method was used in 40 patients.

Intervention: Traditional veinectomy was made (crossectomy, phlebectomy, Cocet procedure). Extravasal correction of valve insufficiency was made by Vedensky lavsam correctors. Sclerosurgery included crossectomy with intraoperational introduction of sclerosis remedy into the trunk of the great saphenous vein through catheter and we used post operational compression vein - afflux sclerotherapy. After out patients surgical treatment all patients were transported home in 2 or 4 hours.

Measurements: Doppler ultrasound and angioscanning control were used a three year patient observation.

Results: 400 patients required sclerotherapy. 106 patients were observed during two years and 72 patients during more than three years. There were no recurrences. No detected imperfect correction of deep venous valve insufficiency (in 5 cases (3.1%) from 163 patients with more than 3 year observations) as a result of anatomical inferiority of valves. In nine cases inferiority with post thrombotic disease, when correction was made out of thrombotic zones in consequence of insolvent valves we observed steady remission.

Conclusions: One day surgery is a real effective and economically convenient method of popular treatment of chronic venous disease of the lower limbs.

OP/20.1

RE-EXPLORATION OF THE SAPHENOFEMORAL JUNCTION IN THE TREATMENT OF RECURRENT VARICOSE VEINS

Viani MP, Pinto A, Poggi RV.
Dept of Surgery, Ospedale di Abbiategrasso - Milano, Italy

Objective: To attest the efficacy of re-exploration of the sapheno-femoral junction (SFJ) in recurrent varicose veins (RVV).

Design: Retrospective study of a consecutive series.

Patients: 55 patients presented RVV from the groin investigated with cw-doppler and phlebography or color duplex scanning after a previous stripping of long saphena (LS) (50 cases) and re-exploration of SFJ for recurrence (5 cases). All the patients underwent re-exploration of the SFJ by a single surgeon (MPV) through a lateral approach.

Intervention: After skin incision 1 cm proximal to the inguinal fold and identificated the femoral artery, the dissection is shifted medially to visualize the femoral vein (FV) between the inguinal ligament and its bifurcation in superficial and the deep. The SFJ is easily identified and ligated flush with the FV. The FV is completely isolated medially and laterally to exclude the presence of tributaries and a drain is placed. The procedure was completed with stripping of LS in 11 cases (20%) and ligation of incompetent perforating veins in 16 (29%).

Measurement: Intraoperative finding was compared to preoperative one. All patients were reviewed. Median follow-up was of 2.5 years.

Result: 35 (63%) patients had intact major tributaries emerging from the FV, 20 (36%) had an intact SFJ, and 11 (20%) had an intact LS in the thigh. 1(1,8%) patient was submitted to reintervention due to hemorrhage from the stump of the SFJ. 4 (7,2%) infections of the inguinal incision were registered. In 3 patients the drainage was mantained for 12 days due to lymphorragia that spontaneously evoved. During the follow-up 48 (87,2%) patients were asymptomatic, 2 (3,6%) presented a new recurrence in the groin 12 and 24 months after the intervention and 5 (9%) presented a recurrence due to incompetent perforating veins of the leg after a period ranging 1 to 24 months.

Conclusions: the intervention has proved to be simple and effective. Complete isolation of the FV to identify every tributarie and ligation of the SFJ flush with the FV is essential to avoid further recurrences.

P213

COMPLICATIONS OF AMBULATORY PHLEBECTOMY (MULLER METHOD)

Dortu J.A. , Dortu J. , Constancias-Dortu I.

2 rue des Glières . 74000 . Annecy . France

Ambulatory Phlebectomy , technic offered since 1966 by Robert Muller (Neuchatel,Switzerland) , constitue today a major therapeutic weapon against the varicose disease .

Phlebectomy must answer to four principles :

- ambulatory
- innocuity
- efficacity
- esthetism .

Incidents or complications of this technic must be well known to answer to these four principles .

Most of them will spontaneously desappear . Some of them will easily cured with a specific treatment .

Among these incidents , the unesthetic consequences of the ambulatory phlebectomy could be a disappointment for certain patients .

However, notable adverse complications are extremely rare and would never appear with a good practice .

P014

LEONARDO'S VEIN OR POSTERIOR ARCH VEIN: ANATOMO-CLINICAL AND STATISTICAL STUDY FROM ITS TREATMENT BY AMBULATORY PHLEBECTOMY (MULLER METHOD).

Dortu J.A., Dortu J, Constancias-Dortu I.
2 rue des Glières. 74000. Annecy. France

Objective: To have a new approach about a not very well known vein which is very often implicated in the varicose disease.

Design: Retrospective review of 1560 successive ambulatory phlebectomies for varicose veins of the lower limb.

Conclusions: Leonardo's veins have four principal interests:
1. Anatomical interest
   because of its connections with some of the most dangerous perforating veins of the legs (Cockett, Dodd, Boyd).
2. Physio-pathological interest
   because this vein is one of the most important ways of shunt for the reflux between the long saphenous vein and the short saphenous vein.
3. Chirurgical interest
   because this vein is one the wrong way routes of the stripper in surgery of the long saphenous vein and treatment of this vein by the ambulatory phlebectomy is very elegant.
4. Statistical interest
   because of the high frequency of its pathology.

PI/9.4

INVERTING STRIPPING VERSUS CONVENTIONAL STRIPPING OF THE LONG SAPHENOUS VEIN

PD COLERIDGE SMITH, CM BUTLER, KM SUMMERVILLE, JH SCURR

Dept Surgery, UCL Medical School, The Middlesex Hospital, London W1N 8AA

Previous studies suggest that it is desirable to strip the long saphenous vein to prevent recurrence of varices, but this may cause haematomas, skin bruising and pain. Inverting stripping has been suggested as an alternative method which may reduce the extent of these complications.

The aim of this study was to compare the results of conventional vs. inverting stripping of the long saphenous vein in a randomised, double-blind study. 67 patients with primary incompetence of the long saphenous vein were identified by colour duplex ultrasound imaging. They were randomised to undergo either conventional (Codman stripper) LSV stripping or inverting stripping (Oesch technique), with saphèno-femoral junction ligation and avulsion of varices. Blood loss due to stripping was assessed and post-operative pain recorded using a visual analogue scale (VAS). 1 week later the extent of thigh and calf haematomas was measured using duplex ultrasound imaging, by an observer blinded to the type of the operation. 59 patients completed the study.

RESULTS Operation	Number of patients	Blood loss* (ml)	Pain* VAS - mm	Total no of Haematomas	Bruising* skin area cm2
Conventional	31	50 (30-75)†	32 (22-48)‡	12	128 (52-251)‡
Inverting	28	25 (18-50)†	25 (20-40)‡	10	167 (80-275)‡

*Median (inter-quartile range). †p=0.0034, ‡N.S. Mann-Whitney u-test.

The blood loss was less in the inverting stripping group, but similar post-operative pain and bruising were observed in the two groups. Marginally fewer patients had haematomas following inverting stripping.

P190

CROSSECTOMY AND FOLLOWING SCLEROTHERAPY, IN OUTPATIENTS FOR TREATMENT OF PRIMARY VARICES OF LOWER LIMBS.

Lombardi A
Studio Flebologico Lamarmora, via Lamarmora, 24 51021 Firenze, ITALY

Objective: Treatment of primary varices of the lower limbs in outpatients by crossectomy and following sclerotherapy saving v; saphena.

Design: Considerations in follow-up of personal casuistic.

Patients: 190 (210 legs) with primary varicose veins.

Intervention: Study of patients by Doppler cw, plethysmography LRR, ECO (in the last two years). I performed crossectomy by local anaesthesia (subcutaneous infiltration) in outpatients. After two months I performed sclerotherapy or phlebectomy if necessary in remaining varicose veins.

Results: Very good reaching objectives a) saving saphena, b) no ospedalisation c) disappearing varices, d) aesthetics, e) good compliance of patients.

Conclusions: Advanced technologies in vascular diagnostics demonstrate how v. saphena is rarely varicose and its importance in venous haemodynamics so that we must try to save it whenever it is possible.

P220

"LIMITED" STRIPPING OF LONG SAPHENOUS VEIN IN THE TREATMENT OF VARICOSE VEINS:PRESERVATION OF THE CALF PART OF LSV FOR CARDIOVASCULAR SURGERY

Zbroński R, Pardela M, Urbaniak A, Sławski M, Matuszewska-Zbrońska H,

2nd Department and Clinic of General Surgery and Vessels Zabrze, Poland

Objective:To evaluate patency of calf part of LSV remaining after its "limited" stripping for using in the future cardiovascular surgery.
Design:Retrospective rewiev of a consecutive series.
Patients:54 patients/74 lower legs/ with isolated insufficiency of sapheno-femoral junction.
Intervention:All patients were undergone operation "limited" stripping of LSV according to the Negus procedure /Phlebology 86,1,33/.
Measurements:Three and twelve month after operation the patency of calf part of LSV was evaluated using Doppler technique.
Results:The patency of calf part of LSV has been remained in 61 legs /83%/ after three months and respectively in 53 legs /71,6%/ after twelve months.
Conclusions:After operation "limited" stripping of LSV the calf part of LSV may be used for cardiovascular surgery in the future.

P221

RESULTS OF LIMITED STRIPPING OF LONG SAPHENOUS VEIN AND ADDITIONAL SCLEROTHERAPY IN THE TREATMENT OF PRIMARY VARICOSE VEIN.

Zbroński R, Pardela M, Urbaniak A, Kozłowski A, Sławski M, Kowalczyk S, Lemieszewski A,

2nd Department and Clinic of General Surgery and Vessesl, Zabrze, Poland

Objective:To evaluate few invasive operation procedure /limited stripping of long saphenous vein/ with additional sclerotherapy.
Design:Retrospective rewiev of a consecutive series.
Patients:54 patients /74 lower legs/ with isolated incompetence of sapheno-femoral junction.
Intervention:After clinical examination using Doppler procedure the patients undergone operation limited stripping of LSV. After three months we have performed injection sclerotherapy /if patient have needed it/.
Measurements:Clinical outcome by scoring system documented at 12 months and after 36 months operation.
Results:In 62 lower legs we obtained very good result, in 10 good and in 2 lower legs bat result after 12 months and respectively in 60 lower legs we obtained very good results, in 8 good and in 6 bad results after 36 months.
Conclusion:This connection therapy is simple safe and obtained better cosmetic results comparing with Babcock operation.

P085

PREVENTIVE OPERATION OF LOWER LIMBS VARICOSE VEINS COMBINED WITH OBLITERATION IN ELDERLY PATIENTS.

Zbroński R, Pardela M, Kozłowski A, Urbaniak A, Lemieszewski A, Sławski M, Kowalczyk S,
2nd Department and Clinic of General Surgery and Vessels Zabrze, Poland

Objective:Assessing a method of treating lower limbs varicose veins, which produces good functional results, minimal complications and is feasible in outpatient conditions.

Design:Retrospective assesment of a homogenous group of patients.

Patients:20 patients with primary varicose veins.

Intervention:The patients were operated by a single surgeon using local anaesthesia. Classical crossectomy was carried out and long saphenous vein was ligated, followed by 2 ml of 3% Aethoxysclerol injection in distal part of LSV. The patients was mobilised immedia-tely after surgery.

Measurements:The condition of the limbs was assessed after 12 months according to the grades /objective scale/:very good, good, poor.

Results:15 cases were evaluated as very good, 4 as good and 1 poor. The last case was operated in the classical method.

Conclusion:Preventive combined treatment of lower limbs varicose veins in people of 65 and more years of age allows to decrease significantly functional distress with minimalised complications.

P223

CLINICAL RESULTS OF LONG AND SHORT SAPHENOUS VEIN STRIPPING FOR PRIMARY VARICOSE VEINS

Kosage T, Soma T, Mo M, Yamazaki I, Ichikawa Y, Kondo J, Matsumoto A.

Dept of Cardiovascular Surgery, Saiseikai Yokoham-shi Nanbu Hospital
First Dept of Surgery, Yokohama City University, Yokohama, Japan

Objective: To review the long term clinical results of long and short saphenous vein stripping (LSS) as radical operation for primary varicose veins.

Design: Review of a consecutive series of patients by a questionnaire

Patients: LSS or long saphenous vein stripping (LS) were performed in 120 patients from 1987 to 1990. 80 patients (66.7%) answered the questionnaire. 82 legs with LSS and 36 legs with LS were assessable.

Measurements: Existence of recurrent varicose veins, leg pain after operation and sensory loss. Degree of satisfaction is divided into five categories (very satisfied, satisfied, neither, unsatisfied, very unsatisfied)

Result:

	LSS (%)	LS(%)	
Recurrence	14.6 (12 legs)	16.7 (6 legs)	NS
Leg Pain	15.8 (13 legs)	16.7 (6 legs)	NS
Sensory Loss	28 (23 legs)	25 (9 legs)	NS

Degree of satisfaction was not different between LSS and LS.

Conclusions: Long and short saphenous vein stripping does not increase the rate of complication, compared with only long saphenous stripping. Although, recurrence rate was not different between LSS and LS, LSS is useful for unknown origin either long or short saphenous veins.

P225

PERSONAL EXPERIENCE IN SURGICAL VARICOSE VEINS TREATMENT

E. Liguori, F. Maramao, C. De Amicis.

VASCULAR SURGERY DEPARTMENT ARMY GENERAL HOSPITAL OF ROME

Varicose veins are one of the commonest maladies of the civilized world. They can produce discomfort, aching, tiredness and cosmetic distress. They can develope complications such as phlebitis and ulcerations. These factors may be appropriate indications for surgery.
However the surgical treatment isn't a definitive and total method to care the varicose veins disease.
Some new collateral varicose veins can occur after a long or short postoperative period and these cases aren't due to an uncorrect surgical procedure.
Therefore, over a period of the last three years, in the Vascular Surgery Department of the Army General Hospital of Rome we have performed, as treatment of choice, the thigh stripping of the long saphenous vein when this is incompetent. We have examined all limbs clinically and with a doppler and color doppler study. When we have detected calf perforator incompetence, we have ligated this veins. 274 patients for a total of 307 limbs underwent correction of their saphenous incompetence and some collateral phlebectomy. The operation consists in a saphenous femoral crossectomy and downward stripping of the long saphenous vein was undertaken to just below the knee but wasn't performed safenous calf avulsion.
The follow up shows a good hemodinamic results. There isn't nerve damage. This procedure is performed in day hospital or better in "one day surgery".

P228

SECOND CROSSE EXISTENCE - PROPOSED SURGICAL PROCEDURE

Ghilardi F, Riba U, Vercelli A

Centro Studio Malattie Vascolari, Torino, Italy
CIDIMU, Torino, Italy

Objective: The ascertained existence of a 2nd cross insufficiency at 1/3 middle thigh (through clinical and instrumental study) and in presence of a sound but insufficient valve apparatus in osteo saphenous femoral vein, demonstrated by Eco-color-Doppler study, a surgical procedure was carried out to preserve neighbouring saphenous vein and correct varicose syndrome.
Design: Retrospective review of a consecutive series.
Patients: 56 consecutive patients operated on Eco-color-Doppler follow up control 6 months later.
Intervention: Insufficient perforans and channels at 1/3 middle thigh (2nd crosse) ligating and distal saphenectomy.
Measurements: Eco-color-Doppler follow up control 6 months after operations.
Results: In addition to correcting the varicose syndrome, disappearance of the femoral osteo-saphenous reflux occurred in all patients with sound but non sufficient valve apparatus. No function recovery of the 1st crosse if valve rims were not pointed out under Eco-Doppler study.
Conclusions: A new surgical procedure is proposed, aimed at eliminating reflux in the 2nd crosse, this being placed at middle 1/3 of the thigh, associated with distal saphenectomy. Proposed procedure allows correction of the varicose condition and re-establishes as well femoral osteo-saphenous continence in cases with anatomically sound valve apparatus.

P230

A COMBINED TREATMENT OF VARICOSE SYNDROME

Scuderi A, Nigri H

Inspemoc Instituto Sorocabano De Pesquisa Molestias Circulatorias R Santa Clara, 494 - Sorocaba 18030-421 (SP) Brazil

The varicose disease is a complex syndrome with several symptoms and signs. It comprised from an aesthetically topic and light symptoms to a very serious manifestations like pain, edema, eczema, ulcers and others. Due these facts the treatment of this syndrome is very complex also. Generally, the physician needs to accomplish more than one therapeutic method to help the patients. Following the authors, the treatment of the varicose illness depend on the kind and severity of the symptoms and lesions. Normally they make a combined treatment using a clinic and surgical procedures. The clinical treatment consist of a dietetic and hygienic cares (rest in Trendelenburg positions and others), several drugs (venotrophic, antiphlogistic, antiallergic antibiotic, anticoagulants and others), topical medicines (salves and other) and physiotherapy (compression therapy like medical stockings and 'pumps' and others). The sclerotherapy is a very important therapeutical method to reduce or remove the varicose veins. In the majority of the cases the authors do the sclerotherapy in the small veins (varicosities and telangiectasis). The most common sclerosing solution is the Polidocanol at 0.25% and 0.5%. The authors prefer the surgical method in large and medium size veins. The surgery is performed always with aesthetically cares from the partial surgery of small branches (with local anaesthesia with the stripper of the saphena vein). The recovery time depends on the degree of the syndrome and on the kind of activities of the patient. Exceptionally, the authors can do the sclerotherapy in large or medium varicose veins (bisecting varicosities, old age, other pathologies, etc). In this case the sclerosing solution is the polidocanol at 1%. 2% or 3% depending on the size of the vein.

V/1.12

AN AESTHETIC SURGERY FOR THE VARICOSE SYNDROME

Scuderi A, Nigri H

INSPEMOC - Inst. Sorocabano de Pesquisa Mol Circulatorias, Rua Santa Clara, 494 Sorocaba, 18030-421 (SP), Brazil

The main target in the treatment of Varicose Syndrome are the physiological results. However the aesthetically aspects are becoming a very important topic. The patients do not accept big incisions in their body, specially, in their legs. Moreover, the big incisions require a long time of recovery and increase the incidence of lymphatic and nerve lesions. In this video, the authors show their experience with the treatment of the Varicose Syndrome by a surgery with very small incisions (point like). This technique can be utilized to treat from a small vein to a lot of large varicose veins. The anaesthesia can be local (small and/or few veins) or epidural (a lot of veins and/or if is necessary to remove the saphena veins). The patients remain in the clinic just a day. The majority of incisions do not require stitches, only "micropore". The authors put the medical elastic stocking just after the end of the surgery directly over the skin.

The results: This technique decrease the complications as infections, lymphatic and nerve troubles. The recovery time is reduced. The patients go back walking at home and can return to their activities faster.

P232

CRYOSURGERY OF VARICOSE VEINS

Maruszyński M, Płachta H,

2 nd Clinic of Surgery, Central Clinical

Hospital of MMA

Warsaw, Poland

Objective: To review clinical experience in cryo-stripping and cryosclerosis of primary varicose veins (VV).

Design: Retrospective review of a consecutive series with early and middle-term follow-up.

Patients:400 patients with (VV) - except telan-giectasia.

Intervention:A metal cryoprobe with -80°C tempe-rature reached on its surface in a few seconds was used.

Measurements: Retrospective clinical studies and Doppler examination:early (2-3 weeks) and middle -term follow-up (6 month - 1 year) checking effe-ctiveness,complications,side-effects and cosme-tic outcome.

Results: 88.5% very effective and good cosmetic outcome(no symptoms,no complaint),11.5% "recur-rents" of (VV),20% transient haematomas and sug-gilation, 3% hyperpigmentation, 2.5% transient of superficial sensibility. No difference depen-dent on the type of (VV) was found.

Conclusion: Cryosurgery is a very effective, safe and well tolerated method of (VV) surgical treat-ment. Cosmetic outcome are very good. Almost all patients can be treated on the out-patient or day case bases.

P233

INFLUENCE OF LOCATION OF VARICES ON THE CHOICE OF THERAPY

Robb G M

New Westminster Vein Clinic, 501 - 625 5th Avenue, New Westminster, BC Canada V3M 1X4

Objective: To confirm the clinical impression that anterior femoral branch varices developing during pregnancy usually respond well to sclerotherapy although the long term results of limited surgery are probably better in the long term.

Design: A retrospective study of results of treatment in varicose veins in five thousand patients over several years. Particular attention was directed to 300 patients presenting with anterior femoral branch varices only.

Results: In this group prior pregnancies were an important factor in 95%. In the early years covered, treatment was surgical using a small incision avulsion technique on an ambulant basis. In cases selected saphenofemoral ligation was not done. This gave excellent cosmetic results. In later years equally good results were obtained by sclerotherapy although it was felt that the long term results were better with surgery. (15% versus 35% recurrence).

P234
RESULTS OF INCOMPETENT GASTROCNEMIUS VEIN HIGH-LIGATION COLOUR FLOW DUPLEX EVALUATION

Juhan C, Barthelemy P, Alimi Y, Di Mauro P, Boucherit A

Service de Chirurgie Vasculaire - Hôpital Nord - Chemin des Bourrely - 13915 Marseille Cedex 20

This study was carried out in order to know whether high ligation of incompetent gastrocnemius vein (GV) always induced the disappearance of GV reflux, and in case of persistent reflux, to find the possible causes. The high ligation of incompetent GV was performed on 106 limbs (65 patients). In the series associate procedures were done on the long saphenous vein: 88 limbs, the short saphenous vein: 39 limbs and the mid calf gastrocnemius perforator vein: 37 limbs. Post operative colour Duplex evaluation was done after a mean follow-up of 16 months (6 to 36 months). Among the 106 limbs, 48 had an additional duplex scanning at 2 months. Beyond 6 months after surgery 62 limbs (58%) were free of reflux in the GV area. Forty four limbs (42%) still presented a reflux in the GV, in spite of the interruption of at least one incompetent GV trunk during the operation. Three categories were established for the types of reflux encountered:
1) Persistence of a direct popliteo-gastrocnemial reflux due to the omission of ligation of a GV trunk in patient with multiple incompetent GV (16 cases)
2) Persistence of an indirect popliteo-gastrocnemius reflux through newly developed short incompetent venous branches (11 cases)
3) Persistence of a reflux in the trunk of a correctly ligated GV, associated to a mid calf gastrocnemius perforator vein incompetence (17 cases).
No correlation was observed between clinical and colour duplex scanning results.
Improvements in pre-operative investigations and surgical strategy are suggested in order to reduce the recurrence of post-operative GV incompetence.

P197
HYPNOSIS IN PHLEBOLOGY

O C Kobau, e Tkach

Unidad Sanitaria N°2 - Clínica de várices computarizada Pedro Morán 5063 (1419) Buenos Aires

Objective: the incorporation of hypnosis as another element of therapeutics in phlebologic consult represents a reopening of a scarcely studied field of our speciality. It will contribute to remove the fear of pain, not only for patients' convenience, but also for the benefit of doctors, who are always exposed to harmful secretion of adrenaline. This occurs whenever a major or minor treatment takes place, if it is accompanied by pain.
Development: a) Brief historical review; b) Description of mechanisms to be used, from quick exploration of patient's suggestibility, suggestion, automatism, inhibition of areas related to motor analyser without eliminating the functioning of other cortical areas, applied language, relaxation in connection with words and erroneous resistance as regards hypnosis, caused by cultural mistakes of the past; c) Use of techniques.
Results: the application of hypnosis tends to avoid: a) pain of puncture in sclerosant injection, of drainage, ulcer cures, anaesthetic infiltration and tissues expansion caused by volume of infiltrated liquid, that consequently changes also the plane of varicose course to be treated during drainage of microsurgery; b) accidental risk of anaesthesia and sensation of residual sleepiness. It is possible to work in an atmosphere of lower emotional tension and improve the relationship doctor-patient. We obtain 75% of positive results, since this method is not applicable for aged or intellectually low patients.
Conclusions: the systematic use with variable degree of hypnotic depth in patients who attended our clinic for over ten years only takes us a few minutes per consult. Knowledge, devotion and training, added to prestige and trustworthiness, will help order patient's decision, as well as make his personality stronger, co-ordinate his psychical energy by correcting wrong mental points of view (caused by previous treatments or erroneous information) and soften reasons of anguish, anxiety and pain, bringing the patient into calm emotional stability.

P198

AMBULATORY SURGERY OF VARICOSE VEINS BY THREE DIFFERENT TECHNIQUES

Sias F., Licheri S., Secci L., Loi R., Daniele G.M..
University of Cagliari, Italy, Department of Surgical Sciences and Organs Transplantation - S. Giovanni di Dio Hospital - Cagliari, ITALY.

Objective: To value the role of the ambulatory surgery of varicose veins of the lower limbs performed by three different tecniques, in regard to the anatomo-pathological situation.
Design: Review of personal experience from January 1991 to January 1995.
Patients: 116 patients (147 legs) with chronic venous insufficiency, ranging in age from 24 to 80 years; 24 men and 92 women; ASA classification: 1 (79%), 2 (19%) and 3 (2%).
Intervention: Under local (91 cases), unilateral superselective spinal (48 cases) and blended or general (8 cases) anaesthesia were performed, for functional problems, the following operations: radical or short safenectomy (53 cases), crossectomy with Muller's phlebectomy (47 cases), CHIVA (47 cases). An additional procedure (sclerotherapy) was necessary in 18% of the patients.
Measurements: Clinical outcome was studied by clinical and ultrasound doppler examination at 6,12 and 36 months after operation.
Results: No complications related to anaesthesia were observed. In the CHIVA group there were 25,5% of superficial veins thombosis.
Conclusions: The absence of general complications, excellent patient acceptance, good aesthetic and functional results and, not last, economical advantages lead the Authors to believe the ambulatory Surgery of varicose veins as the treatment of choice and show the adequacy of the differentiated treatment.

P235

20 YEARS OF THREAD METHOD OF SAPHENECTOMY

DIAZ AUGUSTO.-DIAZ KAROLINNA

INSTITUTO DE ANGIOLOGIA.- QUITO.-ECUADOR

The thread method of saphenectomy is a no-traumatic and sthetic phleboextraction procedure that was used in my country for more than 20 years with satisfactory results and that associated to perforant veins subaponneurotic ligation and scaled varicotomies,becomes in a complete surgical treatment for essential varicose veins of the lower limbs.
With this technique we had treated varicose veins dilatations of internal and external saphenous axis,anarchical,perforant varicose veins of 3th, 4th degree, and for this review we have no notice well defined ethiologic varicose veins well as post-thrombotic, post-traumatic and remainder varicose veins.
For easy diagnosis we have no used invasive or no invasive methods. The surgical technique is as follows:regional and local anesthesia,introduce into vein light thick piece of silk suture that serves to perform phleboextraction by invagination;later,subaponeurotical ligature of perforant veins and subcutaneously ligature of anarchical veins.
It was studied a total of 12,000 surgical procedures.

P199
COMPLICATIONS OF AMBULATORY PHLEBECTOMY
Review of 385 patients

OLIVENCIA J A
Iowa Vein Center, Des Moines, Iowa, USA

Objective: To review complications of Ambulatory Phlebectomy on an outpatient basis.

Design: Retrospective review of 385 consecutive patients.

Patients: 385 consecutive patients.

Measurements: Age and gender distribution, localisation, complications and associated illnesses.

Results: Review of complications.

Conclusions: Technical recommendations, wrapping technique, early ambulation and accurate diagnosis are critical for satisfactory results.

P200
AMBULATORY PHLEBECTOMY IN THE ELDERLY
Review of 100 cases

OLIVENCIA J A
Iowa Vein Center, Des Moines, Iowa, USA

Objective: Review Ambulatory Phlebectomies performed on elderly patients.

Design: Retrospective review of 100 consecutive cases.

Patients: 100 patients (140 limbs).

Measurements: Age and gender distribution, localisation, anaesthesia used, micro-incision technique and hooks used, pre-operative evaluation, intra-operative management, associated illnesses.

Results: Review of complications, wrapping technique and technical recommendations.

Conclusions: Staged shorter procedures on a closely monitored patient during surgery. Pre-operative low molecular weight Heparin, proper wrapping and early ambulation are very important for satisfactory results.

P147

SCLEROTHERAPY IS NOT AN INNOCUOUS THERAPY

AUTHOR: SALVADOR NIETO, M.D., F.I.C.A.

ADDRESS:

- SALVADOR NIETO FOUNDATION. Av. Santa Fe 2679, 2nd. "D"; (1425) Buenos Aires, Argentina. Telefax: (541) 826-8519

OBJECTIVE

To show the potential activity of sclerosing agents far beyond injection point by transteleangiectatic phlebography and emphasize the convenience of being careful with the practice of sclerotherapy.

INTRODUCTION

Nowadays sclerotherapy is so common that a lot of practitioners or, what is the worst, non practitioner people, use this therapy daily. Lacour said that "unfortunately, the use of these therapeutic agents arises the appearance of a great quantity of pseudo-specialists that have extended their employment on varicose veins of all kind and size in an irrational way". In hands of specialists the risk is minimum but it would be fantastic if specialists preach more upon actual indications, technical care, medical capability and, most important, adverse- and sometimes dangerous- effects when laymen execute it.

MATERIAL AND METHODS

12 patients with teleangiectasias were considered. For this study were selected, arbitrarily, patients that had teleangiectasias at the distal area of posterior region of the thigh. Sclerotherapy was performed at that point but a contrast agent (iodine) was used instead of sclerosing agents. Under TV control X ray plaques were taken while the worker was performing the injection. This study reproduced exactly what was usually done in sclerosing sessions, except the use of contrast agent, to watch its pathway.

RESULTS

Once injected, immediately, the contrast agent was generously observed in both deep and superficial venous system, even popliteal, femoral and iliac veins. All the studies showed similar images.

CONCLUSIONS

Transteleangiectatic phlebographies have displayed how far, and fast, sclerosing agents can spread and it has to be assessed what is the activity of those agents there when, as happens in varicose veins, blood flow is very slow.

P236

METHOD OF TRANSIENT LIGATURES

Jorge A. Fernandez

Larrain Hospital (Argentina)

Foot Venous Surgery Through Small Incisions

We describe a technique which is characterised by the use of venous ligatures which are used only during the duration of the operation. The great advantage is that no foreign materials remain in the tissues, thus avoiding complications (granulomas, infection, etc.) This technique is applied in superficial surgical intervention in the legs and feet. The results of seven years' experience in aesthetic venous surgery are described.

The Treatment of
Varicose Veins
and Telangiectasias

Injection Sclerotherapy

Phlebology '95, D. Negus et al. (eds.). Phlebology (1995) Suppl. 1: 504-507

V/2.1

Laser and Non-Coherent Pulsed Light Treatment of Leg Telangiectasias and Venules

Mitchel P. Goldman, MD

Dermatology Associates of San Diego County, Inc

Two LASER and pulsed light systems are currently available which specifically effect leg veins without significant adverse sequelae. The Flashlamp pulsed dye laser (FLPDL) acts within the thermal relaxation times of blood vessels to produce specific destruction of vessels less than 0.2mm in diameter.[1] The Photoderm VL[R] was developed to treat vessels up to 3mm in diameter, 2mm beneath the epidermis.

The main chromophore in the blood vessel is oxyhemoglobin.[2] The oxyhemoglobin absorption spectrum has three major bands calculated for 50um vessels 0.2mm deep, the largest at 418nm and two smaller peaks at 542 and 577nm. A restriction of the application of the 577nm wavelength in the treatment of vascular lesions has been its penetration to only approximately 0.5mm in depth from the dermal-epidermal junction.[3] Increasing the wavelength of the FLPDL to 600nm has also been shown histologically to give deeper thermocoagulation than with a 585nm FLPDL.[4] Calculated energy absorption for oxygenated and deoxygenated hemoglobin is very different for vessels 1mm in diameter and 2mm beneath the dermal-epidermal junction. In these deeper and larger vessels, oxygenated hemoglobin peaks at 920nm and deoxygenated hemoglobin has a plateau "peak" between 660 - 920nm. Thus, using a light source between 600 - 900nm has the advantage of both penetrating to the proper depth of leg telangiectasia and venules while being absorbed by oxygenated-deoxygenated hemoglobin. Thus, longer wavelengths may be desirable when treating deeper blood vessels (Figure 1).

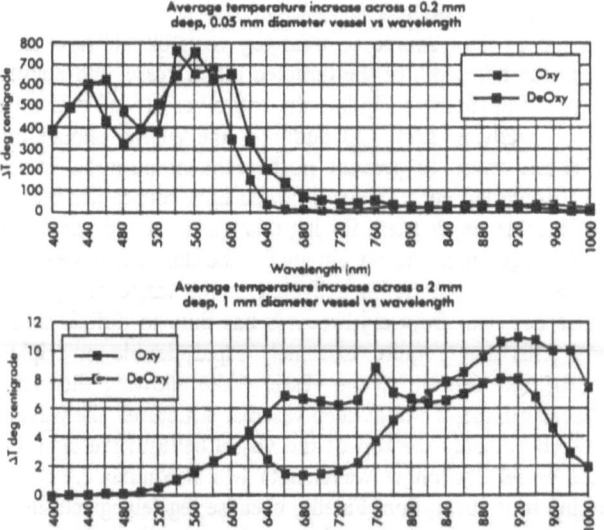

Figure 1: Average temperature increase across a cutaneous vessel as a function of wavelength for two cases: a shallow capillary vessel (similar to those found in a port-wine vascular malformation) and a deeper (2mm) larger (1mm) vessel typical of a leg venule. The calculated curves are generated assuming that the main light absorbing chromophore in the blood is either oxygenated or deoxygenated hemoglobin. Note the dramatic shift in the optimal wavelength as a function of vessel depth and diameter. Also note the difference between oxygenated and deoxygenated hemoglobin. (Courtesy of Shimon Eckhouse Ph.D., ESC Inc., Newton MA) (Reproduced with permission from: Goldman MP, Fitzpatrick RE: Cutaneous Laser Surgery. Mosby, St. Louis. 1994, p23.)

Besides matching the wavelength of the laser to the target tissue, it is ideal to limit the laser energy absorption and contain it within the targeted structure. This is accomplished by pulsing the laser. Limiting the duration of laser exposure ensures that highly specific laser energy is delivered in a period of time less than that required for the cooling of the target vessel, that is, less than the thermal relaxation time.

For superficial cutaneous blood vessels, vessels, thermal relaxation times range from 0.1 to 10msec depending on the size and type of the vessel,[2,5] and average 1.2msec.[2] (Table 1)

Table 1: Thermal Relaxation Time of Cutaneous Vessels	
Vessel Diameter	Thermal Relaxation Time
20um	140 usec
50 um	1.2 msec
100 um	3.6 msec
300 um	5 - 10 msec

In addition to thermal damage produced by the absorption of the 577nm wavelength, photo acoustic "shock-wave" damage resulting from rapid absorption of energy by the oxyhemoglobin molecules has also been demonstrated.[3,6,7] This "shock-wave" promotes the undesirable purpuric response commonly seen with the FLPDL which may take 1 - 3 weeks to completely resolve. The advantage of using longer pulse durations that are still within the thermal relaxation times for the targeted blood vessels is that larger diameter blood vessels may be treated. The average diameters of blood vessels in the upper dermis is 40 - 60um; deeper dermis is 100 - 400um, and subcutaneous tissue is 1 - 3mm. In addition, longer pulses may prevent the "shock-wave" effect, thus preventing purpura.

PHOTODERM VL^R LIGHT SOURCE

An ideal LASER or pulsed light source to treat leg veins should have a wavelength which can penetrate to the full depth of the targeted blood vessel and deliver sufficient energy to the target vessel to thermocoagulate the entire vessel wall without damage to perivascular tissues or overlying skin. In addition, the energy should be delivered without causing a "shock-wave" to prevent post-treatment purpura. In an effort to maximize efficacy in treating leg veins, a novel pulsed light source has been developed. This new technology is more appropriate for leg telangiectasia and venules since these vessels are substantially larger, more deeply situated in the skin, and have thicker walls.

To treat larger vessels, energy must be delivered in the range of approximately 3-30msec pulses so absorbed heat from erythrocytes has time to diffuse throughout the vessel circumference. This may require extremely rapid double or triple pulses to maximize absorption of light fluence to the vessel while allowing thermal cooling of epidermal and perivascular tissue. To treat deep vascular structures, the wavelength must be long enough to reach not only the top of the vessel beneath the skin, but also the entire diameter of the vessel. A longer wavelength will also minimize coupling to the epidermis with resulting heat absorption. Finally, because leg telangiectasia and venules have thicker walls than the ectatic vessels of port-wine vascular malformations, higher energy fluences must be given.

The Photoderm VL^R can deliver energy fluences of 20 - 80 J/cm^2 over 2 - 20msec and may be repetitively pulsed with delays between pulses of 50 - 1,000msec. These parameters will effectively thermocoagulate vessels up to 3mm in diameter, 2mm below the dermal-epidermal junction, as well as telangiectasia 0.2 - 0.6mm in diameter. Histologic evaluation demonstrates that the Photoderm VL^R does not cause vessel rupture as also evidenced by the lack of clinical purpura. Instead, the vessel wall is more slowly heated, producing destruction.

The Photoderm VL^R delivers its energy through a "footprint" whose size and dimensions can vary based on the requirements of the target vessel. The present machine has a standard "footprint" or spot size of 8mm x 35mm. This permits efficient treatment of large sections of vessels.

At this writing, the Photoderm VL^R is in its final stages of development. Vessels up to 3mm in diameter have been successfully obliterated without significant adverse sequelae. Patients report less pain as compared with the FLPDL and treatment sessions are well tolerated. The only adverse effects noted are slight epidermal burning without resulting scarring in patients with pigmented skin, Fitzpatrick type 3 or greater. In addition, intradermal nevi, lentigos, and other pigmented lesions within the treated area may disappear after treatment.

CONCLUSIONS

LASER or pulsed light treatment of leg veins has the advantage of being relatively non-invasive (no needles) and non-toxic (no injection of sclerosing solution). Decreased inflammation also prevents the development of telangiectatic matting. Complete thermocoagulation prevents extravasation of red blood cells which limits post-treatment hyperpigmentation. However, high pressure reflux from incompetent perforator, reticular, or varicose veins must first be treated either surgically or with sclerotherapy before laser or photo thermocoagulation can be expected to be effective.

REFERENCES

1. Goldman MP and Fitzpatrick RE: Pulsed-dye laser treatment of leg telangiectasia: with and without simultaneous sclerotherapy. J Dermatol Surg Oncol 16:338, 1990.
2. Anderson RR and Parrish JA: Microvasculature can be selectively damaged using dye lasers: A basic theory and experimental evidence in human skin. Lasers Surg Med 1:263, 1981.
3. Nakagawa H, Tan OT, and Parrish JA: Ultrastructural changes in human skin after exposure to a pulsed laser. J Invest Dermatol. 84:396, 1985.
4. Goldman L, Kerr JH, Larkin M, et al: 600 nm flash pumped dye laser for fragile telangiectasia of the elderly. Las Surg Med. 13:227-233, 1993.
5. Anderson RR and Parrish JA: Selective photothermolysis: precise microsurgery by selective absorption of pulsed radiation. Science 220:524, 1983.
6. Gange RW et al: Effect of pre irradiation tissue target temperature upon selective vascular damage induced by 577nm tunable dye laser pulses. Microvasc Res. 28:125, 1980.
7. Goldman MP et al: Pulsed-dye laser treatment of telangiectases with and without sub-therapeutic sclerotherapy: clinical and histologic examination of the rabbit ear vein model. J Am Acad Dermatol 23:23, 1990.

Phlebology '95, D. Negus et al. (eds.). Phlebology (1995) Suppl. 1: 508-510

OP/13.3

Sclerotherapy of Internal Pudendal Veins

R. Jacques

Clinique Medic Ami, Rouyn-Noranda, Quebec, Canada

INTRODUCTION

By reviewing 2,000 files in my private practice, I recorded the following prevalences:

- incompetence of the short saphenous veins 3%

- incompetence of the internal pudendal veins 6%

- incompetence of the great saphenous veins 20%

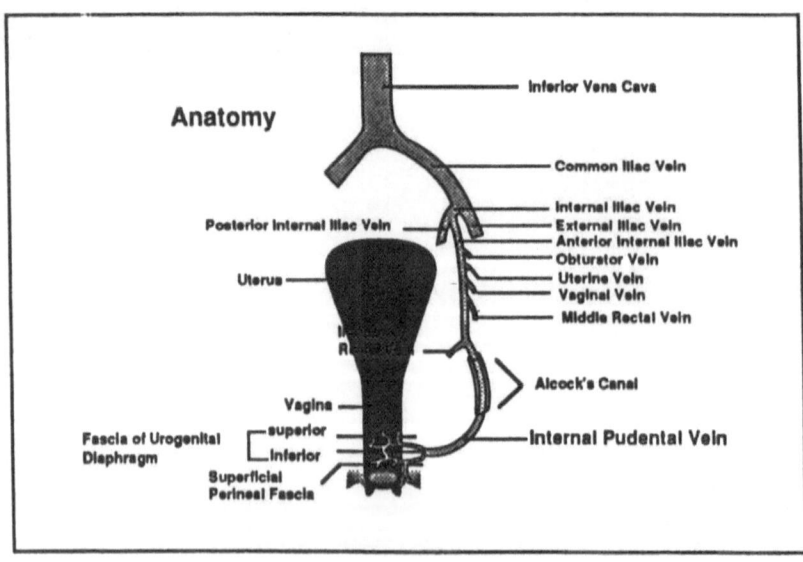

Type of Presentation

In most cases we are facing three clinical types of presentation; the reflux of blood coming from the internal pudendal vein will dilate the veins in the territory of:

- ✓ the great saphenous vein   (40%)
- ✓ the anterior saphenous vein   (20%)
- ✓ the posterior saphenous vein   (40%)

Suspect

- ✓ A woman who has been pregnant;

- ✓ If she suffered of vulvar varicosis during pregnancy;

- ✓ When you see varicosis veins in the territory of the great saphenous vein, the anterior saphenous vein or the posterior saphenous vein, especially if they are of large caliber;

- ✓ With the clinical evidence of a great saphenous vein insufficiency and a negative percussion sign at the crook of the vein.

Clinical Examination

- ✓ Don't forget the examination of the inner face of the thigh, near the vulva.

- ✓ A Trendelenburg Test performed high on the thigh with selective decompression can be useful.

Doppler

✔ A pocket doppler is very useful.

✔ By putting the probe on the most visible branch of the internal pudendal vein, near the vulva and applying pressure on the suspected collateral vein on the lower leg, you can listen for the presence or absence of bloods reflux.

✔ In more difficult cases, a retrograde doppler examination is the method of choice untill the source of reflux is known. When you have a clinical case of great saphenous vein insufficiency, it helps differentiating between a reflux from the crook of the great saphenous vein or from the internal pudendal vein .

Treatment

Sclerotherapy:

✔ **Injection site:** visible veinous branches as close as possible to the vulva.

✔ **Medication:**
First choice: Tetradecyl sulfate

- diffuses better than iodine adding the possibility of a more extensive sclerosis of the internal pudendal vein.

Second choice: iodine

✔ **Concentration and injected volume:**
- 0.5 %
- 0.75 % ⎫ 1/2 cc à 1 cc
- 1 % ⎭

✔ **If there is a recurrence:** you will get good results from a new sclerotherapy and increasing the concentration of the medication by a small amount.

Surgery: Ambulatory phlebectomy.

Conclusion

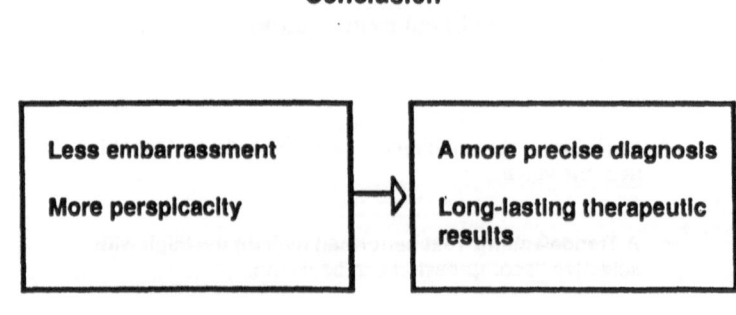

Less embarrassment

More perspicacity

A more precise diagnosis

Long-lasting therapeutic results

Phlebology '95, D. Negus et al. (eds.). Phlebology (1995) Suppl. 1: 511-513

OP/4.4

Echosclerotherapy in the Treatment of Varicose Veins: Short Term Results

P. Pavei, U. Baccaglini, G. Spreafico, P. Sorrentino, C. Castoro, E. Giraldi, V. Fontebasso and E. Ancona

Second Department of Surgery, University of Padua, Italy

INTRODUCTION

Traditional sclerotherapy can be incomplete when treating fat legs or unpalpable small veins or because the doctor prefers not to inject in some area at risk (for example groin or popliteal fossa).
In these areas, in fact, there is the risk of doing an extravascular or intrarterial injection.
The aim of Echosclerotherapy is to reduce these complications and also make it possible to do sclerotherapy in unpalpable veins (1,2). The first doctors who discussed ecosclerotherapy were Knight and Vin in 1989 at the Mondial Phlebological Congress which took place in Strasburg. After this, other authors published their results and defined the indications of this technique (3,4,5,6,7,8).

MATERIALS AND METHODS

An Echodoppler Esaote Biomedica Au 530 with a 10 Mhz anular probe was used. The puncture of the vein was done under echographic guide: the direction of the needle was adjusted in accordance to the modification of the echographic image (depression of the vein or movement to the right or the left).An echodoppler examination was done before every session both in the orthostatic and supine position in order to evaluate the diameter of the vein and its depth . This data is important when choosing the needle and of sclerosing agent. In addition echodoppler can be used to control the efficacy of sclerotherapy evaluating the reduction of vein diameter, the presence of reflux, flux or absence of flux (9,10,11,12).The treatment was considered completed when reflux was no longer detectable.
The puncture was done looking at a transversal section of the vein.
The puncture and the injection was done with the patient in the supine position. The sclerosing agents used were 3 % Soduim Tetradecil sulfate and 8-12% Iode-iodine solution alone or in association, and the quantity of 1 cc or more was used. From 1 to 3 injections were done in the venous trunk of the thigh each time. At the end of the session an elastic-adesive bandage was applied. This bandage can be omitted if there are no important varicosities.

PATIENTS

From May 1993 to October 1994 at the Second Department of Surgery (Padua University- Italy)58 patients, 7 men and 51 women, aged from 26 to 72 years (an average of 54)underwent echosclerotherapy:
46 patients had long saphenous varicose veins; 4 patients had short saphenous varicose veins; 3 had calf perforating veins; 3 Hunter perforating veins and 2 patients had varicose veins determined by a posterior thigh perforator.Patients underwent from 1 to 7 sessions of echosclerotherapy (an average of 3). In 10 patients 3% Sodium Tetradecil sulfate was used; in 10 patients 8-12% Iode Iodine solution was used and in 38 patients both substances were used. Each time more than 1 cc of the agent was utilized. Clinical outcome and duplex data of vein occlusion and absence of blood flow were determined at 1, 3, 6 and 12 month follow up after echosclerotherapy.

RESULTS

Patient follow up varies between 3 and 17 months (an average of 9 months).
We obtained the following results:
1. In 26 patients (45%)complete sclerosis was obtained (19 long saphenous veins, 3 short saphenous veins, 2 calf perforating veins, 2 calf perforators and 2 Hunter perforators.
2. In 24 patients (41%), good sclerosis was obtained but an open stump measuring 2 cm remained near the sapheno-femoral junction.
3. In 2 patients (3.5%), stumps measuring respectively 4 and 10 cm remained but the duplex examination showed no reflux.
4. 2 patients interrupted the treatment.
5. In 4 patients (7%) no results were obtained with this treatment.
The first three group of patients were considered good results: they had, in fact, no clinical evidence of varices and no reflux. That's to say that in 50 out of 58 patients (86%) we obtained a good sclerosis. Our treatment failed in 6 patients (10.5%).
No major complications were observed in any patient. We had only one phlebitis of the great saphenous vein which was due to incorrect bandaging.

CONCLUSIONS

Echosclerotherapy is a technique which allows us to obtain good results even with unpalpable veins and makes sclerotherapy safer, especially when the injection is done near the sapheno-femoral junction, in the popliteal fossa or in other areas at risk. Moreover, echographic evaluation is useful in order to control the effects of the sclerosing agents, to detect eventual recanalization and to decide when the treatment is complete.
At present echosclerotherapy is only used in some situations:
- sclerosis of small unpalpable saphenous varicose veins;
- sclerosis of perforating veins;
- sclerosis of the short saphenous vein.
In literature all the authors report good results with echosclerotherapy (70-80%) without major complications (1,2,8).
Our results (86% of good sclerosis) can also be considered good in that they fall into this range.
Long term results are still not available, but it can nevertheless be supposed that they will be similar to those obtained with a traditional correctly-done sclerotherapy.

REFERENCES

1. Miserey G., Reinharez D., Ecalard P., *Sclerose sous echographie dans certaines zones a risques.* Phlebologie 92 1991, 44: 185-96.
2. Grondin L., Soriano J., *Duplex echo-sclerotherapy: the quest for the safe technique.* Phlebologie 92, John Libbey Eurotext, 1992: 824-825.
3. Cales B., Creusot O., *Apport de l'echosclerose dans le traitment de l'insuffisance veineuse superficielle.* Phlebologie 92, Eds. Raymond Martimbeau, M. Zummo, John Libbey Eurotext, 1992: 817-819.
4. Ermini S., Capelli M., Turchi A., *La sclerose echo-guidée.* Phlebologie 92, Eds Raymond Martimbeau, M. Zummo, John Libbey Eurotext, 1992: 820-821.
5. Forrestal M., Foley D., *A single treatment protocol for sapheno-femoral or sapheno-popliteal incompetence by duplex guided sclerotherapy.* Phlebologie 92, John Libbey Eurotext, 1992:822.
6. Grondin L., Soriano J., *Echosclerotherapy, a Canadian study.* Phlebologie 92, John Libbey Eurotext, 1992:828-831.
7. Schadek M., *Echo-sclerose de la grande saphene.* Phlebologie 1993, 46: 4 673-682.
8. Vin F., *Echosclerotherapie de la veine saphene externe.* Phlebologie 1991, 44: 1 79-84.
9. Schadek M., *Doppler and echotomography in sclerosis of the saphenous veins.* Phlebologie 1987, 2: 221-240
10. Schadek M. Allaert F., *Echotomographie de la sclerose.* Phlebologie 1991, 44: 1 111-130.
11. Avramovic A., De Simone J.C., Avramovic M., *Evaluation de la sclerose de la crosse et du tronc de la saphene interne a l'echotomographie.* Phlebologie 89, A. Davy, R. Stemmer Eds. John Libbey Eurotext 1989: 834-835.
12. Bishop C.R.,Fronek H.S., Fronek A., *Real time color duplex scanning after sclerotherapy of the greater saphenous vein.* J. Vasc. Surg. 1991, 14 (4): 505-510.

Phlebology '95, D. Negus et al. (eds.). Phlebology (1995) Suppl. 1: 514-517

OP/11.3

Echo-Sclérose et Phlébectomie Ambulatoire Associées dans le Traitement des Varices Extra-Sapheniennes

A. Vannuzzi[1] and A. Pieri[2]

[1] Ecole Européenne de Phlébologie, PARIS, France
[2] U.O.de Cardiologie III du Prof. G. de Saint-Pierre, Module de Diagnostic Angiologique, Az. Hôp. de Careggi, FLORENCE, Italy

Echo-Sclerosis and Phlebectomy Combined in the Treatment of Non-Saphenous Varices

SUMMARY

The authors describe their experience with echo-Doppler controlled sclerotherapy associated with phlebectomies and discuss the indications.

ECHO-SCLEROSE ET PHLEBECTOMIE AMBULATOIRE ASSOCIEES DANS LE TRAITEMENT DES VARICES EXTRA-SAPHENIENNES.

A. Vannuzzi */**, A. Pieri **

* Ecole Européenne de Phlébologie - PARIS (F) ** U.O.de Cardiologie III du Prof. G. de Saint-Pierre - Module de Diagnostic Angiologique. - Az. Hôp. de CAREGGI - FLORENCE (I)

La sclérothérapie échoguidée est une technique efficace utilisée maintenant depuis des années et présentant, dans le traitement des varices, de réels avantages par rapport à la sclérothérapie classique.

Elle consiste à pratiquer la sclérothérapie en se servant d'un appareil échodoppler permettant d'avoir un contrôle visuel à la fois de la veine, de l'aiguille et de la progression du liquide sclérosant à l'intérieur du vaisseau. L'injection peut se faire en coupe longitudinale (la plus utilisée par les auteurs) ou en coupe transversale: la première assure une meilleure visualisation de l'aiguille et du produit injecté mais nécessite, plus encore que l'autre, un contrôle constant par aspiration du sang dans la seringue afin d'être sûr de se trouver dans la veine et non dans un plan parallèle; la seconde donne une image moins nette de l'aiguille mais montre bien l'introflexion que celle-ci produit sur la paroi de la veine en y pénétrant.

Le premier avantage qu'offre la technique est la possibilité de déterminer le point exact du siège du reflux et donc, d'y pratiquer l'injection avec plus de précision garantissant ainsi un meilleur résultat, plus rapide et plus durable; cet avantage est particulièrement évident quand il faut traiter des perforantes de cuisse difficilement repérables par la seule palpation. Le recours à l'écho-sclérose permet en outre de traiter en toute sécurité des zones reconnues comme dangereuses et garantit un résultat plus valable même dans les récidives post-chirurgicales qui présentent toujours des difficultés de traitement tant chirurgical que sclérosant.

La possibilité de bloquer les reflux dans les sièges plus profonds permet d'obtenir une réaction inflammatoire moindre et par conséquent une pigmentation cutanée moindre également. En présence de varices de gros calibre, il est néanmoins possible qu'il y ait pigmentation, surtout quand celles-ci sont situées dans la cuisse. Pour pallier à l'éventualité d'une telle complication, il semble logique d'associer phlébectomie ambulatoire et écho-sclérose.

La phlébectomie ambulatoire permet d'enlever de grosses varices avec des incisions minimes mais, n'arrivant pas à bloquer, de façon constante, en profondeur, le siège du reflux qui en est la cause, elle comporte un risque de récidives. En revanche, grâce à l'utilisation simultanée des deux méthodes qui se complètent l'une l'autre, on peut aboutir, au cours d'une seule et même séance, à un résultat tant esthétique que fonctionnel, à savoir la disparition complète des varices visibles sans trace de pigmentation et la sclérose du siège du reflux qui les a causées.

Pour ce qui est de notre expérience, les indications principales pour associer les deux techniques sont les suivantes: A) Varices en écharpe de grandes dimensions à la cuisse causées par un reflux dans la saphène accessoire antérieure (fig.1); B) Varices postérieures de jambe dues à l'incontinence de la perforante du creux poplité (fig.2); C) Varices de gros calibre dans le territoire de la marginale externe alimentées par les perforantes latérales et postérieures de cuisse (fig.3).

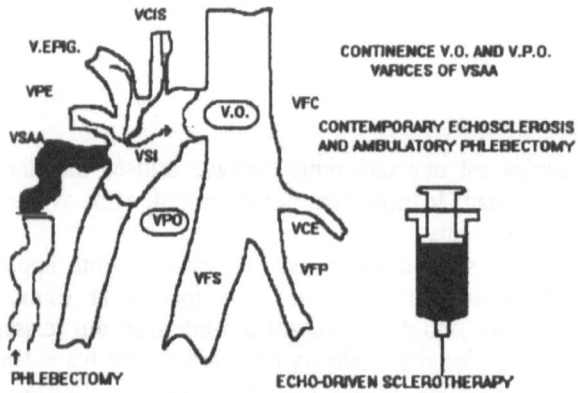

Fig. 1 - Légende: VSI=Veine saphène interne, VO=Valvule Ostiale, VPO= Valvule pré-ostiale, VFC=veine fémorale commune, VFS=veine fémorale superficielle, VFP=veine fémorale profonde, VSAA=veine saphène accessoire antérieure, VPE=veine honteuse externe, V.EPIG=veine épigastrique superficielle, VCIS=veine circonflexe iliaque superficielle.

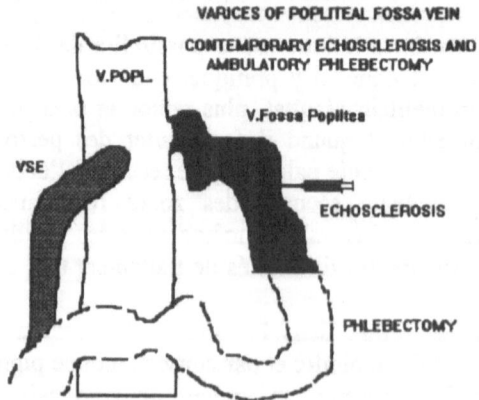

Fig. 2 - Légende: V.POP= veine poplitée, VSE=veine saphène externe, V. FOSSA POPLITEA=veine perforante du creux poplité.

Une autre indication possible sont les récidives post-chirurgicales dans lesquelles se sont formées des neocrosses de petit calibre difficiles à reprendre d'un point de vue chirurgical. Le marquage echo-doppler pré-opératoire est d'une importance capitale pour repérer les sources de reflux qui devront être traitées en premier par écho-sclérose en utilisant les médicaments habituels (trombovar, variglobin) à des doses ou concentrations légèrement plus basses que d'habitude, dans la mesure où la varice qui dérive du point de reflux (déjà marqué préalablement) est enlevée tout de suite après l'injection en pratiquant une phlébectomie. Celle-ci est effectuée de façon classique avec des incisions d'un millimètre en utilisant un bistouri ophtamologique ou, même, en cas de varices pas trop volumineuses, avec une aiguille de 19 G au lieu du bistouri. En décollant délicatement les varices avec le crochet de Muller avant de les enlever, on réduit au minimum le traumatisme des tissus et les teleangectasies réactives sont extrêmenent rares. Au terme de la séance, on appliquera un pansement le long du trajet de la phlébectomie et on fera porter au patient un monocollant de contention de 2ème classe. Il sera immédiatement invité à marcher et après une semaine, au moment où on enlèvera les steristrip, on procédera à un contrôle echodoppler. Lors du contrôle suivant, trois mois après, il sera possible d'effectuer une dernière echo-sclérose dans la perforante au cas où elle ne serait pas complètement fermée.

Par conséquent il semble que l'association de deux méthodes telles que l'écho-sclérose et la phlébectomie ambulatoire, très valables par ailleurs séparément, puisse présenter de réels avantages dans des situations particulières qui devront être évaluées cas par cas durant l'examen écho-doppler.

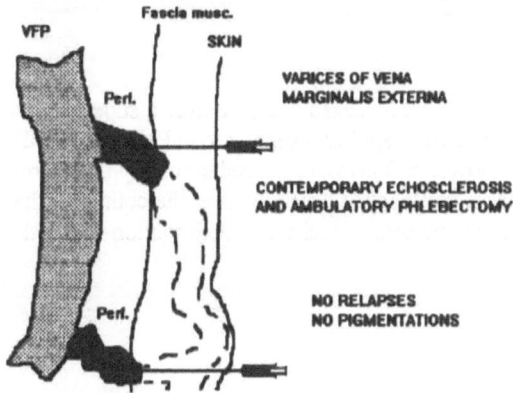

Fig. 3 - Légende: PERF:=veine perforante.

REFERENCES

1 - Calés Gracia B.: Echo-sclerose-Phlébectomie ambulatoire. 2 Techniques à associer. Phlebol., 1993: 46 (4), 665-670
2 - Dortu J.: La phlébectomie ambulatoire. Annales du congrès mondial de phlébologie de Strasbourg - Sept. 1989, art. 387.
3 - Knight R.H., Zymunt J.A.: Echosclerotherapy. Proceedings of the Phlebology Society of America, 1989.
4 - Knight R.H.: Traitment of superficial venous disease with accurate sclerotherapy. Proceedings of the Canadian Society of Phlebology, Whistler, 1991
5 - Muller R.:Introduction à la phlébectomie ambulatoire. Annales du congrès mondial de phlébologie de Strasbourg - Sept.1989, art. 385.
6 - Schadeck M.: L' Echo-Sclérose. La nouvelle donnée. Actes du XI Congrès Mondial de l' Union Internationale de Phlébologie, Montreal 1992, 826
7 - Schadek M.: Duplex et Phlébologie. Ed. Gnocchi, Napoli 1994, pagg. 115-126
8 - Vannuzzi A., Pieri A., : Echosclerotherapy: newer indications of an old technique. Abstracts Book of Anglo '93, 106
9 - Vin F., Schadeck M.: La maladie veineuse superficielle. Masson Ed., Paris, 1991
10 - Vin F.: Echosclerotherapy of recurrent varicose veins after surgery. Acts du XI Congrès Mondial de l' Union Internationale de Phlébologie, Montreal 1992, 827

Phlebology '95, D. Negus et al. (eds.). Phlebology (1995) Suppl. 1: 518-520

OP/4.1

Compressing Sclerosed Veins - A New Technique

A. Chan

Department of Surgery, Lions Gate Hospital, North Vancouver, British Columbia, Canada

INTRODUCTION

Thrombotic masses in varicose veins result from inadequate compression during the early post-injection period.[1] These are unsightly, increase skin pigmentation, and may lead to telangiectatic matting or recanalization.[2] The importance of removing these coagula has been stressed.[3] A method of minimizing their formation is presented.

RATIONALE

This technique is based upon the premise that localized pressure over a superficial vein is more effective than circumferential pressure applied evenly around the limb. This is demonstrated by the fact that a much lower cuff pressure is needed to stop flow through a vein when a length of a firm material is first applied precisely over the course of the vein (Fig. 1). Note that the vein is obliterated by actual mechanical deformation rather than by general increase in tissue tension.

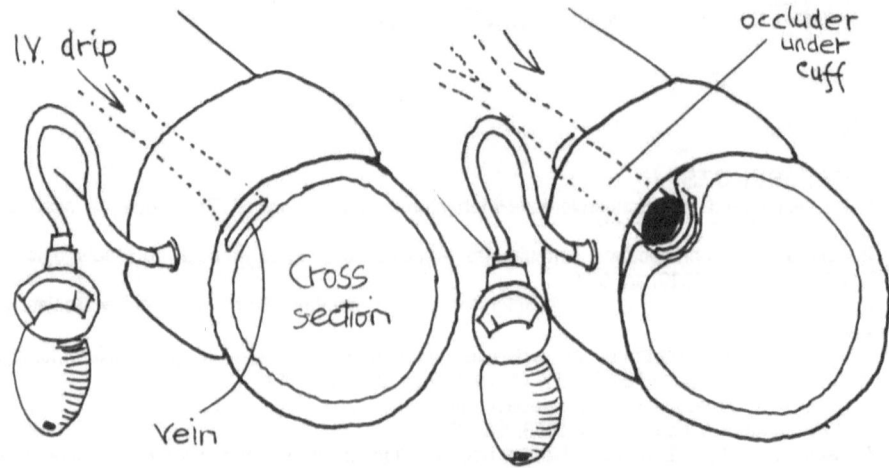

Fig. 1

MATERIALS

8 mm diameter cotton rope was selected as the occluding material because it is a) readily available from sewing supply stores; b) non-allergenic; c) porous and absorbant; d) easily manipulated to follow the course of a vein. 3M micropore tape and 3 inch tensor bandages complete the list.

METHOD

The outline of the vein to be treated is first traced with a skin marker. After the injection treatment is completed, the rope is applied to the skin along the course of the mark. To make the rope firmer and more mouldable, it is taped to the skin at one end while the other end is rotated along its axis to exaggerate the twist (Fig. 2).

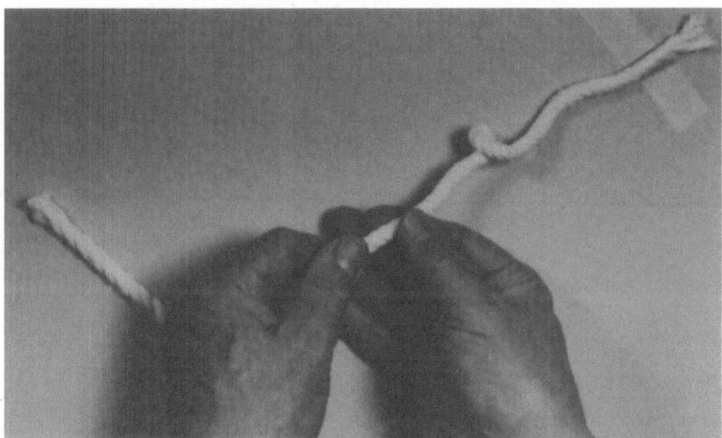

Fig. 2

The rope is secured to the mark sequentially with short strips of micropore tape applied transversely. Various configurations of the rope are used to mimic the vein (Fig. 3).

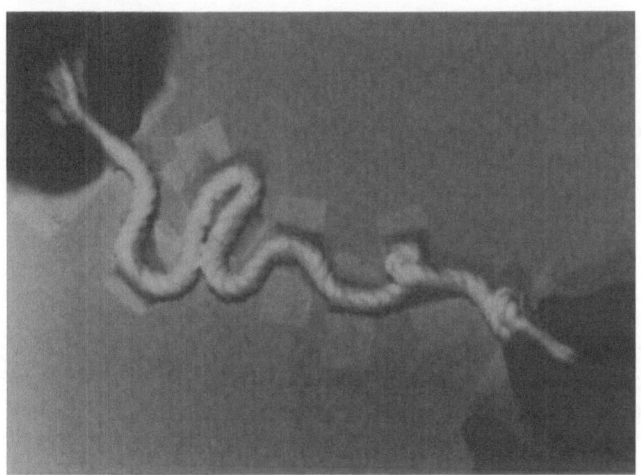

Fig. 3

520

The tensor bandage is then applied snugly over the "roped" vein. After a brisk walk, the patient spends the rest of the day recumbent. On the following day, normal activities are resumed with the bandage on. After 48 hours, the patient is seen and the compression is removed, with no further bandaging.

RESULTS

When it was peeled away, the rope was seen to leave its mark in the form of a gutter, indicating effective compression (Fig. 4). Over 80 patients have been treated with this method. The author found it well tolerated and effective in reducing thrombotic masses and their undesirable side effects. Pressure necrosis of the skin did not occur and secondary clot extraction was not required in this group. Experience with this method has led the author to use sclerotherapy for all uncomplicated varicose veins regardless of size (after sapheno-femoral disconnection when indicated), instead of surgical removal.

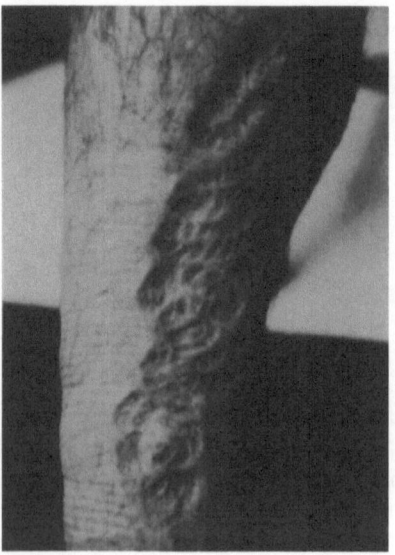

Fig. 4

DISCUSSION

The cornerstore of this technique is the selective compression of the treated vein only. After a little practice, this method adds less than 10 minutes to the session, while reducing the time the bandage is worn significantly.

REFERENCES

1. Scurr JH, Coleridge-Smith P, Cutting P, Optimum compression following sclerotherapy. Ann R Coll Surg Engl 1985;67:109-111.
2. Goldman MP, Sclerotherapy. Mosby Year Book Inc. 1991:170-171.
3. Orbach EJ: The importance of removal of postinjection coagula during the course of sclerotherapy of varicose veins, Vasa 1974;3:475.

Phlebology '95, D. Negus et al. (eds.). Phlebology (1995) Suppl. 1: 521-523

OP/5.3

Segmental Sclerosing or Stripping in Outpatient Treatment

U. Käseberg

Phlebologische Chirurgie Dr. Käseberg, Von-der-Tann-Str. 20, 83022 Rosenheim, Germany

Insufficiency of the long and short saphenous vein is the classic indication for surgical therapy of varicose veins. Basic principles are crossectomy, ligation of incompetent perforating veins in combination with elimination of all insufficient vein trunks.

The better understanding of haemodynamics and morphological foundations and better operating techniques are responsible for the increased success of outpatient treatment.

Requirements for a successful ambulant operative therapy are: a short operating time, an economical use of material and personnel, and a low risk for patients.

Patients are expecting good functional and cosmetic results.

Method

The method of intraoperative segmental sclerosing (IOSS) was developed especially for outpatient vein surgery (1).

It is an alternative treatment to the well known vein stripping.

We wanted to avoid the problems of vein stripping which are uncontrolled haemorrhages, pain and injury to the nervous saphenous.

The principle of our operation technique is characterised by a correct crossectomy, ligation of the insufficient v. communicants and segmental sclerosing of the varicose trunk and lateral system by air block catheter technique.

Small amounts of sclerosing agent reaches directly via ramified branch system.

The segmental procedure harmonises with recirculation systems in upper and lower limb. (2)

522

Sclerosing technique

We take sterile catheters 1.5 x 2.5 x 50 cm with luer connection intraluminal injection should be done orthograd.

We don't need more than 3ml 3% Aethoxysklerol for sclerosing the long saphenous vein. 3-5 small injections are necessary to interrupt the pathological flow in valve destroyed veins.

The greatest danger is the uncontrolled injection in a part of deep veins especially in lower leg veins, because of the many venae communications. It is necessary beforehand to ligate the perforated veins carefully.

Results

We have been performing segmental catheter sclerosing in outpatient treatment since 1965.

More than 11000 operations have been carried out.

Our control group consists of 1000 patients, who were operated on 1992 to 1995.

The preoperative stages were:

st.	IV varicose veins with secondary symptoms	32% (CVI III)
st.	III varicose veins	58% (CVI I-II)
st.	I - II varicose veins	10% (CVI I)
insufficiency of short saphenous vein	5%	
recidive varicose veins	12%	

The postoperative treatment time was from 1 to 5 months.

Incision of local thrombotic areas, sclerosing of insufficient branch veins and ligation of non recognised perforators were necessary.

Operated patients were controlled after 3 months and then again after 6 months and after 1 year.

In 80 patients of the control group invaginated stripping of the long saphenous vein in upper limb was performed.

Stripping in lower leg was not done because of the danger of the nervous saphenous injury.

The results have been identical as those achieved by segmental sclerosing.

In the short observation time we have found a recanalisation of saphenous vein in 23 cases (2.3%)

Noticeable were only a few complications:

large postoperative haemorrhages or haematomas	0.5%
post surgical wound infection	0.3%
nervous saphenous injury	0.6%
lymphical fistules	0.3%

Conclusions

* Method of segmental intraoperative sclerosing (IOSS) is an alternative operating technique in outpatient treatment of varicose veins.
* Postoperative complications were extremely rare.
* The operation can be carried out with local anaesthetics.
* An operating time of more than one hour was rare.
* We didn't need a longer operating time, for patients in stage IV with secondary symptoms or ulcerations.
* The segmentally sclerosing is a minimal invasive treatment, it is simple and economical.
* All stages of disease are possible to treat in outpatient treatment.
* Floride ulcerations are not contraindicated.
* A disadvantage is the extended phase of postoperative treatment because of the local thrombosis of sclerosing parts of veins.
* An exactly applied bandage in postoperative care is very important.

* Segmental intraoperative sclerosing is a useful and gentle alternative for a ambulant phlebosurgery.
* It is possible to have good results by combining intraoperative sclerosing with a short stripping of long saphenous vein and with microsurgery (3).
* IOSS utilised the most well known treatments, surgery and sclerotherapy.

REFERENCES

1. Ermisch H. 1984 intraoperative Sklerotherapie statt Stripping Bericht über 2700 Eingriffe z. ärztl. Fortbild 78, 327-328
2. Hach, Wundele; Der Rezirkulationskreis der primären Varicose Springer Verlag 1994, 27-31
3. Varady 2, Mikrochirurgische Phlebextraktion nach Varady in der Varizenchirurgie Phlebologie 1988 1; 8 - 9.

Phlebology '95, D. Negus et al. (eds.). Phlebology (1995) Suppl. 1: 524-525

PI/9.1

Consensus Conference on Sclerotherapy of Varicose Veins of the Lower Limbs

U. Baccaglini, G. Spreafico, C. Castoro, P. Sorrentino and E. Ancona

Second Department of General Surgery, University of Padua, Italy

Sclerotherapy of varicose veins of the lower limbs is a technique widely used in clinical practice for the past several years now.

The available scientific literature reports conflicting data concerning the results of sclerotherapy. In addition, phlebologists have different opinions on this topic.

These differences make it difficult to clearly define both the role of sclerotherapy in the treatment of varicose veins and the criteria of choice between sclerotherapy and other treatments.

A Consensus Conference on Sclerotherapy of Varicose Vein of the Lower Limbs was organized keeping all these points in mind; the first meeting was held in Padova on the 24th September 1994.

The experts who participated in the CONSENSUS CONFERENCE and the subjects to be discussed were chosen after an enquiry amongst experts personally known, and according to the advice given by several international Phlebological Societies. In fact 35 Representatives or Secretariats of Phlebological Societies were contacted and asked to indicate two experts from every Country and to supply a list of subjects for discussion.

A bibliographic research of the literature classified in the *Index Medicus* from 1966 to 1994 was carried out, and 672 articles were collected and sent for review to the experts. Many papers were unknown to the experts and different opinions were expressed on the same papers. The scientific weight of the majority of papers was considered quite low. Therefore it was recognized that the available literature was greatly limited from the statistical and scientific points of view, and so it was of little use in reaching a consensus.

In trying to reach a consensus both research and personal experience were relied upon. Due to limits in literature, personal experience of the experts has therefore become significant.

Four topics were chosen for discussion: INDICATIONS, TECHNIQUES, RESULTS AND ECHOSCLEROSIS. After the first meeting, ECHOSCLEROSIS was incorporated into TECHNIQUES.

In order to better define the state of art of current sclerotherapy, information obtained from a research done with questionnaires sent to phlebologists in various Countries was also taken into consideration. 6 Countries returned approximately 700 questionnaires.

In conclusion the Consensus Conferenece was based upon 3 points: the literature, the personal experience of the experts and the results of the survey on current practice.

Research and personal experience were approached in a structured manner.

In the Consensus Conference there were:

a) opinions where everybody definitevely agreed or where it was thought that it was probably right. These are definitive or probable opinions, and they can be taken forward to some kind of recommendation.

b) opinions where nobody knew. These are potential research proposals.

c) opinions where nobody agreed. These are areas of active disagreement and here it is more difficult to move forward to research. It must however be stressed that when there was a disagreement, this did not necessarily mean that this was not recommended.

It's a myth to think that a consensus is absent or present. A consensus tends to be graded, there is a whole range of opinions.

The process, and thus the results of the consensus, are dependent on the people who are involved in it. Angiologists, phlebologists, dermatologists, surgeons, vascular surgeons and epidemiologists participated in the Consensus Conference.

Three Consensus Conference meetings were held. The first one in Padova on the 24th September 1994, the second one again in Padova on the 11th - 12th February 1995 and finally the third one in Mogliano Veneto (Venice) on the 13th - 14th May 1995.

All the presentations and discussions of the various sub-committees and Plenary sessions were recorded, transcribed and reviewed by the Chairmen.

A survey was carried out, amongst the participating experts, by means of questionnaires, concerning the most controversial issues which were brought up during the discussions.

A consensus paper was drafted twice and discussed in the Plenary session. It summarizes the principle issues which were dealt with and indicates the points of agreement and disagreement; controversial points are also indicated.

The topics dealt with in the Consensus Conference document are: Consensus Process, Indications, Pre-treatment Assessment, Techniques, Results, Training, Recommendations.

The definitive Consensus Conference document will be availabe in September 1995.

The following is the list of those who participated in the Consensus Conference:

L. Abenhaim	Canada	G. Lipari	Italy
E. Ancona	Italy	O. Maleti	Italy
U. Baccaglini	Italy	L. Norgren	Sweden
E. Baggio	Italy	P. Parpex	France
E. Beverdam	NL	H. Partsch	Austria
K.G. Burnand	UK	P. Pavei	Italy
C. Castoro	Italy	P. Raymond-Martimbeau	USA
F. Chleir	France	M. Schadeck	France
A. Cornu-Thenard	France	U. Shultz-Ehrenburg	Germany
A. Davy	France	JH. Scurr	UK
E. Einarsson	Sweden	P. Sorrentino	Italy
FGR Fowkes	UK	G. Spreafico	Italy
C. Garde	France	I. Staelens	Belgium
H. Gerlach	Germany	R. Stemmer	France
L. Grondin	Canada	F. Vin	France
K. Hoerdegen	Switzerland		

Phlebology '95, D. Negus et al. (eds.). Phlebology (1995) Suppl. 1: 526-529

V/2.2

L'Echosclérose des Varices

G. Gachet

Voiron, Isère, France

Echosclerotherapy of Varices

SUMMARY

Echosclerotherapy (sclerotherapy under echodoppler ultrasound control) is not a revolutionary method but a combination of two completely established techniques. Echodoppler allows safe access to those points of reflux which were not possible to treat previously without risks, and also to treat those varices which are invisible to the naked eye.

This technique does not seek to support other procedures but is useful as a complement to them. The technique is simple, ambulatory, aesthetic, without risk of skin damage, economical and without significant complications in the hands of an angiologist experienced in echo-doppler and sclerotherapy. The obvious simplicity of the method, without fear of serious accident, makes it ideal for those with little previous experience of vascular pathology.

L'ECHOSCLEROSE DES VARICES

Docteur Gilles GACHET
Voiron - Isère - FRANCE

La sclérothérapie est une méthode ancienne consistant à injecter à l'aide d'une seringue un produit sclérosant obturant ainsi la varice que l'on souhaite éliminer. Les avantages sont nombreux : méthode simple, peu douloureuse,ambulatoire, esthétique car sans cicatrice et d'un coût économique très faible.

Le seul danger réside dans le risque d'injecter le produit sclérosant en dehors de la varice souhaitée.

L'utilisation de l'échographie bidimensionnelle a permis de réduire ce risque en donnant des "yeux" à la méthode. Les 2 types de sondes échographiques, linéaires et sectorielles sont utilisables.

TECHNIQUE

L'établissement de la cartographie variqueuse est l'étape préliminaire à tout traitement. L'examen clinique et l'échodoppler se font en position debout. La première étape consiste à éliminer un reflux profond qui pourrait être une cause de récidive variqueuse après traitement des reflux superficiels. L'examen recherche ensuite les incontinences ostiales, tronculaires et collatérales des saphènes puis des reflux variqueux non systématisés. La cartographie est alors réalisée, ce qui permettra ensuite de choisir la ou les meilleures méthodes de suppression des reflux (car on peut les associer).

Si la sclérose est choisie, la cartographie permet d'établir la chronologie de suppression des reflux, en commençant toujours par les ostiums, puis les troncs, et enfin les collatérales des saphènes.

L'écho-sclérose se fait patient couché ce qui évite les malaises vagaux , les mouvements intempestifs du patient et diminue les hématomes par reflux sanguin dans l'orifice crée par l'aiguille dans la varice. La visualisation de la varice peut se faire dans le sens longitudinal ou bien transversal. L'échographie permet de repérer les meilleures zones de sclérose (environ 2 cm en dessous des ostiums) et les meilleurs passages pour y accèder en évitant les artères et les veines profondes. Pour l'échosclérose de la saphène externe, il faut rechercher une artère petite saphène accompagnant cette varice et pouvant être une cause d'accident par injection intra-artérielle. Si cette artère inconstante est présente, l'injection devra se faire dans la partie basse de jambe et il faudra éviter le creux poplité. L'aiguille pénètre la peau selon un angle d'environ 45° en dessous de la sonde, sans la toucher, afin d'éviter une contamination. Le repérage de l'aiguille est facilité dans les tissus par sa mobilisation. Dans la varice, la pointe de l'aiguille est facilement repérable car il s'y produit une aggrégation érythrocytaire instantanée au contact du produit sclérosant. Il apparait alos une opacité punctiforme hyper-échogène. La seule visualisation de la pointe de l'aiguille dans la varice ne doit pas suffir pour injecter le sclérosant ; il faut obtenir, comme pour toute I.V. un bon reflux dans la seringue. L'injection de sclérosant est matérialisée par l'apparition de volutes dans la lumière de la varice (ce phénomène peut être amplifié en injectant 1 cc d'air mélangé au produit utilisé). L'utilisation de produits sclérosants douloureux lorsqu'ils sont injectés en dehors des vaisseaux permet également de s'assurer de la bonne position de l'aiguille dans la veine.

La séance se termine par la mise en place suivant les cas d'une contention pendant quelques jours qui pourra être amovible,cohésive ou collée. Afin d'éviter les TVP, la marche est ensuite indiquée.

Le contrôle de la réussite de l'écho-sclérose se fait lors de la séance suivante. La sclérose étant peu échogène, la vérification se fait selon les critères classiques de diagnostic de thrombose veineuse superficielle, c'est à dire l' absence de flux circulant au doppler et surtout incompressibilité de la varice en échographie. En cas d'échec, une nouvelle tentative pourra être réalisée avec une concentration plus forte, un volume plus important ou un produit plus puissant. On constate parfois une sclérose partielle qu' il est alors facile de compléter par une nouvelle séance.

INDICATION

D'une manière générale, il est préférable de faire opérer les très grosses varices avec de gros reflux car on s'expose à des échecs ou bien à des fortes réactions liées aux grandes quantités de produit utilisé.

L'allergie au sclérosant est bien sûr également une contre-indication. De même il faut être prudent avec les patients porteurs d'une thrombophilie avec déficit constitué de la coagulation.

Par contre, les bonnes indications sont représentées par : Les reflux saphène externe et interne peu importants sur varices peu dilatées, les saphènes antérieures refluantes, les veines de GIACOMINI incontinentes, les varices périnéales, les perforantes

de petit calibre, les échecs de CHIVA, les Varices
résiduelles après chirurgie des troncs sapheniens,
les néogénèses variqueuses, notamment les têtes de
méduses toujours difficiles à réopérer et qui sont
elles mêmes favorisées par les interventions
chirurgicales multiples, les lymphoedèmes qui sont
également favorisés par les interventions chirurgicales
multiples notamment au niveau du creux inguinal en
délabrant les canaux et les ganglions drainant le
membre inférieur, les varices complexes avec
localisation aberrante ce qui est souvent le cas dans
le creux poplité où les variations anatomiques de
la saphène externe sont nombreuses ; enfin, on
rajoutera à cette liste , les reflux ostiaux saphène
des femmes jeunes désirant des grossesses futures,
les personnes âgées et d'une manière générale, les
patients inopérables.

COMPLICATIONS

Ce sont celles de la sclérose classique : réactions
inflammatoires à traiter par AINS et contention,
pigmentation s'estompant avec le temps si l'on évite
l'exposition au soleil, thrombi variqueux nécessitant
parfois une thrombo-aspiration, echec ou reperméation
secondaire nécessitant une 2ème sclérose à des doses,
des concentrations ou des produits plus fort et parfois
en cas de nouvel échec, un éveinage chirurgical, les
TVP spontanées sont très rares surtout si l'on demande
aux patients de beaucoup marcher, avec une contention,
les injections extra-vasculaires avec, suivant les
organes atteints, nécrose cutanée, TVP, ischémie,
dysesthésies ou déficit neurologique. Toutefois, les
complications graves sont très rares et nettement
moins fréquentes que lors de scléroses à l'aveugle.

CONCLUSION

L'échosclérose n'est pas une méthode révolutionnaire
mais la combinaison de 2 techniques confirmées et
complémentaires l'une d'exploration, l'autre de
traitement des varices. La première permet d'accèder
en toute sécurité à des points de reflux qu'il n'était
pas possible de traiter sans risque auparavant et
donc d'étendre la sclérose à la plupart des varices
et non aux seules visibles à l'oeil nu.
 Cette technique n'a pas la prétention de supplanter
les autres procédés mais elle représente un outil
supplémentaire dans l'arsenal thérapeutique. Elle
ne s'oppose pas aux autres méthodes puisqu'elle peut
être utilisée en complément de celles-ci. C'est une
technique simple, ambulatoire, esthétique, non
délabrante, économique et sans grand danger entre
les mains d'un angiologue rompu à l'écho-doppler et
aux sclérothérapies.L'apparente simplicité de la
méthode ne doit pas, sous peine d'accidents graves,
la faire pratiquer par des néophytes en matière de
pathologie vasculaire.

Phlebology '95, D. Negus et al. (eds.). Phlebology (1995) Suppl. 1: 530-532

P239

A Comparative Study of Three Sclerosing Agents in the Treatment of Telangiectasias

J. Gawrychowski, B. Lazar-Czyzewska and A. Romanski

Varicose Vein Clinic, 41-800 Zabrze ul. Gen. de Gaulle'a 60, Poland

INTRODUCTION

Sclerotherapy is advisable for the treatment of superficial varicosities and telangiectasias (3,4, 5,8). None of the sclerosing agents is ideal and all can cause damage to the skin (1,2,3,6). Most frequently we find hyperpigmentation (24-35%), secondary telangiectasias (11-16%), superficial necrosis and/or ulcers (0-3%) and dark permanent skin changes (1-3%) (1,2,6,7).

Since there is still controversy regarding superiority of a specific agents, the aim of our study was to compare effectiveness of the three most commonly used sclerosing agents, namely Sclerodine, Sclerodex (Omega Laboratoires, Canada) and Aethoxysclerol (Kreussler, Germany).

MATERIAL AND METHODS

We reviewed the results of sclerotherapy in three consecutive series of patients treated between 1992-1994 with three different sclerosing agents. Sclerotherapy was performed by the standard technique. 324 patients, all women (aged 18-67, mean 36.5 yrs) with telangiectasias were divided into 3 groups treated by: (1). 0.25% Sclerodine (n=25); (2). Sclerodex (n=121) and (3). 0.25% Aethoxysclerol (n=178).

The results were retrospectively analyzed and compared between the three groups of patients 6 months after the completion of sclerotherapy. The main features estimated were: cosmetic improvement, incidence of side effects and relief of pain.

RESULTS

There were no significant differences in pain relief between the groups. The results of cosmetic effect in patients treated with various agents are summarized in table 1.

Tab. 1. Comparison of cosmetic results of three groups of patients treated by various agents.

Kind of sclerosing agent	all	Sclerotherapy - cosmetic results				
		excellent	good	fair	no change	worse
0,25% Sol. No Sclerodine %	25 / 7,7	4 / 16	11 / 44	7 / 28	2 / 8	1 / 4
Sclerodex No %	121 / 37,3	26 / 21	51 / 42	12 / 9,9	11 / 9,1	2 / 1,7
0,25% Sol. No Aethoxysderol %	178 / 55	57 / 32	130 / 73	7 / 3,9	3 / 1,7	0 / 0
All No %	324 / 100	87 / 27	192 / 59	26 / 8	16 / 5	3 / 1

Tab. 2 Complication of sclerotherapy comparison study of various sclerosing agents

Complication	Sclerosing agent					
	0,25% Sol. n=25 Sclerodine		n=121 Sclerodex		0,25% Sol. n=178 Aethoxysclerol	
	L	%	L	%	L	%
hyperpigmentation after 6 months	8	32	19	15,7	11	6,2
secondary telangiectasias	3	12	14	17	27	15
necrosis and/or ulcer	0	0	4	3,3	0	0
scars	4	16	8	6,6	0	0
superficial phlebitis	3	12	3	2,5	3	1,7
clots	7	28	15	12,4	4	2,2

Postsclerotherapy hyperpigmentation occurred most frequently after Sclerodine (32%), less frequently after Sclerodex (15.7%) and least frequently after Aethoxysclerol (6.7%). Secondary telangiectasias was almost equal (12%, 17%, 15%). Clots and superficial phlebitis were frequent after Sclerodine (28%, 12% respectively) and Sclerodex (12.4%, 2.5%) vs. Aethoxysclerol (2.2%, 1.7%). Necrosis and/or ulcers were found only after Sclerodex therapy (3.3%), but scars mostly after Sclerodine (16%) and Sclerodex (6.6%) (table 2.).

CONCLUSIONS

Aethoxysclerol is a safe sclerosant and gives less hyperpigmentation than Sclerodine and Sclerodex ($p < 0,001$).

REFERENCES:

1. Arramovic A., Arramovic M.: Complications of sclerotherapy: a statistical study. Phlebologie 92. Eds P. Raymond - Martinbeau, R. Prescot, M. Zummo. John Libbey Eurotext, Paris 1992, pp. 853-855
2. D'Agata V., Danieli D., Marzotto P., Pervone L., Sacran N., Dall'Antonia F.: Local complications in course of the sclerotherapy. Phlebologie 92. Eds P. Raymond - Martinbeau, R. Prescot, M. Zummo. John Libbey Eurotext, Paris 1992, pp. 863-865
3. Carlin M.C., Ratz J.L.: Treatment of telangiectasia: comparison of sclerosing agents. J. Dermatol Surg. Oncol. 1987, 13, 1181-1184
4. Goldman P.M.: Polidocanol (Aethoxyschrol) for sclerotherapy of superficial remmles and telangiectasias. J. Dermatol Surg. Oncol. 1989, 15, 204-209
5. Groot W.P.: Sclerotherapy of large veins. J. Dermatol Surg. Oncol. 1991, 17, 589-595
6. Natali J.: Les complications de la scletotherapie dans le traitement des varices. Phlebologie 92. Eds P. Raymond - Martinbeau, R. Prescot, M. Zummo. John Libbey Eurotext, Paris 1992, pp. 853-855
7. Tomban G., Gasbarro V., De Anna D., Morri A., Rosa D., Piva P., Fuga G., Domini S.: Sclerotherapy of veins: complications and their treatment. John Libbey Eurotext, Paris 1992, pp. 856-858
8. Weiss R.A., Weiss M.A.: Resolution of pain associated with varicose and telangiectatic leg veins after compression sclerotherapy. J. Dermatol Surg. Oncol. 1990, 16, 333-336

Phlebology '95, D. Negus et al. (eds.). Phlebology (1995) Suppl. 1: 533-535

P240

Sclerotherapy of Telangiectasias: Results and Complications

J. Gawrychowski, A. Romanski and B. Lazar-Czyzewska

Varicose Vein Clinic, 41-800 Zabrze ul. Gen. de Gaulle'a 60, Poland

INTRODUCTION

Many patients requiring treatment of superficial varicosities and telangiectasias on the legs are seeking relief from pain and discomfort in addition to cosmetic improvement (3,7). Complaints, particularly before or during menses have been associated with superficial varicosities on the thigh. In these patients, sclerotherapy is strongly recommended (7). However, this treatment can cause damage to the skin such as residual hyperpigmentation (24-35%), secondary telangiectasias (11-16%), superficial necrosis and ulcers (0-3%), dark permanent scarring (1-3%), allergy (1,5%) as well as pain lasting more than 5 min. (4,5%) (1,2,4,5,6).

The aim of the present study was to determine the role of sclerotherapy of telangiectasias in reducing leg pain and discomfort and to assess cosmetic results and side effects of this therapeutic procedure.

MATERIAL AND METHODS

The records of 324 patients treated only for telangiectasias during 1992-1994 were reviewed. All patients were female aged 18-67, mean 36.5 years. 55 patients (17%) had venous telangiectasias only (vessels 0.1-1 mm in diameter). In the remaining 83% of patients, venous telangiectasias were accompanied by dilation of larger vessels (1-3 mm). 198 patients (61.1%) suffered from pain and heaviness. Sclerotherapy was performed by the standard technique. We used 0.25% Sclerodine, Sclerodex (Omega, Laboratories Canada) and 0.25% Aethoxysclerol (Kreussler, Germany). Patients received 1-6 sessions of sclerotherapy per region (thigh and cruris - anterior, posterior, lateral and medial as well as ankle - lateral and medial). Cosmetic improvement were classified as follows: <u>excellent</u> -complete disappearance of treated veins; <u>good</u> - greater than 50% reduction in size and/or cleaning of treated veins; <u>fair</u> - slight or minimal improvement; <u>no change</u> and <u>worse</u>. Local complications observed were as follows: hyperpigmentation (a brownish discoloration along the course of a treated vein or around the injection site), secondary telangiectasias, necrosis and/or ulcers, dark scarring, superficial phlebitis and clots inside the vessels. Results were retrospectively analyzed 6 months after the completion of the sclerotherapy and three main features were estimated: cosmetic improvement, incidence of side effects and relief of pain.

RESULTS

Cosmetic improvements were found as follows: excellent (complete disappearance of treated veins) in 87 (27%) patients, good (greater than 50% reduction in the size and/or cleaning) in 192 (59%), fair (slight or minimal improvement) in 26 (8%), no change in 16 (5%), worse in 3 (1%) patients (tab. 1).

Tab.1. Cosmetic improvement of sclerotherapy - results.

Patients		Cosmetic improvement of sclerotherapy				
		excellent	good	fair	no change	worse
No	324	87	192	26	16	3
%	100	27	59	8	5	1

Tab. 2. Sclerotherapy - kinds of complication (out of 324 of patients)

kind of complication	L	%
hyperpigmentation (after 6 months)	38	11.7
secondary telangiectasias	44	13.6
necrosis and/or ulcer	4	1.2
scars	12	3.7
superficial phlebitis	9	2.8
clots	26	8

Local complications found were as follows: secondary telangiectasias in 13.6%, post-sclerotherapy hyperpigmentation in 11.7%, clots in 8%, scars in 3.7%, superficial phlebitis in 2.8%, necrosis and/or ulcer in 1.2% (tab. 2).

164 (82.8%) out of 198 of patients suffering from pain and/or heaviness experienced significant relief of symptoms.

CONCLUSION

Sclerotherapy of telangiectasias is not only a cosmetic procedure but also an efficient method of relieving of associated leg pain.

REFERENCES:

1. Arramovic A., Arramovic M.: Complications of sclerotherapy: a statistical study. Phlebologie 92. Eds P. Raymond - Martinbeau, R. Prescot, M. Zummo. John Libbey Eurotext, Paris 1992, pp. 853-855
2. D'AgataV., DanieliD., Marzotto P., Pervone L., Sacran N., Dall'Antonia F.: Local complications in course of the sclerotherapy. John Libbey Eurotext, Paris 1992, pp. 863-865
3. Goldman P.M.: Polidocanol (Aethoxyschrol) for sclerotherapy of superficial remmles and telangiectasias. J. Dermatol Surg. Oncol. 1989, 15, 204-209
4. Groot W.P.: Sclerotherapy of large veins. J. Dermatol Surg. Oncol. 1991, 17, 589-595
5. Natali J.: Les complications de la sclerotherapie dans le traitement des varices. Phlebologie 92. Eds P. Raymond - Martinbeau, R. Prescot, M. Zummo. John Libbey Eurotext, Paris 1992, pp. 853-855
6. Tomban G., Gasbarro V., De Anna D., Morri A., Rosa D., Piva P., Fuga G., Domini S.: Sclerotherapy of veins: complications and their treatment. Phlebologie 92. Eds P. Raymond - Martinbeau, R. Prescot, M. Zummo. John Libbey Eurotext, Paris 1992, pp. 856-858
7. Weiss R.A., Weiss M.A.: Resolution of pain associated with varicose and telangiectatic leg veins after compression sclerotherapy. J. Dermatol Surg. Oncol. 1990, 16, 333-336

Phlebology '95, D. Negus et al. (eds.). Phlebology (1995) Suppl. 1: 536-538

P129

Physical Sclerosis by Foreign Body After the Method of C. Sánchez et al

C.F. Sánchez[1,2], E. Altmann-Canestri[2,3], U.P. Tropper[1,2], R. Barceló[1,2], E. Tkach[1,2] and D. Clar[1,2]

[1] *Phlebological Clinic,* [2] *Jonh F. Kennedy University,* [3] *Buenos Aires University*

INTRODUCTION

We refer to Toprover's physiopathological groundwork: to create in a vein a phlebitis that will obliterate the vessel [1] and we develop a new technique from there [2].

In order to determine the behaviour of catgut as a physical method of sclerosis, S.A. Nesterov had carried out histopathological studies in animals [1]. Our team investigated the subject carrying out histopathological studies of biopsies made on patients treated by the method of physical sclerosis with catgut, according to our technique.

In this study we also compared the results obtained with regular and chromic catgut.

In both cases we observed that the inflammation produced by the presence of catgut and its consequences (sclerosis of the muscular layers —inner, connective and adventitious—) is the same. The difference lies in that the intensity of the sclerosis obtained is significantly superior with chromic catgut.

In this way, a definite closure of the treated vein is ensured, and therefore we chose this material for our procedure.

The excellent results obtained in the physical sclerosis by foreign body induced us to develop this new technique, specially indicated in those cases where ambulatory phlebectomy is difficult, to wit:

- Dorsum of foot (Fig. 2)
- Lower internal side of thigh, due to lymphorrhagia (Fig. 1)
- Impossibility of extraction
- Overworked area, with danger of induration.

REASONS FOR EACH ONE

Dorsum of foot: Even the most skillful surgeon will find serious difficulties. Phleboexeresis with crochet is hardly practicable because the scarce subcutaneous cellular tissue contains a great number of nerve ends and lymph vessels and, therefore, only minimal maneuvers may be tried so as not to damage them.

On the other hand, sclerosis is more difficult because the communicating veins may carry the sclerosant rapidly to the Lehars's venous sole with unpleasant consequences.

All these reasons induce us to indicate the sclerosis by foreign body as "preferable" in this location. The procedure finds here a great endorsement of its advantages: the sclerosis produced by the catgut has practically no effect on neighboring structures, so that its application avoids damaging noble elements. Besides, its application is simple and its esthetic results unsurpassable.

Lower internal side of thigh: Lymphorrhagia is a hazard clearly illustrated in laborious operations on the lower internal side of the thigh. Let us remember that the lymphatic complex practically flows together massively in this area. Therefore, in this location the same considerations given for the dorsum of foot are valid, that is, a gentle maneuver is made with crochet and exeresis is done if it appears easy; if it fails, the catgut threads are left to sclerose the area.

Impossibility of extraction: when a varix is hard to "catch" with crochet and to extract with Kocher's tweezers one should not insist. *Treatment is a service to the patient and not a personal challenge.* This is a good time to decide for this procedure that will render the expected service and avoid the mentioned difficulties.

Overworked area with danger of induration: when a varicose process has been difficult to extract and affected veins still remain in the area, the possibility of producing a painful induration grows in direct proportion to the number of maneuvers we perform. Here is a new opportunity to complete the work by leaving the catgut.

METHOD

Group of 127 patients, 32 men and 95 women aged between 27 and 59.

TECHNIQUE

A- Marking
We mark the area to be treated with indelible ink, specially points of maximum protrusion (perforans insufficiency).
We can also do this maneuver with Doppler.

B- Position of the patient
Patient decubitus is the most adequate position for the procedure. If possible, we will use C. Sánchez's safety angle for physical sclerosis and ambulatory phlebectomy (60°), placing a polyurethane pillow shaped for this purpose. This simple object is easily handled in the doctor's office and renders best results in order to avoid hemorrhagia (Fig. 3).

C- Antisepsis: as usual

D- Surgical field: as usual.

E- Anesthesia
We use lidocaine 0.5%, which ensures satisfactory anesthesia with ample safety margin.

F- Placing of catgut
• **Preparation of the needle**
To proceed in placing the catgut, we resorted to needles big enough to leave long pieces with only one entrance and exit.
Therefore, the peculiarity of this technique is the instrument to be used. The needle must be 10 to 12 cm long, thin and with excellent sharpness. This condition we will solve in two ways:
1) Veterinary surgery uses suturing needles of that size and extraordinarily lanceolate. The size of the needle will allow, in a short time, to pass a chromed catgut through each

section in which we wish to perform sclerosis. The problem is that the threading of the needle implies catgut superposition which, added to the broadening of the needle at the site of the eye, produces some trauma when penetrating the skin.

2) This inconvenience may be lessened if we use a discardable trocar (Abbocath-T 14G) for spinal anesthesia, whose pavilion we remove by burning it. Through this end we insert 3 to 4 cm of catgut pressing it inside the needle with the tweezers. When the procedure is finished, we dip the needle and catgut into alcohol until the moment of use.

• **Procedure**

We introduce the needle and catgut through the proximal end of the marked area; we advance through the subcutaneous cellular tissue trying to "thread" the vein and when going out we leave a small distal end for later withdrawal. We perform this procedure several times following the location of the varices (Fig. 4). When the varicose complex is greater and certain parts are not visualized, we perform the procedure in a radiating shape (Dr. Barceló's "sun").

Generally only distal ends of catgut are exposed but, as the trajectory follows the varix, when it is short, both ends are exposed. A careful bandage is applied and after 3 days the catgut threads are removed.

CONCLUSIONS

We have been implementing this technique for 15 years with excellent results. Included as a routine complement of ambulatory phlebectomy, it has allowed to overcome the difficulties present at certain locations (foot and internal side of thigh), solving most of their inconveniences (impossibility of extraction) as well as avoiding unpleasant indurations resulting from excessive maneuvers.

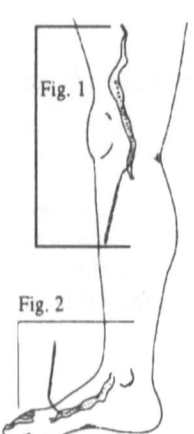

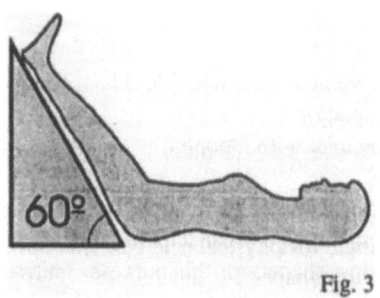

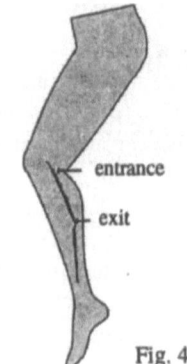

Fig. 1: Risky situation in surgery. Lower internal side of thigh.
Fig. 2: Risky situation in surgery. Dorsum of foot.
Fig. 3: Position of the patient in Dr. Sánchez's safety angle for physical sclerosis and ambulatory phlebectomy.
Fig. 4: Threading of a varicose vein with chromic catgut. Entrance at proximal end and exit at distal end of varicose vein.

REFERENCES

1. Toprover G S. Los principios del tratamiento quirúrgico de la dilataciones varicosas de las extremidades inferiores. Vestnik Khirourguii 1961; 12: 49-56.
2. Sánchez C F, Pereira da Silva L, Altmann-Canestri E, Pace F, Tkach E, Tropper U P et al. Técnica de esclerosis por cuerpo extraño del Prof. Dr. César Sánchez y col. Revista Panamericana de Flebología y Linfología 1994; 15: 31-40.

Phlebology '95, D. Negus et al. (eds.). Phlebology (1995) Suppl. 1: 539-541

P130

Sclerus vs Thrombus

F. Chleir and F. Vin

Hopital Notre Dame de Bonsecours, Paris, France

An analogy has always been drawn between the reaction following injection sclerotherapy for varicose veins and superficial thrombophlebitis. However we know that the symptoms following sclerotherapy are far less than those associated with superficial venous thrombosis. We were aware that the absence of a reaction following injection of a sclerosing drug maybe due to lack of efficacy. When sclerotherapy is conducted improperly superfical thrombophlebitis may occur with associated painful symptoms. We therefore undertook a study comparing the results of sclerotherapy with those of spontaneous superficial thrombophlebitis by studying clinical, biological, echographic and histological data. We compared our own results with those of a literature review.

CLINICAL STUDY

Following sclerotherapy a sensation of discomfort along the course of the vein may be experienced. No inflammatory phenomenon is observed, either visible or on palpation. Following treatment we find that the vein has been converted to a palpable indurated venous cord which is painless and without signs of inflammation. In the long term the indurated cord may persist or disappear totally. By comparison thrombosis is usually painful with an angry inflamed appearance. The inflammatory response occurs not only within the affected vein but also in the surrounding area of skin and subcutaneous tissue. Later we find a dilated partially compressible varicose vein. In the long term recanalisation of the vein results without retraction or disappearance of the varicose vein.

Tournay, Wallois (1), Griton (2) describe the same difference between successful sclerotherapy and superficial thrombophlebitis.

There is clearly a substantial difference even when considering only the clinical symptoms between a successful sclerotherapy treatment and superficial thrombophlebitis.

BIOLOGICAL STUDY

There are a number of specific and sensitive markers that my be used in assessing blood coagulation. We were interested to know whether any of these modern methods of investigation could differentiate between the results of sclerotherapy and superficial thrombophlebitis. Currently available markers for activation of blood platelets include

4 P Factor and Beta Thromboglobulin. We were unable to use these because of the difficulties in obtaining reliable blood samples. Also available are markers of activation of coagulation. These include Fibrinopeptide A released during formation of thrombin, and Fragment 1 + 2, a marker of prethrombin activation. Both of these have short half lives and wide ranges of normal values. For these reasons we considered them unsuitable for use in our investigation.

D-Dimer is the only usable marker in routine clinical practice. We measured this by enzyme linked immunosorbent assay (ELISA). We have investigated ten patients, studied eight days following sclerotherapy and ten patients suffering acute superficial thrombophlebitis. Following sclerotherapy we have never observed elevation in markers of thrombosis. However in patients with superficial thrombophlebitis, inconsistent results were obtained but approximately 50% of cases showed significant modification and markers of thrombosis.

Wupperman (3) has previously shown that most of the plasma coagulation factors and platelets tests are unchanged following injection sclerotherapy. Raymond-Martimbeau (4) studied D-Dimer and Fibrinopeptide A after injection of Sodium Iodine 30 minutes, 48 and 72 hours after treatment. She observed no evidence of systemic thrombus formation. Our findings therefore are in keeping with the results of other investigators published in the literature.

ULTRASOUND IMAGING INVESTIGATION

Ultrasound imaging has been undertaken in longitudinal section. Following sclerotherapy greatest echogenicity is seen near to the wall of the vein with less echogenicity in the centre of the vessel. The sclerus appears to evolve from the wall. In transverse section compression of the vein with the probe results in flattening of the centre of the vein whilst the region near the vein wall remains uncompressed.

In veins affected by superficial thrombophlebitis echogenicity is greatest in the centre of the vein with less echogenicity in the zone near the vein wall. The thrombus evolves from the centre of the vein with the upper limit of the thrombus showing a convex top. Observation during scanning in transverse section shows that the central, echodense region is not flattened but the zone around the outside of the thrombus is easily compressed.

Schadeck (5), Laroche, Dauzat (6), Fauvel, have also found distinct ultrasound imaging appearances for veins following sclerotherapy and those affected by superficial thrombophlebitis.

Ultrasound imaging shows that sclerosis of vein is a process commening in the vein wall where a thrombosis starts in the blood contained in the vein.

HISTOLOGICAL STUDY

We have taken five sections of vein affected by superficial venous thrombosis and five veins eight days after injection of a sclerosing solution. Precise comparison is not possible since the interval between the start of a superficial venous thrombosis and investigation may be difficult to determine. Histological specimens have been studied by Professor Bruneval in the pathology Laboratory of Broussais.

Following sclerotherapy we found endothelial proliferation without disappearance of the vein wall. In the lumen of the vein there was an accumulation of leucocytes but relatively little fibrin. Superficial thrombophlebitis is characterised by presence of large amounts of clot in the venous lumen with little fibrin and many red cells and platelets.

Wupperman (3), Mancini (7), Staubesand, Seydevitz (8) have found similar results with essentially a parietal lesion and formation of a mass of fibrin, platelets and leucocytes in the venous lumen following sclerotherapy. Wenner in 1984 showed us that there is an extended sclerosis without thrombus formation when efficient compression is applied following sclerotherapy. Staubesand reported endothelial lesions with closure of the varicose vein without thrombus formation at one week after treatment.

CONCLUSION

This study of clinical, biological, ultrasound imaging and histological data allowed us to show that a parietal mechanism is responsible for vein destruction following sclerotherapy. However thrombosis results in a series of events in the lumen of the vessel which presumably lead to the clinical symptoms of thrombophlebitis. The results of sclerotherapy include a lesion arising in the vein wall with secondary stimulation of the coagulation mechanism. In contrast thrombus accumulates from circulating blood and coagulation proteins and subsequently invades the venous wall. There are substantial differences between the results of successful sclerotherapy and superficial thrombophlebitis which explain the efficacy of the former and morbidity arising from the latter.

REFERENCES

1. P Wallois. Sclérothérapie des saphènes. Phlébologie, 1995, 48, n°1, 17-23
2. Ph. Griton. La sclérothérapie de la veine saphène interne: indications, techniques et resultats. Actualités Vasculaires Internationales, n° 16, Novembre 1993.
3. Th Wuppermann. Méchanisme de la sclérothérapie des varices. Phébologie, 1991, 44, n° 1, 23-29.
4. P Raymond-Martimbeau, P Leclerc. Effects of sclerotherapy on blood coagulation: a prospective study. XI Congrés Mondial de Phlébologie. Montréal, 1992, John Libbey Eurotext. 441-448
5. M Schadeck, F A Allaert. Echotomographie de la sclérose. Phlébologie, 1991, 44, n°1,11-130
6. J P Laroche, M Dauzat, Gestion de la thrombose veineuse profonde des membres inferieurs par l'utilisation des explorations ultrasoniques. XI Congrés Mondial de Phlébologie. Montréal, 1992. John Libbey Eurotext. 655-656.
7. S Mancini, F Lessueur, F Mariani. La sclérose de la veine grande saphène: étude expérimentale chez l'homme sur l'action sclérosante de la solution iodo-iodurée et le polidodécane (Histologie et microscopie éléctronique) Phlébologie, 1991, 44, n°2, 461-468.
8. J Staubesand, V Seydewitz. Eludes ultra-structurelles des varices sclérosées. Phlébologie, 1991, 44, n°1, 16-22.

Phlebology '95, D. Negus et al. (eds.). Phlebology (1995) Suppl. 1: 542-549

OP/4.5

Sclerotherapy of Saphenous Veins - Results Over 10 Years

P. Wallois

106 Avenue de Suffren, Paris, France

SUMMARY

The author reviews a personal series of patients 10 years after sclerotherapy. Of 102 questionnaires sent, 42 answers were 'usable'. These showed that 50% of the patients were 'satisfied'. A further 79 patients were re-examined 10 years after the initial sclerotherapy: 15 had had no further injections, 31 had further injections once and 15 had required 2 or more series of injections.

Docteur Pierre WALLOIS

INTRODUCTION - MATERIEL

A l'origine de cette étude, une demande pressante d'éléments d'appréciation objectifs des résultats de la sclérothérapie. Ceux qui sont actuellement proposés reposent essentiellement sur des mesures et des analyses prati-
quées au moyen d'appareils d'investigation, signal doppler, écho-doppler,
doppler couleur. Le reproche qu'on peut leur faire c'est de n'avoir analysé
que les résultats à court terme et sur un élément analysable : une image inter-
prétée de la varice traitée.

Il nous a semblé parfaitement légitime de rechercher l'évolution
à plus long terme, et de la varice traitée, et de l'état clinique de celui
qui en était porteur.

Nous avons fait porter notre étude sur ce qui représente pour nous
le noeud de la pathologie variqueuse, les troncs saphéniens, en les séparant
bien de toute la pathologie annexe, plus ou moins dépendantes, les varices
distales, c'est-à-dire, les collatérales du réseau saphénien, ou les varices
isolées.

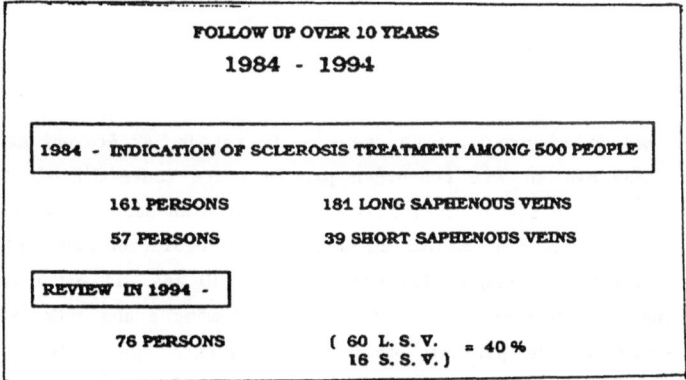

Pour que notre étude soit valable, il fallait qu'elle porte sur
un nombre d'années suffisantes : c'est pourquoi nous avons relevé tous les
cas de saphènes traitées, internes et externes, en 1984 et nous les avons

réétudiés en 1994, soit après une période de 10 ans.

En cette année 1984 environ 500 personnes étaient venues consultéer nous n'avons retenu pour la sclérose des saphènes que 161 personnes avec 181 saphènes internes et 37 personnes avec 39 saphènes externes.

En 1994, nous avons revu à la suite d'une démarche spontanée de leur part 79 personnes, 60 qui avaient été traitées pour une insuffisance du tronc de la saphène interne 16 de la saphène externe. Cela représente un pourcentage de 76/198, c'est-à-dire, 40% environ.

Nous avons d'autre part fait un courrier à tous ceux que nous pouvions joindre en leur demandant de nous renvoyer, à l'aide d'une enveloppe prétimbrée, un questionnaire extrèment simple avec trois options :

1) Je n'ai pas eu besoin d'autres soins

2) Je me suis fait suivre par un autre médecin

3) Je me suis fait opérer.

```
CONTACT WITH OTHERS BY MAIL

102 LETTERS


61 ANSWERS AND 42 USABLE
```

102 lettres ont été envoyées, nous avons eu 61 retour, 42 étaient utilisable, les autres nous avaient été renvoyées par la poste faute d'avoir trouvé le correspondant (16) ou par la famille nous avisant d'un décès (4 cas). Si bien que l'on peut dire que nous sommes au courant de l'évolution chez 119 patients sur 198. Ce qui nous a semblé intéressant après 10 ans. Il est bien évident cependant que la vérification par le médecin lui-même a une autre valeur que celle rapportée par le patient, celle-ci ne viendra donc qu'en corollaire à l'étude principale faite sur les patients qui ont été revus.

Avant d'en venir aux résultats proprement dits de l'étude, il convient cependant de dire un mot sur la sélection et sur le type de traitement pratiqué.

SELECTION ET METHODE DE TRAITEMENT

La base de la sélection a été l'appréciation clinique des éléments
constitutifs du tronc saphénien, essentiellement à partir de la palpation,
c'est-à-dire, le diamètre du tronc perçu et la résistance à la pression, jugée
par la qualité, force et rapidité, du retour au volume initiale de la veine
au moment où le doigt relâche attentivement sa pression. L'analogie de ces
critères pourrait être rapportée à une mesure d'un diamètre perçue à l'écho-
doppler en deça d'un maximum de 8 à 9 mm. Des saphènes plus grosses ont été
acceptées sous la pression du patient, certaines ont merveilleusement tenu,
mais c'est cependant dans les plus grosses saphènes qu'on retrouve les échecs
à long terme, les reprises plus fréquentes, et la nécessité de recourir à
une autre solution.

Le type de traitement est assez facilement défini, malgré des sensi-
bilités individuelles dont il faut savoir se méfier. On peut retenir que la
dose et le produit le plus souvent utilisés ont été le Sotradecol en solution
à 3 % 1 cc en moyenne. Souvent dès la première injection, le résultat a été
obtenu. L'injection est habituellement faite de préférence dans la moitié
supérieure du tronc variqueux pour les saphènes internes, sans rechercher
absolument l'injection à la jonction saphéno-fémorale et dans le quart supérieur à proximité du pli poplité dans la saphène externe, là où elle est parfaitement perçue, sans recherche non plus un point idéal proche de la jonction.
La surélévation de membre au-dessus de l'horizontale permet au liquide de
sclérose d'imprégner suffisamment la zone sus-jacente.

Tel est le schéma de traitement qui a été le plus habituellement
pratiqué : la suppression du tronc étant la plupart du temps suivi d'une diminution immédiate importante des varices collatérales qui ne demandent plus
qu'un traitement d'appoint, avec des doses 5 à 6 fois moins importantes, réalisé avec un minimum de séances, de une à trois en moyenne.

LES RESULTATS

Pour la clarté de l'exposition, nous n'avons retenu que le nombre
de saphènes, sans plus tenir compte du nombre de patients, certains pouvant
avoir une varicose bilatérale.

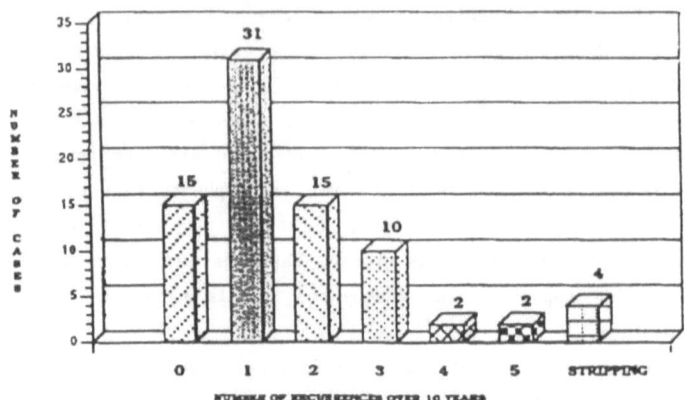

On peut voir sur ce tableau relatif à la saphène interne que sur 79 saphènes suivies 15 n'avaient jamais eu besoin d'être traitées à nouveau, 31 n'avaient nécessité qu'une seule reprise, avec un résultat obtenu dès la 1ère séance ; pour 15 autres, il y avait eu 2 fois reperméabilisation, rapidement neutralisée en 1 ou 3 séances.

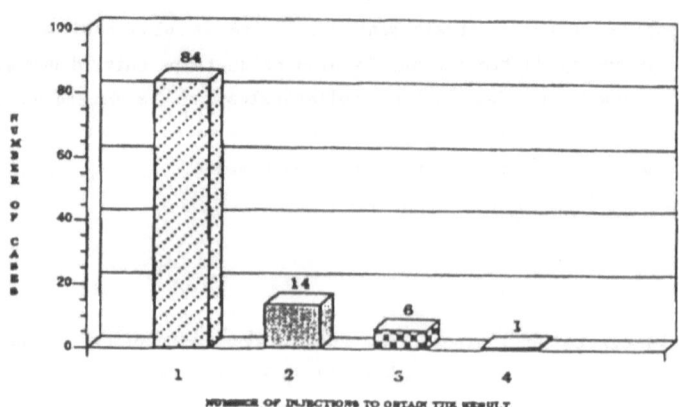

Il faut ajouter que 84 reprises ont été traitées avec un résultat de plus en plus solide avec 1 seule injection, que la dose utilisée était égale ou même sensiblement inférieure à celle utilisée pour obtenir le premier résultat et que le nombre d'injections nécessaires n'a jamais été très important : 14 fois 2 injections, 3 fois 6, 1 fois 4, au-delà un changement d'orientation a été décidé. Si l'on tient compte des 4 premières colonnes on peut considérer comme bons résultats ceux n'ayant pas nécessités plus de 3 reprises en 10 ans, c'est-à-dire, 71 cas sur 79, les 8 autres se répartissent en moins bons résultats 4 cas avec 4 à 5 récidives, les mauvais résultats ayant provoqué de notre part un changement de décision et l'envoi au chirurgien (4 cas).

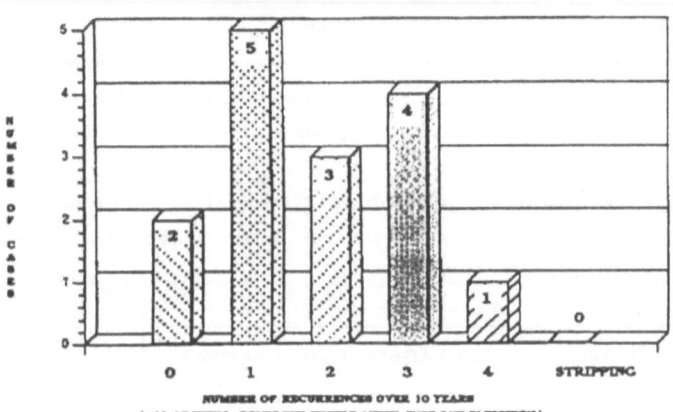

Pour la saphène externe, les résultats sont encore plus satisfaisants, ce qui confirme la notion bien connue que la saphène externe réagit encore mieux que l'interne à la sclérothérapie : sur 16 cas revus, 2 n'ont eu besoin d'aucune reprise, 5 ont vu le résultat définitivement obtenu avec une seule réinjection, 3 avec 2 corrections, 4 avec 3, 1 avec 4. Ce qui est intéressant de constater ici c'est que le nombre d'injections nécessitées chaque fois pour l'obtention du résultat a été d'une seule injection. J'ajoute que les doses utilisées ont été comme pour la saphène interne, soit exactement les mêmes, soit plutôt inférieures, la réapparition de la varice étant en général de moindre volume que lors de la première intervention.

Cette étude est principalement destinée à la sclérothérapie des saphènes ; c'est pourquoi nous avons voulu dissocier leur traitement qui a toujours été pour nous le noeud du problème de tous les gestes complémentaires que nous avons été amené à pratiquer sur les varices restantes. Cela ne se différencie en aucune manière de l'attitude que nous avons dans les suites d'intervention chirurgicale, soit suites immédiates, soit suivies à long terme: il faut toujours surveiller l'évolution d'une varicose traitée quelque soit son traitement, chirurgie ou sclérose.

Le nombre de séances complémentaires est forcément plus important que le nombre de gestes essentiels que nous avons pratiqués sur les saphènes. Nous les avons comptabilisés et ramenés à leur correspondance relativement au nombre d'années et au nombre de patients.

Après la suppression obtenue et maintenue comme on l'a décrit de la saphène interne. On comptabilise 602 séances d'entretien, ceci sur 60 patients et 10 années. Ce qui donne une séance par an et par individu. Le chiffre de 602 semble important. La réalité, une séance par an et par personne correspond à ce qu'il est raisonnable de conseiller pour un suivi chez un variqueux.

Pour la saphène externe, il a été comptabilisé 70 séances complémentaires, beaucoup moins relativement ce qui n'est pas étonnant vu le territoire beaucoup plus limité dépendant de la saphène extere. Ramener au nombre de séances de surveillance cela donne 1 séance tous les 2 ans. La saphène externe ici encore se comporte comme une bonne varice pour la sclérose.

Avant de conclure, je voudrais dire quelques mots des résultats de l'enquête que nous avons entreprise par lettre auprès des patients que nous n'avons pas revus.

---

42 USABLE ANSWERS AGAINST 102 LETTERS

21 NO NEED USING OTHER TREATMENTS
11 TREATED BY OTHER DOCTORS
10 OPERATED BY STRIPPING OR OTHERS
50 % SATISFIED

DISCUSSION

Nous avons été un peu déçus par le faible nombre de réponses, 62 retour dont 16 non parvenus au destinataire et retournés par la poste et 4 avis de décès soit 42 dossiers utilisables.

21 n'avaient pas eu besoin d'autres soins soit 50 % des réponses: mais comment les interpréter ? Satisfaction totale ou négligence.

11 avaient été traitées par un autre médecin

9 avaient subi une intervention (8 strippings et 1 phlébectomie ambulatoire).

Il faut ajouter 2 ou 3 réponses désagréables telles qu'imputation à la sclérose d'une tendinite, un reproche violent de récidive après 5 ou 6 ans, et l'avis d'un décès par infarctus 6 semaines après saphénectomie chirurgicale.

Les réponses sont données pour ce qu'elles sont, si j'en fais état, c'est pour montrer mon souci de faire un travail sérieux : ne compte que le travail réalisé chez les patients revus par moi-même.

CONCLUSION

Que conclure de ces chiffres ? Je les abandonne aux jugements des uns et des autres. On pourra aussi bien leur trouver des arguments pour, que contre la sclérothérapie des saphènes. Certains s'en empareront pour la condamner une fois de plus. J'attends de leur part une étude statistique portant sur le suivi à 10 ans de leurs opérés dans les différentes techniques proposées. Pour moi, si on les considère en eux-mêmes, ils justifient totalement la méthode. Les récidives sont peu nombreuses, elles se contrôlent très facilement, le plus souvent dès la première injection et avec, ce qui n'est pas la moindre raison de satisfaction, un soulagement immédiat des patients qui retrouvent dès la première séance le bien-être auquel ils étaient habitués. Cet aspect de la question ne doit pas nous échapper, on ne peut pas avoir le regard uniquement fixé sur le tuyau veineux, il faut aussi voir son effet pathologique : certaine reperméabilisation donne passage à un certain reflux qui n'est ressenti d'une manière pathologique que s'il revient important. Avant qu'il le soit, ou bien les patients ne jugent pas utiles de consulter, ou bien, au vu de la constatation clinique, le médecin lui-même peut décider de s'abstenir, sous réserve d'un contrôle poursuivi, en gardant, à la moindre manifestation, la possibilité d'intervenir, facilement et efficacement sur le processus.

Toutes ces facilités donnent à la sclérothérapie un avantage de souplesse et d'adaptabilité qui lui confère son rôle irremplaçable et bénéfique si on sait la maintenir dans les limites de ses possibilités. C'est cependant une technique de traitement qui a ses impératifs : d'une part elle nécessite pour chaque cas un travail préalable en vue d'établir un protocole de traitement auquel il faudra se tenir du début à la fin, jusqu'au résultat obtenu, et d'autre part, il faut bien savoir qu'on ne peut pas se lancer dans la pratique de la sclérothérapie sans un véritable apprentissage, sans en avoir acquis la maîtrise parfaite, condition de l'efficacité et aussi de la sécurité ; faute de quoi, on ne peut pas se permettre d'émettre appréciation ou jugement.

Phlebology '95, D. Negus et al. (eds.). Phlebology (1995) Suppl. 1: 550-552

P132

Postsclerotherapy Hyperpigmentation: Incidence, Clinical Features and Therapy

M. Izzo, F. Mariani, F. Binaghi and M. Amitrano[1]

Institute of Internal Medicine, Center of Medical Angiology, University of Cagliari, Italy
[1] Medicine Department, USL-4, Avellino, Italy

INTRODUCTION.

Residual hyperpigmentation that occur following sclerotherapy is an unaesthetic minor complication. In the up to-date literature[1] the frequence of this complication is estimated between 10% and 30% with a highest incidence for smaller and superficial veins and is only partially due to operator (injection technique, fragile vessels, elastic compression after sclerotherapy etc.). Even though a greatest number of these pigmentations whithin two years fades gradually[2], they are responsible of a patients refusal of sclerotherapy. Because the bioptic evidence is a pigmentation due to hemosiderinic pigment, some Authors were stimulated to search new therapies[3-4-5-6].

METHODS.

Using some acids solvents of hemosiderin like an amphiphila solution of trichloroacetic acid at 25% (TCA)[7-8], or an association of trichloroacetic acid, retinoic acid (RA) at 0,05% with hidroquinone (HQ) at 2% (Kligman triplet modified) were treated 50 female patients (mean age 39,7 years), who showed on the lower extremities some residual pigmentations after sclerotherapy and persisting from 6 months to 5 years. The sclerosing agents were sodium salicilate (SAL), chromated glycerin (GLY) and polidocanol (POL).
The way of the treatment was a local application once a week: on the short lasting and less evident pigmentations we applied the only amphiphile trichloroacetic solution (amphiphile is an hydrophilic solution changed chemically in a lipophilic one). For the treatment of the long-lasting pigmentations or of these of larger size was used the Kligman triplet

modified (trichloroacetic acid, retinoic acid and idroquinone in liquid solution)[9-10].

These solutions were painted with a little pencil or with cotton wool and 15 minutes after we rinse with running water (only when occurred a buring discomfort or a whitefrost appearance the rinsing was done before 15 minutes). The application were repeated at intervals of 7-10 days during a time between 4 weeks and 3 months.

40 patients (80%) were treated with the amphiphile TCA and 10 (20%) with the triple association.

Every 2 applications for each patient was take a photographic documentation and a clinical evaluation was performed for dermatologic local adverse reactions as an inflammation or an excessive cutaneous exfoliation.

RESULTS.

38 patients (76%) revealed a total fading of the pigmentation (30 patients of the group treated with TCA only, and 8 patients treated with triple association).

12 patients (24%) showed a significative reduction of their pigmentation (10 patients treated with TCA and 2 patients after treatment with triple association).

Complications: only two patients because of a local inflammation and excessive exfoliation were treated every 15 days for a time under 5 minutes.

CONCLUSIONS.

Indications, technique of sclerotherapy and sclerosing agents seem to be determinant causes in the development of hyperpigmentation. The incidence of this adverse sequelae is 8% in our series of 600 patients treated with sclerotherapy (POL, SAL, GLY) from 1989 to 1993; we found a 2% incidence of untreated pigmentation persisting after 1 year. The results obtained with topic application of TCA and TCA+RA+HQ are very satisfactory for the majority of patients.

BIBLIOGRAPHY.

1)Bergan J., Goldman M.P. Varicose veins and telangiectasias. Ed. QMP, St. Louis - Missouri 1993; 259-60.

2)Goldman M.P. Sclerotherapy. Ed. Mosby Year Book, St. Louis - Missouri 1991; 219-24.

3)Goldman M.P., Kaplan R.P. and Duffy D.M. Postsclerotherapy hyperpigmentation: a histologic evaluation. J.Dermatol.Surg.Oncol. 1987; 13:547.

4)Bernier E.C. and Escher F. Treatment of postsclerotherapy hiperpigmentation with trichloroacetic acid, a mild and effective procedure. Acts of the 2nd Annual International Congress of North American Society of Phlebology, New Orleans - february 1989.

5)Richter G.W. et al. Commentary on hemosiderin. Blood 1965; 25:370-4.

6)Ackerman Z. et al. Overload of iron in the skin of patient with varicose ulcer. Arch.Dermatol. 1988; 124:1376-8.

7)Izzo M. et al. Postsclerotherapy hyperpigmentation: the state of the art. Acts of the 2nd International Congress of Aesthetic Phlebology, Ed. Class International, Genoa (Italy) 1993; 47.

8)Izzo M. et al. L'iperpigmentazione postscleroterapica. IX° Congresso Internazionale di Medicina Estetica, Roma - giugno 1993; 121.

9)Izzo M. Iperpigmentazione postscleroterapica. IV° Congresso Internazionale della Società Italiana di Flebolinfologia, Siena - settembre 1993; 46.

10)Izzo M. et al. Iperpigmentazione postscleroterapica: markers di rischio predittivi. XV° Congresso Nazionale di Medicina Estetica, Roma 1994; 87.

Phlebology '95, D. Negus et al. (eds.). Phlebology (1995) Suppl. 1: 553-555

OP/4.2

Safety Angle in Sclerotherapy

C.F. Sánchez[1,2], E. Altmann-Canestri[2,3], U.P. Tropper[1,2], N. Rosli[2], A.L. Pedrazzoli[2] and J.O. Pacheco[4]

[1] *Phlebological Clinic,* [2] *Jonh F. Kennedy University,* [3] *Buenos Aires University,* [4] *"Prof. Dr. Luis Güemes" Hospital*

INTRODUCTION

This is a simple, practical method, easily reproducible, in order to trace the substances injected during sclerotherapy. With this objective, the sclerosing substance is replaced by a radiopaque substance, tracing its behavior by means of phlebography and kinephlebography, for each of the different positions the patient may adopt during sclerotherapy.

METHODS

Our work team has performed phlebographic and kinephlebographic research studies on the behaviour of 10 c.c. of radiopaque substance injected for ten minutes into the superficial venous system of a group of 53 varicose patients, obtaining the following results [1,2,3].

Fig. 1

Fig. 1. Patient standing: the radiopaque substance flows from the superficial venous system to the deep venous system through the perforans very **rapidly**.

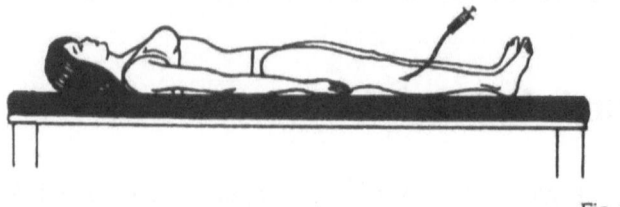

Fig. 2

Fig. 2. Patient lying: the radiopaque substance flows from the superficial venous system to the deep venous system through the perforans **slowly**.

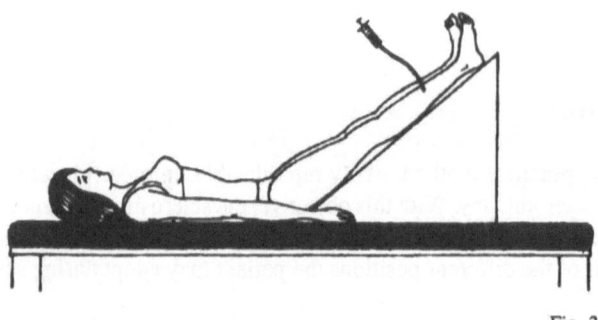

Fig. 3

Fig. 3. Patient lying with the leg raised at a 45° angle with the plane of the stretcher: the radiopaque substance flows **slowly** through the superficial venous system and only passes into the deep venous system at the crook of the internal saphenous vein.

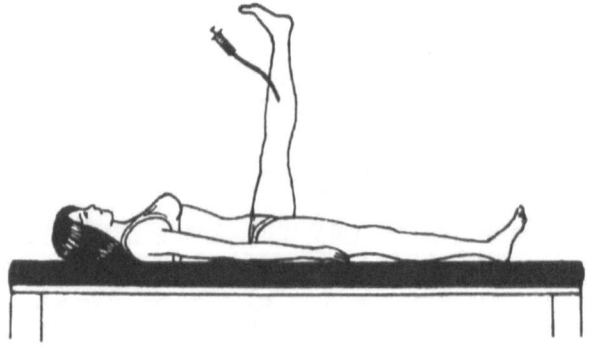

Fig. 4

Fig. 4. Patient lying with the leg raised at a 90° angle with the plane of the stretcher: the radiopaque substance flows **rapidly** through the superficial venous system and only passes into the deep venous system at the crook of the internal saphenous vein.

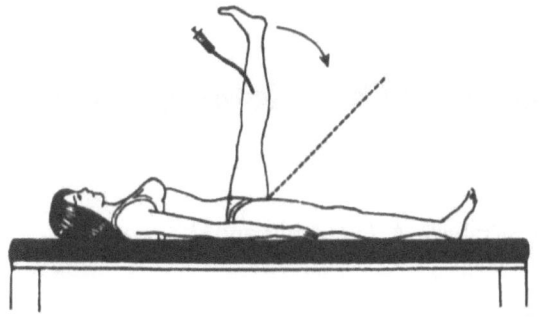

Fig. 5

Fig. 5. Patient lying with the leg raised at a 90° angle with the plane of the stretcher: the radiopaque substance flows **rapidly** through the superficial venous system and only passes into the deep venous system at the crook of the internal saphenous vein. If the leg is lowered slowly, the flow slows down and only below 45° it passes into the deep venous system also through other perforans.

CONCLUSIONS

Comparing the radiopaque substance with the sclerosant, we may state that, in order not to damage the deep venous system when injecting large quantities, it is necessary to do this at the 45° safety angle, already described and demonstrated.

We do not use the right angle because in this position the blood flow is so rapid that it prevents the sclerosant from making contact with the venous endothelium long enough to cause its destruction.

In this position, which we call safety angle, the following is phlebographically and kinephlebographically verified:

1) While the vein is empty and collapsed, smaller quantities and concentrations are able to produce the same effect as when the patient is standing.

2) The superficial system pressure is close to zero and the deep system is 80 mmHg. This means that, to pass into the deep system, the sclerosant must first fill the vein and also keep a pressure supposed to be 80 mmHg.

3) If, for any reason, the sclerosant should pass to the perforans, the high speed of the blood flow at the deep system, produced by the elevation of the limb, would prevent its contact with the vascular wall.

REFERENCES

1. Sánchez C, Altmann-Canestri E, Tropper U. Escleroterapia. Técnica de la Esclerosis. Manual de Escleroterapia y Flebetomía Ambulatoria. Buenos Aires: Editorial Celcius, 1992: 167-171.
2. Sánchez C F, Pisoni M, Altmann-Canestri E, Leguizamón O, Cuccarese M, Ferreti J et al. Ángulo de seguridad en Escleroterapia. Revista Panamericana de Flebología y Linfología 1991; 2: 30-33.
3. Sánchez C F, Rosli N. Escleroterapia. In: Altmann-Canestri E, Sánchez C F, Tropper U et al. Tratado de Flebología y Linfología. Buenos Aires: Fundación Flebológica Argentina, 1995: 110-114.

Phlebology '95, D. Negus et al. (eds.). Phlebology (1995) Suppl. 1: 556-558

OP/11.5

Sclerotherapy with Chromic Glycerol in Chronic Venous Ulcer Treatment

O. Andoniades, R. Mirabella, A. Avramovic and M. Conzi

Department of Surgery, Section Phlebology and Limphology, Bernardino Rivadavia Hospital associated with the University of Buenos Aires, Argentina

INTRODUCTION

Chronic venous ulcer of the lower limbs is a pathology with a prevalence of approximately 2 % in our country [1]. This complicated condition is the cause of many problems in the workplace.

Most patients have been suffering from their condition for several months, even for years. Many of them have been operated on without success and many others are not able to have, or just do not want surgery.[1,3,4]

In a selected group of such patients we applied the unique treatment of sclerotherapy in disabled veins around, inside and under the ulceration, with an adequate compression.

With this method we achieved a reduced period with no additional bedrest necessary.[1,2]

METHODS

In a first stage we eliminated the ulcer infection by administration of appropriate antibiotics, according to cultures taken from the bottom of the ulcer. The patient was taught to carefully clean, wash and scrub the wound and leg with salt water and soap once a day (the ulcer for five minutes and the surrounding skin for ten minutes).[1,2,5]

The ulceration had to be covered with sterile vaseline and gauze and the limb had to be bandaged from the foot to the knee with an elastic bandage for twenty four hours a day. The patient was allowed to walk all day long.

Once a week we administered injections of chromic glycerol 1.11 % enriched with 15 % glucose in the ulcer and its surrondings, up to five mililiters per session.

PATIENTS

We chose thirty two patients aged between 32 and 74, 21 females, 11 males, with the certain diagnosis of venous ulcer produced by pure venous insufficiency with a diameter between 1 and 2 inches, who were divided in two groups of 16 each.

MEASUREMENTS

The healing time was measured in weeks, comparing the two study groups and evaluating sclerotherapy on trophic lesions.

HEALING RATES. RESULTS

Eleven patients treated with this method achieved complete healing in between 7 and 12 weeks.(FIG.1)
Four patients showed improvement but their lesions had completed healing until the 18th week.
We lost contact with the remaining patient.
The control group also had good results, but over a longer period: between 16 and 24 weeks.

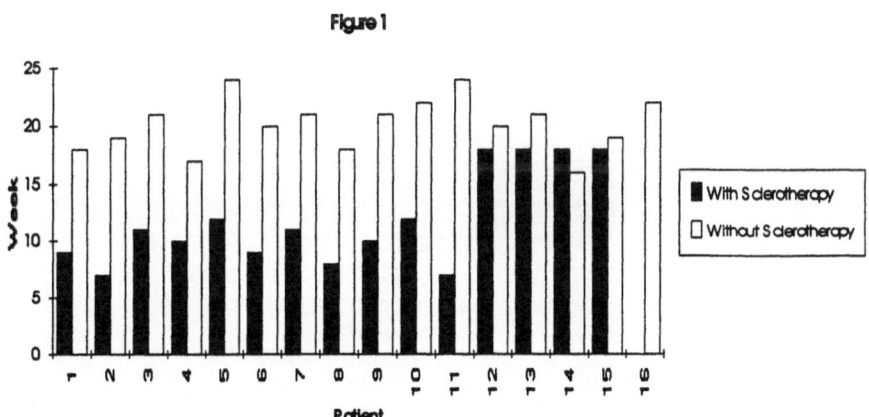

Figure 1

CONCLUSIONS

Sclerotherapy in areas with trophic damage and open ulcer is a useful and safe method for application in chosen cases. It certainly helps in reducing the healing period.

ADVERSE REACTIONS

We did not observe any case of nechrosis nor injuries in the area of treatment. Just a slight edema and pain, wich is a normal effect of sclerotherapy.
The absence of allergic reactions and / or secondary effects confirms the use of chromic glycerol as a drug of choice for the treatment of this pathology.

REFERENCES

1. Fillipin D., Efecto de la escleroterapia en la úlcera venosa.
   Prensa Médica Argentina. 1985; 92:28-30.
2. Ouvry P.A., Davy A., Guenneguez H.. La Scleroterapie dans l'ulcere d'origine veineuse. Phlebologie 1985; 38:37-42.
3. Parpex P., La Sclerose des varices dans l'ulcere de jambe.
   Phlebologie 1986; 39:335.
4. Sigg K., Varizen ulcus cruris und thrombose. Springer - Verlag, Berlin. 1968; 44 - 47.
5. Vigoni M., Phlebologie. Maloine Editor, Paris. Cap 3°. Les ulceres variqueux. 1983; 127.

Phlebology '95, D. Negus et al. (eds.). Phlebology (1995) Suppl. 1: 559-560

P135

Retrospective Clinical Study of the Sclerosis of Primary Varicose Veins in the Foot

Aurora Avramovic and M. Avramovic

Clinica Flebologico Privado, Riobamba 747, Buenos Aires (1025), Argentina

OBJECTIVE

To try to show the possibility of undergoing sclerosis of varicose veins of medium and great calibre in the foot, as a complement of a saphenectomy or a sclerosis of the long saphenous in a superficial venous insufficiency.

DESIGN

A retrospective study of 106 cases treated during 1992-1994 was carried out. The patients were of both sexes and between 35 and 61 years of age. The 106 patients presented superficial venous insufficiency of different degrees with varicose veins in the back part of the foot.

METHOD

After carrying out the treatment for the varicose veins in the thigh and the leg in a semi-sitting position, the sclerosis of the varicose veins in the foot is begun. This treatment is performed in a standing position, subsequent to an Echo-Doppler examination and to the elimination of any other pathology, by injecting 1 cc. of 3% Hydroxipolietoxidodecano with 2 cc. of 25% Hypertonic glucosed solution. The contents of the syringe should be distributed in various sections of the varicose vein and it is necessary to make sure that the injection is well inserted in the vein so as to assure a firm fibrosis and reduce any inflammatory reaction. Elasto-compression should be performed during seven days so as to prevent an edema.

MEASUREMENTS

In follow-up sessions at 15 days, 45 days, 3 months, 6 months and 12 months, possible recanalization is controlled and complementary sclerotherapy is carried out.

RESULTS

86 patients reacted positively to fibrosis and total sclerotherapy and did not present complications. 11 patients presented hyperreaction and they underwent extraction of the thrombus; 6 patients presented edema in the foot that retrograded in 30 days with elasto-compression and Benzopironas; and 3 patients presented scleroresistance.

CONCLUSION

The sclerosis of varicose veins in the foot resulted postively in 90% of the cases. For a lasting result, it is necessary for the medical doctor to receive training as well as to correctly manage both the doses and the concentrations.

REFERENCES

1. Stemmer R., Sclerose des Varices et Compression. Phlebologie 1991; 44:49-67.
2. Sigg K., Varizen-Ulcus Cruris und Thrombose. Springer-Verlag. Berlin 1968; 77-85.
3. Spano V., Microcirugia de las Varices del Pie. Flebologia. Argentina 1993; 6:265-276.
4. Tournay et Collaborateurs, La Sclerose des Varices. Expansion Scientifique Francaise. Paris 1985.
5. Wallois P., Les Varices du Pie. Phlebologie 1986;39:273-275.

Phlebology '95, D. Negus et al. (eds.). Phlebology (1995) Suppl. 1: 561-563

V/2.3

Segmental Sclerosing Instead of Stripping - An Alternative in Phlebosurgery (11,000 Interventions)

E. Ermisch and U. Käseberg

Phlebologische Chirurgie Dr Ermisch, Lessingstr. 7, 02763 Zittau, Germany

Superficial venous incompetence of the lower limbs is characterised by raised venous pressure and reverse venous flow within venous trunks and their collaterals. Long standing venous disease may result in trophic skin changes and possibly venous ulceration.

The aim of surgical treatment is to prevent the progress of venous disease. Since 1965 we have been developing surgical techniques suitable for use in treating outpatients. We have developed a mini-surgical procedure under local anaesthesia and conducted 11,000 interventions of this kind which we report below.

Usually hospitals use sapheno-femoral ligation combined with stripping methods under regional or general anaesthesia [1]. We consider this method unsuitable for outpatient treatment as walking ability is not regained for several hours after the operation. Disadvantages of this operative approach include damage to tissues and including neurological injury with permanent post operative consequences.

In the years 1911 to 1929 Unger [2] published a number of articles describing intraoperative sclerotherapy using uretheric catheters. We have revised, updated and technically improved this technique. [3]

OPERATIVE TECHNIQUE
1. Clinical and phlebological tests are performed to determine the extent and function of the pathology in the deep and superficial veins. [4]

2. After ligating the sapheno-femoral or sapheno-popliteal junction, the incompetent residual trunk is treated by sclerotherapy. This is achieved by an air-bloc-catheter inserted into the vein. In this way, the whole of the incompetent system can be destroyed by 3-4 small segmental incisions. Intact valves in competent tributaries block further spread of the sclerosing agent. The thrombotic process is therefore limited to the diseased veins with preservation of normal veins.

3. Sclerosing agent can not achieve obliteration of the communicating veins so surgical ligation of these vessels is carried out. This segmental sclerosing treatment is painless and requires only 0.2 to 0.5mls of 3% aethoxysclerol in the form of a foam. The maximum quantity used for the entire operative procedure in one extremity is just 2-2.5ml.

   After the wound has been closed and an appropriate bandage applied, the patient leaves the operating table, dressed himself and is taken home by car. The patient receives written instructions to avoid sitting and standing, to keep the bandage tight but comfortable and to keep walking as much as possible during the first seven post operative days.

We now review 30 years of experience with 11,000 operations of this type.

The long-term results are the criterion of efficacy. A comparison has been undertaken between 1,290 limbs after stripping and a group of 2,700 limbs after sapheno-femoral junction ligation combined with sclerotherapy. We agree that recurrence of varices is unavoidable in any series and in our groups was 5.3% for the stripping group and 6.6% for the sclerosing group.

The evaluation of the sclerosing group shows the following results:

2.8% Lapses of the short saphenous vein as a result of incomplete junction ligation

2.4% Recurrence due to residual perforating veins.

5.3% Recurrence in lateral tributaries of the saphenous system

0.4% Recurrence of ulceration due to incomplete removal of the haemodynamic problem

The examination of recurrences confirms that a large number of new secondary symptoms may arise from incomplete removal of incompetent veins. The outcome may be improved if residual varices identified during follow up are treated by subsequent sclerotherapy.

Economic considerations are always important. The costs for outpatient treatment are 5.6 to 1.3 times lower that for operative interventions in hospital or clinics and show at least the same operative short and long term results.

REFERENCES

1. Babcock W R. (1907): Extraction for the removal of varicose veins. J.amer.med.Assoc.1907. Bericht Chirurg 1910. 1339. New York med.J. 1907 86, 253

2. Unger, F. (1911): Einiges über Krampfadern. Bln. klin. Wschr.
   Unger, F. (1927): Zur Technik der Varizenbehandlung, Bln. Ges. Chir.
   (1927): Ref. Zbl. Chir. Nr. 51, 3273, 3287
   (1929): Bln. Ges. Chir
   (1930): Zbl. Chir. 3, 159

3. Ermisch, H. (1984) Intraoperative Sklerotherapie statt Stripping, Bereicht über 2700 Eingriffe. Z. ärztl. Fortbild. 78, 327-328.

4. Hach, Wunderle: Der Rezirkulationskreis der primären Varikose. Springer-Verlag 1994, 27-31.

Phlebology '95, D. Negus et al. (eds.). Phlebology (1995) Suppl. 1: 564-566

P136

Toward a Phlebology without Pigmentations

C.F. Sánchez[1,2], U.P. Tropper[1,2], R. Barceló[1,2], D. Raifman[2,3], F.E. Kira[2,3] and C.D. Lesnik[2,3]

[1] Phlebological Clinic, [2] Jonh F. Kennedy University, [3] Medical Center

INTRODUCTION

The treatment of varicose pathologies with sclerotherapy and ambulatory phlebectomy constitutes a remarkable progress in phlebological medicine, these therapies being nowadays preferred in view of their advantages. But these valuable allies, which bring so many successful solutions for varicose diseases, occasionally produce pigmentary sequela. It is therefore imperative to find a solution for this unesthetic complication, either by procedures which may prevent residual pigmentation or by methods to eliminate it once it has appeared.

Stimulated by this concern, in order to elucidate its mechanism of action, our team carried out biopsies of post-sclerotherapy stains to find out their etiology and location in order to carry out prevention and treatment.

To this effect, we performed biopsies of 3, 6 and 9 month old stains in 27 patients, subjecting them to coloring with hematoxylin/eosin, Prussian blue and Fontana-Mason.

No significant modifications were observed with hematoxylin/eosin, except a slight infiltration of lymphocytes in the dermis, which increased with the age of the stain.

The samples subjected to coloration with Prussian blue showed the presence of hemosiderin in the dermis, its concentration being higher as the stain grew older.

The Fontana-Mason coloration, specific for determination of melanin, showed that its deposits in melanocytes increased in proportion to the amount of hemosiderin in the dermis and its time of evolution.

In short, the conclusions of this study show that the hemosiderin deposited in the dermis by extravasation from the injected veins, acts by exciting the melanocytes in the basal layer of the skin, which respond by increasing the amounts of melanin. This fact produces the hyperpigmentations occasionally observed after sclerotherapy. Their close relation to their aggravation with time, teaches us, in the first place, to take the necessary preventive measures to avoid them, and if they do appear after all, to establish precocious treatment, since the time elapsed will not lessen them, but on the contrary, it will worsen them even more.

PREVENTION

The first and most effective step is prevention. In predisposed patients, if a greater irritation is followed by a greater pigmentation, we should avoid irritation when performing sclerotherapy as well as ambulatory phlebectomy.

Sclerotherapy

NO to high concentrations in small veins.
NO to exposure to sunlight after sclerotherapy.
YES to slow injections so as not to produce trauma in the vein.
YES to the foam method to avoid injecting outside the vein.
YES to precocious thrombectomy in superficial veins.
YES to adequate compression to avoid hematoma.

Ambulatory phlebectomy

NO to inopportune maneuvers. When the extraction is difficult, C. Sánchez's method of sclerosis by foreign body should be applied.
YES to avoiding formation of hematomas:
  • using C. Sánchez's safety angle during surgery
  • applying an adequate compressive bandage
  • in case a hematoma appears in spite of these measures, creams with heparins or heparinoid substances should be indicated.

PRECOCIOUS TREATMENT

As soon as a pigmentary stain is noticed, treatment must start, especially in patiens with previous pigmentary antecedents. For this purpose we indicate application of creams with demelanizing agents. There is a variety of substances which interfere with the syntesis of melanin. Of all these, the most widely used is Hydroquinone 2% or 4%, which sometimes may have a slight irritating effect. Creams may be formulated including acid Vitamin A and Hydrocortisone which, besides its metabolic effect, would diminish any erythema or burn. It is not convenient to prepare a large quantity of cream because the Hydroquinone decomposes (color alterations are observed) and it turns ineffective.

SPECIFIC TREATMENT

Once the stain is definitely established, we proceed to its depigmentation by desquamation of the melanin-loaded epithelium, which we carry out by a physical method —cryotherapy— or chemical method —trichloroacetic acid— or a biological method [1, 2, 3].

Cryotherapy

Mechanism of Action: It produces desquamation of superficial layers of the skin, preserving the basal layer untouched and inhibiting melanocytes by low temperature.
Materials: We use pieces of dry ice, easily obtainable in ice-cream shops.

566

Technique: A small piece of dry ice is held with tweezers and applied to the surface under treatment for 4 seconds. A momentary bleaching will be produced in the area, which will be insensitive, hardened and stiff as if frozen. After 5 or 6 hours, a blister may appear with a clear and then turbid content. When the blister breaks, it will be replaced by a scaly scab which will spontaneously fall after two weeks, leaving instead a rosy, fine epithelium under regeneration. The scab must not be pulled out, either by the patient or the doctor; it must be left to fall by itself [1, 2, 3].

Trichloroacetic acid

Mechanism of action: it coagulates proteins in the skin surface, its destructive power being proportional to its concentration and time of exposure. In a 20/30% concentration it exfoliates; in a 50/75% concentration it is caustic.
Materials: It is available as hygroscopic crystals. Simple contact with a few drops of water is enough to dissolve it. Pharmacies will prepare it in different concentrations.
Technique: the skin surface is cleaned and grease is removed by alcohol. The area to be depigmented is marked and then a swab is introduced in the trichloroacetic acid solution and applied to the marked area. The patient will experience a tolerable pain, the skin will turn erythematous and then white due to protein coagulation of corneous albumins. Then this action is neutralized applying a swab moistened with an alkaline solution (bicarbonate, milk of magnesia, etc.). After 3 or 4 days, the area will show a brownish cover similar to onion skin, formed by the exfoliated corneous layer which falls spontaneously and must not be pulled out. This application may be repeated if necessary. The patient must avoid exposure to the sunlight until all irritation disappears [1, 2, 3].

Biological method - Acid vitamin A or retinoic acid

Mechanism of Action: its action on the epidermis is to diminish cellular cohesion on one hand, facilitating elimination, and on the other, to increase the activity of the germinative layer, accelerating the renewal of epidermic cells.
Materials: Acid Vitamin A is commercially available as cream in three concentrations: 0.05%, 0.1% and 0.2%. It may also be prepared as cream, gel or lotion, specially if smaller concentrations are desired.
Technique: apply on the area, avoiding folding places, in very small quantities and massaging thoroughly. As it may have a slight irritative effect (erythema, slight peeling, burning), it is convenient to start with lower concentrations, applying it every other day to allow a gradual acceptance of the skin, since these slight disturbances disappear with use.

REFERENCES

1. Sánchez C, Altmann-Canestri E, Tropper U. Escleroterapia. Pigmentación residual post-escleroterápica. Manual de Escleroterapia y Flebetomía Ambulatoria. Buenos Aires: Editorial Celcius, 1992: 270-279.
2. Sánchez C F, Rosli N. Escleroterapia. In: Altmann-Canestri E, Sánchez C F, Tropper U et al. Tratado de Flebología y Linfología. Buenos Aires: Fundación Flebológica Argentina, 1995: 161-166.
3. Viglioglia P A and Rubin J. Discromías. Cosmiatría. Buenos Aires: Ediciones de Cosmiatría, 1979: 160.

Phlebology '95, D. Negus et al. (eds.). Phlebology (1995) Suppl. 1: 567-570

OP/5.1

The Complex Treatment of Lower Leg Varicosities in Outpatient Clinic

K. Twardowska-Saucha, M. Pardela, D. Czaczka and W. Saucha

Varicose Vein Clinic "Medservice", Zabrze, Poland

INTRODUCTION

The lower extremity venous system diseases present itself as a serious problem of Polish patients; however often underestimated by physicians. It is the matter of common knowledge that more than 20% of population suffer, from varicose veins and its complications [1]. Until nineties the treatment of the disease was carried on by several different specialists often fragmentarily and inefficiently. It frequently caused prolonged disability for work and was very expensive. The general tendency of Polish surgeons was the treatment by means of partial phlebotomy of visible varices combined with more or less accurate LSV stripping. It was caused by the fact that varicose vein were considered as not a serious problem and treated by not very experienced surgeons, in many cases without the proper diagnose. Such a treatment produced extremely high recurrence ratio ranging to 60% of cases consequently leading to many severe complications.

A lot of patients with complications such as phlebitis, oedemas, hyperpigmentation or leg ulcers were treated by physicians who focused only on the symptomatic treatment of the disease without the awareness of its complex origin. Thus, the treatment lasted very long and was highly ineffective.

Sclerotherapy was applied in Poland earlier, but due to the lack of necessary diagnostic equipment, wrong diagnostic procedures and the lack of the obliterating drugs of new generation the required therapeutic effect was not achieved. In addition, the knowledge of injecting procedures was inadequate. Therefore, frequent complications occurred or the treatment was even highly ineffective. Last but not least the important cause of bad condition of Polish "varicose patients" was insufficient knowledge as far as prophylaxis is concerned, both among physicians and their patients.

In 1991 "Medservice" developed the idea of founding the centre of complex treatment of lower leg varicosities in outpatient procedure.

METHODS AND RESULTS

Before opening the Clinic, the following procedures were tested: diagnostics, compression sclerotherapy and leg ulcer healing in ambulatory, and surgical and anaesthetic methods at the hospital ward. During two-year studies the complex scheme of treatment was performed. The procedure focused on the adequate diagnosing by means of non invasive methods including CW-Doppler, D-PPG and basic physical tests. It was also of great importance to choose the most effective and at the same time possibly low-traumatizing methods of treatment. The staff of the clinic went through many courses in recognised European and Canadian centres. Our aim was to deal with the majority of patients in day care mode [2,3].

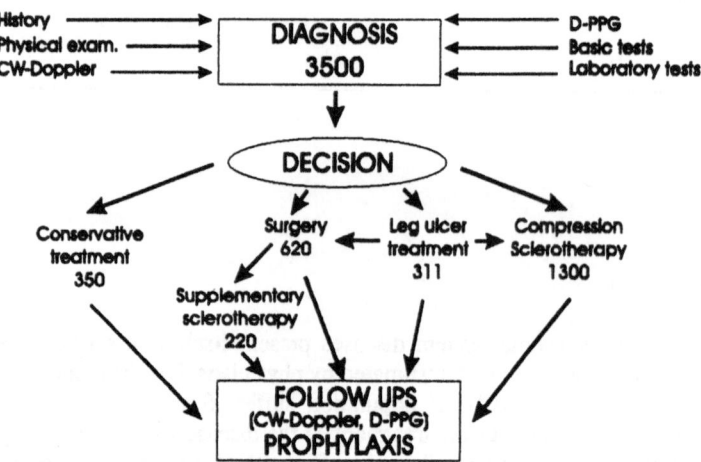

Figure 1. The complex treatment organization scheme and number of treated patients.

Until 1995 year 3500 patients were admitted and treated in the sections presented in Fig.1. In many cases patient underwent more than one type of treatment to achieve satisfactory results. 814 patients did not required any treatment or did not come back after the first visit. 105 patients were referred to the hospital word (extreme obesity, severe pulmonary or cardiovascular insufficiency, positive HBs antigenemia).

Conservative treatment was performed mainly in patients suffering from thrombophlebitis or elderly individuals who did not accept surgery or sclerotherapy and in few patients with severe diseases, making more invasive intervention impossible. Medical compression stockings, non-steroid anti-inflammatory drugs, vasoactive drugs and fractionated heparin were of frequent use in these cases.

Leg ulcer healing usually demanded 4-10 sessions in weekly intervals. Ulcers were cleaned-up, hydrocolloids or Unna's rigid paste bandage were often applied as well as defibrinating ointments or granulation stimulants. Good reaction for oral pentoxyfilline and diosmine was often observed. All the patients except those with severe arterial insufficiency were treated by compression ranging from 2nd to 3rd degree. The complete ulcer healing was achieved in 72% in a period from 2-12 weeks; cleaning -up the ulcer, marked progress of granulation, significant decrease of pain and oedema accompanied by distinct decrease of ulcer area (more than 50%) were observed in 23% of patients. Only 5% presented weak progress or no reaction at all.

In many cases it was possible to complete healing process successfully by means of surgical treatment or incompetent perforators sclerotherapy.

Generally 1520 patients were treated by means of sclerotherapy. There were two main groups of patients.

1. - cases in which compression sclerotherapy was the treatment of choice due to the presence of big varicose tributaries, insufficient perforating veins or reticular veins and teleangiectasia with competent LSV and SSV ostia,

2. - cases in which supplementary compression sclerotherapy was applied after surgical procedure performed in the cases of LSV or SSV insufficiency (this usually started from 2 to 3 months after surgical treatment).

Compression sclerotherapy was performed using hypertonic dextrose and saline solutions, detergents or iodide sclerosants according to individual indications. The modified air-block technique was of use in many cases in order to achieve a firm sclerosis with lower solution concentrations. Compression ranging from 20 to 35 mmHg., reinforced by local sponge pads was applied immediately after injections. From 2 to 12 sessions (mean 4) in 1-2 week distance were necessary to complete the treatment. Good and very good functional and cosmetic results were observed in more than 85% patients; however in some patients it took even 9 months to complete the resorption process.

Patients with incompetent LSV and/or SSV were also treated in outpatient procedure by means of limited or even extensive surgery. All of them were thoroughly consulted by our internist prior to surgery. New operation instruments (modified hooks and RYVA Set) made operations both radical and aesthetic. The following techniques were used: limited and complete stripping of LSV, ligation and/or stripping of SSV, selective dissections of insufficient perforators. Special indent knives RYVA-Varix-Set made the excision of large tributaries and even spacious varicose plexes possible through a 5mm cut. Due to modified phlebectomy hooks smaller traumatization of tissues was achieved. Thanks to these hooks the distal fragment of LSV was easily accessible during stripping.

All the patients were premedicated by means of intravenous Diazepam. Pentazocine (Fortral) was applied to some of them before or during the operation.

The anaesthesia was obtained by femoral block and infiltration of the local anaesthetics. The operations usually took from 25 to 50 minutes. There were no intraoperative complications. Patients were applied compression of 2nd degree immediately after operation. The procedure was well tolerated by the patients and also well accepted due to its gentleness, reduction of the incision number (only 2 to 5), their small diameter and short recovering period. No serious complaints were presented in the postoperative period. All the patients were discharged from the Clinic in the period from 2 to 4 hours after surgery. Quick mobilising of the patients resulted in avoiding thrombotic complications without antithrombotic agents. The follow-ups performed on 7th and 28th day after operation revealed regular wound healing. Patients who had completed 6-months observation period revealed gradual regression of venous insufficiency symptoms confirmed by the prolongation of To values (refilling time) and increase of Vo (venous pump power) in D-PPG.

In order to improve cosmetic effects many patients underwent supplementary sclerotherapy two or three months after they had been operated on.

In less than 10% of cases not very serious complications were observed i.e.:

- in ulcer care - elongated treatment, allergies, pain during first two weeks of treatment, or troublesome infections.

570

- in compression sclerotherapy - incomplete effect or no reaction at all, matting, hyperpigmentation or small allergic reactions,
- in surgery - prolonged wound healing period, extensive or long-lasting bruises, sometimes not serious infections of the wound area (especially in groin and ankle region).
Serious complications such as necrosis of small area of the skin, superficial thrombophlebitis or extensive infected haematomas were really rare (less than 2%).
The recurrence ratio did not exceed 6% in surgery, 19% in sclerotherapy and 37% in chronic leg ulcer managing in a period of two years. Part of them probably could be avoided if the patient continued the proper prophylaxis.

CONCLUSIONS

The analysis shows that complex treatment including conservative, obliterating and surgical procedure allows us to receive satisfactory functional and cosmetic results in more than 90% of the patients in outpatient varicose vein management.

Non-invasive diagnostic methods give the physician the advantage of accurate and handy information on each stage of treatment and recovery period.

This complementary treatment is safe, convenient and very well accepted by the patients.

It is also of great importance, that outpatient procedure is much cheaper than traditional, hospital treatment.

Thus it could be considered to be the method of choice in varicose veins treatment.

REFERENCES

1. Pudlik Z., Politowski M.: The frequency of the occurrence of varicose vein and their complications in Poland. Zdrowie Publiczne 1971, 3, 189-199
2. Schubert H.J.: Arguing for day care varicose vein surgery. Phlebology 1992, John Libbey Eurotext, vol.1, pp.271
3. Twardowska-Saucha K., Czaczka D., Fiutek Z., Zbroński R.: The outpatient procedure in venous disorders of the lower extremity treatment in Poland. Proceedings of the European Congress of the International Union of Phlebology, Budapest 6-10 Sept. 1993, pp.101-102

Phlebology '95, D. Negus et al. (eds.). Phlebology (1995) Suppl. 1:571-573

OP/4.3

Sclerotherapy of the Long Saphenous Vein - A Prospective Duplex Controlled Comparative Study

M. Zummo[1], M. Forrestal[2]

[1] Private practice, Montreal, Canada, [2] Private practice, Chicago, USA

INTRODUCTION

In 1976 and later, Cloutier[1] and then Cloutier & al.[2,3] established that the long saphenous vein (LSV) could be efficiently sclerosed by injecting an iodine - sodium iodide solution at distance from the sapheno-femoral junction (SFJ) at two different points, described as «summation effect». During the injection, the vein is sequestrated at its proximal end by applying external pressure with the index. This digital pressure applied on the SFJ slows the emptying of the sclerosing solution in the deep venous system thus allowing a longer contact period of the solution with the LSV endothelium.

Later, one of us[4,5] showed that there is a direct relationship between the rate of success of the method and the importance of the reflux. Using a Doppler probe, one can quickly establish the degree of reflux in the LSV by quantifying the height of the reflux. The more distal the reflux can be heard in the LSV, the lesser are the chances of successful obliteration of that same vein.

This study was designed as a prospective longitudinal clinical study and had two purposes:
- To assess the efficiency of a technique derived from Cloutier's method of sclerosing the long saphenous vein using ultrasound guidance;
- To establish if compression of the vein after sclerotherapy treatment makes a difference in the outcome.

Patients consulting in our private clinics with primary varicose veins due to LSV incompetence proven by Duplex scanning and continuous wave (CW) Doppler examination were included in the study. Post-surgical recurrences and patients with a history of DVT were excluded.

Each patient was examined, insonated and duplex scanned in the upright position. The extent of the reflux was assessed by CW Doppler ultrasound. The veins were scanned with one of the following duplex scanner: Ultramark-4 from ATL with a 5-7-10 MHz annular probe (Mtl) and either an Acuson X-P10 color flow duplex with a 7 MHz probe or a Diasonic DRF-400 with a 10MHz probe (Chi). Deep and superficial veins were tested for patency, compressibility and competence. The veins examined were the LSV at the junction and at distance from the junction, the common femoral vein, the superficial and deep femoral veins.

When the LSV were proven incompetent, they were injected using a technique similar to that of Cloutier. The patient was injected, under ultrasound guidance, either in the sitting (Mtl) or recumbent (Chi) position with the inferior limb externally rotated. The patients were injected with either an iodine - sodium iodide solution (Sclerodine®) in progressive concentration (3%, 6%, 6%) (Mtl) or a 3% sodium tetradecyl sulfate solution (Sotradecol®) (Chi). Only the group of patients from Chicago was systematically fitted with graded compression stockings placed before getting up from the examination table (30-40 mmHg). The stockings were worn daily for at least one week.

The diameters of the common femoral vein, SFJ and of the LSV at 30 MM from the junction were measured before the injections at the initial visit and at the follow-up visits. After the injections, the aspect of the lumen of the LSV was also noted and classified as follow:

- Excellent sclerosis when the vein was un-compressible, with no detectable flow in the lumen and the lumen took a flaky or chalky aspect;
- Total occlusion when the lumen had completely disappeared;
- Sub-total occlusion when the lumen was still present near the SFJ with complete disappearance of the lumen at distance from the SFJ.
- Some patients had only a C.W. Doppler control at the follow-up visit and were recorded as CW Doppler neg.;

RESULTS

The two populations are comparable in terms of age: Mtl 46.1 ± 11.4, Chi 46.3 ± 10.3 with an almost identical distribution. Most subjects are female: Mtl 100%, Chi 86.3%. The population from Mtl is composed of 91 subjects and 166 limbs and that of Chi of 51 subjects and 74 limbs.

The distribution of right versus left LSV is almost inverted when comparing the groups from Mtl (R=55.5%; L=44.5%) to Chi (R=47.3%; L=52.7%). The importance of the reflux was assessed using *Hach's classification*:

Grade 1: Reflux heard at mid thigh	Grade 3: Reflux heard at mid-leg
Grade 2: Reflux heard at knee level	Grade 4: Reflux heard at ankle

No subject had a reflux of grade 1. As expected and from our experience, most had a reflux of grade 3 (Fig.1).

In both groups, the diameter of the LSV at 3 cm from the SFJ is diminished by a little more than 50% (p < 0.01) of its initial size (Table 1) when sclerosis succeeded. At the SFJ, that reduction is approximately of 30%. These values could be excellent indicators of the success or failure of this procedure. This remains to be proven as data are still coming in. At three months and in 50% of cases, the lumen of the sclerosed veins will be imaged as flaky or chalky. The LSV are un-compressible and no flow can be detected into the lumen. Only 5.6% of injected veins will be completely occluded with no visible lumen and 13.2% will be completely occluded, except for the junction itself which remains open for a few millimeters, usually less than 3 to 5 mm. It this later case, the diameter of the SFJ is smaller than prior to the treatment.

Distribution
in regard to reflux

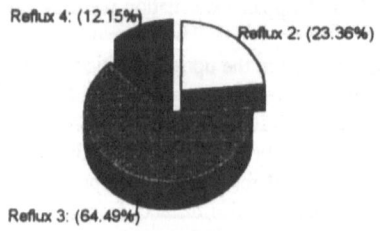

Reflux 4: (12.15%)

Reflux 2: (23.36%)

Reflux 3: (64.49%)

Figure 1

Comparison Between Diameters		SFJ (in MM)	Reduction	At 3 cm (in MM)	Reduction
Before treatment	Mtl	9.1 ± 2.6		6.4 ± 1.8	
	Ch	8.7 ± 2.1		6.7 ± 2.3	
After treatment	Mtl	6.4 ± 2.7	29.7%	2.9 ± 1.5	54.7%
	Ch	6.1 ± 2.1	33.0%	2.7 ± 1.3	59.7%

Table 1

Data are available at three months. The results from the two groups are almost identical. The success rate is 91.5% in Mtl and 90.8% in Ch. In both groups, some LSV showed reflux at the initial evaluation, remained opened after the treatment but reflux had disappeared. This has also been described by other authors[6,7,8]. This phenomenon was present in 2.8% of patients in Mtl and 6.2% in Chi. In both groups, the patients were asymptomatic after the treatment.

CONCLUSION

Our results compare to the results from other authors. There does not appear to be any difference in response after 3 months between the group that was fitted with elastic stockings and the other group that was not. Compilation of results at 12 months is still underway but more than 83% (N=12) of patients have responded favorably. No accident or complication were reported. The technique is well tolerated and appreciated by the patients as an alternative to the traditional stripping. There is no convalescence after the injection treatment and the patient can go back to work immediately in most cases.

Sclerotherapy of the SFJ under ultrasound guidance is a safe and a highly efficient method and is an interesting alternative to surgery.

REFERENCES

[1] Cloutier G. Sclérose des crosses des saphènes internes et externes: Nouvelle approche. Phlébologie 1976; 3: 227-232.

[2] Cloutier G, Sansoucy H. La sclérose des crosses saphènes internes et externes avec compression: Résultats. Phlébologie 1980; 4: 731-735.

[3] Cloutier G, Zummo M. La sclérose des crosses avec compression: Résultats à long terme. Phlébologie 1986; 1: 145-148.

[4] Zummo M. La sclérose des varices selon Cloutier. Phlébologie 1991; 1: 37-43.

[5] Zummo M. Insuffisance ostiale de la veine saphène interne: Critères pour le choix du traitement. In: Davy A, Stemmer R, editors. 10ième Congrès Mondial U.I.P.. Strasbourg: John Libbey Eurotext, 1989: 791-793.

[6] Vin F. Contrôle d'efficacité des traitements sclérosants en phlébologie. Phlébologie 1990; 4:673-680.

[7] Shadeck M. Sclerotherapy of the Great Saphenous Veins: A 48- months follow-up. J. Dermatol. Surg. Oncol. 1990; 16(1): 87 (Abstract).

[8] Kuefner G. Normalization of Long Saphenous Flow after Ultrasound-guided Injections: An Acceptable Failure. Dermatol. Surg. 1995; 21(1):91 (Abstract).

Phlebology '95, D. Negus et al. (eds.). Phlebology (1995) Suppl. 1: 574-576

OP/11.1

Duplex Scanning in the Mechanism of the Sclerotherapy: Importance of the Spasm

M. Schadeck and F.A. Allaert

Ecole Européenne de Phlébologie, Hôpital Notre-Dame de Bon-Secours, 66 rue des Plantes, 75014 Paris, France

INTRODUCTION

The arrival of ultrasonic techniques in phlebology has enabled the evolution of sclerotherapy of large veins to be observed, over periods from a few seconds to a few weeks or months (1).

After Doppler (2,3,4) studies, ultrasound imaging studies (5,6) were made to evaluate the results of the sclerotherapy of larger veins.

The use of probes of higher frequencies - between 10 and 13 MHz - has enabled this knowledge to be conisderably added to. It is now possible to understand better the successive stages of the sclerosing mechanism effect on larger veins. Just after the injection, we can observe a spasm which occurence depends of the technique used.

The goal of this study is to evaluate the role of this spasm occuring after the injection of a sclerosing agent in the formation of sclerosis.

Once again the purpose of the of the sclerosis is the transformation of the vein in complete fibrosis.

METHODS

For obvious practical reasons, this study is made at the level of the sapheno femoral junction because of its anatomical situation.

It is a comparative study of 3 groups undergoing three different therapeutic protocoles, delivered at random.

Inclusion criterias were patients with incompetence of the great saphenous vein (GSV), with a diameter lesser than 6 mm.

The Duplex scanning confirmes the reflux of the GSV and makes measurement of its diameter.

30 patients recieved a sclerotherapy, each group of 10 according to 3 types of technique.

In the 3 groups, the injection is made under echographic guidance in the terminal segment of the GSV according to the same technique. The two first groups are treated with Sotradecol 3% and the third one with Aetoxysklerol 4%. At the first session the dose is of 2cc in one point. At the second session, four weeks later, the dose is of 3cc in two different points. Only the first group undergoes an alternative compression (7) (AC) made by the echoprobe on the vein. This AC is a vertical up-and-down movement, compressing the vein against the underlying aponeurosis every second, and then relaxing this pressure. This device is applied during two to three minutes.

The diameter of the GSV is measured before the injection in standing then in supine position, then 3 minutes after the injection.

RESULTS

The first group (I) has a mean of age of 51,1 years, 7 females and 3 males.
    The mean of diameter of the vein is of 5,1 mm.
The second one (II) has a mean of age of 51,2 years, 9 females and 1 male.
    The mean of diameter of the vein is of 4,9 mm.
The third one (III) has a mean of age of 47,9 years, 7 females and 3 males.
    The mean of diameter is of 4,9 mm.
Currently, we observe after injection only little decrease of cross section area of the vein.

In this study, an important spasm occurs in 100% of cases in the groups I and III and the average reduction of the size after spasm was 76,5% compared to to cross section area measured on the patients in standing position.

In the second group, the reduction of the cross section area of the vein was of 17,5 %.

The GSV were closed in 90% of the cases in the groups I and III, and in 70% in the group II.

DISCUSSION

The spasm generaly occurs about 30 seconds after the end of the injection only on that part of the venous wall impregnated with the sclerosing product. The vein can then take on different forms, cup shave when the parietal area impregnated is very narrow or tubular when the product has spread homogeneously over a few centimetres.

The analysis of this study shows similar good results in the groups I and III where we observe an important spasm of more than 75%.

576

In the first group, we can explain the occurence of the spasm because the AC permits the product to spread little by little on the borders and finally reach the aponeural side of the vein. In the group III, only the properties of the product (Aetoxysklerol 4%) permit alone the transformation.
In the last group (II), the absence of occurence of the spasm do not permits the same quality of results than in the other groups (I and III).
This spasm because of the important reduction of the size of the vein, permits to increase the concentration of the product in the lumen, and, with an homogeneous repartition of this product on the wall.
The post-injection spasm is the first stage in the mechanism of the sclerotherapy, and precedes the different modifications of the vein at the level of the wall and of the lumen.
It is so a very good way to the transformation of the vein in fibrosis wich is the goal of the sclerotherapy.

CONCLUSION

Echotomographic study of the sclerosis mechanism has been rich in discoveries, particularly in the first stage of the sclerosis.
The spasm appears of a great importance in this mechanism. It favors the best results and above all the transformation of the vein in fibrosis.

REFERENCES

1. Schadeck M. Etude par Doppler et échotomographie de l'évolution des veines saphènes sous traitement sclérosant. Communication au 19ème Congrès de Pathologie Vasculaire, Paris, 14 March 1985.
2. Schultz-Ehrenburg U., Tourbier H. Dopplerkontrollierte Verodungs behandlung der Vena saphena magna. Phlebol u. Proktol 1984; 13: 117-22.
3. Tourbier H. and Schultz-Ehrenburg U. Sclerotherapy of stem varicosis. Doppler finding long-term results and new indication. Phlebology '85, D.Negus & Jantet G., 1986 John Libbey and Co Ltd, pp. 137-140 .
4. Zummo M. Insuffisance ostiale de la veine saphène interne: critères pour le choix du traitement. Phlébologie '89, A.Davy, R.Stemmer eds, 1989 John Libbey Eurotext Ltd , pp. 791-793.
5. Schadeck M. Doppler and echotomography in sclerosis of the saphenous vein. Phlebology, (1987) 2, 221-240.
6. Avramovic A., De Simone J., Avramovic M., Brizzio E. Evaluation de la sclérose de crosse et du tronc de la saphène interne grâce à l'échotomographie. Phlébologie '89, A.Davy, R.Stemmer eds, 1989 John Libbey Eurotext Ltd, pp. 794-796.
7. Schadeck M. Duplex and Phlebology. G.Gnocchi Ed., Napoli, Sept.1994, pp. 91-114.

Phlebology '95, D. Negus et al. (eds.). Phlebology (1995) Suppl. 1:577-580

OP/11.4

A Comparison Study of Sclerotherapy vs Phlebectomy

P. Ilieff

Phlebological Praxis, Steinbühlweg 15, 4123 Allschwil, Switzerland

I believe that vein stripping is not appropriate in every case. Modern methods of investigation allow less radical treatment. The combination of selective surgery with supplementary compression sclerotherapy may be as effective as extensive surgical treatment.

AIM

The aim of this study is to compare two methods of venous treatment, sclerotherapy and phlebectomy by studying duration of treatment, differences between age groups, patient compliance and side effects.

METHOD

ANOCOVA statistical analysis was used to compare two groups of patients: Sclerotherapy and Phlebectomy. A total of 1456 patients with chronic venous insufficiency and varicose veins have been treated in the study.

For the sclerotherapy group the number of patients was 690, mean age 55.9 years, the oldest 89, the youngest 22 years, total right legs of 481, total left legs 513. For the phlebectomy group the respective parameters were n = 766, mean age 50.5 years, oldest 84 years, youngest 22 years, R legs 355, L legs 411.

The entry criteria were as follows:
1. Varicose veins with complications
2. Chronic venous insufficiency, stage I - III
3. Sapheno-femoral junction ligated or competent.
4. Same compression technique for both groups
5. Same medical personnel treating both groups

Exclusion criteria:
1. Incompetence of the sapheno-femoral junction valve
2. Diseases that could affect patient mobility
3. Inflammatory skin processes
4. Periphery arterial diseases greater than Fontaine stage II
5. Pregnancy
6. Deep vein thrombosis in the preceding 12 months
7. Sensitisation to sclerosing agents or to local anaesthetic solutions
8. Patients with cardiac arrhythmies
9. Patients taking regular analeptics, beta-blockers or chronic alcohol intoxication
10. Poor compliance in wearing elastic stockings

Indications for Sclerotherapy or Phlebectomy
The selection of treatment was done empirically, except for patients with certain contra-indications for surgical intervention. The morphology picture of the varicose veins was essential for the final decision. The decision was independent of the degree of chronic insufficiency. Perforator insufficiency was regarded as an indication for surgical treatment.

RESULTS

The following rules based on the work of Sigg, were followed during sclerotherapy.
1.   Puncture with an open needle with the patient standing
2.   Subsequent elevation of the leg
3.   Injection of sclerosing agent to empty veins
4.   Application of compression bandaging
5.   Immediate mobilisation
Compression is the key to successful sclerotherapy. Incorrectly applied compression may lead to phlebitis with pain and inflammation of the vein. There is also a risk of skin pigmentation. An incision is made in these circumstances and the clot removed. The sclerosing therapy was carried out with Variglobin 2% - 4%.

Outpatient phlebectomy was carried out under local anaesthesia with 0.25% Scandicain-Adrenalin (Mepivacain) (with addition of 0.9% NaCl and 1.4% Natrium bicarbonate). Blood pressure and pulse were recorded before and after treatment. The use of Epinephrine and Trendelburg's position were intended to minimise the bleeding. On completion a bandage was applied for one week (changed after the first 24 hours). One week later the bandages were exchanged for compression stockings, which were worn for six weeks. The final outcome was evaluated by clinical examination at the end of this period.

Complications of sclerotherapy were allergy to bandages: 3, syncope: 0, variglobin allergy: 2, phlebitis: 6, poor compliance with compression: 2, skin pigmentation: 5.

Complications of phlebectomy included: allergy to bandages: 5, syncope: 0.

STATISTICAL EVALUATION OF SCLEROTHERAPY

The aim of this evaluation is to investigate the possible difference in the duration of compression between the left and the right leg and to investigate the possible difference in numbers of incisions due to age, leg and % of variglobin.

There was a total of 304 patients who had both legs treated and 386 patients who had either the left (n=209) or the right (n=177) under the therapy. There was therefore 690 patients, 513 left legs and 481 right legs. The average age of the group was 55.9 years with coefficient of variation 23%. The age of the youngest and the oldest patient were 22 and 89 years, respectively.

For the analysis of incision, only patients who had variglobin 2% or 4% were included. There were a total of 625 patients (n=508 for 2% and n=117 for 4%). Out of the 625 patients, 345 had either the left or right leg, whereas in the remaining 280 patients, both legs were treated.

Statistical methods used
The duration (in days) of compression was logarithm transformed, since the original duration appeared to be skewed and with long upper tail. For the 386 patients who had either left or right leg treated, the comparison of duration of compression was between leg and between patients. The comparison was pooled based on the weighted average, the weight being proportional to the square of the standard error of the estimated between leg difference.

The possible influence of the age, leg and variglobin concentration of the chance of incision was examined by means of logistic regression.

Statistical results
Duration of compression
The geometric average of the compression days for the left leg is 10.11 days with between patients coefficient variation 50%. The (geometric) average for the right leg is 9.6 days with between patients coefficient of variation 53% (p=0.87, N.S..).

Incision
For the 345 patients who had only one leg treated, the logistic regression reveals that there is no significant leg or age effect (p=0.7 and 0.36). However, the 2% VA patients had a higher incidence of incision than that of 4% patients (p=0.0037).

Since there was no leg effect, we combined the results of the left and the right leg for the patients who had both legs under treatment. When either the left of the right leg required incision, the patient was considered to be requiring incision.

The result of logistic regression of the 625 patients revealed that there was no significant age effect (p=0.62N.S.) and there was a significant difference between the 2% and the 4% group (p=<0.0001). The observed percentage of patients requiring incision for the 2% and 4% group were 15.8% (80/508) and 33.3% (39/117), respectively. The chance of incision for the 2% group was therefore significantly lower than the 4% group.

EVALUATION OF THE PHLEBECTOMY

The aim of this evaluation is to investigate the possible difference in duration of compression between the left and the right leg, and the possible inference due to patient characteristics, to investigate the possible factors (e.g. sex, age and amount of local anaesthesia), that might increase the chance of complication during the surgery.

There was a total of 766 patients, a total of 411 left legs and 355 right legs. The number of female patients was 669 and male patients 97. The average age of the female patients was 50.7, whereas male patients 50.4.

Statistical methods
For the evaluation of the difference in duration of compression between the left and the right leg, an analysis of covariance (ANOCOVA) was carried out on the logarithm of the duration in days. The model included age, gender and leg as explanatory variables. The difference of duration between legs was calculated after adjustment for age and gender.

Statistical Results
The (geometric) average of the duration of compression of the left leg was 8.9 days between patient coefficient of variation 30%. The (geometric) average of the duration of compression of right leg was 8.8 days with between patient coefficient of variation 25%. There was no significant difference between left and right leg (p=0.32). Every 10 years increase in age is expected to prolong the duration of compression by 43% (se 0.8%). The age effect on duration is significant (p=<0.0001).

DISCUSSION

* There was a difference in duration of compression between the two treatment groups. The phlebectomy patient group is expected to have duration of compression one day less than a sclerotherapy patient.
* There was an age effect in duration of compression for both treatment groups. Increase in age leads to an increase in duration of compression. A 25 year increase leads to 1 day more compression.
* Under sclerotherapy treatment, patients with 4% VA had significantly higher chance of incision. However, the effect of age and leg was not evident.
* Both methods confirm that both types of treatment are safe and guarantee high compliance.

580

REFERENCES

1. Harry B Abramowitz, Jerusalem, Israel: The Use of Photoplethysmography in the Assessment of Venous Insufficiency: a comparison to venous pressure measurements from Surgery St Louis. Vol. 86, No 3 pp. 434-441, September, 1979.
2. Arnost Fronek: Noninvasive Diagnostics in vascular disease.
3. Fischer, G Früh: Resultate der Varizenoperation beim Vorliegen eines primären Lymphödems. Phlebol 1991, 56-69.
5. Alina Fratila et al: Stellenwert der perkutanen microchirurgischen Phlebextraktion nach Varady in der Varizenchirurgie. H+G Zeitschrift für Hautkrankeheiten 65 (5) 487-491. Grosse Verlag Berlin 1990.
6. Hans Jörg Leu, Novaggio: Komplikationen der Varizen-Chirurgie - eine Literaturübersicht. Ars Medica 4/94.

Phlebology '95, D. Negus et al. (eds.). Phlebology (1995) Suppl. 1:581-583

P140

Combination of Aetoxisclerol with Glucose for the Treatment of Varicose Veins and Telangiectasias

F. Zuccarelli

Head of Phlebology and Angiology Department, Hôpital Saint Michel, Paris, France

GENERAL CHARACTERISTICS OF AETOXISCLEROL

AETOXISCLEROL is hydroxypolyethoxydodecane, to which is added absolute ethyl alcohol to ensure stability of the substance with respect to heat. Its international non-proprietary name is poliodocanol. It is marketed under the name AETOXISCLEROL. Hydroxypolyethoxydodecane is a hydrophilic ether chain (polyethylene hydroxide) combined with a liposoluble or hydrophobic alcohol (dodecylic alcohol).

The surfactant action of AETOXISCLEROL, responsible for its therapeutic activity, is derived from this dual hydrophilic and hydrophobic property.

PRINCIPLE OF ACTIVITY

The principle site of actions of AETOXISCLEROL are the lipids on the cell surface, which induces deterioration of the intima. Animal trials have studied intravenous injections of the compound into intact veins. The smooth, healthy intima in this type of vein does not allow any possible contact with endothelium, as in the case of the deteriorated intima of a varicose vein.

Healthy veins have a high blood flow rate and the wetting agent is therefore rapidly eliminated by the blood stream by combining with serum lipids in a higher proportion. These elements explain the mechanism of action of AETOXISCLEROL which only acts on pathological endothelium observed in varicose veins and which is inactive on endothelium of healthy vessels.

When injected into a varicose vein, emptied of its blood, the sclerosing agent penetrates into the intima of the vein and almost immediately induces the formation of desquamation thrombus surrounding the sclerosing agent. The thrombus adheres to the lesion and is integrated into the thickened intima before undergoing the classical process of connective tissue organisation, transforming it into a fibrous cord. The larger the thrombus, the lower its fibril content and the greater the risk of revascularisation. It therefore seems particularly useful to obtain thrombi as small as possible in order to ensure more extensive fibrous organisation. It is also generally accepted that the sclerosing effect is not only related to the concentration of the agent in the vein, but also depends on the sclerosant-endothelium contact time. This concept led us to use a technique designed to increase this contact time.

INDICATIONS

AETOXISCLEROL is essentially indicated in the treatment of varicose branches and reticular varices at concentrations ranging between 3% and 0.5%.

It does not appear to be sufficiently active, even at 3%, to be used systematically in the saphenous veins. On the other hand, in varicosities or telangiectasias, it was used with a certain reticence due to the fear of excessive reactions, even at dilute doses, with a risk of developing pigmented areas or paradoxical reactions of post-sclerosis capillaritis.

PARTICULAR CASE OF TELANGIECTASIAS

These lesions, very frequently unsightly and poorly tolerated by our patients, can be eliminated by sclerosis which remains a simple and reliable method, which has still not been replaced in this indication.

To treat the telangiectasias, or the veins supplying them, it is firstly essential to sclerose the saphenous reflux vessels, when present, and then use mild sclerosing agents to treat the branches of the varicosities not eliminated by eradication of the reflux from larger veins. In this indication, AETOXISCLEROL is generally used at the concentration of 0.50% or 0.25%, or even 0.10%, depending on the size of the vein, but, as we have mentioned it has often been considered to induce excessive reactions.

These paradoxical reactions appeared to be exclusively related to the marked fluidity of the substance which, despite every precaution during injection (which is easy and straightforward), meant that an excessive volume of substance was injected at each session and into each site. One way of avoiding this problem was to increase the number of injection sites and inject only a few drops of substance into each site, but the repetition of these injections, resulting in a large number of injections, was sometimes poorly tolerated by the patient and carried a risk of inducing local concentration phenomena due to the limited distance between two injections into the same vessel.

By modifying the viscosity of AETOXISCLEROL, we considered that much better results could be obtained in this type of indication, especially in view of the concept that sclerosis is not only related to the concentration of the sclerosing agent, but also to the sclerosant-endothelium contact time.

The need to dilute AETOXISCLEROL, from 0.50 to 0.25% with distilled water in the treatment of telangiectasias, decreased its sclerosant activity proportionally. In contrast, by adding 66% glucose as diluent, this loss of efficacy is compensated by a longer persistence of the active ingredient in contact with the venous endothelium, which induces less severe and more effective reactions. The disadvantage of the excessive toxicity of AETOXISCLEROL in small venous dilatations is consequently overcome by adding 66% glucose as diluent.

The milder reactions obtained greatly decrease the risk of neovascularisations around the injection sites. The quantity of substance injected, the sudden turgescence of the vessels, the size of blanching during injection, which must be less than a circle 2 cm in diameter, on average, and the force of injection have been considered to be responsible for the pathogenesis of these networks. By making AETOXISCLEROL thicker, the addition of glucose overcomes most of these drawbacks, as the thicker the substance, the smaller the volume injected and the lower the force of injection. By increasing the injection time in this way, it is easier to ensure a more limited blanching halo; furthermore as the substance is milder, the turgescence of the vessels and the spastic effect of the injection are reduced. This combination is very useful in this aesthetic indication, in which the result must also be aesthetic, as it considerably decreases the risks of paradoxical reactions. Moreover, the number of thrombi is markedly reduced due to the gentler nature of the sclerotic reaction.

EXTENSION TO OTHER TYPES OF VEINS

The advantage of the altered viscosity of AETOXISCLEROL in the treatment of telangiectasias has led to the use of this property in sclerosis of varicose branches or reticular varices and the systematic addition of glucose retains the same beneficial properties as those observed in the treatment of varicosities.

IN PRACTICE

For the treatment of varicose branches or redicular varices, the AETOXISCLEROL solution is prepared in a 2 cc syringe with systematic addition of 0.5 cc of half-strength glucose.

For telangiectasias, we use a more balanced dosage with equal parts of 0.50% AETOXISCLEROL solution and glucose diluted to one half in water. We usually prepare 1 cc of 0.50% AETOXISCLEROL and add 1 cc of 30% glucose solution, prepared by diluting 0.5cc of 66% glucose with 0.5 cc of distilled water.

We sometimes use 1/3 0.5% AETOXISCLEROL - 1/3 66% glucose - 1/3 distilled water. The relative proportions of the various substances is subsequently determined on the basis of personal experience.

CONCLUSION

The addition of dilute 66% glucose improves the efficacy of AETOXISCLEROL in the treatment of telangiectasias by allowing a longer sclerosant-wall contact time and by decreasing the intensity of the reaction, which eliminates the risk of peripheral neovascularisation and thrombi, which must be avoided in a treatment which is essentially designed to be aesthetic.

Phlebology '95, D. Negus et al. (eds.). Phlebology (1995) Suppl. 1: 584-586

OP/5.6

Tactics and Results of Sclerotherapy and Surgery of Primary Varicose Veins

P.B. Dimakakos[1], K. Katsenis[1], M. Papasava[1], A. Papageorgiou[1] and D. Mourikis[2]

[1] Vascular Surgical Department of the 2nd Surgical Clinic and [2] X-Ray Department, University of Athens, Athens, Greece

INTRODUCTION

Recovery time cost, relapse, cosmetic result and fast restoration to previous way of life should be the features of surgical treeatment of primary varicose veins [1], one of the most frequent operative procedures of general surgery [2].

The preservation of healthy veins however, for use of arterial grafts, constitutes the responsibility of the surgeon and the indication for removal should be done with rigorous criteria. Cosmetically, the result improves with sclerotherapy. As a principle, the removal of only the diseased venous parts, according to functional and not the anatomical findings [3] only, is well-grounded.

The morphology, stage of disease and haemodynamic parameters, require an individualized therapeutic scheme.

Our tactics with the above principles were applied in an one-day surgery program on an ambulatory basis.

METHODS

Over 2300 lower extremities suffering from primary varicose veins were operated and studied clinically, by Doppler ultrasound examination, Duplex scanning, ambulatory venous pressure and phlebography or varicography.

The clinical findings are shown in Table 1 and the tactics applied in Table 2.

Table 1. Clinical findings

Group I	Incompetence of SFJ	1660 (71.75%)
Group I	Incompetence of SPJ	186 (8.3%)
Group I	Accessory veins	302 (13%)
Group I	Incompetent perforators only	162 (7%)

Table 2. Surgical instructions

Preoperative	Postoperatively (hospital)	Postoperatively (home)
General examinations	Feeding immediately	1-4 postop.day: 4 hrs walking
Marking of veins	Active exercise of the legs	4th postop.day: Removal of sutures and 2hrs walking
Epidural anesthesia	2nd postop.hour: Free walking	12th postop.day: Elastic stockings, return to previous normal life
Operation	3rd postop.hour: Changing of bandages. Discharge	

All patients were operated under epidural anesthesia without the administration of heparin or antibiotics. The postoperative complications and results 3 months and 6½ years later are shown in Tables 3-5.

Table 3. Postoperative complications

Subesthesia	159 (9.2%)	Lynmphedema	2
Arterial damage	1	Deep vein thrombosis	2
Hematoma	24	Pulmonary embolism	-
Oedema of the ankle	223	Inflammation (redness of skin)	335 (17%)

Table 4. Results after 12 weeks

Satisfied	1112	
Improvement	110	} 91.2%
Improvement + Sclerotherapy	350	
Not satisfied	150	9.8%

Table 5. Results after 6½ years (n=1055)

Sclerotherapy	105	(9.9%)
Requested sclerotherapy	20	(1.8%)
Elastic stockings	156	(14.5%)
Re-operations	6	(0.6%)
Change of profession	12	(1.2%)
Without symptoms	576	(72.1%)

CONCLUSIONS

Combined surgical treatment and sclerotherapy give the best results, while elastic stockings, professional orientation, regulation of body weight and physiotherapy maintain the good results.

REFERENCES

1. Goren G, Yellin AE. Ambulatory stab evulsion phlebectomy for truncal varicose veins. Am J Surg 1991, 162:166-74.
2. Kitslaar PJ EHM, Rutgers PH. Varicose veins and the vascular surgeon: From nuisance to challenge. Eur J Vasc Surg 1993, 7:109-12.
3. Varady Z. Moeglichkeiten der Varizenoperation. Angio 1986, 8:223-6.

Phlebology '95, D. Negus et al. (eds.). Phlebology (1995) Suppl. 1: 587-590

P193

Le Traitement Ambulatoire des Varices: Association de Sclérose et Chirurgie

D. Paolicelli, N. Crocetti and R. Di Pinto

Centre Phlébologique, v. Dandolo 74, Rome, Italy

Ambulatory Treatment of Varicose Veins by a Combination of Sclerotherapy and Surgery

SUMMARY

Analysis of the results of 139 patients undergoing ambulatory treatment for varicose veins permits the following conclusions:

- the treatment is completely ambulatory and is very acceptable to patients; it is socially acceptable and practically without risk.

- it provides satisfactory control of symptoms and good long term results, with some subsequent attendances for injection sclerotherapy to control recurrent varices.

- recurrent varices most often appear most frequently shortly after treatment and in patients with a strong predisposition (to varicose veins).

588

LE TRAITEMENT AMBULATOIRE DES VARICES:
ASSOCIATION DE SCLEROSE ET CHIRURGIE

D.Paolicelli,N.Crocetti,R.Di Pinto

Centre Phlébologique - v.Dandolo 74 - Rome - Italie

INTRODUCTION

Dans les dernières cinq années,dans notre centre phlébologique,nous a-
vons adopté,pour le traitement des varices,une ligne thérapeutique qui
tient compte de la prédisposition du patient au développement de la
maladie variqueuse,de la grossesse et de la possibilité de poursuivre
un traitement ambulatoire et des résultats durables et   facilement
corrigibles en cas de recidive.Sur cette base nous avons   rassemblé
les patients en 3 groupes principaux:
   1)Patients pour lequels nous prévoyons un maximum évolutif de la maladie variqueuse à cause de la considerable prédisposition familiale
et de l'éxistence d'une incontinence de la saphène dès le jeune âge et
avant la grossesse.Dans ces cas nous effectuons,généralement,un stripping par invagination de la saphène suivit par des microphlébectomies
des varices collaterales.
   2)Patients qui ont développé les varices à un âge moins jeune   et
après les grossesses.Dans ces cas nous procédons à une crossectomie en
anesthésie locale associé à sclérose et/ou à microphlébectomies   des
varices collatérales.et des saphènes.
   3)Patients non plus jeunes ou agés qui possèdent des nombreuses varices avec ou sans incontinence de la veine saphène.Ces cas la   sont
traités,exclusivement,par sclérose.

METODES

Cet étude à été fait pour évaluer les résultats obtenus sur les   patients du deuxième group de 1 à 5 années après la fin du traitement.Le
nombre et le type des interventions et le nombre et le sex des   patients operés sont exposés dans le tableau 1.
   L'interruption de la jonction saphèno-fémorale ou de celle sapheno-
poplitée a été éxécuté sous anesthésie locale avec Mepivacaine au 2%.
La dose utilisé a été de 2 à 5 ml.Une semaine après   l'intervention

nous avons entrepris la sclérose de la saphène et des autres varices pour laquelle nous avons utilisé du Trombovar au 3% et 1%,de l'Atossi-sclerol de 3% au 0.25% et du Scléremo au 0.70% selon le calibre des varices.Le traitement a été conduit du haut en bas et une contention élastique a été systématiquement appliqué jusqu'à 1 mois après la fin du traitement.Quand cela est terminé,nous préscrivons une thérapie a-vec des flavonoides.

TABLEAU 1

	n.intervention	n.patients opéré	crossec.int.	crossec.ext.
Femmes	105(75.5%)	78(74.3%)	78(74.3%)	27(25.7%)
Hommes	34(24.5%)	27(25.7%)	17(50.0%)	17(50.0%)
Total	139	105	95(68.3%)	44(31.7%)

Aux controles nous avons évalué l'apparition de récidives,la réappari-tion de la symptomatologie et,avec un examen doppler,la récanalisation de la veine saphène.

RESULTATS

Le 89.9% des patients n'avait aucune symptomatologie aux controles. L'examen doppler nous a montré que le 61.2% des veines saphènes trai-tées étaient récanalisées.Dans le 93.0% des ces cas nous avons trouvé un reflux très lent que nous avons jugé hémodynamiquement insignifiant. Dans le 4.7% des ces cas nous avons relevé un reflux important et dans le 2.3% il n'y avait pas de reflux.Dans le 33.1% des cas nous avons trouvé des varices recidives(seulement le 26% des recidives donnait u-ne symptomatologie).Le traitement des varices recidives a necessité de un minimum de 1 séance jusqu'à un maximum de 4 séances de sclérose avec une moyenne de 2.1 séances (on a exclu les séances esthétiques). En ce qui concerne les complications nous avons observé un hématome post-opératoire dans 4 cas(2.9%).Dans 6 cas(4.3%)nous avons remarqué douleur post-operatoire durée à peu près 20 jours.Nous n'avons pas eu des complications majeures.

CONCLUSIONS

L'analyse du follow-up nous entraine a quelques considerations:
-Le traitement est totalement ambulatoire,bien accepté par les pa-tients;il a un bas coût social et les risques sont pratiquement nuls.
-Il nous donne un satisfaisant contrôle des symptômes et un résultat

durable à condition que le sujet se soumet à des contrôles périodiques et à quelques séances de sclérose des récidives.

-Les varices récidives se manifestent plus souvent dans un temps assez bref et dans des sujets avec une forte prédisposition.

Phlebology '95, D. Negus et al. (eds.). Phlebology (1995) Suppl. 1: 591-592

P143

How Compression and Ambulation Robs You of Success in the Sclerotherapy of Large Varicose Veins

D. Brian McDonagh MD

1101 Perimeter Drive, #615 Schaumberg, Il60173, USA

Sclerosants are phlebitogenic agents: they are designed to work by irritating and injuring a segment of vein wall. The objective of the sclerosant is to initiate an inflammatory reaction which shrinks that segment of vein in a process of fibrosis, thereby obliterating the vein lumen. The goal (similar to vein ligation surgery) is to secure the multiple points of origin of reflux without surgery or anaesthesia.

But why has sclerotherapy been more effective in small veins rather than large vein disease? Why does it work well in some patients' large varicose veins and not in others? Why has sclerotherapy been apparently 'limited' in large vein disease?

Historically sclerotherapy became more effective when an analytic diagnostic and therapeutic approach replaced random injections of varicosities. This coincided with the addition of compression and ambulation to the treatment program. Since that time the quality of outcomes plateaued and surgical ligation was often justified to correct the major avenues of reflux.

Over two years ago, I examined each component of sclerotherapy for both its positive and negative potential in contributing to the outcome of sclerotherapy. For example, the impact of injections varied according to the exposure time from both the concentration and the volume of sclerosant. Other treatment principles include injecting first from deep to superficial, from large to small and from proximal to distal. These principles address the problems of reflux and the direction of abnormal blood flow, yielding more rapid therapeutic effects.

However the greatest contribution to improved results came from depressurising the large calibre varicose veins. We are aware how compression alone has a positive and rapid therapeutic influence on the red, swollen and ulcerated leg. Clinically it has a powerful anti inflammatory effect by denying the space necessary for the development of inflammatory oedema. But by the same rationale it has the capacity to attenuate the irritating effects of a sclerosant and deny you clinical success.

Compression exists in two important forms: external and internal. Externally by compression stockings or bandaging and internally (venous hypertension) when the patient stands or walks. The internal pressure stretches the vein wall and inhibits inflammatory intra-abdominal pressure, then the post injection inflammatory reaction will continue unopposed in the vein(s) without blood or reflux. I use a sliding scale which varies from 12 to 24 hrs for large vein disease, to 2 hours for medium sized veins/venules and 20 minutes for small vein disease. Large and medium vein patients remain A.R.P. (anti-reflux position) in the office for 45 minutes before going home to complete their time; this initial A.R.P. time establishes the initial therapeutic response as seen on ultrasonic imaging. The chances of a full thickness vein wall response and selective fibrosis is greatly increased. There will also be less opportunity for intravascular haematoma formation or painful thrombophlebitis. Overall, the patient response has been much more rapid. These principles also apply to medium and small vein disease where progress is enhances, resulting in less treatment sessions.

SUMMARY

1. Previously medium and small vein disease has responded better and more predictably than large vein disease because of lower intraluminal pressure/flow rates.
2. Compression and venous hypertension attenuate the effects of sclerosants.
3. All veins respond equally to sclerotherapy when depressurised.
4. All large vein disease patients must receive Duplex ultrasound 'mapping', drawn in the patient record by the ultrasonographer before treatment, and again before the patient is finally discharged, or it will be impossible to monitor outcomes. It is also necessary to differentiate recurrence from progression of vein disease when the patient presents with more veins in the years ahead.
5. Now that we can see large vein disease for the first time using Duplex ultrasound, all prior judgements and claims of therapeutic efficacy are suspect.

Phlebology '95, D. Negus et al. (eds.). Phlebology (1995) Suppl. 1: 593-595

OP/11.6

The Effect of Sclerotherapy on Restless Legs: 4 Year Follow-Up[1]

A. Kanter[2]

[1] Reprinted by permission of the publisher from *The Effects of Sclerotherapy on Restless Legs: 4 Year Follow-Up*, Kanter A, Dermatol Surg 1995;21:328-331. Copyright 1995 by Elsevier Science Inc.
[2] *In Private Practice, Vein Centers of Orange County & Torrance, USA*

INTRODUCTION

Restless Legs Syndrome (RLS) is a common, chronic, familial disorder of unknown etiology characterized by relentless leg discomfort when stationary, compelling leg movement to obtain temporary relief. It is classified as a sleep disorder due to its frequent nocturnal occurrence in association with periodic leg movements, and resultant sequelae of sleep deprivation. [1] Pharmacotherapy maintenance with currently available agents is often limited by side effects.

Anecdotal reports associating vein disease with "restless legs" have appeared in the European literature [2-6]. We have received anecdotal reports of coincidental relief from RLS symptoms in patients following sclerotherapy for varicose vein disease. Medical education for both sleep disorders and venous disorders is often lacking for primary care providers, thus leading to under-recognition, misdiagnosis, and inappropriate treatment.

To more fully explore the connection implied by these reports, we prospectively evaluated the immediate and long-term effects of sclerotherapy on RLS in patients seeking treatment for concomitant vein disease, while documenting the incidence and characteristics of ICSD-defined RLS [7] in this population of patients.

METHODS

All patients presenting to our vein treatment center for lower extremity vein evaluation over a 2-year period completed a questionnaire regarding RLS symptoms. If respondents qualified for the ICSD definition of chronic RLS by the author's interview, they were further questioned regarding the diurnal/nocturnal occurrence of symptoms, bed-partner complaint of patient leg movements, family history for RLS, and duration of RLS.

A thorough phlebological clinical exam for all patients included greater and short saphenous vein axis interrogation with CW-Doppler (standing), with stratification into Doppler-positive (DP) and Doppler-negative (DN) groups based on the presence or absence of reflux.

594

Subjects electing to undergo treatment for their vein disease received sclerotherapy in affected vessels with sodium tetradecyl sulphate (Sotradecol; Elkins-Sinn, Cherry Hill, NJ). Excluded from this phase of the study were patients with previous vein treatment by either surgery or sclerotherapy, and patients taking drugs known to affect either RLS or the central nervous system. Sclerosant concentrations of 0.16% and 0.5% were used to treat DN (Goldman-Duffy Types 1-3) veins, and 1% and 3% (French method) for DP (Goldman-Duffy Type 4) saphenous-related veins. Post-treatment ambulation and compression with full-length 30-40 mm gradient stockings was employed after each weekly treatment session, until all targeted vessels were sclerosed visually for DN, and by CW-Doppler for DP patients. Change in RLS symptoms was solicited and recorded at each office visit, and by telephone interview at six months and annually thereafter.

RESULTS

Incidence and Characteristics

Of 1,397 patients interviewed, 312 (22%) had true RLS, with a DN/DP ratio of 3:2. 46% reported a positive family history, 94% reported daytime symptoms, including 39% who experienced both day and night symptoms. The mean duration of symptoms was significantly longer in the DP group (10.9 years) than in the DN group (7.4 years, p=.0042).

Treatment Effect

The treatment group consisted of 113 patients electing to undergo sclerotherapy, and did not differ from the main group in demographics, vein type, or RLS characteristics. 111 (98%) reported relief from RLS symptoms, defined as either complete resolution or sustained marked improvement. No patient complained of worsening of symptoms.

The number of treatments required for relief is shown in Figure 1. The mean ± SD of treatments required for relief in those patients responding was 2.2 ± 1.7. The cumulative recurrence rate over four years shown in Table 1 was 9% (8/86), 33% (28/86), 45% (39/86), and 49% (42/86) at 1, 2, 3, and 4 years respectively.

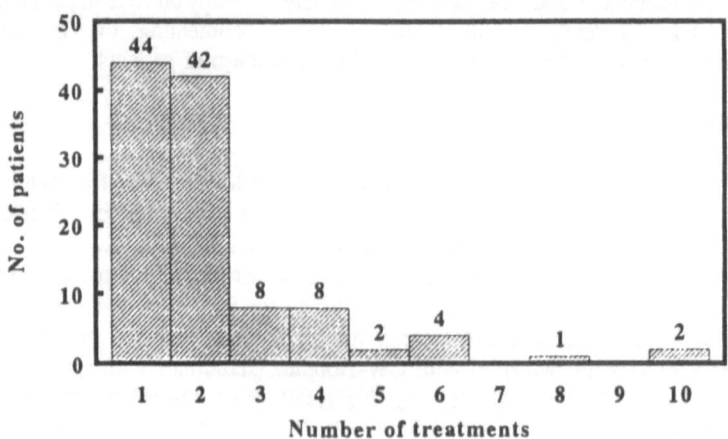

Figure 1. Number of patients whose RLS symptoms resolved at given number of treatments

Table 1. Recurrence rate

	1 Year	2 Years	3 Years	4 Years
Doppler-negative group	7/49	18/49	25/49	27/49
Doppler-positive group	1/37	10/37	14/37	15/37
Total	8/86	28/86	39/86	42/86
Total recurrence	9%	33%	45%	49%

CONCLUSIONS

ICSD-defined RLS is common in patients with both saphenous and non-saphenous vein disease, usually responds promptly to sclerotherapy, and tends to recur over time. Periodic sclerotherapy of selected patients with concomitant RLS and varicose vein disease is an alternative to chronic maintenance pharmacotherapy to treat RLS symptoms. Such patients may have predominantly diurnal rather than nocturnal RLS symptoms, possibly representing a sub-population of patients with generally milder symptoms within the spectrum of severity for RLS.

Updated medical education curricula for both venous and sleep disorders should improve recognition, accuracy of diagnosis, and appropriateness of treatment recommendations for RLS sufferers.

REFERENCES

1. Walters AS, Hening W. Clinical presentation and neuropharmacology of restless legs syndrome. Clin Neuopharmacol 1987;10:225-37.
2. Tournay R, Wallois P. Les varices de la grossesse et leur traitement principalement par injection sclerosants. Paris L'Expansion Scientifique Francaise;1948:91.
3. Oesch A, Widmer LK, Claus L. Leg problems of venous origin. Observations on 2,036 subjects of the Basel study 3. Schweitz Med Wochenschr 1973;103:1755-8.
4. McEwan AJ, McArdle CS. Effect of hydroxyethylructosides on blood oxygen levels and venous insufficiency symptoms in varicose veins. Br Med J 1971;2:138-41.
5. Pulvertaft TB. Paroven in the treatment of chronic venous insufficiency. Practitioner 1979;223:838-41.
6. Hobbs JT. The enigma of the gastrocnemius vein. Phlebology 1988;3:19-20.
7. Lawrence KS. International Classification of Sleep Disorders: Diagnostic and Coding Manual. Lawrence KS; Allen Press, Inc. 1990:69-71.

Phlebology '95, D. Negus et al. (eds.). Phlebology (1995) Suppl. 1: 596-599

OP/4.7

Mythes et Réalité de la Sclérose des Varices des Membres Inférieurs

H. Partsch[1], U. Baccaglini[2] and R. Stemmer[3]

[1] *Univ. Hautklinik, Wilhelminenspital, Wien*
[2] *Chirurgia Generale II, Ospedale Guistinianeo, Padova*
[3] *Strasbourg*

Le but principal de la Conférence de Consensus était trouver l'accord le plus large possible sur les techniques de sclérose des varices des membres inférieurs. Dans les publications faisant date on relève la technique de l'Ecole Française de Phlébologie inspirée par Raymond Tournay (1), celle de l'Ecole Suisse de Phlébologie formée par K. Sigg (2) et l'Ecole Irlandaise de Phlébologie personnifiée par G. Fegan (3). Or les conceptions techniques de la sclérose des varices des membres inférieurs sont fort différentes d'une école à l'autre. Le tableau N° 1 synthétise quelques unes de ces différences:

	EC. FRANCE	SIGG	FEGAN
IND. CHIR	OUI	NON	NON
AIRBLOCK	NON	OUI	NON
PROX-DIST	OUI	NON	NON
DIST-PROX	NON	OUI	OUI
COMPRESSION	GR. VAR.	TOUJOURS	TOUJOURS

Tabl. N° 1 Ecoles de sclérothérapie.

Il ne tient pas compte d'autres différences typiques, comme la taille des aiguilles, la ponction et l'injection, la technique de compression et sa durée, etc.

Il est donc apparu rapidement qu'il était quasiment impossible de trouver un consensus sur des positions aussi éloignées. Le questionnaire adressé aux phlébologues (voir communication: La formation en sclérothérapie) a posé aux phlébologues des questions simples sur les différentes techniques de base de leur sclérose et leur a demandé s'ils se prévalaient d'une école de Phlébologie: Sigg, Fegan, école française ou autre.

Les résultats parvenus ont été dépouillés et pour les mêmes raisons invoquées dans la communication sur la formation du médecin, n'ont pas été calculés du point de vue statistique, car pour certains l'échantillon de réponse risquait de ne pas être considéré comme représentatif.

RESULTATS

L'appartenance à une école de sclérose a eu une réponse positive de 77 % des questionnaires retournés, selon le tableau N° 2 (en **gras** les pourcentages maxima).

	A	CH	CND	D	F	GB	I	SC	USA	DIV	TOTAL
n	63	54	13	324	294	23	60	31	166	11	1039
EC. FR.	14	18	**77**	25	**81.3**	8	29	0	21	21.5	**38.0**
SIGG	25	**52**	0	27	0.3	4	22	0	5	7	15.0
FEGAN	1.6	0	0	0.3	0	48	0	17.5	12	**36**	4.4
AUTRES	22	15	15	**28**	12.7	16	18	17.5	32.5	29	19.6
S / R	37.4	15	8	19.7	15.7	24	31	65	29	7	23

Tabl. N° 2: Ecoles de sclérose (%)

Il était prévisible que l'école française soit massivement représentée en France et celle de Sigg en Suisse (à un degré moindre). Fegan obtient le meilleur résultat en Grande Bretagne, mais n'est pas connu en France et en Italie. 65 % des Scandinaves ne se réfèrent à aucune école. Nos collègues américains sont le moins liés à une école de sclérose, un tiers se proclame d'une autre école et 29 % n'a pas répondu à cette question. L'école française a été citée par 38 % de l'ensemble des réponses, soit pratiquement deux fois plus que les deux autres écoles réunies.

Une autre différence fondamentale est donnée par la stratégie thérapeutique. L'école française propose de commencer le traitement à partir de l'aine, soit le point de reflux le plus haut situé, puis de traiter les points de fuite du haut vers le bas. Par contre aussi bien les techniques de Sigg que celle de Fegan recommandent de traiter à partir de la périphérie vers la racine du membre. Ces deux techniques ont été recensées parmi toutes les 1039 réponses, mentionné dans le tableau N° 3, et sans tenir compte d'une appartenance à une école de sclérose. On peut conclure à la prépondérance nette de la technique dite de haut en bas (proximal - distal) qui recueille 52 % des suffrages, contre 26 % pour les deux techniques de bas en haut (distal - proximal). 22 % des médecins ont omis de répondre à cette question.

	A	CH	CND	D	F	GB	I	SC	USA	DIV	TOTAL
n	63	54	13	324	294	23	60	6	166	11	1039
PRO-DISTAL	43	27	92	50	70	12	41	45	55	25	**52**
DISTAL-PRO	38	45	8	31	7	67	44	0	22	67	**26**
S / R	19	28	0	19	23	21	15	49	23	8	**22**

**Tabl. N° 3: Pourcentage des techniques de haut en bas (pro-distal)
et de bas en haut (distal-pro) suivant les pays.**

Ce résultat de 52 % de proximal - distal est à comparer avec les 38 % d'école française qu'il dépasse nettement, alors que les 26 % de distal - proximal dépassent les 19,4 % recueillis par les techniques cumulées de Sigg et Fegan.
Cette divergence a conduit à des recherches plus poussées. Comme indiqué dans la tableau N°1, la technique de Sigg et Fegan réfute toute indication chirurgicale et la sclérose de l'école irlandaise s'arrête à mi-cuisse. Il était donc significatif de rechercher les indications

de la sclérose des crosses des veines saphènes internes dans les trois écoles (possibilité de répondre oui pour les deux thérapeutiques à la fois). Le tableau N° 4 mentionne les indications de traitement des médecins ayant admis leur appartenance à une des trois écoles ou ayant indiqué une technique personnelle (PERS.)

	FRANCE	SUISSE	IRLANDAISE	PERS.
	414	165	46	213
CHIRURGIE	81.7	73.9	69.5	78.9
SCLEROSE	39.6	21.2	39.1	17.4
S / REP	5.0	12.1	6.5	12.2

Tabl. N° 4: Indications pour la crosse de la saphène interne (%).

Il ressort de ces chiffres que les indications de la chirurgie des veines saphènes internes est devenue une indication fréquente, même si elle est réfutée par Sigg et Fegan (en **gras**). Elle a pris le pas sur la sclérose de la crosse pour l'école suisse. Elle est même traitée par sclérose par les médecins se réclamant de la technique de Fegan (pourcentages en **gras**), technique arrêtant pourtant la sclérose à mi-cuisse. Les médecins se réclamant d'une technique personnelle penchent plutôt vers la chirurgie des crosses de saphène interne.

La notion de la direction du traitement a été mise en corrélation avec les écoles de sclérose, comme l'indique le tableau N° 5.

	FRANCE	SUISSE	IRLANDAISE	PERS.
n	414	165	46	213
PROX-DISTAL	75	39	39	51
DISTAL-PROX	6	51	49	32
S/ REP	21	10	12	17

Tabl. N° 5: Direction du traitement (%)

Uniquement 6 % des adeptes de l'école française sclérosent du bas vers le haut, dont un tiers utilise les deux techniques à la fois. Les techniques suisse et irlandaise comportent par contre 39 % d'erreurs qui semblent aller totalement à l'encontre des positions de Sigg et Fegan. La technique personnelle favorise la sclérose de haut en bas.

Une autre caractéristique de le technique de Sigg est l'utilisation constante de l'air block qu'il pratique lors de chaque injection. Le questionnaire en a fait état et les réponses selon les écoles sont indiquées dans le tableau N° 6.

	FRANCE	SUISSE	IRLANDAISE	PERS.
n	414	165	46	213
TOUJOURS	5	16	2	9
PARFOIS	39	40	33	37
JAMAIS	41	27	54	35
S / REP.	15	17	11	19

Tabl. N° 6: Utilisation de l'air block (%)

Cette technique d'injection est réfutée par les élèves de l'école irlandaise, rarement utilisée en France et uniquement appliquée systématiquement par 40 % des médecins se réclamant de l'école de Sigg. La réponse « parfois » semble s'appliquer aux télangiectasies. Les préférences données par les adeptes de technique personnelle se rapprochent des tendances de l'école française.

Un autre point important des techniques de Sigg et Fegan est la compression systématique après toute sclérose. Les résultats des questionnaires figurent dans le tableau N° 7. Cette mesure est certainement celle qui est la plus respectée par les adeptes des techniques suisse et irlandaise.

	FRANCE	SUISSE	IRLANDAISE	PERS.
n	414	165	46	213
TOUJOURS	49	88.5	83	75
GR. VARICES	43	3	13	17
RARE	7	6.5	0	4
S/ REP.	1	0	4	4

Tabl. N° 7: Compression après sclérose (%).

L'école française fait bande à part en ne comprimant dans 43 % de scléroses que les varices de gros calibre, alors que les écoles suisse et irlandaise ainsi que les individualistes préfèrent comprimer systématiquement. Mais toutes les écoles font une part de plus en plus importante à la compression par bas élastiques en remplacement des bandes, techniques qui n'étaient pas usitées lors de la description princeps des différentes écoles. Le pansement fixe posé par le médecin semble prendre le pas sur les pansements amovibles.

CONCLUSIONS

Toutes ces constatations presque convergentes démontrent que les préceptes des écoles ou familles de sclérose sont de plus en plus modifiés et adaptés à une technique strictement personnelle. Le point essentiel reste la conduite du traitement, le début proximal recueillant la majorité des adhésions. En deuxième lieu on peut citer le choix du sclérosant et le type et la durée de la compression.

Le sclérothérapeute individualise sa propre technique en empruntant à chaque école la part qui lui semble la mieux apte à obtenir les meilleurs résultats dans son environnement professionnel et social. Ces patchworks de techniques individuelles vont donc à l'encontre des cadres plus ou moins rigides des écoles de sclérose. Il convient donc de se libérer de cette notion. Cet abandon a été proposé à la conférence de consensus sur la sclérothérapie des varices des membres inférieurs.

Le mythe des écoles de sclérose avec leur chapelet de dogmes techniques a ainsi succombé à la réalité d'une adaptation individuelle des différentes techniques, mais subordonnées au principe thérapeutique général des varices qui s'impose aussi bien au sclérothérapeute qu'au chirurgien: priorité à l'élimination des reflux pathologiques.

BIBLIOGRAPHIE:

1. R. Tournay et coll.: La sclérose des varices. Paris, Expansion Scientifique Française, 4ème éd. 1985;
2. K. Sigg: Varizen, Ulcus cruris und Thrombose. Berlin, Springer, 4. Aufl. 1975.
3. G. Fegan: Varicose Veins. Compression sclerotherapy. London, Heinemann, 1967.

Phlebology '95, D. Negus et al. (eds.). Phlebology (1995) Suppl. 1: 600–603

OP/11.7

La Formation en Sclérothérapie

H. Partsch[1], U. Baccaglini[2] and R. Stemmer[3]

[1] Univ. Hautklinik, Wilhelminenspital, Wien
[2] Chirurgia Generale II, Ospedale Guistinianeo, Padova
[3] Strasbourg

Une Conférence de Consensus sur la sclérose des varices des membres inférieurs s'est tenue à Padova. Les préparatifs ont démontré l'absence de toute publication sur la formation pratique des phlébologues, mais aussi de toute donnée sur le recrutement des médecins la pratiquant et de leur activité.

Pour combler cette lacune un questionnaire a été élaboré et tous les Présidents des Sociétés Phlébologiques adhérentes à l'Union Internationale ont été priés de le répercuter sur les membres de leurs Sociétés respectives. Les réponses reçues se répartissent ainsi:

A	Österreichiche Arbeitsgemeinschaft für Phlebologie	67
CH	Schweizer Gesellschaft für Phlebologie	54
CND	Société Canadienne de Phlébologie	13
CH	Deutsche Gesellschaft für Phlebologie	324
F	Société Française de Phlébologie	294
GB	Venous Forum RSM	23
I	Assoziazzione di Flebologia ambulatoriale	60
SC	Scandinavian Society of Phlebology	31
USA	North American Society of Phlebology	166
DIV	divers pays	11
TOTAL		1039

Le nombre de ces réponses ne représente qu'une fraction variable des membres des Sociétés, qui n'ont pas saisi l'importance de cette enquête. Malheureusement bon nombre de questionnaires est revenu plus ou moins incomplet. De ce fait les chiffres qui seront indiqués par la suite ne prétendent nullement être pleinement représentatifs des principaux membres de la phlébologie mondiale, mais donnent tout de même des tendances générales. Ces bases partielles n'ont pas permis un dépouillement statistique incontestable.

RESULTATS

Le tableau N° 1 suivant indique le pourcentage des spécialités d'origine indiquées par les sclérothérapeutes (en **gras** souligné figure le chiffre maximum par spécialité).

	A	CH	CND	D	F	GB	I	SC	USA	DIV	TOTAL
n	63	54	13	324	294	23	60	31	166	11	1039
DERM	**60.3**	27.9	15.4	46.3	6.1	-	6.7	-	17.5	18.2	24.8
ANGIO	-	14.3	7.7	-	**83.3**	-	10.0	-	-	-	24.5
OPR	7.9	31.3	**53.8**	19.5	9.1	-	20.0	3.2	17.5	-	15.7
CHIR	20.6	9.4		20.0	0.4	21.7	30.2	**58.2**	15.0	9.1	15.0
CVASC	6.3	-		4.0	0.4	**78.3**	20.0	32.2	5.4	72.7	7.2
SP DIV	-	-	7.7	0.6	0.7	-	-	3.2	**40.4**	-	7.6
INT	3.2	**11.2**	-	5.3	-	-	1.7	-	1.8	-	2.8
ORTHO	-	-	-	1.9	-	-	1.7	3.2	-	-	0.7
GYN	-	1.8	-	1.2	-	-	**1.7**	-	0.6	-	0.6
PHLEB	1.7	3.7	7.7	-	-	-	-	-	-	-	0.5
S/Rép.	-	-	7.7	0.6	-	-	-	-	1.8	-	0.6

Tab. 1 Sclérothérapie et spécialité médicale
DERM = dermatologues ANGIO = angiologues OPR = omnipraticiens
CHIR = chirurgiens généraux CVASC = chirurgiens vasculaires
SPDIV= spécialités diverses INT = médecine interne ORTHO = orthopédistes
GYN = gynécologues PHL = phlébologues S/Rép. = sans réponse

Ce tableau suggère quelques remarques par pays:

L'Autriche compte parmi ses sclérothérapeutes 60,3 % de dermatologues , pourcentage le plus important de tous les autres pays. L'enseignement en angiologie y est trop récent pour que des jeunes angiologues figurent déjà dans cette liste.

La Suisse par contre compte déjà quelques angiologues, l'enseignement officiel de cette spécialité date d'environ 6 ans. Les groupes des dermatologues et omnipraticiens sont équilibrés autour de 30 %, mais celui des internistes atteint 11.2 %, l'angiologie étant rattachée aussi bien à la dermatologie qu'à la médecine interne.

Le Canada se caractérise par le plus important pourcentage de médecins généralistes, 53.8 % et un groupe moins important de dermatologues.

L'Allemagne avec ses 46.3 % de dermatologues se classe immédiatement derrière l'Autriche. Les omnipraticiens sont pratiquement aussi nombreux que les chirurgiens.

La France occupe une place à part avec ses 83.3 % d'angiologues médicaux, conséquence d'un enseignement spécialisé depuis 1966. Les dermatologues par contre sont très peu nombreux. A noter le nombre minime de réponses venant de chirurgiens.

En Grande Bretagne seuls les chirurgiens vasculaires et généraux traitent des varices, avec un maximum de 78.3 % de chirurgiens vasculaires pour tous les pays.

L'Italie compte 50 % de chirurgiens parmi ses sclérothérapeutes, mais aussi 10 % d'angiologues, spécialité enseignée par certaines universités italiennes.

La Scandinavie fait la part du lion à la chirurgie, avec plus de 90 % de réponses et confirme ainsi cette tendance qu'elle partage avec la Grande Bretagne.

Les Etats-Unis indiquent 40.4 % de spécialités diverses, car non spécifiées dans les réponses. Les dermatologues, omnipraticiens et chirurgiens forment des groupes identiques, avec la restriction précédente.

Finalement le total général indique une prééminence des dermatologues et des angiologues, ces derniers surtout du fait de la France.

Le tableau N° 2 indique l'activité des sclérothérapeutes recensés (en **gras** les chiffres maxima).

	A	CH	CND	D	F	GB	I	SC	USA	DIV
n	63	54	13	324	294	23	60	31	166	11
VUS	36.1	47.5	94,6	63.8	79.3	20.4	22.5	**5.1**	27.3	**118.9**
TRAIT	17.3	28.6	**93.0**	35.9	77.4	7.3	17.4	1.7	24.4	46.6
%	47.9	60.2	**98.4**	56.3	97.6	35.8	77.3	33.3	89.4	39.2

Tab. N° 2: Activité moyenne des sclérothérapeutes
VUS = nombre de malades veineux vus par semaine
TRAIT = nombre de malades traités par semaine
% = pourcentage de malades traités

Un chirurgien roumain voit en moyenne le plus de malades par semaine, mais il additionne tous les malades vus par son équipe hospitalière entière. En pratique individuelle les Canadiens voient 94.6 malades par semaine et en traitent 98.4 %, chiffres qui devraient s'expliquer par la densité faible de phlébologues dans ce vaste pays. L'angiologue français est consulté par environ 79 malades vus par semaine. Ce chiffre français élevé s'explique par l'enseignement angiologique de ce pays, les malades venant consulter directement l'angiologue pour se faire traiter (97.6 %).

Le sclérothérapeute scandinave ne voit que 5.1 malades par semaine et ne sclérose en moyenne que 1.7 malades pendant le même période.

Le tableau N° 3 est consacré aux sclérothérapeutes ayant indiqué le maximum d'activité par pays (en **gras** le maximum)

	A	CH	CND	D	F	GB	I	SC	USA	DIV
VUS	300	300	300	**700**	250	90	100	20	300	**1000**
par	CHIR	DERM	ANG	DER	ANG	CHIR	CHIR	CVASC	OPR	CHIR
TRAIT	200	190	310	**400**	250	44	100	19	300	300
par	CHIR	ANG	ANG	DER	ANG	CHIR	CHIR	CHIR	OPR	CHIR

Tab. N° 3: Activité maximale des sclérothérapeutes
VUS = nombre de malades veineux vus par semaine
TRAIT = nombre de malades traités par semaine

Mis à part le cas roumain déjà cité, le record absolu en pratique individuelle revient à un dermatologue allemand consulté par 700 malades par semaine, l'activité la plus faible est celle d'un chirurgien scandinave avec 20 malades par semaine. Le nombre maximal de traitements se chiffre à 400 pour un dermatologue allemand, contre 19 scléroses par semaine effectuées par le chirurgien scandinave le plus actif. Ces chiffres extrêmes posent au moins le problème de la qualité de soin et celui de l'expérience minimale.

Un autre aspect du questionnaire concernait l'enseignement de la sclérothérapie. La variabilité entre le nombre d'options de réponses suggérées par les questionnaires nationaux était importante, de sorte qu'il a fallu se limiter à 4 réponses possibles dont les résultats figurent dans le tableau N° 4. Chaque médecin pouvait cocher plusieurs réponses (**en gras** les pourcentages maxima).

	A	CH	CND	D	F	GB	I	SC	USA	DIV	TOTAL
n	63	54	13	324	294	23	60	31	166	11	1039
HOP/UNI	47	39	15	38	48	**65**	41	62	11	61	38.0
COLLEG	24	32	**38**	27	33	0	4	0	31	0	27.2
COURS	10	13	31	23	12	15	**44**	10	35	23	21.5
AUTOD	19	16	16	12	7	20	11	**28**	23	16	13.3

Tab. N° 3: Formation des sclérothérapeutes
HOP/UNI = enseignement hospitalier ou universitaire
COLLEG = apprentissage par collègue
COURS = enseignement par cours spécialisés
AUTOD = médecin autodidacte

Les chirurgiens de Grande Bretagne et de Scandinavie ont des points communs: ils apprennent la sclérose des varices dans leurs hôpitaux et de ce fait ne demandent pas les conseils de leurs collègues en dehors des cliniques. La France dispense un enseignement officiel de l'angiologie, mais 33 % font tout de même appel à l'expérience de leurs collègues, ce qui relativise la qualité de l'enseignement de la sclérothérapie.

Les Canadiens aiment voir travailler leurs collègues (38%) et fréquentent assidûment les cours de sclérothérapie (31%), mais y sont dépassés par les collègues italiens (44 %).

La Scandinavie, les Etats-Unis et la Grande Bretagne comptent au moins 20 % d'autodidactes, les Français en déclarent le moins avec 7 %.

CONCLUSIONS

La diversité importante de toutes ces réponses suggère qu'il serait souhaitable que l'enseignement des maladies vasculaires touchant les artères, les lymphatiques et les veines soit officialisé sous forme de spécialité d'angiologie dans tous les pays développés. La formation en sclérothérapie devrait se développer avant tout dans les Centres Universitaires dispensant l'enseignement officiel, mais aussi faire appel à des Polycliniques privées pour combler une carence éventuelle. Les angiologues seraient ainsi mieux formés en sclérothérapie, leurs malades accéderaient à des soins de qualité et en conséquence la Santé Publique diminuerait ses dépenses de soins et d'invalidité.

Phlebology '95, D. Negus et al. (eds.). Phlebology (1995) Suppl. 1: 604-605

PI/9.1

Consensus Conference on Sclerotherapy of Varicose Veins of the Lower Limbs

U. Baccaglini G. Spreafico, C. Castoro, P. Sorrentino and E. Ancona

University of Padua, Italy, 2nd Department of General Surgery

Sclerotherapy of varicose veins of the lower limbs is a technique widely used in clinical practice for the past several years now.

The available scientific literature reports conflicting data concerning the results of sclerotherapy. In addition, phlebologists have different opinions on this topic.

These differences make it difficult to clearly define both the role of sclerotherapy in the treatment of varicose veins and the criteria of choice between sclerotherapy and other treatments.

A Consensus Conference on Sclerotherapy of Varicose Vein of the Lower Limbs was organized keeping all these points in mind; the first meeting was held in Padova on the 24th September 1994.

The experts who participated in the CONSENSUS CONFERENCE and the subjects to be discussed were chosen after an enquiry amongst experts personally known, and according to the advice given by several international Phlebological Societies. In fact 35 Representatives or Secretariats of Phlebological Societies were contacted and asked to indicate two experts from every Country and to supply a list of subjects for discussion.

A bibliographic research of the literature classified in the *Index Medicus* from 1966 to 1994 was carried out, and 672 articles were collected and sent for review to the experts. Many papers were unknown to the experts and different opinions were expressed on the same papers. The scientific weight of the majority of papers was considered quite low. Therefore it was recognized that the available literature was greatly limited from the statistical and scientific points of view, and so it was of little use in reaching a consensus.

In trying to reach a consensus both research and personal experience were relied upon. Due to limits in literature, personal experience of the experts has therefore become significant.

Four topics were chosen for discussion: INDICATIONS, TECHNIQUES, RESULTS AND ECHOSCLEROSIS. After the first meeting, ECHOSCLEROSIS was incorporated into TECHNIQUES.

In order to better define the state of art of current sclerotherapy, information obtained from a research done with questionnaires sent to phlebologists in various Countries was also taken into consideration. 6 Countries returned approximately 700 questionnaires.

In conclusion the Consensus Conferenece was based upon 3 points: the literature, the personal experience of the experts and the results of the survey on current practice. Research and personal experience were approached in a structured manner.

In the Consensus Conference there were:

a) opinions where everybody definitevely agreed or where it was thought that it was probably right. These are definitive or probable opinions, and they can be taken forward to some kind of recommendation.

b) opinions where nobody knew. These are potential research proposals.

c) opinions where nobody agreed. These are areas of active disagreement and here it is more difficult to move forward to research. It must however be stressed that when there was a disagreement, this did not necessarily mean that this was not recommended.

It's a myth to think that a consensus is absent or present. A consensus tends to be graded, there is a whole range of opinions.

The process, and thus the results of the consensus, are dependent on the people who are involved in it. Angiologists, phlebologists, dermatologists, surgeons, vascular surgeons and epidemiologists participated in the Consensus Conference.

Three Consensus Conference meetings were held. The first one in Padova on the 24th September 1994, the second one again in Padova on the 11th - 12th February 1995 and finally the third one in Mogliano Veneto (Venice) on the 13th - 14th May 1995.

All the presentations and discussions of the various sub-committees and Plenary sessions were recorded, transcribed and reviewed by the Chairmen.

A survey was carried out, amongst the participating experts, by means of questionnaires, concerning the most controversial issues which were brought up during the discussions.

A consensus paper was drafted twice and discussed in the Plenary session. It summarizes the principle issues which were dealt with and indicates the points of agreement and disagreement; controversial points are also indicated.

The topics dealt with in the Consensus Conference document are: Consensus Process, Indications, Pre-treatment Assessment, Techniques, Results, Training, Recommendations.

The definitive Consensus Conference document will be availabe in September 1995.

The following is the list of those who participated in the Consensus Conference:

L. Abenhaim	Canada	G. Lipari	Italy
E. Ancona	Italy	O. Maleti	Italy
U. Baccaglini	Italy	L. Norgren	Sweden
E. Baggio	Italy	P. Parpex	France
E. Beverdam	NL	H. Partsch	Austria
K.G. Burnand	UK	P. Pavei	Italy
C. Castoro	Italy	P. Raymond-Martimbeau	USA
F. Chleir	France	M. Schadeck	France
A. Cornu-Thenard	France	U. Shultz-Ehrenburg	Germany
A. Davy	France	JH. Scurr	UK
E. Einarsson	Sweden	P. Sorrentino	Italy
FGR Fowkes	UK	G. Spreafico	Italy
C. Garde	France	I. Staelens	Belgium
H. Gerlach	Germany	R. Stemmer	France
L. Grondin	Canada	F. Vin	France
K. Hoerdegen	Switzerland		

P122

HYALURONIDASE IN THE PREVENTION OF EXTRAVASATION NECROSIS: A DOSE-RESPONSE STUDY

Zimmet, S.

Austin Regional Clinic, Austin, Texas, USA

Objective: To conduct a dose-response study using hyaluronidase in the prevention of necrosis following intradermal 23.4% sodium chloride.

Design: Randomised blinded parallel dose design.

Subjects: 65 female Sprague-Dawley rats.

Intervention: Hyaluronidase was administered in doses ranging from 18.75 to 900 units (in volume of 3 mL) immediately following the intradermal instillation of 0.25 mL of 23.4% sodium chloride. A control group had no therapy.

Measurements: The incidence of ulceration was compared for each group three days after the intervention.

Results: A statistically significant protective effect was found in the treated vs. untreated groups (p=.01 Scheffe comparison). A dose response curve was generated (SPSS CNLR non-linear regression routine). Maximal protection (0.67) was achieved with 75 units of hyaluronidase and was not improved upon by higher doses.

Conclusion: In the event of extravasation with 23.4% sodium chloride, in the model studied, one can expect maximal protection with a dose of 75 units of hyaluronidase. Maximal benefit yielded avoidance of ulceration in 67% of treated subjects.

P203

COMPRESSION SCLEROSURGERY- RADICAL TREATMENT OF VARICOSE VEINS.

Konstantinova G., Alekperova T.

Russian Centre of Out-Patients Surgery, Moscow, Russia.

OBJECTIVE: To develop simple and effective low-traumatic method for out-patients treatment primary varicoses.
DESIGN: Retrospective review of a three years experience.
PATIENTS: 350 patients with primary varicoses.
INTERVENTION: Sclerosurgery treatment was conducted in specialized Centre of out-patients surgery. Sclerosurgery included crossectomy, Coccet procedure and catheter sclerotherapy of long or short saphenous veins by aethoxyscleroly. Just after operation all patients were made active and transported at home.We used compression sclerotherapy of varicose tributary in 1-1.5 months after operation, additionaly.
MEASUREMENTS: All patients were included in every six months follow-up with Doppler-ultrasound and Duplex-scanning control.
RESULTS: 117 patients (33.4 per cent) required postoperative compression sclerotherapy. In 29 cases we had hyperpigmentation in operation area, which dissapeared in 5-9 months period. We had not reccurent varicoses during three years follow-up.
CONCLUSION: Sclerosurgery is effective complex method for out-patients treatment of the primary varicoses.

P124

TREATMENT OF LESIONS BY ESCAPE OF VARIGLOBIN DURING SCLEROTHERAPY OF CROSSE S.I.

Dr Rebuffo E M and Zannini G
Phlebology Surgery, Turin, Italy

Objective: Evaluating lesions, healing at short medium term and patients' clinical conditions.
Design: Retrospective review of a consecutive series.
Patients: 8 patients out of 70, all greatly obese.
Intervention: During sclerotherapy of S.I. crosse, total or partial escape of Variglobin at 8%-12% of concentration is possible; the area in infiltrate, inflamed, hard, with a diameter of 3-4 cm near the groin; 8 mg of Betamethasone is directly injected into the nodule, and after 4 days another 8mg of Betamethasone; the resolution is quick, the nodule disappears and the patient does not feel pain any more.
Measurements: Clinical state is evaluated and thanks to a plicomether the necrotic area is controlled after 48-72 hours and after 7-14 days.
Results: 7 patients out of 8 had rapid resolution of both lesions and pain just after 48 hours. 1 patient had a light residual fibrosis.
Conclusions: This technique rapidly resolves lesions and allows sclerotherapy to be pursued without any discomfort for the patient.

P125

UNUSUAL VARICOSE VEINS . TREATMENT BY "SWISS METHOD" OF SCLEROTHERAPY.

DESOGUS A. I., CARTA M., POLO M.
Department of General and Emergency Surgery, "Marino" Hospital, Cagliari, Italy

OBJECTIVE: review preliminary clinical experience in varicose veins (V.V.) treatment characterised by an anomalous course independent by a saphenous reflux.

DESIGN: retrospective review of 229 cases of primary varicose veins.

PATIENTS: 9 of them (8 fem, 1 male; age 2 to 51 yr.) were characterised by anomalous course of V.V., not maintained by saphenous reflux, usually located in posterior or posterolateral surface of the leg. Clinical balance included physical and doppler C. W. examination; only in two cases descending flebography has been performed.

INTERVENTION: standardised surgical thecnique (Muller) and local anaesthetic procedure was performed (day surgery) in two cases. In 7 pt. we used the so called "swiss method" of sclerotherapy, developed by Sigg, with "air block" technique and immediate elastocompression. No injection was made in a leg with oedema.

MEASUREMENTS: clinical outcome by scoring system and anatomo-physiological response by C. W. doppler, documented 2, 6, 10 months following therapeutic procedure.

RESULTS: In our experience sclerotherapy, according to a less traumatic approach, plays a precise role. Sclerotherapy is the choice treatment, with safe, rapid, aesthetic and stable results.

P126
RADIOFREQUENCY BIPOLAR ELECTROCOAGULATION AND SCLEROTHERAPY: AN EFFECTIVE PROCEDURE

Lifsitz S, Baza R
Surgical Department, Hospital Penne, Bahia Blanca, Argentina

Objective: To avoid the undesirable brown pigmentation which appears when sinuous veins showing raised patterns, are treated exclusively with sclerotherapy.

Design: Retrospective review of a consecutive series.

Patients: 50 patients, (64 legs) with sinuous varicose veins of small and medium diameter showing a raised pattern image.

Intervention: This procedure proposes the combination of radiofrequency bipolar electrocoagulation and the injection of low doses and low percentage of sclerosant material.

Measurements: Slides were taken before and 15 days after the technique was performed.

Results: 87% of the patients treated showed a surprisingly good and quick result, without sequelae of undesirable pigmentation.

Conclusions: This procedure provides an effective result in the treatment of sinuous varicose veins of small and medium diameter with raised pattern image.

P128
TREATMENT OF LESIONS FOLLOWING EXTRAVASATION OF CHEMOTHERAPICS BY LASER HeNe/IR

Balzarini A, Pirovano C, Martino G, Zanolla R
Physical Therapy and Rehabilitation Department, National Cancer Institute, Milan, Italy

Objective: To evaluate the therapeutic efficacy and the optimisation of the Laser HeNe/IR application times in the treatment of the lesions following extravasation of chemotherapics.

Design: Randomised study (three groups), open clinical trial.

Patients: 60 patients with extravasation of chemotherapics.

Intervention: 20 patients (group A) were treated by Laser HeNe/IR 20' a day for 2 weeks; 20 patients (group B) by Laser HeNe/IR 10' a day for 2 weeks; the last group of 20 patients (group C) only with pharmacological therapy.

Measurements: Daily local examinations, evaluation of the pain intensity by Keele Scale and of the articular functionality by Range of Motion (ROM), photographs of the lesions at the beginning and at the end of treatment. The follow-up lasted 2 months.

Results: In the patients treated by Laser we observed an immediate reduction of the lesions dimensions, the oedema, the erythema and the pain, starting from the first days of therapy. At the end there was a complete recovery, above all in the early treated cases (23 patients). In the late treated ones we noticed, at the control after 2 months, a tendency to vascular sclerosis. No statistically significant differences were evident between A and B groups as regards the treatment duration. In group C there was persistence of signs and/or symptoms in 7 cases, with the need of prolonging therapy and, in 5 of them, with presence of vascular sclerosis at the final control.

Conclusions: The therapeutical scheme B has showed the same efficacy as scheme A, but has resulted more favourable in terms of time/patient, allowing a larger number of treatments.

P133

ESTHETIC SCLEROTHERAPY OF THE FACE
Technique, Indications and Limits
Report on a series of 86 cases

Christian DANIEL, M.D., Angiology and Phlebology Unit
2, rue de la Réunion, 92500 Rueil-Malmaison, FRANCE

Esthetic sclerotherapy of the face not only concerns varicosities but also bluish varicose veins and pseudo-angiomatous forms.

The technique and results are presented for 86 cases with a follow-up of six months to seven years, and by comparison with the literature.

In this series, the micro-sclerosis technique gave 74 % satisfactory results (total disappearance and patient satisfaction) without contra-indications or notable adverse effects. Moreover, the cost-efficacy ratio of this technique is markedly superior to that of other techniques such as Laser and Electrocoagulation.

Nonetheless it is important to emphasize the training required to perform micro-sclerosis and the extreme caution in choosing the doses and quantities of sclerotics to be used. Together with appropriate training, this technique also requires experience and dexterity.

P134

"PREVENTION OF THE POST - SCLEROTHERAPEUTIC PIGMENTATION"

Avramovic Aurora, Avramovic Miguel

CENTRO FLEBOLOGICO PRIVADO
RIOBAMBA 747 BUENOS AIRES (1025) ARGENTINA

OBJECTIVE: To prevent the formation of post-sclerotherapeutic pigmentation, which according to our statistics, is the most frequent local complication (14%) that may bring about legal - medical problems.

DESIGN: A decrease of residual pigmentation during the last 4 years, by applying the correct technique.

METHOD: Before undergoing sclerotherapeutic treatment
1) Observe the type and color of skin, 2) Prepare de skin, 3) No infections must be present, 4) Correctly feel the varicose veins. 5) Use the correct sclerosing substance, 6) Know well the pigmentation capacity of each sclerosing substance.

DURING SCLEROTHERAPY: 1) Go from the vein of greater caliber to the smaller. 2) Cathetize the vein. 3) Inject in different parts. 4) Inject slowly. 5) Correct position of the patient. 6) Sessions each 15 or 21 days.

AFTER SCLEROTHERAPY: 1) Avoid venous estasis (elastocompression). 2) Avoid the melanocitic process (antinflammatories, inhibitors of the biosynthesis of thyroxine.)

RESULTS: Using these preventive methods, we have reduced the formation of post - therapeutic pigmentation by 50% with respect to the former statistic.

CONCLUSION: The prevention is useful to reduce aesthetic problems produced by sclerotherapy. Research must be continued in order to discover other ways to totally eliminate this complication.

P137

TREATMENT OF LESIONS FOLLOWING EXTRAVASATION OF CHEMOTHERAPICS BY LASER HeNe/IR

Balzarini A, Pirovano C, Martino G, Zanolla R.

Physical Therapy and Rehabilitation Department, National Cancer Institute, Milan, Italy.

Objective: To evaluate the therapeutic efficacy and the optimization of the Laser HeNe/IR application times in the treatment of the lesions following extravasation of chemotherapics.

Design: Randomised study (three groups), open clinical trial.

Patients: 60 patients with extravasation of chemotherapics.

Intervention: 20 patients (group A) were treated by Laser HeNe/IR 20' a day for 2 weeks; 20 patients (group B) by Laser HeNe/IR 10' a day for 2 weeks; the last 20 ones (group C) only with pharmacological therapies.

Measurements: Daily local examination, evaluation of the pain intensity by Keele Scale and of the articular functionality by Range of Motion (ROM), photographs of the lesions at the beginning and at the end of treatments. The follow-up lasted 2 months.

Results: In the patients treated by Laser we observed an immediate reduction of the lesions dimensions, the oedema, the erythema and the pain, starting from the first days of therapy. At the end there was a complete recovery, above all in the early treated cases (23 patients). In the late treated ones we noticed, at the control after 2 months, a tendency to vascular sclerosis. No statistically significant differences were evident between A and B groups as regards the treatment duration. In the group C there was persistence of signs and/or symptoms in 7 cases, with the need of prolonging therapy and, in 5 of them, with presence of vascular sclerosis at the final control.

Conclusions: The therapeutical scheme B has showed the same efficacy than the A one, but has resulted more favourable in terms time/patient, allowing a larger number of treatments.

P145

DOES NEOVASCULARIZATION OCCUR POST SCLEROTHERAPY

Forrestal M, Zummo M,

Private Phlebology Practice, Chicago, USA, Montreal, Canada

Objective: To evaluate patients who have been treated by sclerotherapy of the LSV for the type of SFJ recurrence known as neovascularization, a documented phenomenon following surgical treatment.

Design: Prospective longitudinal clinical study utilizing DUS imaging pretreatment and at 3, 6, and 12 months post-treatment. Further yearly studies are planned.

Patients: 80 patients with primary varicose veins and SFJ reflux.

Intervention: All patients treated with sclerotherapy of the SFJ and LSV by one of two methods. 40 patients treated via DUS guided injection of 3% Sotradecol. 40 patients treated with progressive concentration injections of iodine-sodium iodide.

Measurements: DUS evaluation of the post sclerotherapy treated SFJ and proximal LSV for presence of neovascularization recurrence at 3, 6, 12 months and yearly.

Results: The patients tolerated sclerotherapy well with minimal complications. Early fibrosis as demonstrated on DUS by noncompressible proximal LSV segment without Doppler flow was noted at 3-6 month FU. Full one year studies will be presented.

Conclusions: Preliminary data have not revealed neovascularization in this series. Long-term followup is needed and underway to gain insights into whether or not sclerotherapy by its different mode of action will stimulate the neovascularization recurrence described following surgery.

P146
COMPARISON OF HYPERTONIC SALINE AND HYPERTONIC COMBINATION OF DEXTROSE AND SALINE IN TREATING VENULECTASIA AND TELANGIECTASIA.

Zummo M
Private practice, Montreal, Canada

Objectives: To elucidate the effects of Sclerodex™ (NaCl/Dextrose) and an equivalent formulation without dextrose with regards to adverse reactions, complications profile and discomfort.

Design: Prospective longitudinal "*paired*" comparison study.

Patients: Between 20 and 40 female outpatients, aged 20-70 years, with primary varicose veins (VV) of the lateral thighs will have their venulectasia and telangiectasia injected.

Intervention: Patient with VV of the lateral thighs are injected on one side with a 20% dextrose (200 mg/mL) / 10 % NaCl solution (100 mg/mL) with an osmolarity of 4.2 mOsm/mL and on the other side with an 11% (110.4 mg/mL) NaCl solution, osmolarity 4.19 mOs/mL

Measurements: Disappearance of VV is evaluated by comparing macro-photographs taken before and two to four weeks after the treatment. Comparison of pain, cramps, redness, pigmentation, neo-vascularization, thrombus formation, eschar formation, hypersensitivity to either sclerosing agents is evaluated.

Results: The study is still under way.

Conclusion: Up to this stage, there is no significant difference between the two sclerosing agents.

V/2.4
ECHOSCLEROTHERAPY OF THE GREATER SAPHENOUS VEIN - THE CATHETER BANDING TECHNIQUE

Louis Grondin, Ron Young, Lilly Wouters
The Grondin Medical Centre
3rd Floor, 1504 15 Avenue S.W.
Calgary, Alberta   T3C 0X9   Canada

Introduction: After having performed over 2,300 junctional echo-sclerotherapy treatments in the past three years in three distinct phases, we have observed a consistently good response for the lesser saphenous vein and a variable response for the greater saphenous vein. This observation has led us to hypothesize that the presence of variable deep venous pressure is the cause of this variable response to sclerotherapy for the sapheno-femoral junction.

Objective: To increase the initial closure rate of junctional sclerotherapy.

Design: Randomized parallel group, open clinical trial.

Patients: 40 patients with similar clinical and echographic findings have been treated with and without the catheter banding technique.

Intervention: Following the open catheter technique, a Dexon band is applied to the greater saphenous vein while the patient is in a tilt down position. The patient is subsequently observed for closure of the junction one month after injection.

Results: The two groups of 20 patients each, were well matched for age, sex and severity of disease. The data retrieved showed no statistically significant improvement in outcome with the banding technique.

Conclusion: For sclerotherapy of the greater saphenous vein at the sapheno-femoral junction, we were unable to demonstrate improvement in immediate closure rate with the application of a circumferential Dexon band following the injection.

612

OP/11.2
ECHOSCLEROTHERAPY - A THREE YEAR FOLLOW-UP

Louis Grondin, Ron Young, Lilly Wouters
The Grondin Medical Centre
3rd Floor, 1504 15 Avenue S.W.
Calgary, Alberta T3C 0X9 Canada

Objective: To evaluate the short and long term effectiveness of a new method of echosclerotherapy in the treatment of greater and lesser saphenous vein insufficiency.
Design: Retrospective review of a consecutive series.
Patients: Three distinct echosclerotherapy techniques were utilized in a patient population of approximately 140. Complications, effectiveness, as well as advantages and disadvantages of each technique were observed over a three year period.
Intervention: The open catheter echosclerotherapy technique previously described by Grondin in San Francisco in 1990, and in Montreal in 1992.
Measurement: A scoring system that evaluates compressibility, reflux severity, and maximum vein diameter in the proximal segment of the saphenous trunk is used before treatment, one month after treatment, six months, one year, two years, and three years after treatment.
Results: Initial results of 89% closure rate of the lesser saphenous vein and 82% closure rate of the greater saphenous vein were observed. No intra-arterial complications were observed. The most significant complications were three deep venous thrombosis, two of which appear to have been re-exacerbation of an underlying (chronically active) deep venous thrombosis.
Conclusion: The open catheter echosclerotherapy technique, previously described, seems to be a safe, reliable, effective, and less costly approach to primary saphenous disease, specifically in cases of lesser saphenous vein incompetence.

V/2.5
DUPLEX ULTRASOUND GUIDED SCLEROTHERAPY: SAFETY AND COMPARISON OF TECHNIQUES.

Grondin L., Young R., Wouters L.

The Grondin Medical Center, Calgary, Canada.

The term echosclerotherapy was first introduced to the medical literature in the latter part of the 1980's by Dr. Robert Knight. He described this procedure as the use of ultrasound to guide the sclerotherapy act in the treatment of large truncal varices. Since its introduction numerous publications have appeared in both the European and North American literature regarding its hypothetical usefulness. Unfortunately, numerous complications have been reported consisting mainly of intra-arterial injections following this technique.

In 1990 we modified the technique described by Dr. Knight, by using continuous doppler and ultrasound monitoring during the act of sclerotherapy. Furthermore, we have substituted the closed needle technique for an open, disposable catheter technique. To date we have treated upwards of 2,400 patients and have not caused any extravasation or intra-arterial injections.

This presentation will describe our technique with a video projection.

P139
MICROSCLEROTHERAPY OF THE FACIAL TELANGIECTASIS

Louis Grondin, Ron Young, Lilly Wouters
The Grondin Medical Centre, Calgary, Canada

A significant number of patients consult with facial or nasal telangiectasias. Several techniques are available in the treatment of this condition, namely electrocauterisation and laser therapy.

Sclerotherapy remains an effective, affordable and relatively pain-free procedure which should be offered to the patient first.

In this presentation we will discuss the technique, medication and general guidelines for this therapy.

V/2.7
THE SCLEROTHERAPY : WHAT TO DO AND WHAT TO AVOID.

Schadeck M., Mako S.

4, avenue de Melun. 94190 - Villeneuve-St-Georges FRANCE.

Objective: To show the importance of a strict clinical examination, completed, if necessary by a Duplex Scan examinatoin before each stage of sclerotherapy.
For each patient with varicose disease, even moderate (telangiectasiae), the research of an incompetent saphenous vein is systematic.
The sclerotherapy of telangiectasiae and small varicosities needs previously the research of a reflux or of feeding branch. Then it is made carefully with an appropriate equipment, using needles of 3/10, a low concentration of the sclerosing agent (Aetoxisclerol 0,25 - 0,30%) a small quantity of agent (<0,3 to 0,5 cc).
The injections must remain strictly intravascular and is interrupted at the slightest extravasation.

Conclusion: The respect of these simple rules allows an efficient and harmless treatment of varices.

614

OP/4.6

POST SCLEROSIS RECURRENCES OF THE GREAT SAPHENOUS VEIN

Schadeck M, Allaert F A

Ecole Européenne de Phlébologie, Hôpital Notre-Dame de Bon-Secours, Paris, France

Objective: to demonstrate the role of the non treated tributaries in the repermeation of saphenous trunks and the occurrence of a new reflux.

Design: randomised parallel study between two groups of 25 patients.

Patients: 50 patients with an incompetent great saphenous vein.

Intervention: a therapeutical protocol identical for the two groups, is realised with the Sotradecol 3% injected:
- at the first session in the terminal segment of the incompetent GSV,
- at the following sessions in the underlying segments.
The first group undergoes in addition a treatment of the different tributaries of the saphenous trunk.

Measurements: the Duplex Scanning checks, after the treatment, the evolution of sclerosis and its extension on the trunk. It localizes the different tributaries at the thigh.

Results: the thrombotic transformations occurring after sclerosis usually lead to a repermeation of the saphenous trunk within one to three months. The existence of non treated tributaries ending at the saphenous trunk facilitate this repermeation. The first group do not present early recanalization.

Conclusions: the treatment at the same time of GSV and of its tributaries avoid the early recanalization and delay its secondary repermeation.

V/2.6

DUPLEX SCANNING IN THE MECHANISM OF THE SCLEROTHERAPY : THE INITIAL STAGE.

Schadeck M.

4, avenue de Melun - 94190 - Villeneuve-St-Georges FRANCE

Objective: To understand the role of the sclerosing agent on the wall immediately after the injection and thus, to try to measure its action.

Design: 20 patients with an incompetent Great Saphenous Vein (GSV) receive a sclerotherapy, each group of 10 according to 2 types of technique.

Intervention: The saphenous reflux is confirmed by duplex (Esaote Idea colour with a probe of 10 MHz). In the 2 groups, the injection is made under echographic guidance in the terminal segment of the GSV:
- at the first session with Sotradecol 3% 2cc, - at the second session, 4 weeks later, with Sotradecol 3% 3cc injected in two points. The difference between the 2 groups consists in an alternative compression realized in the first group and made by the echoprobe at the level of the injection, in a "come and go" movement.

Measurements: The diameter of the vein is measured before the injection in standing then in supine postiion, then 3 minutes after the injection. We observe the endoluminal and parietal transformation, particularly a spasm, and the course of the sclerosing agent

Results: Without the alternative compression manouver the spasm spontaneously occurs in 15% of cases. With this compression manouver, a spasm (mean 75% of reduction of the cross section) occurs in 100% of cases.

Conclusion: The sclerosis is thus better quality and more rapid when made with alternative compression because of a presence of the spasm.

P227
La cryochirurgie des varices

Dr J.M. COGET - Dr J.P. MILLIEN

61 Rue de Turenne 59000 LILLE

Technique moderne d'éveinage des varices, la cryochirurgie a des adeptes car c'est une méthode tout à fait élégante, qui parfois permet de faire des éveinages sans cicatrice basse. Mais elle est aussi très performante au niveau des collatérales permettant par une simple incision de 2 à 3 mm, d'enlever des branches tortueuses, qui par la phlébectomie classique aurait nécessité de multiples incisions. Même si l'investissement est onéreux, la gamme de cryodes permet maintenant de faire des éveinages performants, qui réduisent le nombre de scléroses post-opératoires et rendent la surveillance moins lourde.

P229
VARICOSE VEINS, MINISURGERY AND SIMULTANEOUS SCLEROTHERAPY

Simkin R, Bulloj R, Fuentes A

Malabia 3166 (1425) Buenos Aires, Argentina

The authors present their experience with a combined treatment of superficial system surgery and simultaneous sclerotherapy. The treatment of the superficial system is carried out with the technique of venous minisurgery, the retrogressive stripping with a personal model stripping (Simkin), which permits us to keep the aesthetic sense of surgery and at the same time carry out the sclerotherapy. The drugs used for this kind of technique are the polidecan the sodium salicylate and the chromated glycerine. The fact that these patients remained with a post-operative compressing bandage helped and favoured the results of this combined treatment. This kind of technique is done in this way. First, the surgical treatment -fronatal and retral is carried out: then follows the sclerosing treatment, evaluating the size of the vein to be treated in the same way as if the patient were in a daily consult. The results with this kind of technique make the treatment faster and the results, more encouraging, which reduces the cost of the ambulatory treatment and the patient's time.

616

P141

SCLEROTHERAPY UNDER DUPLEX GUIDANCE IN LONG
SAPHENOUS VEIN RECURRENCY AFTER SURGERY

Ouvry P

9 rue Jules Ferry, 76200 Dieppe France

Objective: to show the technique of injection under duplex guidance
and evaluate the results of sclerotherapy in long saphenous vein
recurrency.
Design: Retrospective review of a consecutive series
Patients: 50 patients with long saphenous vein recurrency proven by
duplex sonography.
Intervention: Every patient had injections of STD or Iodine until
complete suppression of the reflux.
Measurements: Every patient had duplex sonography after the
procedure and after one you follow up.
Results: No problems of tolerance; immediate results are good
(suppression of the reflux for every patient).
Conclusions: Sclerotherapy, improved by injections under duplex
guidance seems to be a very efficient treatment for long saphenous vein
recurrency.

P142

AMBULATORY PHLEBECTOMY OF THE FOOT
Review of 75 cases
OLIVENCIA , J.A.
Iowa Vein Center
Des Moines, Iowa USA

Objective: Review of Ambulatory Phlebectomies
performed on the foot.
Design: Retrospective review of 75
consecutive cases
Patients: 75 patients (105 limbs) in a two
year review
Measurements: Age distribution, gender
distribution, micro-incision
technique, diferent hooks,
anesthesia technique.
Results: Review of complications, wrapping
technique and technical suggestions
Conclusions: Acurate diagnosis, knowledge of
local anatomy, proper surgical
technique, careful wrapping
and early ambulation are essential
for satisfactory results.

PI/6.4
THE AUSTRALIAN POLIDOCANOL (AETHOXYSKLEROL) OPEN CLINICAL TRIAL RESUTLS AT THREE YEARS

P Conrad, Mark Malouf, Sydney, Australia

The ANZ Society of Phlebology has been conducting a clinical trial using Polidocanol (Aethoxysklerol) for compression sclerotherapy of suitable leg varicose veins, venules, and spider veins since September 1991. 120 doctors with various prior sclerotherapy experience using sodium tetradecyl sulphate and/or hypertonic saline as sclerosants, are participating in the non-randomised, non-blinded trial. Strict trial guidelines apply. The results at three years are presented. By June 1994 Polidocanol (Aethoxysklerol) had been used on 24,466 legs. 10,332 legs with varicose veins, 14,134 legs with smaller veins. There were no deaths and no anaphylaxis reported. Pain on injection was very rare. Thirty eight possible allergic/sensitivity reactions were reported. Three deep vein thromboses were reports, 51 legs suffered injection ulcers. Superficial thrombophlebitis and excessive hyperpigmentation were rare. In comparing retrospectively their use of Polidocanol (Aethoxysklerol) with sodium tetradecyl sulphate or hypertonic saline sclerosants, 77% of injecters with previous sodium tetradecyl sulphate experience rated the effects of Aethoxysklerol superior; 75% of the injecters experienced with hypertonic saline reported the efficacy of Aethoxysklerol superior. Similar trends were indicated with respect to the incidence and severity of complications of Polidocanol compared with the other two solutions. Conclusion: Polidocanol (Aethoxysklerol) is safe, as effective, and has fewer and less severe complications than the sclerosants that have been in use in Australia to date. Properly conducted randomised and blinded trials are proceeding.

V/2.12
ABDOMINAL VARICOSITIES TREATED BY SCLEROTHERAPY.

Kanter A.

Vein Center of Orange County (Private Practice) Irvine USA.

Objective: To follow the long-term natural evolution of post-thrombotic central vein recanalization, and describe the results of subsequent sclerotherapy to collateral abdominal veins.

Design: Case study.

Patient: Single case history followed 13 years.

Intervention: Compression sclerotherapy to abdominal varicosities using sodium tetradecyl sulphate.

Measurements: Initial venography, with subsequent clinical, duplex ultrasound, and MRI-Angio exams.

Results: Late resolution of lower extremity edema after spontaneous recanalization of the inferior vena cava was observed, followed by successful single-treatment sclerosis of abdominal varicosities.

Conclusion: Eventual recanalizsation of central vein thrombosis is possible many years after the inital event. Sclerosis of abdominal collateral veins may be performed successfully and safely following careful documentation of central vein recanalization.

P144

COSMETIC SCLEROTHERAPY OF BREASTS AND HANDS

de Groot W P,

Seattle USA

Most cosmetic sclerotherapy is performed for unsightly veins of the legs.

The author has treated a limited number of female patients for enlarged veins of the breasts and hands, which the patients deemed to be cosmetically unacceptable.

Sclerotherapy of the breasts and hands have to be done with great care. In most cases, a result can be obtained which is satisfying to the female patient.

OP/4.8

FADING RATE OF PIGMENTATION INDUCED BY SOTRADECOL FOR SCLEROTHERAPY OF VARICOSE VEINS

Marley W

Private Practice, Bensalem, PA, USA

Objective: To determine the incidence of pigmentation using sotradecol for sclerotherapy of reticular and varicose veins and record the rate of resolution.

Design: Prospective consecutive series of patients treated and evaluated by the same physician.

Patients: 144 previously untreated patients with reticular varicose veins were given a total of 620 sclerotherapy sessions with sotradecol as the sole sclerosant.

Intervention: Patients were treated using the "French" technique of sclerotherapy (i.e. proximal to distal). The concentration of sotradecol ranged from 0.1% to 3.0%. The presence or absence of pigmentation was noted on each subsequent visit at two weeks, six months, 1 year, 2 years and 4 years.

Results: Sclerotherapy induced pigmentation was seen in 25% of patients at two weeks. 16.7% at six months, 9% at one year, 4.8% at two years, and 2.1% at four years. Pigmentation lightened in all patients over time.

Conclusion: Sclerotherapy induced pigmentation fades over time and should not be a major concern when using sotradecol as the sole sclerosant.

V/2.13

COMPRESSION SCLEROTHERAPY.

Bernbach H.R.

Lugano, Switzerland.

Increasing safety and efficacy of the sclero-therapeutic procedure was the aim of Karl Sigg's injection technique. This is shown in detail (thick needle, glass syringe, air block).
Using a strong but safe sclerosing agent (iodine solution, Variglobin) in its different concentrations is the other hallmark of this technique.
We have added a system of strong compression which reduces "thrombus"-formation and by that the rate of recanalisation.
This latter being the main cause of many cases of recurrencies after inadequately performed sclerotherapy.
Our way of treating telangiectasias is shown.

OP/9.5

THE EFFECT OF SCLEROTHERAPY ON THE NUMBER OF CIRCULATING ENDOTHELIAL CELLS (CEC)

Strejcek J*, Buhtová L**, Marecková B***

*Centre for Dermatological Angiology, **1st Dermatological Clinic:
Univ. ***1st Pathological Institute Charles Univ., Prague, Czech Republic

Objective: It is assumed that the mechanism of action of sclerosing agents causes damage to the endothelial resulting in endofibrosis.
Design: We assumed that it may be possible to detect an increase in circulating endothelial cell (CEC) immediately after sclerotherapy.
Patients: 30 normal controls without varices, 30 patients with reticular varicose and telangiectasies, 30 patients with extensive varicosities (truncal with large convolutes).
Measurements: We developed a special method for counting of CEC in peripheral blood and monitored the numbers of CEC after injections of Aethoxyclerol in patients with varicose veins. The number of the circulating cells has be related to:

a) the concentration of Aethoxycerol
b) total amount of injected drug
c) times from injection

Results:

1) Ratio of CEC in normal controls without varices 2.3, retiucalr and telangiectasia 2.45, extensive varicosities 3.8 in 0.9 microlitres
2) Number of CEC increases in dependence to used concentration of Aethoxyclerol
3) In dependence to total amount of injected drug there is also an increase of CEC
4) During the first 30 minutes after sclerotherapy the number of CEC rapidly increases to 28, after 120 minutes is 30. Then the number of CEC slowly decreases, but after 10 days it is still elevated (8 per 9 microlitres).

Conclusions: The method of counting CEC seems to be useful for testing of the extent of endothelial damage caused by sclerotherapy, maybe also for testing how 'strong' the used agent is, or to detect some complications of sclerotherapy (DVT?)

PI/6.3

AETOXISCLEROL 4% IN THE TREATMENT OF THE SAPHENOUS VEINS

Schadeck M

4 avonuede Melun 94190 Villeneuve-St-Georges France

Objective: To evaluate the efficiency of Aetoxisclerol 4% in the treatment of the great saphenous veins versus Sotradecol 3%.
Design: Comparative study of 2 groups undergoing 2 different sclerosing agents, delivered at random.
Patients: 30 patients with an incompetent Great Saphenous Vein (GSV) receive sclerotherapy, each group of 15 according to 2 types of sclerosing agent.
Intervention: The duplex scanning confirms the reflux of the GSV and makes measurement of its diameter GSV with a diameter higher than 6mm are excluded.
In the two groups the injection is made under echographic guidance in the terminal segment of the GSV according to the same technique. The first group is treated with Sotradecol 3% and the second one with Aetoxisclerol 4%.
Measurements: The diameter of the vein is measured before the injection in standing then in supine position, then 3 minutes after the injection. We observe the endoluminal and parietal transformations, particularly a spasm in the second group and the course of the sclerosing agent.
Results: After two sessions there is a significant difference between the two groups. A spasm (a mean of 75% of reduction of the cross section) occurs in 100% of cases in the group 2 (Aetoxisclerol) with excellent and quick results (85% disappearance of reflux). In the group l, (Sotradecol 3%) the spasm occurs only in 15% of cases (65% of good results).
Conclusion: Aetoxisclerol 4% can give an excellent result in the treatment of the great saphenous vein because of the occurrence of a spasm and of its importance in the mechanism of sclerotherapy. It favours the best results in a short time.

PI/6.5

CONTROVERSIES IN SCLEROTHERAPIE

Jean van der STRICHT

HOPITAL MOLIERE-LONGCHAMP. UNIVERSITE DE BRUXELLES
23 avenue du Domaine - II90 BRUXELLES

Controversies about sclerotherapy have already existed for a long time. They generally concern the techniques: material used, position of the patient, the use of a complementary contention and compression, the air -block technique, tricks to avoid complications etc... Such debates are always instructive but they never lead to a real concensus. The techniques vary from one phlebo-logist to another. Standardization of sclerotherapy cannot be imagined and morover is undesirable.

More recent controversies concern the investigation methods, especially Doppler, that are used before or even during a sclerosing treatment. Some of these new methods are really helpful in particular cases. Nevertheless, as a general rule, their necessity is inversely proportional to the experience of the phlebologist.

Real problems remain when talking about indications of sclerotherapy. How far can sclerotherapy be proposed to treat the main incompetent superficial veins ? To what extend is phlebectomy a challenge to sclerotherapy ? More pragmatic: what about sclerotherapy in young people, especially in young girls before pregnancy. Is sclerothe-rapy compatible with obesity ?

Finally a problem that has rarely been brought up: does " excessive sclerotherapy " exist ?

When can venous the drainage of the skin be trea-tened ?

The question is imperative and the answer is far from being easy.

The Treatment of Varicose Veins and Telangiectasias

Pharmacology

Phlebology '95, D. Negus et al. (eds.). Phlebology (1995) Suppl. 1: 622-625

P290

The Effects of Cyclo 3 Fort Treatment on Hemorheological Disturbances During a Provoked Venous Stasis in Patients with Chronic Venous Insufficiency

C. Le Dévéhat, T. Khodabandehlou, M. Vimeux and M. Dougny

Unité de Recherches d'Hémorhéologie Clinique, Pavillon J. Renault, 1 avenue Colbert, Centre Hospitalier, 58000 Nevers, France

INTRODUCTION

The presence of multiple microcirculatory alterations has already been pointed-out in chronic venous insufficiency "CVI" (1-6). In previous studies, we reported impairements in RBC rheologic behaviour which appear in the early uncomplicated stages of this disease (7-8). These impairements were shown, on the other hand, to be exaggerated when using a venous stasis stimulus in patients. In fact, changes induced in this way proved useful to better understand the hemorheological disturbances in such patients. In this work, we have investigated hemorheological effects of Cyclo 3 Fort in patients suffering from a venous incompetence. Because of the concomitant responsability of venous stasis in initiating or worsening venous circulation (7-8), we have performed measurements at rest and after a 10-minute provoked venous occlusion.

MATERIAL AND METHODS

PATIENTS : 49 patients suffering from a CVI of the lower limbs were recruited for this study. They were without varicose veins, did not present other concomitant pathology and had no medication known to modify the rheological properties of blood. 7 healthy volunteers without CVI or other pathology and without any medication have also participated in this study. Each patient gave informed consent. The project was approved by the institutional review board of human investigation of the regional university. Table 1 presents the characteristics of patients compared to those of the controls.

	Healthy subjects	Placebo group	Cyclo3 Fort group
sex ratio (F/M)	1/6	25/0	22/2
age (years)	35.3+-8.4	58.3+-11.7	58+-10.7
BMI (kgs/m^2)	23.44+-3.2	26.8+-4.6	29.8+-4.7

TABLE 1 : Clinical characteristics of patients and healthy subjects.

Furthermore, the patients were characterized by normal hemodynamic parameters as evaluated by a strain gauche plethysmograph. These results already discussed in our previous work (7) are not shown here.

STUDY PROCEDURE : Patients were studied in supine position. Hemorheologic measurements were performed on blood samples drawn from a foot vein before and following the end of a provoked venous stasis (PVS). This venous stasis was provoked by a cuff inflated at 100 mmHg during 10 minutes around the knee. The study was based on a double blind randomised treatment with Cyclo 3 Fort versus placebo. Both Cyclo 3 Fort and placebo were administrated at a daily dose of 2 capsules during 4 weeks. Investigations were performed before beginning the treatment "D0" and at the end of treatment "D28".

HEMORHEOLOGIC INVESTIGATIONS : Red blood cell aggregation measurements were performed by Sefam erythro-aggregometer (9). The apparatus is based on the analysis of the backscattered light through the blood suspension in a Couette flow. The derived parameters are red blood cell aggregation time "t" corresponding to the rate of "rouleaux-formation" and the disaggregation shear stress "σ d". Measurements were performed at a standard hematocrit of 0.40 and at a temperature of 37°C.

Plasma viscosity "PV" was measured by a capillary viscometer (KSPV4,Myrenne) at 37°C, fibrinogen level by a thrombin clotting time technique and albumin level by a colorimetric technique.

STATISTICAL ANALYSIS : Before and after treatment, differences between before and after PVS were analysed by Student "t" test for paired data. The effects of a 4-week treatment have been assessed by comparing the parameters'variations resulted from the PVS, between D0 and D28 by Student "t" test in each group.

RESULTS

To better take note of the effects of Cyclo 3 Fort versus placebo, results are expressed as relative variation (Δ%) between before and after the PVS at each time for each group. (Table 2)

Hematocrit :At D0 in the placebo group, the PVS leads to a significant variation which is maintained at D28. In the Cyclo 3 Fort group, Δ % resulted by the PVS is not significant neither at D0 nor at D28.

Fibrinogen : In contrary to healthy subjects, patients showed at D0 a significant increase in fibrinogen level as a result of PVS. These increase persisted in both groups at D28.

Albumin : In placebo group at D0 the PVS leads to a significant increase which becomes more pronounced at D28. In patients with Cyclo 3 Fort, the deleterious effect of PVS is reduced after 4 weeks (3.42 % at D0 versus 2.12 at D28).

Plasma Viscosity : At D0 before stasis, both groups of patients had significantly higher values of this parameter than that in healthy subjects. In placebo group, the increase of this parameter as a results of PVS at D0 was as important as D28. In Cyclo 3 fort group, the increase is significantly less pronounced at D28 than D0.

Red blood cell aggregation : (Table 3)

At Day0 before stasis, RBC aggregation parameters are significantly impaired in both groups of patients when compared to those in healthy subjects. Impairements become significantly more pronounced after stasis. In placebo group at D28, the PVS is showed to be still accompanied by disturbances in RBC aggregation parameters.

Whereas Cyclo 3 Fort exhibited an inhibitory effect on RBC hyperaggregation tendency caused by PVS. In fact, one can notice that the variations of parameters as a result of PVS are significantly less important at D28 when compared to D0 in Cyclo 3 Fort group.

| | Healthy subjects n = 7 | | Placebo group n = 25 | | | | Cyclo 3 Fort n = 24 | | | |
| | | | D0 | | D28 | | D0 | | D28 | |
	BVS	AVS	BVS	AVS	BVS	AVS	BVS	AVS	BVS	AVS
Hematocrit (%)										
mean	42.3	42.3	40.36	42.48	40.1	42.56	41	42.12	39.9	41.25
± SD	± 3.4	± 5.1	± 3	± 3.8	± 2.6	± 3.7	± 3.25	± 4.8	± 2.7	± 4.6
	NS		$p < 0.001$		$p < 0.001$		NS		NS	
Δ%	0.23		5.02		6.10		2.63		3.08	
Fibrinogen (g/l)										
mean	2.68	2.7	3.03	3.3	2.87	3.06	3.14	3.28	3.19	3.3
± SD	± 0.2	± 0.2	± 0.5	± 0.6	± 0.5	± 0.5	± 0.6	± 0.5	± 0.5	± 0.5
	NS		$p < 0.001$		$p < 0.001$		$p < 0.02$		$p < 0.01$	
Δ%	0.87		9.22		7.2		3.55		3.76	
Albumin (g/l)										
mean	45.6	48	47	48	47	49.75	44.9	46.9	44.9	45.9
± SD	± 5.1	±4.9	± 3.5	± 4	± 5.5	± 5.9	± 3.2	± 3.4	± 3.4	± 4.6
	NS		$p < 0.05$		$p < 0.02$		$p < 0.05$		NS	
Δ%	5.87		2.33		6.05		3.42		2.12	
Plas. Visc. (mPa.s)										
mean	1.3	1.31	1.33	1.38	1.32	1.38	1.34	1.42	1.33	1.36
± SD	± 0.03	± 0.06	± 0.07	± 0.1	± 0.1	± 0.1	± 0.06	± 0.08	± 0.08	± 0.07
	NS		$p < 0.01$		$p < 0.001$		$p < 0.001$		$p < 0.02$	
Δ%	0.42		4.85		5.13		5.66		2.63	

Table 3 : Hematocrit, fibrinogen, albumin, plasma viscosity expressed as mean +- SD as well as their relative variations (Δ%) between before (BVS) and after (AVS) venous stasis before (D0) and after (D28) treatment.

| | Healthy subject n = 7 | | Placebo group n = 25 | | | | Cyclo 3 Fort n = 24 | | | |
| | | | D0 | | D28 | | D0 | | D28 | |
	BVS	AVS	BVS	AVS	BVS	AVS	BVS	AVS	BVS	AVS
RBC agg.time (mPa)-1										
mean	21.2	20.7	19.2	16.13	20.14	17.13	16.01	13.62	16.5	14.87
± SD	± 3.5	± 4.8	± 5.5	± 5.6	± 7.04	± 6.4	± 4.2	± 4.1	± 4.8	± 4
	NS		$p < 0.001$		$p < 0.001$		$p < 0.001$		$p < 0.02$	
Δ%	3.62		15.5		14.7		14.4		8.6	
RBC disagg.shear stess (mPa)										
mean	79.25	83.84	91.66	102.6	92.3	102.3	95.57	111.5	95.7	102.54
± SD	± 6.7	± 14.5	± 17.5	± 22.6	± 22.3	± 27.6	± 13.3	± 17.7	± 14.8	± 17.7
	NS		$p < 0.01$		$p < 0.01$		$p < 0.001$		NS	
Δ%	5.9		10.9		11.22		16.0		3.1	

Table 3 : RBC aggregation parameters and their relative variations between before (BVS) and after (AVS) venous stasis at D0 and D28.

It is also interesting to note that the relative variation of RBC disaggregation shear stress at D28 in Cyclo 3 Fort group is similar to that in healthy subjects.

DISCUSSION

From our data, venous insufficiency can be characterized by hemorheologic disturbances mainly those concerning RBC aggregation accompanied by plasmatic proteins changes. The main emphasize is however on the aggravation of disturbances in stasis circumstances. Indeed, a fluid exudation occuring during stasis results to increases in local hematocrit, plasma fibrinogen level and decrease in albumin level. These changes lead to increase the local RBC hyperaggregation tendency which occurs mainly in the venules being the region of lowest shear rate, even in the normal circulation. This RBC hyperaggregation tendency, in turn, will tend to perpetuate stasis and aggravate the situation. This shows the particular benefits which could be expected from any pharmacological intervention decreasing red cell aggregability, thus helping to break the vicious circle. In regard to Cyclo 3 Fort, its benefic effects were mostly revealed through its inhibitory action on deleterious effects of venous stasis. In fact, a 4-week treatment is found to significantly reduce the RBC hyperaggregation tendency caused by stasis. These results indicate the protective role of Cyclo 3 Fort against stasis. Data also emphasize on the usefulness of a venous occlusion as a test allowing to assess the hemorheological efficacy of Cyclo 3 Fort.

REFERENCES

1. C. Le Dévéhat, T. Khodabandehlou : Concepts hémorhéologiques de l'insuffisance veineuse et implications du traitement par Cyclo 3 Fort. Precepta Medica, N°6, 106-108, 1990.
2. TR. Cheatle, C. Quashie, B. Villemur, P. Carpentier : Two-dimensional laser doppler perfusion imaging and microcirculatory function in patients with venous skin damade. Phlebology 10, 32-36, 1995.
3. SK Shami, SJ Chittenden, JH Scurr, PD Coleridge Smith : How is skin blood flow altered in venous disease. Phlebology 92, John Libbey Eurotext, 88-90, 1992.
4. HAM Neumann : TcPO2 values in patients with and without pericapillary cuffs in chronic venous insufficiency and porthyria cutanea tarda. Phlebology 92, John Libbey Eurotext, 178-178, 1992.
5. G. Belcaro, M. Grigg, A. Rulo, A. Nicolaïdes : Blood flow in the perimalleolar skin in relation to posture in patients with venous hypertension. Annals of vascular surgery vol 3, N°1, 5-7, 1989.
6. G. Cjuffetti, E. Mannarino, R. Paltriclia, V. Malagigi et al : Leucocyte activity in chronic venous insufficiency. Intern. Angiol. vol 13, N°4, 312-316, 1994.
7. C. Le Dévéhat, T. Khodabandehlou, M. Vimeux, M. Dougny : The effects of Cyclo 3 Fort treatment on hemorheological disturbances during a provoked venous stasis in patients with chronic venous insufficiency. Clin. Hemorh. vol 14, suppl.1, 553-563, 1994.
8. J. Remacle, C. Michiels, T. Arnould : Stasis as a cause for the appearance of venous pathology. Phlebology 92, John Libbey Eurotext,251-253, 1992.
9. M. Donner, M. Siadat, JF Stoltz : Erythrocyte aggregation approach by light scattering determination. Biorh. 25, 367-375, 1988.

Phlebology '95, D. Negus et al. (eds.). Phlebology (1995) Suppl. 1: 626-628

P292

Chronic Venous Insufficiency of the Legs
The Treatment with Mesoglicano (Prisma)

F. Rück, M. Agostini and G. Giansiracusa

Surgical Department, General Hospital, Lovere, Italy. Chief of Department L. Beluffi

INTRODUCTION

Venous diseases present a high percentage of morbidity, and occupy the fifth-sixth place, for frequency, among morbid affections.
It has a main aethiologic agent, venous congestion, the alteration of the vein parietal permeability, being veins subjected to a hypertension regimen caused by the communicans valvulare failure.
The medical treatment has always been a problem. The discovery of the importance of the association endothelium-blood led to revalue the substances produced by endothelium for the treatment of phlebopathies.
The Mesoglicano (Prisma) is a product extract from mammels aorta and consists for 83% of "heparan sulphate" and for 17% of "dermatan sulphate", molecules belonging to the "glicosamino-glycous" family. They fulfil an antithrombotic and profibrinolitic action, an improvement of the endothelial selective permeability at the endothelial level.

MATERIALS AND METHODS

During the year 1993, the Angiology Service of Surgical Department of Lovere General Hospital treated no. 300 patients, 80% women and 20% men, with an age from 30 to 75 years.
The patients, at the beginning of the treatment presented:

- aedema	in 75% of the cases	- sensation of heaviness in 95% of the cases
- pain	in 95% of the cases	- heat in 80% of the cases
- paresthesia	in 70% of the cases	- functional impotence

The seriousness of the symptomatology was valued as follows:

0	= absent
+	= light
++	= medium
+++	= severe

The patients underwent the treatment with:
- 50 mg Mesoglicano tablets, two tablets per day for 4 weeks.

RESULTS

All patients underwent a control after two weeks and after a month in order to check the efficacy and the tolerance to the Mesoglicano. The two week treatment resulted to be very efficacious. The general symptomatology was improved and the importance of each symptom was reduced.

Symptom	Basic	After 14 days	Final
Aedema	+++	++	0
Sensation of heaviness	+++	++	+
Functional impotence	++	+	0
Paresthesia	++	+	0
Pain	+	//	0

The Mesoglicano proved to be well tolerated by 90% of the patients. The other 10% of the patients suffered from cutaneous reaction and gastric intolerance.

DISCUSSION

The results of this treatment confirm the efficacy of the Mesoglicano given "per os" to the patients suffering from cronic venous insufficiency of the legs and it is proved to be well tolerated. The system by which Mesoglicano acts, consists in reducing the venous hypertension activating the fibrinolysis, in re-establishing and controlling the endothelium permeability; it doesn't alter the emocoagulative aspect as it acts at endothelium and blood level only.

REFERENCES

1. Andreozzi G.M. et al.: "Effects of mesoglycan sulfate on the arterial elastic module". Angiology, 38, 593, 1987.
2. Bartha K., Kovacs T., Lerant I.; Papp B.; Csouka E., Kolev K., Machovic R. : Interaction of antithrombin III and thrombin-antithrombin III complex with cultured aortic endothelial cells, Thromb. Res. 1987, 47, 541.
3. Bracale G.C. et al.: "L'attività terapeutica del mesoglicano per os nelle flebopatie. Valutazione clinica ed emocoagulativa". Minerva Medica, 77, 1919, 1986.
4. De Gaetano G. : "L'interazione delle piastrine con la parete vascolare". In : Piastrine, trombosi e aterosclerosi, Masson Editori, 53,58,1986.
5. Rosenberg R.D.: "Role of heparin and heparinlike moleculesin thrombosis and atherosclerosis". Federation proceedings, 44, 2, 1985.
6. Rück F. : Acts V European American Symposium on venous diseases. Vienna 1990

7. Rück F. : Acts XI World Congress Union. Internationale de Phlébologie. Montreal 1992.
8. Thomas D.P. et al.: "The antithrombotic action of heparan sulphate".
9. Vercelloni M. et al.: "Studio clinico sull'efficacia del mesoglicano nel trattamento delle flebopatie degli arti inferiori." La Clinica Terapeutica, 115, 173, 1985.
10. Vittoria A. et al.: "Azione del mesoglicano sul sistema fibrinolitico nell'uomo dopo somministrazione di 24-48-72 mg per via orale. Studio dell'azione profibrinolitica esercitata alla dose di 48 mg per os in confronto con placebo". Atti del IX Congresso Nazionale della Società Italiana di Patologia Vascolare, Copanello (CZ) 6-9 Giugno 1987.

Phlebology '95, D. Negus et al. (eds.). Phlebology (1995) Suppl. 1: 629-631

P294

The Influence of Rutosides on Increased Capillary Permeability in Chronic Venous Insufficiency as Measured by Video Capillaroscopy

S. Bort, U. Hahn, M. Hahn, T. Klyscz and M. Jünger

Department of Dermatology, University Hospital Liebermeisterstr. 25, D-72072 Tübingen, Germany

INTRODUCTION

It is well known that patients suffering from chronic venous insufficiency develop microangiopathy in which the capillaries become elongated, dilated and tortuous. In severe chronic venous insufficiency the capillaries have a glomerulus-like appearance (1, 2). In areas with marked skin changes like induration and hyperpigmentation the number of capillaries is reduced. The introduction of fluorescence videomicroscopy made it possible to study the transcapillary diffusion of the low-molecular-weight tracer Na-fluorescein. Fluorescence videomicroscopy opens the way to visualizing and quantifying the diffusion of Na-fluorescein through the capillaries. With this kind of investigation it was determined that the mean fluorescent light intensity increases significantly in mild and severe chronic venous insufficiency, indicating increased transcapillary diffusion (3, 8).

The aim of this study was to investigate the influence of O-(ß-hydroxyethyl)-rutoside (HR) on the transcapillary and interstitial diffusion of Na-fluorescein out of the capillaries of the retromalleolar skin in patients with chronic venous insufficiency. O-(ß-hydroxyethyl)-rutoside is a bioflavonoid which reduces the postcapillary filtration of macromolecules (5) as well as experimentally induced edema (7).

METHODS AND PATIENTS

This clinical trial was carried out in a randomized, double-blind, placebo-controlled design with a parallel group comparison. 50 patients with chronic venous insufficiency were admitted into a double-blind, placebo-controlled study. The primary objective of the study was to determine the efficacy of O-(ß-hydroxyethyl)-rutoside (HR) compared to a placebo in reducing the capillary permeability of low-molecular-weight substances.

After a compression therapy of at least 14 days 25 patients were treated with a placebo and 25 patients were treated with 500 mg HR twice daily. Before the treatment period and immediately afterwards the skin in the medial malleolar region was examined using intravital fluorescence microscopy. The measurements were performed in a skin area without any macroscopically visible trophic disturbances. After a bolus injection of Na-fluorescein (0.3 ml 20% solution per liter of calculated blood volume) the dynamic filling of the capillaries and the leakage of the fluorescent dye through the capillary wall into the interstitial space was recorded with a high-resolution video camera for 30 min and stored on video tape. The fluorescent light intensity was measured by video densitometer with a window representing an actual skin area of 1.0×1.0 mm^2. The fluorescent light intensity of Na-fluorescein was analyzed by a special computer system and was expressed in arbitrary units. These light intensities differed from individual to individual because of the natural variation in the light transmission properties of the skin. In order to obtain comparable values, we expressed the intensities as a percentage of the maximal individual intensity of each patient and each examination. The absolute fluorescent light intensity was corrected by the natural fluorescence of the skin itself, measured before the dye reached the capillaries.

RESULTS

The results are given in Table 1 for patients treated with O-(ß-hydroxyethyl)-rutoside and patients treated with placebo. The mean values were calculated and compared with the Wilcoxon test. It is clear from the table that there was no statistically significant difference in the mean fluorescent light intensity (analyzed over a period of 30 minutes) between the two groups).

Table 1

	verum group n = 25	placebo group n = 25
light intensity before therapy (arbitrary units)	2436 ±229	2457 ±146
light intensity after therapy (arbitrary units)	2457 ±119	2501 ±126

DISCUSSION AND CONCLUSION

Electron microscopy showed a marked restoration of the ultrastructure of the myocytes after compression therapy (6). The number of capillaries is also increased after compression therapy (4). It is possible that the compression therapy performed by each patient for at least 2 weeks before taking O-(ß-hydroxyethyl)-rutoside or placebo had a certain effect on transcapillary permeability. Therefore, O-(ß-hydroxyethyl)-rutoside in addition to compression therapy does not significantly reduce capillary permeability in patients suffering from chronic venous insufficiency.

REFERENCES

1. Fagrell B.
    Local Microcirulation in Chronic Venous Incompetence and Leg Ulcers
    Vasc. Surg. 1979, 4: 217-225

2. Fagrell B.
    Micorcirculatory disturbances - the final cause for venous leg ulcers?
    Vasa 1982, 11 (2): 101-103

3. Jünger M., Bort S., Hahn U., Klyscz T., Geiger H., May B., Schiek A.
    In vivo Nachweis erhöhter Durchlässigkeit der Hautkapillaren bei
    fortgeschrittener chronischer Veneninsuffizienz (CVI).
    Deutscher Dermatologie-Kongress, 37. Tagung Deutsche Dermatologische
    Gesellschaft,
    Düsseldorf, 14 -18 Juli 1993

4. Galler S., Klyscz T., Hahn M., Jung MF., Steins A., Jünger M.
    Auswirkungen einer standardisierten Kompressionstherapie auf die kutane
    Mikrozirkulation in Stauungsgebiet bei chronisch venöser Insuffizienz (CVI) im
    Stadium I/II nach Widmer.
    Deutscher Dermatologie-Kongress, 38. Tagung Deutsche Dermatologische
    Gesellschaft,
    Berlin 29.04.-03.05.1995

5. Gerdin B., Svensjö E.
    Effect of HR O-(ßHydroxyethyl)-rutosides on increased microvascular
    permeability to macromolecules induced by histamine, bradykinin and fibirn
    degradation products.
    Abstract 2nd World Congress for Microcirculation, July 22-29, 1979, La Jolla

6. Hammersen F., Hesse G.
    Strukturelle Veränderungen der varikösen Venenwand nach
    Kompressionsbehandlung.
    Phlebologie und Proktologie 1990; 19: 193-199

7. Lund F., Fagrell B., Kunicki J., Glenne PO.
    O-(ßHydroxyethyl)-rutoside in postischaemic and stasis oedema of the rat tail.
    IIIe Congres International de Phlebologie, Steuvert & Soon, Apeldoorn, 1970

8. Speiser D. E., Bollinger A.
    Microangiopathy in mild chronic venous incompetence (CVI): Morphological
    alterations and increased transcapillary diffusion detected by fluorescence
    videomicroscopy.
    Int. J. Microcirc:Clin Exp. 1911; 10: 55-66

Phlebology '95, D. Negus et al. (eds.). Phlebology (1995) Suppl. 1: 632-637

P296

Effects of a Micronized Flavonoid Fraction (Daflon 500) on the Clinical, Haemodynamic and Cytokine Alterations of Patients with Varicose Disease

S. Signorelli[1], G. Pennisi[1], M.P. Costa[1], V. Monte[2], M.G. Malaponte[2], M.C. Mazzarino[2] and G.M. Andreozzi[2]

[1] Department of Internal Medicine "A.Francaviglia", Chair of Angiology and [2] Institute of General Pathology, Chair of Immunology, University of Catania, School of Medicine, Italy

The pathogenetic role performed by venous stasis and hypertension has long been emphatized (1,2) on the other hand a less importnace has been given to the performance of plasmatic cells, regarding the chronic venous insufficiency (C.V.I.). We must remember that some time ago the presence of disorders in permeability of venous wall to permit the passage of elements with high weight (i.e. fibrinogen) was found.(1) We know that the relationship between the haemodynamic imbalance (i.e. venous stasis and hypertension) and blood cells merits more attention. In fact Mayses (3) said that the erythrocitary count and the haematocrit of samples of peripheral blood measured in baseline condition and after venous stasis were less in varicose patients than in normal subjects. Thomas (4) found that the count of white cells after a period of orthostatic position reduced more in varicose patients than in healty patients. In other studies(5,6) performed with intravital biomicrhoscopy was possible to obeserve the reduction of "capillary loops" caused by venous stasis and it was possible to hypotize that a leucocitary thrombosis can be responsible for severe changes of skin in varicose patients. On the other hand it is know that the haemodynamic disorders can produce ischemia but it can cause tessutal damages can be caused by means of the endothelium. Between endothelium and margined and trapped leucocytes by venous stasis and hypertension a strong link begins and leucocytes also becomes immunologically activated as the production of cytokines can be demostrats (7). The macrophagic production of citokynes is able to cause a lack of antithrombotic proprierties in the endothelium (8,9) which are the best conditions to create tessutal ischemic damage also in varicose patients.

From this cultural background we performed a study to evaluate the possibility to reduce by means of drugs the production of cytokines and we used synthetic diosmin plus esperidin (Daflon 500 -Servier) that has phlebotonic and anti-inflammatory abilities.(10,11,12)

Case report and metodology.

We studied 16 patients suffering from primitive varicose disease with ages between 42 and 68 years old (mean age 53.2+/-7.2) and a mean period of clinical history of 5 years. In all patients a clinical and instrumental visit was made to find objective and subjective symptons of C.V.I. We took into consideration the presence of heaviness of lower limbs, parestesise, cramps at rest, orthostatic oedema , we created an arbitrary score for this evaluation. 1 egual to absence of symptom, 2 slight symptom, 3 strong symptom and 4 stronger sensation of symptom. We determined , by c.w. ultrasound , the venous pressure in orthostatic position and after muscular activation on the great saphenous vein and on the posterior tibial vein.(13)We gave all patients synthetic diosmin plus esperidine, 1 gr/daily for one month. Both at the first observation and after a month we took blood samples directly from the great saphenous vein to determine the production of interleukin-6 (IL-6) both at rest and after 10 minutes of venous occlusion , performed by inflatted cuff to 60 mmHg.

Results:

Clinical symptoms - parestesia: at the beginning were observed a score of 1 in 35% of the observations,score 2 in 60.5%, score 3 and 4 in 1% and after 1 month we observed score 1 in 93.50%, score 2 was reduced to 4.5%, scores 3 and 4 were not modified. Orthostatic heaviness: score 1 was found initially in 6.25%, score 2 was observed in 68.75%, score 3 in 28% and only in 2% did we find score 4. After the therapy score 1 raised to 91.75%, e score 2 decreased to 6.25%, scores 3 snd 4 decreased to 1%.

Cramps at rest: initially we found in 50.75% score 1, in 41.25% score 2, in 6.25% , score 3 and in 1.75% score 4. After the therapy we found , respectively score 1 increased to 91.75%, while score 2 was reduced to 6.25% and scores 3 and 4 decreased to 1%. Oedema: at the beginning score 1 was found in 12.5%, score 2 in 55.25%, score 3 in 31.25% and we found score 4 only in the 1%. After therapy we observed score 1 in 85%, score 2 in the 11.25, score 3 in 2.5% and score 4 was present in 1%.

Venous pressure: the baseline mean value of orthostatic venous pressure was 80.62+/-18.13 mmHg and decreased to 59.69+/-11.50 mmHg (P<0.001) after the period of therapy. The mean of value of venous pressure by muscular activation was initially 56.88+/-18 mmHg and after 1 mounth the mean of value was decreased to 41.28+/-12.03 mmHg (P<0.5).

Interleukin-6: at the beginning we found a mean of IL-6 of 1.88+/-0.36 nanogr/ml and a mean of 2260+/-432.79 nanogr/ml caused by lipopolysaccarides (LPS 1 microgr/ml), the mean level of IL-6 raised respectively to 395.50+/-39.65 nanogr/ml and 4043+/-594.21 after V.O. After the therapy the means were reduced respectively to 1.68+/-0.20 nanogr/ml and 581+/-176.42 nanogr/ml. The levels of IL-6 by V.O. were reduced to 184.10+/-37.71 nanogr/ml and to 479.40+/-56.75 nanogr/ml. The statistical comparison of resting level between baseline and 1 month of therapy showm a statistical significance (P<0.3); most significant was the statistical evidence showed between the values caused by V.O. In fact we observed P<0.1 in the comparison between baseline and therapy concerning blood samples at rest and we found a most importnat difference (P<0.001) concerning the level caused by V.O. and stimulated by LPS .

Conclusion: this study confirms the efficancy of this drug on clinical symptoms and shows the decrease of the orthostatic venous pressure as a haemodynamic marker for clinical changes. We believe that the results of IL-6 are most significant , in fact we observed a significant increase of the production of this cytokine caused by V.O. and a higher level was caused by LPS stimulus. The cytokinic production was less after the therapy and we observed a lower activation of macrophages caused by LPS and by V.O. These data show that by means of the activation of leucocytes the haemodynamic phase of C.V.I. can go forward to the tessutal phase.The different levels of venous hypertension and experimental venous stasis are be able to transform leucocytes from resting to primed and activated cells (14,15).Diosmin and esperidin significantly modify venous hypertension , clinical symptoms ,and moreover, can change the leucocitary immunological activation that can cause some "local ischemic disorders" leading certainly to the creation of a venous ulcer of the skin.

References:
1) Browse NL, Burnard KG
The cause of venous ulceration.
Lancet 1982;2:243-245

2) Homans J

The aetiology and treatment of varicose ulcers.

Ann.Rev.Immunol. 1984;2:283-318

3) Moyses C, Cederholm-Williams SA, Michel CC

haemoconcentration and the accumulation of white cells in the feet during venous stasis.

J.Microcirc.Clin.Exp.1987;5:311-320

4) Thomas PRS,Nash GB, Dormandy JA

White cell accumulation in the dependent legs of patients with venous hypertension: a possible mechanism for throphic changes in the skin.

Br.Med.J.1988;296: 1693-1695

5) Bollinger A,Haselbach P, Schnewlin G,Junger M

Microangiopathy due to chronic venous incompetence evaluated by fluorescence videomicroscopy.

In: Negus D, Jantet G, eds. Phebology 1985; John Libbey & Co: London 1986

6) Franzek UK,Speiser D, Haselbach P,Bollinger A

Morphologic and dynamic microvascular abnormalities in chronic venous incompetence(CVI).

In: Davy A,Stemmer R eds. Phlebologie 1989; John Libbey Eurotext Ltd: Montrouge France:104-107

7) Engler RL,Dahlgren MD,Peterson MA,Dobbs A,Schmid-Schoenbein GW

Accumulation of polymorphonuclear leucocytes during three-hour myocardial ischemia.

Am.J.Physiol.1986;251. H93-H100

8) Edwards AT, De Friend DJ,Corson RJ,Mc Collum CN

Oxygen-derived free radicals and ischemia -reperfusion in venous ulceration.

Br.J.Surg. 1992;79:369

9)Pober JS

Cytokine-mediated activation of vascular endothelium.

Am.J.Patol. 1998; 133:426-433

10) Labrid C

Inflammation en phlébologie: activité des dérives des flavonoides.

Rev.Gynecologue-Obstetricien 1991; 3:164-172

11) Amato C

Advantage of a micronized flavonoidic fraction (Daflo 500) in comparison with a nonmicronized diosmin.

Angiology 1994;45:n^6 part 2: 531-535

12) Duchene Marullaz P, Amilel M, Barbe R

Evaluation of the clinical pharmacology activity of a phlebotonic agent.Application to the study of Daflon 500 mg.

Int.Angio 1988; 7(suppl 2):25

13) Bartolo M,Antignani PL, Nicosia PM, Todini AR

Non invasive venous pressure measurements.

Int. Angio. 1988; 7:2,182-9

14) Andreozzi GM,Signorelli S,Martini R, Di Pino L

Haemodynamic and cellular effects in ischemia and stasis: clinical and experimental models.

Int.J.Microcirc.Clin.Exp. 1994; 15: (51) : 247

15) Andreozzi GM,Martini R,Di Pino L,Signorelli S

Clinical physiology of the microcirculation.

In: New trends in vascular exploration . Borgatti E,De Fabritiis A eds. Minerva Medica Torino; 1994: 24-36

	BEFORE	AFTER	P
BASELINE	1.88 +/- 0.36	1.68+/-0.20	< 0.3
BASEL + LPS	2260+/-432.79	581+/-176.42	< 0.1
VENOUS OCCL.	395.50 +/-39.65	184.10 +/-176.42	<0.1
V.O. + LPS	4043 +/-594.21	479.40 +/-56.75	< 0.001

Tab.1: means, S.D. of levels of IL-6 in different moments of experiment and statistical analysis of differences between before and after therapy.

The Treatment of Varicose Veins and Telangiectasias

Compression

Phlebology '95, D. Negus et al. (eds.). Phlebology (1995) Suppl. 1: 639-642

V/3.4

Elastic Stockings and Bandages. The Checking of their Compression Effectiveness in the Consulting Room

C.R. Catz

Centro de Ulceras y Varices. Buenos Aires, Argentina

INTRODUCTION

In the daily phlebologic practice it is usual to prescribe the elasto-compression to be worn by the patient in short periods of time,for example after doing sclerotherapy; or in longer periods of time, as prescribed in a recurrent venous insufficiency. The variety of materials we have for that is large,ie, the width of bandages, the percentage of their stretching, their manufactured and control of quality, the different trade-marks, standardization, whether the patients can afford for them or not, etc. Besides we know that each pathology needs a particular elasto-compression with a determined range of pressure. Thus the need of using in the consulting room, a simple device to measure the pressure of the elasto-compression recorded, either by the doctor or patient, and the need of checking the pressure "sold" by the manufacturer or orthopedist arise. Considering those experiences from Dr. Sigg in the 60s. we have designed our device, the so called ball-tensiometer, to carry out this work.

MATERIAL AND METHOD

The sensor device has a ball made of latex whose measures are about 3 cm diameter by 0.5 cm wide. Its main feature is that of being almost flat. This is stuck to a rubler tube, that one used as oxygen connections, long enough so as to permit its handling along the lower limb. The kit ball-tube is placed inside a cellophane or polythene cover; this will allow the pulling of it without harming the ball when this is compressed between the skin and the bandage or the elastic stocking. The other end of the tube es connected to a watch-tensiometer with an insufflator. We ask the patient, either to bandage himself or to put the stocking on as he usually does while we place the ball just on the internal malleolar; the tube remains along the internal face of the tibia. To make the sliding easier we spread some talcum powder on the skin of the limb before placing the device. Undergoing our first practices we did not know the exact bulk of air the ball should have had to record the real pressure carried out by the bandage or stocking. Did it have to be collapsed, half or full inflated? Thus to establish the appropiate bulk of air, we did as follow:

ADJUSTING THE DEVICE: following the same way to control the normal use of a tensiometer, we take a bottle, that one of an adult-arm size, we wrap it up with the handle of a tensiometer (the same we will use to do the control), and we place our ball between the bottle and the handle. We start to insufflate pressure at the handle of the tensiometer, checking at the same time if both watches record the same pressure. Measuring different initial bulk of air in our ball, we conclude that this must have the minimum bulk of air inside it to keep its shape without collapsing its walls.(In practice this bulk of air allows us, if we compress the ball between our fingers, to record 120 mmhg; if we do not get this recording, we can insufflate air to the system with the handle; if we over record this, we let air out until adjusting it). After having practised this above described for many times, we conclude that the ball-tensiometer allows us to record pressure between 0 and 60 mmhg. We find out that those records over the one registered, 60 mmhg alter the results.

METHOD IN THE RECORDING

We make an elasto-compression data card to record there all the pressure obtained along each limb in three different heights: 1/3 inferior of the leg, corresponding to that pressure just above the internal malleolar; 1/3 mid just in the middle of the leg; and 1/3 upper, only 5 cm below the knee. We always work along the interior face of the tibia on which we slide our device pulling from the cellophane cover. Notice that the device alllows us to record the adjusting pressure along the whole limb, but only for this statistical work we have chosen these three locations. We call "descending gradient" when the recorded pressures fromm the malleolar to the knee go down. We call "without gradient" when the recorded pressures from the malleolar to the knee do not change. And we call"reverse altered gradient" when the pressures are larger in an upper level and lesser in a lower one.

RESULTS

Using the ball-tensiometer, 100 measurements were done at random in patients attending external consulting rooms in three different places, La Matanza Hospital, French Hospital and Central Military Hospitall from Buenos Aires, and from private consulting room. Among those 100 measurements, 56 correspond to elastic stockings and 44 to elastic bandages. Checking our data, we find out, as regards elastic stockings,that in 25 cases (45%) a "descending gradient" is recorded; in 7 cases (12.5%) a "without gradient"; and finally in 24 cases (42.5%) a "reverse altered one" is recorded. As regards elastic bandages, we find out that in 11 cases (25%) a "descending gradient"is recorded; only in 2 cases (5%) a "without gradient"; and finally in 31 cases (70%) a "reverse altered one" is recorded.

DISCUSSION AND CONCLUSIONS

With our first practices, we coould soon appreciate the practical advantages of using the ball-tensiometer since we could find out records that did not correspond to the medical prescriptions ordered, as regards gradients and pressures. Considering the elastic stockings we could see that when these were given to the orthopedic houses following the measurements suggested by the manufactures, they improved the recorded results. We should also have to say that sometimes we get to know, question our patients, that some stockings are sold without taking any measures on the anatomy of the patient. Or that on other occasions, some patients wear stockings that have already been worn by other people. These facts should be considered and dealt with by the doctor to improve the results of the elasto-compression. Now considering the elastic stockings, when we practise repeated measurements on the same stocking, similar results are registered, though there are slight variations. Thus when we prescribe the wearing of elastic stocking, once we check the pressure this does on "the limb thath has, at that moment, determined anatomical measures", we can assume that those pressures would keep for a time, until the elasticity of the stocking or the size of the limb changed. As regards elastic bandages, the statistics show poor results, but these can easily be corrected when the profesional takes the measure himself, or when he asks the patient to give a different pressure to the bandage when he does it himself. We have noticed that when told, those patients who bandage themselves learn how to correct it inmediately. In spite of what statistics show, we conclude that the daily bandage, when properly taught, allows us to have the required pressure gradients, though the anatomical conditions may vary. This is different to the stockings that cannot so easily be adapted to these changes. Thus we conclude:

1) The ball-tensiometer satisfies largely our expectancy since it allows us to take measures of the pressures of the elastic bandages and stockings with a slight possibility of getting wrong results, when the system has been well adjusted. We also insist on the simplicity and low cost of this system.

2) That the statistics, although the people taken at random were a few, show a large variability of results that makes the professional think aabout the need to control, first all the materials used by the patients, and then to check the treatment later to guarantee and improve the elatocompressive method.

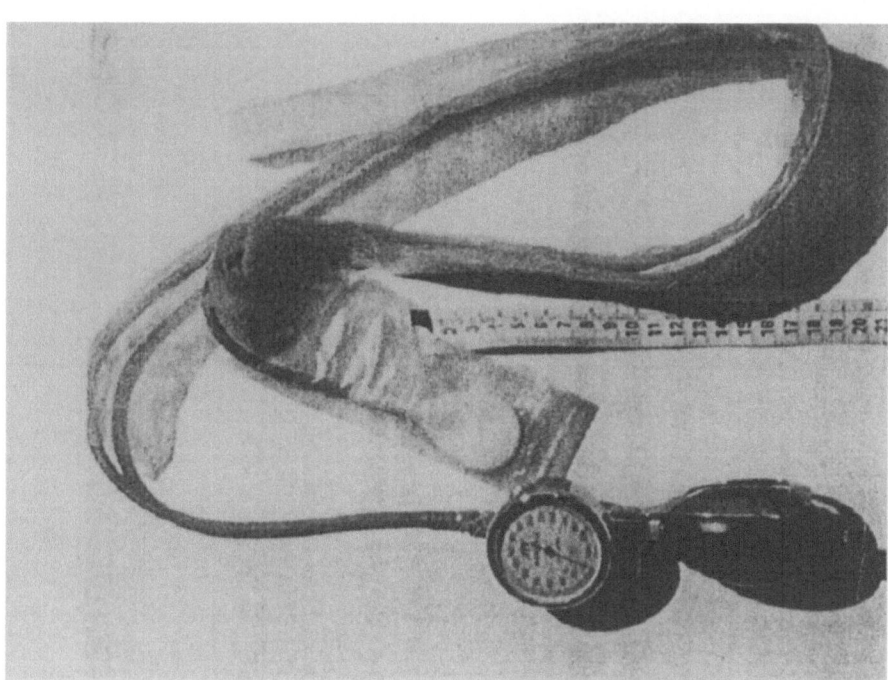

3) that the 45% of those stockings controlled at random has a descending pressure gradient from the ankle to the knee. And that these results are the same after measuring the same legs many times. This shows the need to check the stockings our patients wear; since once the proper pressure is recorded, this one is expected to be the same through time.

4) that only in 25% of the bandages we can expect a proper pressure gradient if the professional do not check and teach the patient how to correct the bandage himself daily; this would greatly improve just with a few indications.

Phlebology '95, D. Negus et al. (eds.). Phlebology (1995) Suppl. 1: 643-646

OP/1.3

The Effects of Long Term Graduated Compression Treatment on Venous Function During Pregnancy

C. Austrell, I. Thulin and L. Norgren

Department of Surgery, Lund University, S-22185 Lund, Sweden

During pregnancy venous wall changes are induced, resulting in a reduced venous tone. Less contractility of the veins might cause functional disturbances such as diminished venous emptying and reflux. An interesting finding is that a venous reflux appearing during the course of the pregnancy may reverse after delivery (1). Varicose veins frequently develop during pregnancy but may disappear spontaneously after delivery.

It has been shown that during standing an orthostatic uterovascular syndrome may occur (2). In a study on 40 women during gestational week 38, 70% had evidence of such symptoms (3). Graduated compression stockings reduced the orthostatic changes of maternal and fetal haemodynamics significantly (4).

In one investigation we have shown that graduated compression treatment during pregnancy utilising stockings of pantyhose model exerting an ankle pressure of 25 mm Hg, reduced to 12 mm Hg at mid thigh, increased the expelled venous blood volume from the leg significantly, while the refilling rate was influenced to a lesser degree (5). The women also reported a marked reduction of subjective discomfort while wearing graduated compression. In another study (60) we investigated 15 women in gestational week 32 to 38 in the supine and standing positions with and without graduated elastic compression. When the woman was raised to a 60° upright position, maternal and fetal orthostatic heart rate increases were recorded. These increases were significantly reduced when graduated elastic compression was applied. On the other hand, no significant variations of the uterine artery blood flow velocity measured by ultrasonography were registered.

In a recent study (7) on a similar group of pregnant women, duplex examinations were performed of the femoral vein in the supine and standing positions with and without elastic graduated compression. Parallel to an orthostatic increase of the heart rate we found a significantly decreased femoral vein flow velocity in the vertical position. This decrease was less pronounced when graduated compression was used. The flow velocity was also significantly higher with graduated compression than without in the standing position.

These findings support the use of graduated compression during late pregnancy.

The aim of the present study was to further evaluate the venous function if stockings were used from an earlier stage of the pregnancy and to compare two levels of compression pressures of the pantyhose stockings.

MATERIALS AND METHODS

Seventy-five pregnant women were included and randomised to treatment with either medium or low pressure graduated compression, 25 and 13 mmHg respectively at the ankle level, reduced by 50% at the thigh. Twenty-five women withdrew their participation at various times (11 with low, 14 with medium pressure compression). In 11 cases the reason was a problem with fitting of the pantyhose stockings, the remaining 14 had complicating disease or early delivery.

Thus, 50 women with a mean age of 31 years (range 23-40 years) fulfilled the study. Nineteen women had their first pregnancy, the remaining 31 their second to fifth pregnancy. Three examinations were performed, the first one as close to gestational week 20 as possible (range 18-25), the second one around week 33 (range 30 - 37) and the final post partum examination was performed with great time variations (median week 18, range 8-32). This last examination was fulfilled by 44 of the 50 women.

Twenty-two of the women were randomised to wear low pressure compression stockings while 28 women had medium pressure compression stockings. For all calculations, the examination results of both legs were used.

For comparison eight women participated as controls without any use of graduated compression during the course of the pregnancy. They were only examined at the first and second occasions and not after delivery. Otherwise they did not differ from the study group regarding age or number of pregnancies.

The venous function was investigated by foot volumetry (8), a dynamic plethysmographic method performed with the patient standing and exercising with 20 knee-bendings during 40 seconds. During this exercise the expelled venous volume (EV_{ml}) was calculated. The initial foot volume was also registered (FV_{ml}). The refilling rate after exercise was calculated ($Q_{ml/min \times 100ml}$).

At the first examination foot volumetry was initially performed without stockings, thereafter with stockings applied. At the second examination the reversed order was used, and at the follow up only examination without stockings was accomplished.

For statistical comparisons a two-tailed Wilcoxon signed ranks test was used and p-values < 0.05 were regarded significant.

RESULTS

The basic foot volume increased significantly in all three groups (controls, low compression and medium compression groups) from the first to the second examination. In both compression groups there was also a significant reduction from the second examination to the follow-up. There was no difference between the first examination and the follow up regarding any of the compression groups (table 1).

A numerical increase of the expelled venous volume in the control group from the first to the second examination was recorded, though not reaching any statistical significance. On the other hand there were corresponding significant increases with both low and medium pressure compressions (p<0.05). In the medium compression group the expelled volume was even significantly higher at the final examination than at the first one (table 2).

While stockings increased the expelled volume during the first examination regardless of the compression class, there was no difference at the second examination when examining with or without stockings.

The refilling rate after exercise did not reach reflux values in any of the groups. The values ranged from 2.9 to 4.2 ml/mm x 100 ml. Even though controls increased the refilling rate from 2.9± 0.3 to 3.5 ± 0.4 ml/min x 100 ml, these changes were not statistically significant.

DISCUSSION

Some potential advantages should be considered using graduated compression treatment during late pregnancy, reduction of discomfort, particularly caused by swelling (3), probably prevention against thromboembolic complications and maybe also against future venous incompetence and the development of varicose veins. Finally, it has recently been shown that this kind of treatment is effective to prevent haemodynamic changes of the so-called oscillating vena cava syndrome (4). That study used a compression pressure at the ankle level of 40 mmHg, and compression with a gradient of 30-40 mmHg has previously been advised (9). In our previous investigations we have used an ankle pressure of 25 mmHg, and have been able to show an improvement of the venous function. Therefore it was found valuable to study if the compression pressure could be still more reduced, in order to accomplish the simplest possible use of the pantyhose stockings. We have previously shown that for less advanced venous incompetence such as primary varicose veins, a low graduated compression pressure is as effective as a higher one (18 vs 30 mmHg at the ankle level). The present study thus compared stockings exerting 25 and 13 mmHg at the ankle level, each pressure reduced by 50% at the thigh.

Both the medium and the low compression pressure stockings were found to improve the venous emptying, measured during exercise in the standing position. At the first examination there was an immediate increase of the expelled venous volume as the stockings were applied. During the second examination there was no difference between the results achieved with and without stockings. At the final examination post partum, venous emptying was still improved, and in the medium compression group it was even higher than during previous examinations.

As reflux values were not recorded it is not relevant to talk about any reduction of reflux, numerically there were even only slight variations between the groups.

Finally, stockings reduced the volume of the feet, which shall be interpreted as a reduction of swelling. It was also the subjective understanding by the participants that discomfort and edema were reduced. In general, the treatment was appreciated, and very few subjectively noticed differences between low and medium compression stockings were reported.

Withdrawals due to fitting problems were similarly common in the two groups (15%), the reason always being that the waist part of the pantyhose stockings slipped down or was not as selfadjusting as was wished as the abdomen increased its size.

Whether an improved or a less deteriorated venous function during pregnancy is predictive of any reduced development of varicose veins or chronic venous incompetence in the future is not possible to say. Theoretically this may be true, and based on other studies also thromboembolic events and haemodynamic disturbances may be prevented by the use of graduated elastic compression. If difficult to put on or wear, this kind of treatment is usually forgotten by patients, and pregnant women may not be different. One of the major problems in the use of high pressure graduated compression stockings is the difficulty to put them on.

The present study supports the use of a lower ankle pressure than 40 mmHg, but as there were few differences between stockings exerting 25 or 13 mmHg at the ankle, it is not possible to conclude which to recommend. As reflux was infrequent one should possibly select a group of women where reflux is present to evaluate the two different pressure levels as regards also the effect of the reflux. Regarding comfort, the low compression stockings were possibly found somewhat easier to put on and wear, but this subjective finding is not conclusive.

Finally modification of the waist part of the pantyhose stockings are important to get the best compliance in future studies.

REFERENCES

1. Struckmann J R, Meiland H, Bagi P, Juul-Jörgensen B. Venous muscle pump function during pregnancy. Acta Obstet Fynecol Scand 1990; 69:209-215
2. Schneider K T M, Bollinger A, Huch A, Huch R. The oscillating vena cava syndrome during quiet standing - an unexpected observation in late pregnancy Brit J Obstet Gynecol 1984; 91:766-780
3. Schneider K T M, Bung P, Weber S, Huch A, Huch R. An orthostatic uterovascular syndrome - a prospective, longitudinal study. Am J Obstet Gynecol 1993; 169:183-188
4. Weber S, Schieder KTM, Bung P, Fallenstein F, Huch A, Huch R. Effect of compression stockings on circulation in late pregnancy. Geburtsh und Frauenheilk 1987; 47: 395-400
5. Nilsson L, Austrell C, Norgren L. Venous function during late pregnancy. The effect of elastic compression hosiery. VASA 1992; 21:203-205

6. Austrell C, Nilsson L, Norgren L. Maternal and fetal haemodynamics during late pregnancy. Effect of compression hosiery treatment. Phlebology 1993; 8:1555-1557

7. Norgren L, Austrell C, Nilsson L. The effect of graduated elastic compression stockings on femoral blood flow velocity during late pregnancy. VASA 1995; in press.

8. Norgren L. Functional evaluation of chronic venous insufficiency by foot volumetry. Acta Chir Scand; suppl 444.

9. Skudder P A, Farrington D T. Venous conditions associated with pregnancy. Semin Dermatol 1993; 12:72-77.

10. Sjöberg T, Norgren L, Einarsson E. Functional evaluation of four different compression stockings in venous insufficiency. Phlebology 1987; 2:53-58.

	First examination	Second examination	Final examination
Controls	966 ± 11	- * - 1004 ± 12	-
Low pressure compression	941 ± 14	- * - 975 ± 17	- * - 950 ± 15
Medium pressure compression	949 ± 11	- * - 974 ± 12	- * - 952 ± 13

Table 1.   Foot volume (FV$_{ml}$) measured at the three examinations.
$* = p < 0.05$

	FIRST EXAMINATION		SECOND EXAMINATION		FINAL EXAMINATION
	without compression	with compression	without compression	with compression	without compression
Controls	12,4 ± 1,0	-	13,9 ± 1,3	-	-
Low pressure compression	13,6 ± 0,6	14,5 ± 0,7	14,7 ± 0,7	14,6 ± 0,7	14,1 ± 0,6
Medium pressure compression	12,2 ± 0,5	13,4 ± 0,5	14,0 ± 0,6	13,6 ± 0,6	13,3 ± 0,6

Table 2.   Expelled venous volume (EV$_{ml}$) at the three examinations.
$* = p < 0.05$ n.s. = non significant

Phlebology '95, D. Negus et al. (eds.). Phlebology (1995) Suppl. 1: 647-649

OP/1.2

Do Ready-Made Stockings Fit Our Patients?

M.C. Mooij and S.H. Oosterwal

Flebologisch Centrum Oosterwal, Polikliniek voor Dermatologie en Flebologie, Overkrocht 10, 1815 KX, Alkmaar, Nederland

INTRODUCTION

Up to now compression is the best treatment for patients with chronic venous insufficiency. Elastic stockings provide this compression most easily. The pressure exerted by the stockings should be linear decreasing towards the heart. This graduation of compression is achieved easily by custom-made elastic stockings. However, so far, ready- made stockings have been prescribed most frequently by physicians, for different reasons. But, to ensure a graduated pressure profile, the circumferences of the limb measured have to match at 6 levels at least with the ready-made sizes. Our experiences made us expect this to be almost impossible. To find proof for this, we compared the sizes (circumferences and lengths) of over 20,000 legs of our patients with the sizes of flat-knitted below-knee stockings that are to be found in the ready-made size chart which is being used for the production of elastic stockings.

METHODS

The circumferences and lengths of the legs of our patients have been measured at the A, Y, B, B1, C and D levels (only below-knee stockings). According to our criteria, the stocking fitted if the size measured did not deviate 0.75 centimeter from the ready-made size at the levels D, C and B1 and 0.5 centimeter at the levels B, Y and A. Measurements had been taken while the patient sat on a chair with his leg on a bench with a height of 40 centimeter. The data recorded have been compared with the data from the ready-made size chart of the Hohenstein Institute and those of the recently developed chart of the European Committee for Standardization in order to find out whether any ready-made stocking could be prescribed to any of our patients.

RESULTS

Whereas a deviation from the standard size value (of 0.75 and 0.5 centimeter respectively) was allowed, to none of our patients a ready-made stocking could be prescribed. The specific combination of circumferences at all levels (at the A, Y, B, B1, C and D levels!) of each individual patient could not be found in the size chart.

Once we changed the deviation into 2 centimeter, the same result was found.

In the charts below the circumferences of the patients have been printed together with their corresponding ready-made sizes. The charts indicate the variations in the sizes measured in relation to a specific B size. This B size (20 cm) is the size we find with our patients most frequently.

The ready-made size chart shows us that A equals B. However, in chart 1 and 2, we see that A does not match the ready-made size. Obviously this may lead to too much pressure and the risks of corns and discomfort with many patients.

In chart 3 the Y does not match with the ready-made size. If the value of the Y measured is higher than the ready-made size value, the risks of irritation and even necrosis, particularly with patients with extensive atherosclerosis, are greater. For most women this size does not cause too much of a problem.

The B1 is widely accepted as being an important value for the efficacy of the elastic stocking. Therefor, a proper fitting of the stocking at this level is essential. Charts 5 and 6 show that most ready-made stockings provide too little pressure at these levels.

In charts 7 to 10 the ready-made sizes differ widely from the sizes of the legs measured. The consequences are an increased pressure at the proximal end of the stocking so that blood- and lymph-flow will be impeded, and a reduction in the desired effect on the venous pumpfunction.

Our results are supported by a study from Wienert, who showed that to 75% of the normal legs ready-made stockings do not fit. In another study, Hohlbaum claimed that to 50% of all the women ready-made stockings do not fit, simply because their D size is bigger than their C size.

We wondered whether it would be possible to create a new ready-made size chart based upon the measurements of our patients. We calculated the average circumferences of the different levels for each B size. Next we compared the sizes of our patients with our new ready-made chart. The outcome shows that to only 5% of our patients, mainly women, ready-made elastic stockings would "fit".

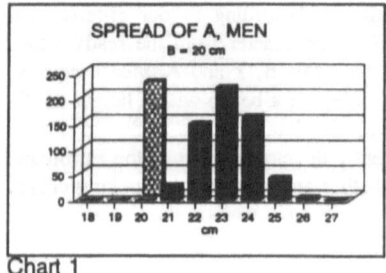

Chart 1

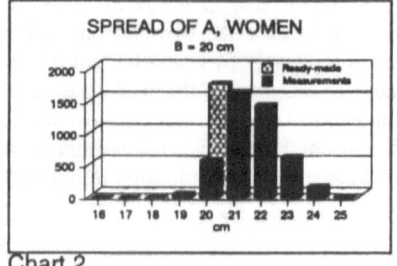

Chart 2

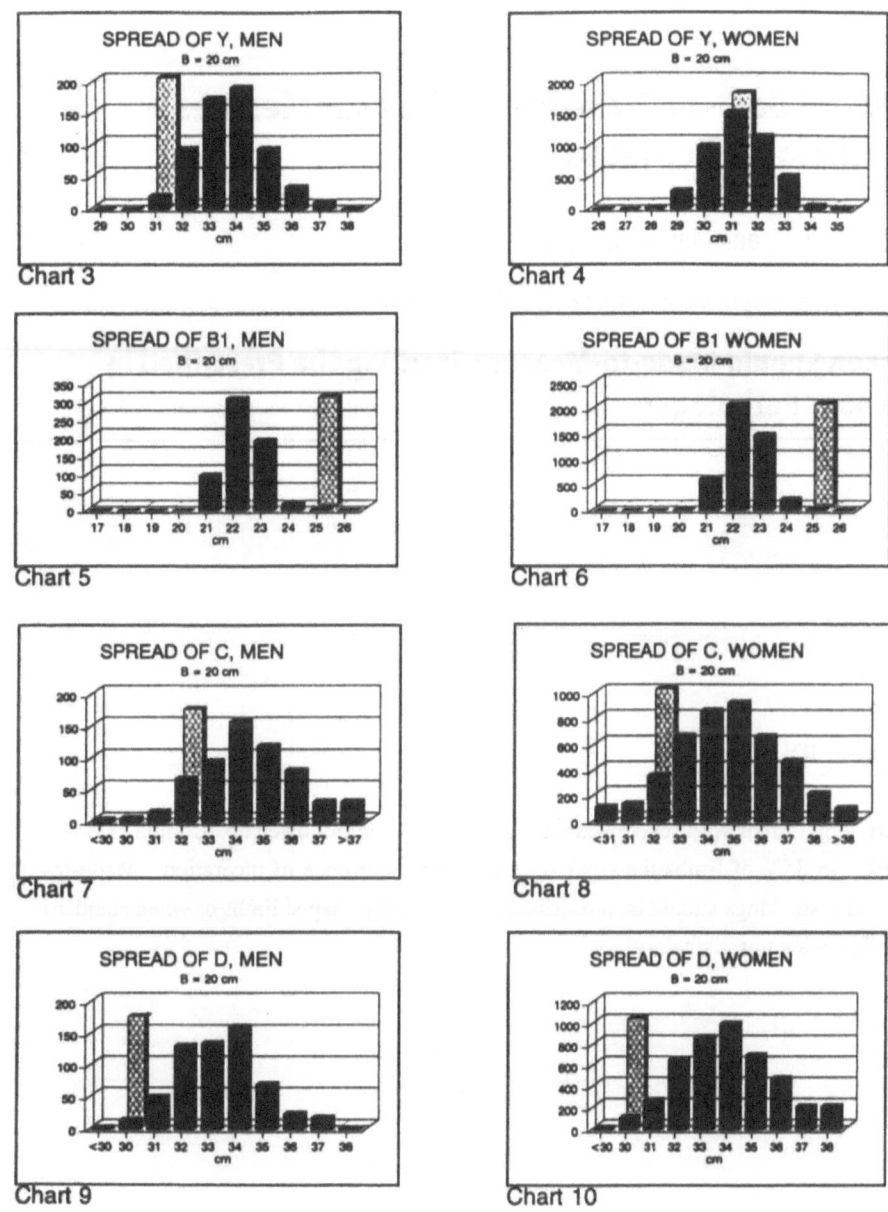

Chart 3

Chart 4

Chart 5

Chart 6

Chart 7

Chart 8

Chart 9

Chart 10

CONCLUSION

It was not possible to find a ready-made, flat-knitted stocking, with a considerable slope, that would properly fit to any of our patients.

Phlebology '95, D. Negus et al. (eds.). Phlebology (1995) Suppl. 1: 650-653

OP/1.1

Quand doit-on Prescrire des Bas sur Mesures en Pathologie Veineuse?

Paul A. Ouvry and Pierre A.G. Ouvry

Service d'angéiologie, Clinique St Pierre, 76200 Dieppe, France

When Should Made-to-Measure Stockings be Prescribed in Venous Pathology?

SUMMARY

Sixty-eight limbs with severe venous insufficiency were followed up. Compliance was 85%. In 85% of limbs the stocking prevented recurrence of ulceration. Made-to-measure stockings should be prescribed in abnormally shaped limbs or when standard stockings are unsatisfactory.

Quand doit-on prescrire des bas sur mesures en pathologie veineuse ?

OUVRY Paul A., OUVRY Pierre A.G.

Service d'angéiologie - Clinique St Pierre

76200 DIEPPE - France

Introduction

Du type de l'insuffisance veineuse dépend le degré de la stase. Encore trop souvent de nos jours, quand la stase est sévère, la contention appliquée est insuffisante. Cela explique la persistance de certains troubles, trophiques et beaucoup de récidives d'ulcères. Une élévation importante de la pression veineuse périphérique exige en fait le port de bandes ou de bas confectionnés avec des fils à allongement court.

Ces bas, à coefficient de rigidité élevé, vont mieux résister que d'autres aux reflux veineux importants. Leur but n'est pas tant de comprimer la jambe que de résister efficacement à l'oedème dès qu'il apparaît.

Ces bas sont fabriqués avec des fils épais sur des métiers rectilignes et comportent donc une couture. Ils sont élastiques dans le sens de la largeur et seulement extensibles dans le sens de la longueur. Cela permet d'appliquer la mesure, donc la pression voulue, à la hauteur voulue.

Aussi leurs mesures seront elles prises, après la réduction de l'oedème, avec le plus grand soin. La cause la plus fréquente des insuffisances veineuses sévères est la maladie post thrombotique.

A ce sujet, une étude prospective récente de H.R Büller permet d'apprécier l'efficacité des bas sur mesures dans la prévention de la maladie post thrombotique. Le risque est réduit de 50%. Nous avons entrepris nous même une étude rétrospective afin de contribuer à cerner ce problème, sans nous limiter à la seule maladie post thrombotique mais en y incluant d'autres formes d'insuffisance veineuse sévère.

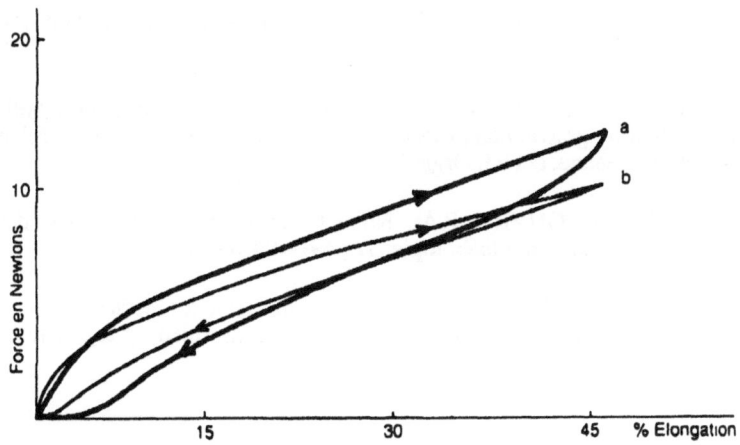

Force elongation curves of elastic knitwear
a) of a made to measure stocking
b) of a standard stocking

Matériel et méthode

68 membres inférieurs atteints d'insuffisance veineuse sévère ont été étudiés chez 42 patients qui ont tous subi un examen écho-doppler systématique.
57% d'entre eux souffraient d'une maladie post thrombotique.
Les autres souffraient d'insuffisance veineuse superficielle ou fonctionnelle ou d'angiodysplasies.

Résultats

L'observance par le patient du port du bas a été bonne dans 85% des cas.
Son port a été efficace dans la prévention des récidives d'ulcères dans 85% des cas.

Discussion

Le bas sur mesures est plus coûteux qu'un bas standard. Le surcoût est-il justifié ? La réponse nous paraît clairement être oui. En effet, le bas sur mesures s'adapte mieux qu'un autre à un membre dont la morphologie s'écarte de la norme. Par ailleurs, son efficacité a bien été démontrée par H.R. BÜLLER dans la préventionde la maladie post thrombotique . Enfin, notre étude montre de très bons résultats dans la prévention des ulcères veineux pour une population d'insuffisants veineux sévères.
Seule une étude prospective permettrait de savoir si la contention standard peut faire aussi bien.

Conclusion

La contention sur mesures est plus longue à prescrire que la contention standard. Elle nécessite l'intérêt du médecin et la formation de l'équipe soignante. Elle est utile en première intention quand le galbe du membre est anormal, en seconde intention sur un membre de morphologie normale après échec ou intolérance de la contention standard.

Références

.   BASSÌ Gl., Stemmer R. Traitements mécaniques fonctionnels en phlébologie. PICCIN, PADOVA 1983

.   BULLER HR, BRANDJES DPM and JW ten CATE. The post thrombotic Syndrome. Prévention by graded compression elastic stockings. International Union of Angiology. SALF. Beaune, oct 93. Organisateur : F. Becker.

.   GUENNEGUEZ H., OUVRY P.A., Nouvelle approche thérapeutique des ulcères post phlébitiques rebelles. Phlébologie, 37 ; 491-499. 1984.

.   STOLK R., SALZ P., A Quick Pressure Determining Devic for medical stocking based on the determination of the counterpressure of air filled leg segments. Swiss Med 10 (1988) N° 42,91-96

Matériel et méthode

68 membres inférieurs atteints d'insuffisance veineuse sévère ont été étudiés chez 42 patients qui ont tous subi un examen écho-doppler systématique.
57% d'entre eux souffraient d'une maladie post thrombotique.
Les autres souffraient d'insuffisance veineuse superficielle ou fonctionnelle ou d'angiodysplasies.

Résultats

L'observance par le patient du port du bas a été bonne dans 85% des cas.
Son port a été efficace dans la prévention des récidives d'ulcères dans 85% des cas.

Discussion

Le bas sur mesures est plus coûteux qu'un bas standard. Le surcoût est-il justifié ? La réponse nous paraît clairement être oui. En effet, le bas sur mesures s'adapte mieux qu'un autre à un membre dont la morphologie s'écarte de la norme. Par ailleurs, son efficacité a bien été démontrée par H.R. BÜLLER dans la préventionde la maladie post thrombotique . Enfin, notre étude montre de très bons résultats dans la prévention des ulcères veineux pour une population d'insuffisants veineux sévères.
Seule une étude prospective permettrait de savoir si la contention standard peut faire aussi bien.

Conclusion

La contention sur mesures est plus longue à prescrire que la contention standard. Elle nécessite l'intérêt du médecin et la formation de l'équipe soignante. Elle est utile en première intention quand le galbe du membre est anormal, en seconde intention sur un membre de morphologie normale après échec ou intolérance de la contention standard.

Références

.   BASSI Gl., Stemmer R. Traitements mécaniques fonctionnels en phlébologie. PICCIN, PADOVA 1983

.   BULLER HR, BRANDJES DPM and JW ten CATE. The post thrombotic Syndrome. Prévention by graded compression elastic stockings. International Union of Angiology. SALF. Beaune, oct 93. Organisateur : F. Becker.

.   GUENNEGUEZ H., OUVRY P.A., Nouvelle approche thérapeutique des ulcères post phlébitiques rebelles. Phlébologie, 37 ; 491-499. 1984.

.   STOLK R., SALZ P., A Quick Pressure Determining Devic for medical stocking based on the determination of the counterpressure of air filled leg segments. Swiss Med 10 (1988) N° 42,91-96

OP/15.7

THE USE OF HOME SEQUENTIAL GRADIENT PNEUMATIC COMPRESSION IN THE TREATMENT OF CHRONIC VENOUS INSUFFICIENCY

Arcelus JI, Caprini JC, Hoffman KN, Finke N, Pollard JC, Size G. Department of Surgery, The Glenbrook Hospital, Glenview, IL and North western University Medical School, Chicago, IL, USA.

Objective: To evaluate the use of home sequential gradient pneumatic compression (SGPC) in the treatment of chronic venous insufficiency of the legs.

Design: Prospective study including a consecutive series of patients.

Patients: Twenty patients with symptomatic chronic venous insufficiency of the legs confirmed by duplex scanning and photoplethysmography (PPG).

Intervention: All patients were instructed to use SGPC at home for four hours a day, in addition to the use of elastic compression stockings, during six months.

Measurements: Clinical improvement was assessed by a scoring system completed by the patients before and after completion of the study. All patients underwent air plethysmography (APG) before compression and 1, 3 and months thereafter. The following APG indices were calculated: venous volume (VV), venous filling index (VFI), ejection fraction (EF) and residual volume fraction (RVF).

Results: The median total score was 20 (IQR 17, 2-20) before treatment and 13 (IQR 10-13) after SGPC treatment (Wilcoxon rank test, p<0.0001). The individual scores given by patients to symptoms such as swelling, pain, decreased physical activity and cosmetic problems were significantly reduced. Median APG indices VV, VFI and RVF did not change significantly. However, EF increased significantly after 3 and 6 months of compression (Wilcoxon rank test, p<0.05) compared to baseline.

Conclusions: The use of SGPC improved the ejection fraction, as measured by air plethysmography, in a group of patients with chronic venous insufficiency of the legs. This favorable hemodynamic response could explain the remarkable clinical response to this therapeutic modality.